			III A	IV A	V A	VI A	VII A	
								$1s^2$ **He** 4.0026
			5 (He) $2s^2$ $2p^1$ **B** 10.81	6 (He) $2s^2$ $2p^2$ **C** 12.011	7 (He) $2s^2$ $2p^3$ **N** 14.007	8 (He) $2s^2$ $2p^4$ **O** 15.999	9 (He) $2s^2$ $2p^5$ **F** 18.998	10 (He) $2s^2$ $2p^6$ **Ne** 20.179
	I B	II B	13 (Ne) $3s^2$ $3p^1$ **Al** 26.98	14 (Ne) $3s^2$ $3p^2$ **Si** 28.09	15 (Ne) $3s^2$ $3p^3$ **P** 30.974	16 (Ne) $3s^2$ $3p^4$ **S** 32.06	17 (Ne) $3s^2$ $3p^5$ **Cl** 35.453	18 (Ne) $3s^2$ $3p^6$ **Ar** 39.948
28 (Ar) $3d^8$ $4s^2$ **Ni** 58.71	29 (Ar) $3d^{10}$ $4s^1$ **Cu** 63.55	30 (Ar) $3d^{10}$ $4s^2$ **Zn** 65.37	31 (Ar) $3d^{10}$ $4s^2$ $4p^1$ **Ga** 69.72	32 (Ar) $3d^{10}$ $4s^2$ $4p^2$ **Ge** 72.59	33 (Ar) $3d^{10}$ $4s^2$ $4p^3$ **As** 74.92	34 (Ar) $3d^{10}$ $4s^2$ $4p^4$ **Se** 78.96	35 (Ar) $3d^{10}$ $4s^2$ $4p^5$ **Br** 79.904	36 (Ar) $3d^{10}$ $4s^2$ $4p^6$ **Kr** 83.80
46 (Kr) $4d^{10}$ **Pd** 106.4	47 (Kr) $4d^{10}$ $5s^1$ **Ag** 107.87	48 (Kr) $4d^{10}$ $5s^2$ **Cd** 112.40	49 (Kr) $4d^{10}$ $5s^2$ $5p^1$ **In** 114.82	50 (Kr) $4d^{10}$ $5s^2$ $5p^2$ **Sn** 118.69	51 (Kr) $4d^{10}$ $5s^2$ $5p^3$ **Sb** 121.75	52 (Kr) $4d^{10}$ $5s^2$ $5p^4$ **Te** 127.60	53 (Kr) $4d^{10}$ $5s^2$ $5p^5$ **I** 126.90	54 (Kr) $4d^{10}$ $5s^2$ $5p^6$ **Xe** 131.30
78 (Xe) $4f^{14}$ $5d^9$ $6s^1$ **Pt** 195.09	79 (Xe) $4f^{14}$ $5d^{10}$ $6s^1$ **Au** 196.97	80 (Xe) $4f^{14}$ $5d^{10}$ $6s^2$ **Hg** 200.59	81 (Xe) $4f^{14}$ $5d^{10}$ $6s^2$ $6p^1$ **Tl** 204.37	82 (Xe) $4f^{14}$ $5d^{10}$ $6s^2$ $6p^2$ **Pb** 207.2	83 (Xe) $4f^{14}$ $5d^{10}$ $6s^2$ $6p^3$ **Bi** 208.98	84 (Xe) $4f^{14}$ $5d^{10}$ $6s^2$ $6p^4$ **Po** (210)	85 (Xe) $4f^{14}$ $5d^{10}$ $6s^2$ $6p^5$ **At** (210)	86 (Xe) $4f^{14}$ $5d^{10}$ $6s^2$ $6p^6$ **Rn** (222)

In several cases, atomic weight is rounded off to four or five significant figures. See inside back cover for listing of 1970 International Atomic Weights. Electron configurations taken from *Theoretical Inorganic Chemistry* by M. Clyde Day and Joel Selbin, Reinhold Publishing Corporation, except numbers 104 and 105, which are predicted by analogy.

[a] Value in parentheses denotes mass number of most stable known isotope.
[b] Name and symbol are not officially accepted. Kurchatovium, Ku, has been proposed by Russian investigators and rutherfordium, Rf, by American investigators for element 104.

64 (Xe) $4f^7$ $5d^1$ $6s^2$ **Gd** 157.25	65 (Xe) $4f^9$ $6s^2$ **Tb** 158.93	66 (Xe) $4f^{10}$ $6s^2$ **Dy** 162.50	67 (Xe) $4f^{11}$ $6s^2$ **Ho** 164.93	68 (Xe) $4f^{12}$ $6s^2$ **Er** 167.26	69 (Xe) $4f^{13}$ $6s^2$ **Tm** 168.93	70 (Xe) $4f^{14}$ $6s^2$ **Yb** 173.04	71 (Xe) $4f^{14}$ $5d^1$ $6s^2$ **Lu** 174.97
96 (Rn) $5f^7$ $6d^1$ $7s^2$ **Cm** (245)	97 (Rn) $5f^8$ $6d^1$ $7s^2$ **Bk** (247)	98 (Rn) $5f^{10}$ $7s^2$ **Cf** (249)	99 (Rn) $5f^{11}$ $7s^2$ **Es** (254)	100 (Rn) $5f^{12}$ $7s^2$ **Fm** (255)	101 (Rn) $5f^{13}$ $7s^2$ **Md** (256)	102 (Rn) $5f^{14}$ $7s^2$ **No** (254)	103 (Rn) $5f^{14}$ $6d^1$ $7s^2$ **Lr** (257)

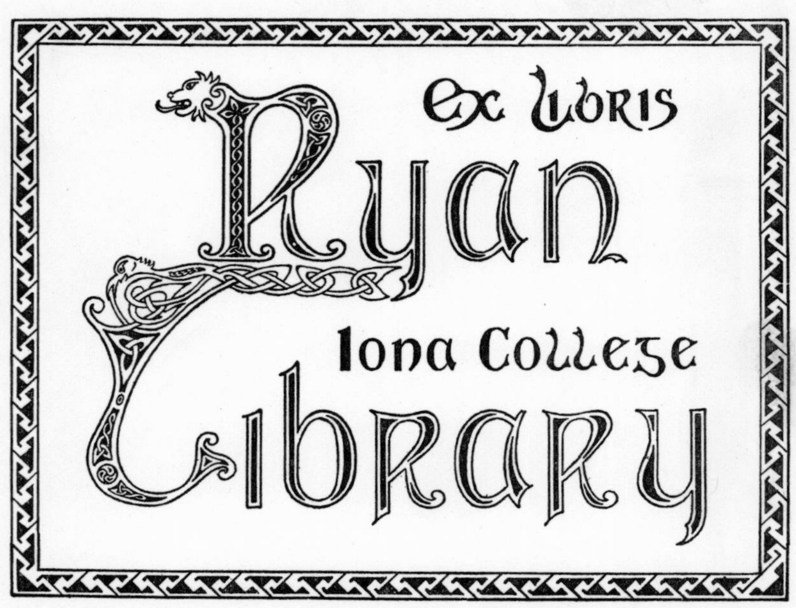
Ex Libris
Ryan
Iona College
Library

a chemical background for the paramedical sciences

a chemical background for the paramedical sciences

second edition

gerald f. grillot

syracuse university

HARPER & ROW, PUBLISHERS
NEW YORK, EVANSTON, SAN FRANCISCO, LONDON

Sponsoring Editor: Joe Ingram
Project Editor: Robert Ginsberg
Designer: Michel Craig
Production Supervisor: Robert A. Pirrung

Formerly published under the title *A Chemical Background to Nursing: and Other Paramedical Programs*

A Chemical Background for the Paramedical Sciences, Second Edition

Library of Congress Cataloging in Publication Data

Grillot, Gerald Francis, 1914-
A chemical background for the paramedical sciences.

First published in 1964 under title: A chemical background to nursing, and other paramedical programs.
Bibliography: p.
1. Chemistry. I. Title. [DNLM: 1. Chemistry–Laboratory manuals. 2. Chemistry–Nursing texts.
QD31 G859c 1974]
QD33.G895 1974 540'.2'461 73-9288
ISBN 0-06-042511-3

contents

preface

This textbook is a revision of a previous edition which was assembled to meet the needs of a course in chemistry being taught to students of nursing at Syracuse University. Some of the simple mathematical operations of chemistry have been introduced into the book so that the student may fully understand the meaning of the chemical equation, the analytical results that are compiled by the hospital laboratory, and the significance of acidity and pH. The structural formulas and the systematic nomenclature of the field of organic chemistry should be sufficiently detailed for the needs of the students in the various paramedical fields.

Based on the suggestions of those teachers who have used the previous edition, the material in this second edition has been expanded. However, I have attempted to arrange this new material in such a fashion that those not desiring to include this additional material can readily delete it. Thus the discussions of the periodic table and of atomic orbitals are placed at the end of Chapter 3, while an elementary discussion of molecular orbitals appears at the end of Chapter 16 following a discussion of aliphatic hydrocarbons. Some teachers may choose to introduce atomic orbitals during their discussion of atomic structure in Chapter 3 and to introduce molecular orbitals while they are considering covalent compounds in this chapter. In the early portion of Chapter 21, carbohydrates are discussed from a limited structural viewpoint. An elementary discussion of stereochemistry is then given so that the structure of this type of substance can be considered in greater detail. In Chapter 5, the gas laws and the kinetic-molecular theory of gases may be deleted without destroying the continuity of the text. Those instructors who wish to develop a more quantitative approach to chemical equilibrium and ionization can include the material on equilibrium constants and the ionization constants of acids and bases at the end of Chapters 8 and 9, respectively. Although methods of determining the pH of an acidic or basic solution are introduced in Chapter 10, methods and calculations for determining the pH of a buffer solution are placed in Appendix B.

Because of their important role in protein synthesis and in genetics, nucleic acids are discussed in a separate chapter, Chapter 29, instead of in the chapter on proteins. Chapter 28 contains an elementary discussion of photosynthesis and its role in the biosynthesis of carbohydrates.

I hope that teachers and students will find the discussions of significant figures, exponents and exponential numbers, logarithms, answers to problems and selected study questions, and the listing of suitable reference texts and films a useful addition to the text.

An ever-increasing number of students in paramedical fields, besides those in nursing, are now being trained to serve and assist the medical profession. It is felt that this textbook should adequately meet in the chemical field the needs of the various paramedical programs. The title of this edition has been selected to emphasize this fact.

Students entering the nursing program at Syracuse University now usually complete their requirement in chemistry during the second semester of their residence. The course taught at Syracuse meets for three hours of lecture, one hour of recitation, and a three-hour laboratory period each week. Because most of those admitted to this program have completed a satisfactory high school course in chemistry, some of the more elementary material in the beginning chapters is omitted. The following chapters are usually covered in this course: 2, 3, 5 (in part), 7 to 10, 15 to 22, 24 to 27.

In some nursing and other paramedical programs, students may be required to complete the conventional year-long course in general chemistry and then take an additional semester in what is frequently called applied chemistry. This text should contain sufficient material to meet the needs of such a course. It is suggested that Chapters 7 to 10 be reviewed, followed by a study of Chapters 14 (radioactivity), 15 to 23 (organic chemistry), and 24 to 34 (biochemistry).

The material in chemistry in this text has been developed for students of the paramedical sciences but should also be quite suitable for students of the biological sciences, particularly if they are enrolled in a two-year junior or community college in a program in which chemistry must be completed in an academic year.

Your attention is called to a suitable testing program for courses based on this text. This is the American Chemical Society cooperative test in inorganic–organic–biological chemistry. Form 1971, which is now available, consists of three subtests, one in inorganic chemistry, one in organic chemistry, and one in biological chemistry. A new form of this examination is now in preparation. For more information concerning these tests, write to: Examination Committee ACS, University of South Florida, Tampa, Florida 33620.

To those graduate students who have assisted me in teaching chemistry to students of nursing, and to the users of the previous edition who have so kindly offered suggestions for improvements in this text, my heartiest thanks. In looking forward to future editions of this text, I will be most grateful for those suggestions that you will make to improve it. Many of my students have aided me in recognizing numerous applications of chemistry to the various paramedical fields, for which I have felt most fortunate and grateful.

GERALD F. GRILLOT

foreword to the student

Chemistry *is the science that deals with the structure and composition of matter and with the transformations that matter undergoes. For many years, this science has been making valuable contributions to medicine. These contributions have become more important as chemical compounds of a single definite composition have replaced as medicinals the herbs and plant extracts of yesteryear.*

The essential chemical reactions that occur in the body are multitudinous and complex. The study of the chemistry of life processes is called biochemistry. *This special branch of chemistry has not only elucidated many of the normal reactions of the body, but it is furnishing us with valuable information concerning the abnormal (pathological) reactions characteristic of certain diseases. Note in particular the frequent references in this text to the pathological chemistry characteristic of* diabetes mellitus.

Other contributions of chemistry have been in the fields of deficiency diseases, vitamins, hormones and their role in the body, alkaloids, nutrition, dietetics, and so forth. Also, the technologist in the hosptial laboratory must have considerable training in the field of chemistry if he is to carry out necessary chemical tests with the accuracy required of him.

The present trend in the education of paramedical students is not only to teach them to carry out the task assigned to them, but also to introduce them to the fundamental theory relating to these procedures. The understanding that the paramedical student thus obtains should result in doing assignments with greater caution and care, which should lead to more dependable results. The knowledge of chemistry that the student obtains should be of great value to the paramedic in making useful decisions in times of emergency, particularly when qualified personnel are not available to assist. Usually these students begin their basic training by studying such fundamental sciences as chemistry, biology, microbiology, physiology, anatomy, nutrition, and dietetics.

The following reasons can be cited for the study of chemistry by the paramedical student:

1. *Chemistry should serve as a background to the study of such related fields as physiology, microbiology, introduction to*

medical science, pharmacology, therapeutics, nutrition, and clinical subjects.

2. *It should assist in recognizing significant relationships between chemistry and other parts of the curriculum and in establishing the habit of using this knowledge of chemistry to obtain a greater understanding of the art.*
3. *It should serve as an aid in providing for the safety and comfort of the patient and for the intelligent handling of equipment and supplies.*
4. *It should stimulate an understanding interest in the contributions that chemistry and outstanding chemists have made toward the development of modern civilization, especially as it relates to the care of the sick, the conquest of disease, and the improvement of health.*
5. *It should provide a foundation in the science that will enable the student to read with a reasonable degree of understanding the literature of medical science as it relates to nursing and the other paramedical fields, thus favoring the continued development of the student.*
6. *It should develop an understanding and appreciation of the methods of science as they find application in the care of the sick and the study of disease.*

G.F.G.

a chemical background for the paramedical sciences

chapter 1

---

introduction to the study of chemistry

1

Science and the scientific method

Science can be defined as a body of facts and relationships that are organized in such a fashion as to be useful in the discovery of new facts and truths.

One of the important basic sciences is *chemistry*. In the foreword to the student, we defined chemistry as the science that deals with the structure and composition of matter and with the transformations that matter undergoes.

Another science closely related to chemistry is *physics*. This science is primarily concerned with energy, energy relationships, and the transformations of the various kinds of energy. The relationship between these two sciences becomes evident when it is recognized that in nearly all chemical transformations, a loss or gain of a stored type of energy called *chemical energy* is involved.

The ancients speculated regarding natural phenomena without recourse to experimentation. The recent rapid development of such sciences as chemistry and physics has resulted from the development and application of the *scientific method*. There are several essential steps in the scientific method which will now be described.

The first step in the scientific method usually involves the collection of facts and data concerning a certain natural phenomenon. In chemistry, this is usually accomplished by performing a series of well-organized experiments. These data and facts are correlated, and a generalized statement called a *scientific law* is formulated. Such a law attempts to describe some behavior of nature. The various laws describing the behavior of gases, such as those of Boyle and Charles (see Chapter 5), were discovered by these means.

By the use of his imagination, the scientist attempts to formulate an explanation for certain behaviors of nature; this is called an *hypothesis*. If newly discovered facts about this behavior are brought to the attention of the scientist and the hypothesis is not in agreement with these facts, it must then be revised or abandoned. If, however newly discovered facts are in harmony with the hypothesis, it becomes a *scientific theory*. In this fashion, the kinetic-molecular theory of gases was developed to explain the behavior of all gases. This was originally formulated as an hypothesis but is now accepted as a theory

Figure 1.1 A modern biochemical research laboratory. (Courtesy Bristol Laboratories, Syracuse, N.Y.)

because all the facts concerning the properties of gases are consistent with it.

A brief history of chemistry

CHEMISTRY IN ANCIENT TIMES

Many skills and arts were known to the ancients: the smelting of iron; the preparation of enamels to cover tiles; the production of gold, glass, copper, bronze, dyes, cements, soap, and beer; etc. The preparation of these substances was based on experience, and the necessary skills must have been discovered accidentally. There was little speculation, if any, concerning why certain changes take place, and the science of chemistry can hardly be considered as having existed in this period. Many special skills were known to the Egyptians, but the practice of many of them was reserved for the priests of the temples.

THE AGE OF THE PHILOSOPHERS (ABOUT 600 TO 200 B.C.)

The Greek philosophers of this period attempted to give plausible reasons for the occurrence of certain phenomena in nature. Because they did not confirm their reasoning by experimentation, many of their explanations seem quite absurd today. The Greek philosopher Empedocles taught that all matter was composed of different proportions of four elements: earth, air, fire, and water. Another Greek philosopher, Democritus, suggested that all matter was composed of minute discrete particles called *atoms*. Dalton revived this latter theory in 1802, and it is still accepted today, although some of its details have undergone modification.

THE AGE OF ALCHEMY

Alchemy flourished during the Middle Ages and lasted for about ten centuries. A number of chemical processes were described during this period, but on the whole, little progress was made in the development of chemistry as a science. The goal of the alchemists was the transmutation of base metals into gold and the discovery of the universal solvent. They spent much time and effort in attempting to find the *philosopher's stone*, a substance that would accomplish the transmutation of metals. Some of the alchemists also attempted to formulate an *elixir of life* that would cure all physical ills and permit men to retain their youth. Unfortunately, many of the alchemists were charlatans and capitalized on public gullibility and superstitions. At the end of this period, they mostly fell into disrepute.

In the early sixteenth century, Paracelsus, a great scholar, induced many alchemists to devote their efforts to the discovery of remedies and drugs for the alleviation of human ills, and he declared that their true goal should be the discovery of new methods of combating disease. This latter period is frequently called the age of *iatrochemistry*.

THE ERA OF THE PHLOGISTON THEORY

Most scientists of this period believed that a substance called *phlogiston* caused burning and escaped during combustion. Because some substances burned to an ash that weighed more than the substance consumed, they assumed that in these cases, phlogiston had a negative weight. The phlogiston theory flourished during the first seven decades of the eighteenth century and was disproved in 1777 by Lavoisier's discovery of the true nature of combustion. However, many of its adherents were slow to abandon the theory. Priestley, the discoverer of oxygen, believed in the phlogiston theory until his death, and his last writings were in its defense.

THE MODERN PERIOD OF CHEMISTRY

Beginning with the discovery of oxygen by Priestley in 1774 and the explanation of the true nature of combustion by Lavoisier a few years later, this period saw the rapid development of chemistry as a science, primarily because of the application of the scientific method to its problems.

The following is a chronological listing (not to be memorized) of some of the contributions of chemistry and related fields of science that have had a great influence on modern medicine and biology:

1774 Priestley discovers oxygen.
1777 Lavoisier discovers the true nature of combustion, the death blow to the phlogiston theory.
1802 Dalton proposes the atomic theory.
1818 Chevreul discovers cholesterol.
1820 Strychnine, brucine, and quinine first isolated.
1822 Dr. Beaumont studies gastric secretion via a stomach fistula.
1828 Wöhler synthesizes urea to overthrow the vitalistic theory.
1830 Nicotine isolated.
1831 Atropine isolated.
1839 Mulder recognizes and suggests name for proteins.
1845 First use of ether in anesthesia.
1856 First synthetic dye, mauve, prepared by Perkin.
1858 Kekulé suggests the tetracovalency of carbon.
1857 Bernard isolates glycogen from the liver.
Pasteur studies ferments (enzymes).
1860 Kekulé and Couper suggest first modern concept of valence and molecular structure.
1861 Traube establishes that biological oxidations occur in cells and not in the liver or blood.
1865 Kekulé deduces the cyclic structure for benzene.
1875 First acid hydrolysis of proteins.
1877 Franz Boll isolates rhodopsin, or visual purple, from the rods of the retina.
1879 Baeyer synthesizes the dye indigo.
1883 Fischer determines structure of sugars.
1896 The Büchner brothers establish that enzymatic action can be extracellular.
1897 Abel and Crawford isolate adrenalin.
1901 Fischer and Fourneau synthesize first peptide, glycyl-glycine.
Loew recognizes presence of enzyme catalase in many tissues.
1902 Fischer and Hofmeister independently propose the peptide hypothesis of protein structure.
Bayliss and Starling discover secretin, a hormone of the digestive system.
1904 Knoop suggests β-oxidation theory of fat metabolism.

1906 Willstätter begins study of chlorophyll.
1907 Sørensen suggests the use of the *p*H notation to indicate acidity and basicity.
Hopkins studies amino acids in nutrition.
1908 Gelmo prepares sulfanilamide.
1909 Ehrlich prepares salvarsan and becomes the father of chemotherapy.
1911 Funk isolates the antiberiberi principle from rice polishings and suggests the name *vitamine* for these substances.
1912 Warburg initiates experimental studies on the metabolism of tissue slices.
1913 McCollum and Davis isolate vitamin A.
1917 Hill, Meyerhof, and Embden establish that lactic acid is a product of muscular action.
1919 Kendall isolates thyroxine.
1920 MacLeod, Banting, and Best isolate insulin and establish its role in carbohydrate metabolism.
1922 Evans and Sure independently isolate vitamin E, the antisterility factor.
1923 Allen and Doisy prepare an active extract of the estrogenic hormones.
1925 Keilin identifies the iron-containing oxidation enzymes called *cytochromes*.
Collip establishes the existence of the parathyroid hormone.
1926 The synthetic antimalarial pamaquin is prepared.
Sumner prepares crystalline urease.
1928 Rose establishes that certain amino acids are essential to growth.
1929 Fleming observes that *Penicillium notatum* produces a substance, which he calls *penicillin*, that destroys staphylococcus organisms.
1930 Northrop prepares crystalline pepsin.
1932 King and Waugh isolate crystalline vitamin C.
1933 Coenzymes isolated and study of their structure begins.
1935 Stanley isolates tobacco mosaic virus.
Dam isolates vitamin K.
Laquer and associates isolate the male sex hormone testosterone from bull testes.
Structure of nicotine adenine dinucleotide phosphate (NADP) established.
1936 Kögl and Tonnis isolate biotin from dry egg yolk.
Krebs suggests the citric acid cycle for the aerobic breakdown of pyruvic acid.
Stanley establishes the nucleoprotein nature of the viruses.
Structure of nicotine adenine dinucleotide (NAD) established.
Preparation of prontosil by Domagk initiates use of sulfa drugs.
Carl and Gerti Cori isolate glucose-1-phosphate, an intermediate in glucose metabolism.

1937 Elvehjem and associates recognize nicotinic acid as the anti-pellagra factor.
1939 Ansbacher and Fernholz find that synthetic menadione has vitamin K activity.
1940 Florey, Chain, and associates at Oxford begin the study of the isolation of penicillin.
1944 Waksman and associates isolate streptomycin from cultures of *Streptomyces griseus.*
R. Woodward and W. Doering synthesize the antimalarial quinine.
1948 Isolation of crystalline vitamin B_{12}.
1952 Isoniazid found effective in the treatment of tuberculosis.
1953 Du Vigneaud synthesizes oxytocin, a hormone produced by the pituitary.
Watson and Crick propose that in DNA, two polynucleotide chains form a double helix.
1954 Du Vigneaud and associates announce the synthesis of vasopressin, a hormone of the pituitary.
1955 Sanger and his students establish the polypeptide structure of insulin.
Franke and Fuchs demonstrate the effectiveness of oral doses of Carbutamide, a sulfa drug, in the control of diabetes.
1960 Robert Woodward reports the total synthesis of chlorophyll a.
1961 Dr. Klaus Hofmann and associates at the University of Pittsburgh synthesize a peptide with the same biological activity as the adrenocorticotrophic hormone (ACTH) produced by the pituitary of the pig.
Dr. Melvin Calvin receives the Nobel prize in chemistry for his elucidation of the pathway of photosynthesis.
1962 Drs. John Kendrew and Max Perutz receive Nobel prize in chemistry for their X-ray studies of the structure of hemoglobin and muscle myoglobin.
1963 Robert Schwyzer and P. Sieber of Ciba, Ltd., announce the total synthesis of the adrenocorticotrophic hormone of the pig pituitary.
The amino acid sequence of the enzyme ribonuclease is established by Hirs, Stein, Moore, and Anfinsen.
1964 Drs. K. Bloch and F. Lynen receives the Nobel prize in medicine for establishing the mechanism of the metabolism of fatty acids and cholesterol.
1966 The synthesis of bovine insulin is reported.
1967 Drs. A. Kornberg and M. Goulian make an exact copy of the DNA molecule that occurs in the bacteriophage QX 174.
1968 Drs. R. Holley, H. G. Khorana, and M. Nirenberg receive the Nobel prize in medicine for establishing the role of the genetic code in protein synthesis.
1969 Both Merck, Sharp and Dohme Laboratories and Rockefeller University announce the synthesis of the enzyme ribonuclease.

Dr. J. Beckwith and associates at Harvard University isolate individual genes from *Escherichia coli.*

1972 The structure of the active portion of the human parathyroid hormone is established and the biologically active portion of the molecule is synthesized in the Ciba-Geigy Laboratories in Switzerland.

STUDY QUESTIONS

1. Give a suitable definition of (a) science, (b) chemistry, and (c) physics.
2. How does a scientific law differ from a civil law?
3. What contributions did the early Egyptians make to chemistry? Why was the chemistry of their day considered an art rather than a science?
4. The Greek philosophers believed that the earth was made of a combination of four elements. What were these four elements?
5. What were the goals of the alchemists?
6. What was the phlogiston theory, and why was it abandoned?
7. When did the modern period of chemistry begin?
8. What contributions to science did Democritus, Paracelsus, Priestley, and Lavoisier make?

chapter 2

some fundamental concepts of chemistry

2

Before proceeding further, the student, depending upon his background, should become acquainted with or review some of the fundamental concepts of chemistry. This chapter discusses matter and energy, varieties of matter (based on composition), chemical and physical changes, atoms and molecules, and symbols and formulas. Atomic theory is discussed in more detail than in Chapter 1.

The *metric system of weights and measurements* is also introduced in this chapter. Nearly all measurements in chemistry employ the metric system, and although the system is not in general use in the United States, it is now the official system of weights and measurements for science and medicine. A study of the metric system, density, and specific gravity is a valuable introduction to pharmaceutical arithmetic.

The student will also learn about heat energy units, such as the calorie and the degree, knowledge which will be needed later when he studies basal metabolic rate (BMR) and calorimetry (see pages 422–425).

Matter

In the introduction, chemistry was defined as the science that deals with matter and its transformations. *Matter* is defined as anything that has mass or occupies space. Although the terms "mass" and "weight" are fundamentally quite distinct, they are frequently substituted for each other. The *mass* of a body is the quantity of matter that it contains. *Weight* is a measure of the gravitational attraction for a body. If an object is taken from the earth to the moon, its mass will not change. Because the moons gravitational attraction is considerably smaller than the earth's, the weight of an object on the moon will be appreciably smaller than its weight measured on the surface of the earth. On the earth's surface, weight is proportional to mass, and the two terms are used interchangeably (Figure 2.1).

Energy

Physics deals with the varieties and the various transformations of energy. *Energy* is the power or ability to do work. The numerous

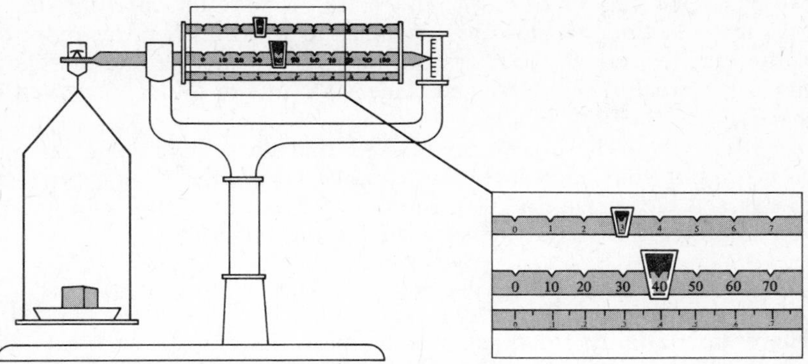

Figure 2.1 A balance, such as the one illustrated, can be used to determine mass. Because weight is a measure of gravitational attraction, if the object on the balance were on the moon, it would weigh one-sixth what it weighs on earth. But since mass is a constant, the mass of the object (43.22 g) is identical on the earth and the moon.

varieties of energy, such as heat, electrical, radiant, mechanical, and chemical energy, may be freely transformed one into another. A definite quantity of one kind is always converted into a certain amount of another type; thus 1 watt-hour of electrical energy is equivalent to 860 calories of heat.

Kinetic energy is the energy of a moving body. A body in a state or position such that it can produce work is said to have *potential energy,* which is, therefore, stored energy. *Chemical energy* is an important type of potential energy. When coal burns, it combines with oxygen, and a certain amount of energy stored in the coal as chemical energy appears as heat and light. Electrical energy is stored as chemical energy when a storage battery is recharged. When the storage battery is discharged, chemical energy is transformed into electrical energy, which may be further transformed into heat, light, or mechanical energy. In the body, the chemical energy in the food eaten is converted into both heat and the mechanical energy of motion. In the firefly, chemical energy, in part, is transformed into light, and in electric eels, chemical energy is transformed into electrical energy.

Chemical reactions in which heat is evolved are called *exothermic reactions,* and those in which heat is absorbed by the reactants are called *endothermic reactions.* In an exothermic reaction, the amount of chemical energy present in the reacting substances is decreasing, whereas in endothermic reactions, heat is being transformed and stored as chemical energy in the products.

The laws of the conservation of matter and energy

In 1775, Lavoisier stated that *matter can be neither created nor destroyed.* This statement is now known as the *law of conservation of matter*. According to this law, the mass of material entering into a chemical reaction will always be equal to the mass of the products of the reaction. In all ordinary transformations, energy, like mass, can be neither created nor destroyed. This is the *law of the conservation of energy*.

Recently it has been established that these two laws are not rigorous but that under certain conditions, matter is converted to energy and vice versa. In 1905, Einstein suggested the following equation to indicate the equivalence of mass and energy:

$$E = mc^2$$

where E is energy (in ergs),* M is mass (in grams), and c is the velocity of light (in centimeters per second). In all ordinary chemical reactions, the amount of mass changed to energy is so insignificant that the loss may be disregarded. In the explosion of a nuclear energy bomb, an appreciable amount of mass from the atomic nuclei is being transformed into a very large quantity of energy. From Einstein's equation, it can be calculated that 1 gram of matter will be equivalent to 21,500,000,000 kilocalories of energy.

Metric units of weight and measurements

The metric system of weights and measurements is a decimal system based on multiples and submultiples of ten. It was introduced in France during the French Revolution and is now used universally except in some English-speaking countries. Official international standards of the metric system are maintained at the International Bureau of Weights and Measures at Sèvres, near Paris. National standards are compared with these, and copies are maintained at our Bureau of Standards in Washington.

The following table indicates the prefixes for the multiples and submultiples of the units of the metric system:

kilo-	1000 units
hecto-	100 units
deka-	10 units
deci-	$\frac{1}{10}$ unit
centi-	$\frac{1}{100}$ unit
milli-	$\frac{1}{1000}$ unit
micro-	$\frac{1}{1,000,000}$ unit

THE METRIC UNIT OF LENGTH

The standard units of length in the metric system is the meter (m). In 1875, representatives of 17 nations established the meter as

*The erg is a unit of energy and is the work done when a force of 1 dyne acts through a distance of 1 centimeter. A dyne is the force required to give a mass of 1 gram an acceleration of 1 centimeter per second. For further details, consult a physics text.

equal to the distance measured at the freezing point of water (0° Celsius) between two marks on a platinum–iridium bar maintained at Sèvres. Because of a demand for a more precise definition of the meter, in 1960 the wavelength of the orange-red line of the spectrum of the krypton atom of mass 86 (see page 33) replaced the meter bar at Sèvres as the formal international standard of length. The meter is defined as 1,650,763.37 times the length of the orange-red line of krypton 86. Other units of length frequently used are the kilometer (km, 1000 m) and the centimeter (cm, $\frac{1}{100}$ m). Ten millimeters (mm) equal 1 cm. One meter is equal to 39.37 inches, and 1 inch equals 2.54 cm.

ILLUSTRATIVE PROBLEM 1
Convert 55 cm to millimeters.

$$55\ \cancel{\text{cm}} \times \frac{10\ \text{mm}}{1\ \cancel{\text{cm}}} = 550\ \text{mm}$$

ILLUSTRATIVE PROBLEM 2
Convert 4 in. to centimeters.

$$4\ \cancel{\text{in}}. \times \frac{2.54\ \text{cm}}{1\ \cancel{\text{in}}.} = 10.16\ \text{cm}$$

THE METRIC UNIT OF MASS

The unit of mass in the metric system is the gram (g). It is equal to $\frac{1}{1000}$ of the weight of the standard kilogram, which is a piece of platinum–iridium stored at the International Bureau of Weights and Measures (Figure 2.2). The gram is approximately the weight of 1 cubic centimeter (cm^3) of water at its maximum density (4° Celsius). One pound is equal to 454 g. Other units of mass are the kilogram

Figure 2.2 *The United States Prototype Kilogram No. 20. The diameter is 39 mm, and the height is 39 mm. (Courtesy National Bureau of Standards, Washington, D.C.)*

(kg, 1000 g, about 2.2 lb) and the milligram (mg, 1/1000 g). Weights in chemistry are usually expressed decimally in grams.

ILLUSTRATIVE PROBLEM 3

A man weighs 154 pounds. Express his weight in kilograms.

$$154 \text{ lb} \div \frac{2.2 \text{ lb}}{1 \text{ kg}} = 154 \not{\text{lb}} \times \frac{1 \text{ kg}}{2.2 \not{\text{lb}}} = 70 \text{ kg}$$

THE METRIC UNIT OF VOLUME

The unit of volume in the metric system is the liter. A liter is the volume of a cube 10 cm on a side, or 1000 cm^3. It was originally intended that a kilogram of water at its maximum density (4°C) would occupy a volume of 1 liter. Due to some discrepancies in the measurement, a kilogram of water occupies a volume of 1000.028 cm^3. This variation is so small that, for practical purposes, one can consider that a gram of water at 4°C occupies a volume of 1 cm^3, or 1 milliliter (ml). One liter is equal to 1.06 liquid quarts, while an ounce of liquid has an approximate volume of 30 ml.

ILLUSTRATIVE PROBLEM 4

What is the volume in liters of a jar that holds 5 quarts of water?

Because there are 1.06 quarts in a liter volume, then a quart will be equal to 1/1.06 liter, and five quarts will be equivalent to

$$5 \text{ qt} \times \frac{1 \text{ liter}}{1.06 \text{ qt}} \quad \text{or} \quad 4.72 \text{ liters}$$

DENSITY AND SPECIFIC GRAVITY

Density is the mass or weight of a substance per unit volume and is expressed as

$$\text{density} = \frac{\text{mass}}{\text{volume}}$$

or as

$$D = \frac{M}{V}$$

Density is expressed as grams per milliliter in the metric system and as pounds per cubic foot in the English system (Table 2.1). A milliliter of liquid mercury weighs 13.6 g; therefore its density is expressed as 13.6 g/ml. Water has a density of 1 g/ml, because 1 g of water at 4°C will occupy a volume of 1 ml.

ILLUSTRATIVE PROBLEM 5

A metal cylinder has a volume of 26 ml and weighs 104 g. What is its density?

$$D = \frac{M}{V} = \frac{\text{grams}}{\text{milliliters}} = \frac{104 \text{ g}}{26 \text{ ml}} = 4 \text{ g/ml}$$

TABLE 2.1 DENSITIES OF COMMON SUBSTANCES IN GRAMS PER MILLILITER (ROOM TEMPERATURE)

Substance	Density	Substance	Density
Hydrogen (gas)	0.00009	Table salt	2.16
Carbon dioxide (gas)	0.00198	Sand	2.32
Balsa wood	0.16	Aluminum	2.70
Cork wood	0.21	Iron	7.9
Oak wood	0.71	Silver	10.5
Water	1.00	Lead	11.3
Eucalyptus wood	1.06	Mercury	13.6
Magnesium	1.74	Gold	19.3

Specific gravity is the ratio of the weight of a substance and an equal volume of water. Because concentrated sulfuric acid is 1.84 times as heavy as an equal volume of water, its specific gravity (sp gr) is 1.84.

ILLUSTRATIVE PROBLEM 6

What is the specific gravity of glycerol if 200 ml of this substance weighs 252 g?

Because 1 ml of water weighs 1 g, then 200 ml of water will weigh 200 g. The specific gravity of the glycerol will be the ratio of the weight of the 200 ml of glycerol and the 200 ml of water, or

$$\text{sp gr of glycerol} = \frac{252}{200} = 1.26$$

In the metric system, density and specific gravity are numerically equal.

The measurement of heat energy

In the measurement of heat energy, units for the measurement of intensity and quantity of heat are necessary. The *calorie* (Figure 2.3), the *kilocalorie*, and the *BTU* (British Thermal Unit) are used in measuring the quantity of heat. A calorie (cal) is the quantity of heat energy required to raise the temperature of 1 g of water 1°C.* The kilocalorie (kcal), or kilogram-calorie, is equal to 1000 cal. The caloric value of foods is generally expressed in kilocalories per gram; thus energy equivalent to 4 kcal is produced from each gram of carbohydrate or protein utilized in the body, while fats produce 9 kcal per gram. The BTU is the quantity of heat required to raise the temperature of 1 lb of water 1°F.

ILLUSTRATIVE PROBLEM 7

Combustion of 5 g of fat will heat a kilogram of water from 20° to 65°C. How many calories of heat are generated by the combustion of a gram of fat?

*In exact measurements, the temperature change is taken from 14.5 to 15.5°C.

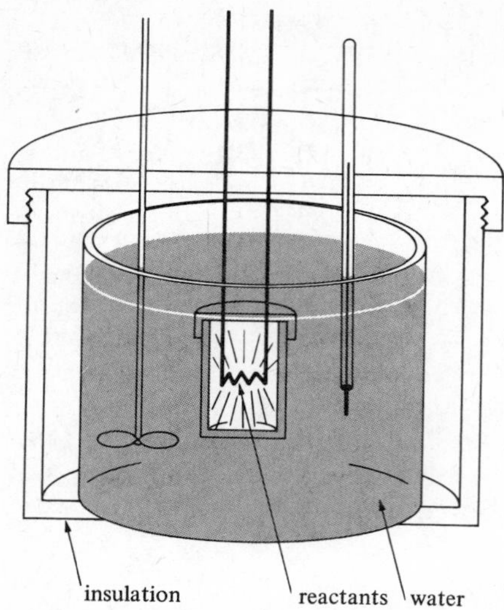

Figure 2.3 A bomb calorimeter.

The 5 g of fat produces

$$1000 \text{ g } H_2O \times (65^\circ - 20^\circ C) = 1000 \text{ g} \times 45^\circ$$
$$= 45{,}000 \text{ cal}, \quad \text{or} \quad 45 \text{ kcal}$$

The 1 g of fat will generate 9,000 cal, or 9 kcal of heat on combustion.

Thermometers are instruments used to measure the intensity of heat. The Celsius thermometer is used to measure temperature in scientific work. The freezing point of water is designated as 0°C (Celsius), while the boiling temperature of water at a pressure of 760 mm is taken as the 100° mark on the Celsius thermometer. The space on this scale between these two marks is divided into 100 equal segments. On the Fahrenheit scale, 32° is the freezing point of water and 212° is its boiling point (at 760 mm pressure). There are 180 degrees between the freezing point and the boiling point of water on the Fahrenheit scale. One Celsius degree is then equal to $\frac{9}{5}$ or 1.8 Fahrenheit degrees (Figure 2.4).

The following formulas are employed to convert Fahrenheit temperatures (°F) to Celsius temperatures (°C) and vice versa:

$$°F = (°C \times \tfrac{9}{5}) + 32$$
$$°C = (°F - 32) \times \tfrac{5}{9}$$

ILLUSTRATIVE PROBLEM 8

Convert 20° C to the corresponding Fahrenheit temperature.

$$°F = (20^\circ \times \tfrac{9}{5}) + 32^\circ = 36^\circ + 32^\circ = 68^\circ \text{ F}$$

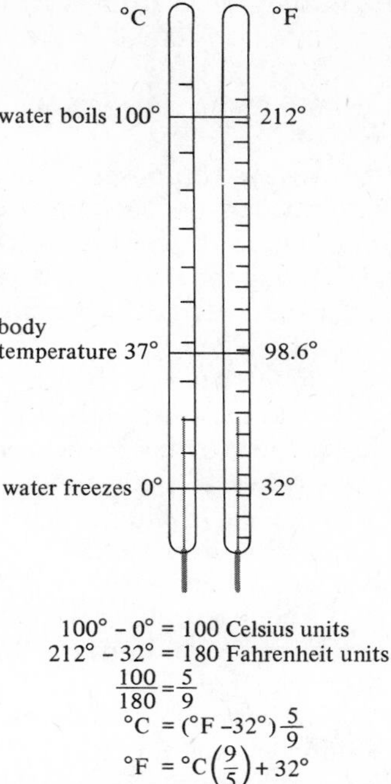

Figure 2.4 A comparison of the Celsius (centigrade) and Fahrenheit temperature scales.

ILLUSTRATIVE PROBLEM 9
Convert 14° F to the corresponding Celsius temperature.

$$°C = (14° - 32°) \times 5/9 = -10°C$$

ILLUSTRATIVE PROBLEM 10
Convert −13° F to the corresponding Celsius temperature.

$$°C = (-13° - 32°) \times 5/9 = -45° \times 5/9 = -25°C$$

Varieties of matter

From the standpoint of composition, there are three varieties of matter: elements, compounds, and mixtures.

The *elements* are elementary particles of matter and cannot be decomposed by chemical means into simpler substances. There are now 103 elements known, of which hydrogen is the simplest and lawrencium is the most complex in structure.*

*Discovery of element 104 is claimed by a Russian group at Dubna and by a group at the Lawrence Radiation laboratory in California. The latter group also claim that they have prepared element 105, to which they have assigned

Elements are the building units from which chemical compounds are formed, and every compound is composed of two or more different elements.

The following are four important characteristics of *compounds:*

1. Compounds are always composed of two or more elementary substances, or elements, and can be decomposed into these elements by chemical means. Iron sulfide is formed by the union of iron and sulfur. Sodium chloride (common salt) is composed of the metal sodium and the gas chlorine. Water is decomposed by electrolysis into the gases hydrogen and oxygen.
2. Compounds are *homogeneous;* that is, all the particles of the compound are identical in appearance and in chemical composition.
3. Compounds have their own characteristic properties, which are different from the properties of their constituent elements. Iron is magnetic and reacts with acids to liberate hydrogen gas; sulfur is soluble in carbon disulfide. But iron sulfide is nonmagnetic, is insoluble in carbon disulfide, and reacts with acids to produce foul-smelling hydrogen sulfide. White crystalline sodium chloride is formed from soft metallic sodium and the poisonous yellow-green gas chlorine.
4. All compounds have a definite and fixed composition. This is known as the *law of definite proportion or definite composition.* Pure iron sulfide (FeS) always contains 63.52 percent iron and 36.48 percent sulfur. Pure water contains 11.19 percent hydrogen and 88.81 percent oxygen.

Mixtures are generally heterogeneous; that is, they are composed of two or more distinctly different particles. In a mixture of iron and sulfur, the iron and sulfur particles can be discerned easily. A mixture has a variable composition, and the various constituents of the mixture can usually be separated by mechanical means. In a mixture of iron and sulfur, the iron particles are removed by a magnet and the sulfur may be dissolved and removed from the mixture by carbon disulfide. By evaporating the resulting carbon disulfide solution, the sulfur is recovered. The compound iron sulfide, if pure, is not attracted by the magnet, nor does it or any part of it dissolve in the carbon disulfide.

Solutions may be considered as a special type of mixture, but unlike most mixtures they are homogeneous. In solutions, all particles dispersed in the solvent are of molecular dimension. A further discussion of solutions appears in Chapter 7.

---

the name hahnium in honor of the German physicist Otto Hahn. The Russian group has assigned the name kurchatovium to element 104 in honor of the Russian physicist I. V. Kurchatov, while the California group call this element rutherfordium to honor the British physicist Lord Rutherford. The International Union of Pure and Applied Chemistry, however, has not as yet assigned official names to elements 104 and 105. See the June 1970 issue of *Scientific American* (page 83).

States of matter

Gases, liquids, and *solids* represent the three states of matter. Gases have neither a fixed volume nor a fixed shape and spread through and occupy the entire container into which they are placed.

Liquids have a fixed volume but no fixed shape. The molecules in a liquid are much closer than in a gas, as evidenced by the low compressibility of liquids compared to the high compressibility of gases. Heat is required to separate the molecules of a liquid and convert it to a gas. This is called the *latent heat of vaporization* of the liquid.

A solid has rigidity; that is, it has a fixed size and shape and possesses mechanical strength. The attractive forces between the molecules are much greater in solids than in liquids; the *latent heat of fusion* is the heat required to convert the solid to the liquid state.

Physical and chemical changes

Physical changes take place in matter without modifying its composition. The melting of sulfur, the evaporation of ether, and the freezing of water are examples of physical changes.

Chemical changes involve the modification of the composition of matter. Some examples are the burning of wood, the rusting of iron, the souring of milk, the rotting of wood, and the union of iron and sulfur to form iron sulfide.

Atoms and molecules

Early chemists were unable to distinguish between atoms and molecules. As chemistry developed as a science, it was recognized that molecules of a compound were made up of a smaller number of particles (atoms) of each of the constituent elements.

If a portion of water is continuously subdivided, eventually a particle, the molecule, consisting of two hydrogen particles and one oxygen particle will be obtained which still possesses the properties of water. Further division of this molecule can lead only to particles of hydrogen and oxygen, and the properties of water will no longer be evident. *A molecule of a compound can be defined as the smallest particle that still possesses the properties of the compound.*

The particles of elements that enter into chemical reaction are called *atoms.* Each oxygen molecule contains two atoms. We are certain of this fact because when oxygen combines with hydrogen to form water, each molecule of oxygen produces two molecules of water, and there must be at least one oxygen atom in each water molecule. Incidently, the properties that are ascribed to oxygen in Chapter 4 are those of the molecule, because individual free oxygen atoms are much more reactive than atmospheric oxygen.

The molecules of the following elements also contain two atoms; that is, they are *diatomic:* fluorine, chlorine, bromine, iodine, hydrogen, and nitrogen. In the case of a number of the elements, such as helium, mercury, and sodium, the atom and molecule are identical (*monatomic* molecules). Ozone is a *triatomic* form of oxygen, whereas phosphorus is *tetratomic* and the crystalline forms of sulfur are *octatomic.*

Symbols

Symbols are abbreviations for the names of the chemical elements. The symbols of some elements are derived from the Latin name of these elements. Some examples are Na for sodium (L. *natrium*), Cu for copper (L. *cuprum*), Ag for silver (L. *argentum*), Fe for iron (L. *ferrum*), Hg for mercury (L. *hydrargyrum*), and Pb for lead (L. *plumbum*).

Some other symbols are derived from the first or first two letters of the English name of the element. Thus

O	oxygen	H	hydrogen
S	sulfur	N	nitrogen
P	phosphorus	He	helium
Ne	neon	Ca	calcium

If this rule were always followed, several elements would have the same symbol; in such cases, some elements have symbols consisting of the first letter and some other characteristic letter of their name.

Mg	magnesium	Mn	manganese
Cd	cadmium	Cl	chlorine

To the chemist, a symbol is an abbreviation of the name of the element and may also designate one atom of the element (see page 24).

Formulas

A *formula* is a combination of symbols of elements that is a shorthand method of indicating a compound and in addition indicates the chemical composition of the compound. The formula FeS is used not only to designate iron sulfide but also to indicate that one atom of iron is combined with one atom of sulfur in this compound. The formula NaCl indicates that common salt is composed of an equal number of atoms of the metallic solid sodium and the yellow-green gas chlorine. Because the formula for sulfuric acid is H_2SO_4, each molecule of this compound contains two atoms of hydrogen, one atom of sulfur, and four atoms of oxygen.

A formula may also be interpreted in a quantitative fashion. In such an interpretation, the formula represents one molecule of the compound. If two molecules of salt are to be indicated, this is expressed as 2NaCl.

The atomic theory

The Greek philosopher Democritus first suggested that all matter was particulate in nature and was made up of hard, indivisible particles. The law of definite composition further suggests that matter must be composed of distinct individual particles, called atoms, which function as units during chemical reactions.

In 1802, the English teacher and philosopher John Dalton revived the atomic theory to describe the structure of matter. In his version of the atomic theory, he made the following assumptions:

1. All matter is composed of tiny indivisible particles called atoms.
2. Atoms of any one chemical substance are identical and have the same weight, size, and shape.
3. The atom is the smallest particle of matter that can enter into a chemical reaction.
4. Atoms are indestructible.
5. Compounds are formed by the union of two or more atoms of different elementary substances.

Most of the above statements are accepted today, although some of them need modification. At present, it is known that the atom is divisible and is made up of smaller particles called *electrons, protons,* and *neutrons.* Furthermore, in radioactivtity (see Chapter 14), atoms spontaneously decompose into smaller particles and new atoms. The atoms of a certain element may not be identical in all details. (See the discussion of isotopes on page 31.)

Atomic weights

Because atoms are too small to be weighed individually (the carbon atom has a weight of 0.00000000000000000000002 g), equal large numbers of atoms of elements may be weighed, and relative weights are assigned to the elements by comparing the resulting weights with some arbitrary standard. These relative weights of atoms are called *atomic weights.* In 1961, the International Union of Pure and Applied Chemistry (IUPAC) adopted the ordinary carbon atom as the standard of atomic weights and assigned to it an atomic weight of 12. If the carbon atom has an atomic weight of 12, hydrogen will have an approximate atomic weight of 1. Thus the carbon atom is approximately 12 times as heavy as a hydrogen atom. Because a sulfur atom is about 32 times as heavy as a hydrogen atom, its atomic weight will be approximately 32. The table of atomic weights of the elements is given inside the back cover.

Originally oxygen was chosen as the standard of atomic weights. Due to some difficulties in defining an exact atomic weight for oxygen, it was abandoned in favor of carbon, which is the accepted standard. The new atomic weights will differ slightly from those previously in use. Thus the atomic weight of oxygen will be 15.994

instead of 16.000, and that of hydrogen will be 1.00797 instead of 1.0080.

The *gram-atomic weight*, or the *gram-atom*, of an element is the atomic weight of the element expressed in grams. For carbon, the gram-atomic weight is 12 g.

Note that the symbol of an element may designate a gram-atomic weight of the element (see page 22).

STUDY QUESTIONS

1. Define the following terms: atom, matter, calorie, element, energy, molecule, kilocalorie, exothermic reaction, compound, mixture, atomic weight, endothermic reaction.
2. State the following laws: the law of the conservation of matter; the law of definite composition; the law of the conservation of energy.
3. Are the laws of the conservation of matter and energy rigorously true? Explain.
4. Contrast weight with mass; density with specific gravity.
5. Discuss the advantages of the metric system. What are the units of length, mass, and volume in this system of measurements?
6. Give an example of chemical energy. Is chemical energy kinetic or potential energy?
7. List several varieties of energy.
8. If 23 g of sodium combines with 35.5 g of chlorine to form sodium chloride, how many grams of sodium chloride will be formed? Why?
9. The freezing point and the boiling point of water are frequently used in the calibration of thermometers. What are these temperatures on the Celsius scale? On the Fahrenheit scale? One degree Celsius is equal to how many Fahrenheit degrees?
10. How do physical and chemical changes differ?
11. Which of the following are chemical changes: burning of coal; melting of butter; solution of sugar in water; digestion of food; souring of milk; breaking of a dish; evaporation of water; ripening of fruit?
12. What are the three varieties of matter?
13. What characteristics do all compounds exhibit?
14. How do mixtures differ from compounds?
15. How do solutions differ from mixtures? From compounds?
16. For what are symbols and formulas abbreviations? Do they have any other significance?
17. List several symbols of elements that are derived from the Latin name of the element.
18. List several symbols of elements that are derived from the first letter of the English name of the element.
19. Find in the table of atomic weights (inside back cover) the symbols for xenon, selenium, potassium, mercury, scandium, tungsten, cadmium, arsenic, boron, beryllium, and lithium.
20. What are the assumptions that Dalton made in his atomic theory? How has his theory been modified?
21. What element has been chosen as the standard for atomic weights? Why?
22. A bromine atom is five times as heavy as an oxygen atom, while a carbon atom is three-fourths the weight of an oxygen atom. What are the atomic weights of these elements?

PROBLEMS

1. Calculate the height of a man 6 ft. tall in (a) meters and (b) centimeters.
2. Convert 1 qt to cubic centimeters.
3. What is the weight of a 150 lb man in (a) kilograms and (b) grams?
4. What is the approximate volume in milliliters of a 3 oz bottle?
5. Convert 4 in. to centimeters; 100 in. to meters; 16 kg to pounds; 22.5 g to ounces and pounds.
6. Express the sum of the following weights in *grams:* a 10 g weight, a 5 g weight, two 2 g weights, a 500 mg weight, two 20 mg weights, a 5 mg weight, a 2 mg weight and a 1 mg weight.
7. Normal body temperature is 98.6°F. What is normal body temperature on the Celsius scale?
8. Dry Ice sublimes at −80°C. Convert this temperature to the corresponding Fahrenheit temperature.
9. Liquid ammonia boils at −40°F. At what temperature does it boil on the Celsius scale?
10. Room temperature is about 75°F. What is room temperature on the Celsius scale?
11. Aluminum melts at 659°C. At what Fahrenheit temperatures does it melt?
12. How many calories of heat are required to raise the temperature of 550 g of water from 20°C to 37°C?
13. A chemical reaction was made to take place in a calorimeter that contained 1 kg of water. The heat of the reaction raised the temperature of the water from 18°C to 26.2°C. Calculate the number of calories of heat evolved. Assume that all the heat of the reaction is absorbed by the water.
14. Twenty-five milliliters of a certain liquid weighs 33 g. What is the density of this liquid?
15. Glycerol has a specific gravity of 1.26. What is the weight of 3 liters of glycerol?
16. A certain volume of glucose solution weighs 19.78 g, while an equal volume of water weighs 15.38 g. What is the specific gravity of this solution?
17. Stainless steel has a specific gravity of 7.8. What is the weight of a stainless steel bar that is 15 cm long, 2.5 cm wide, and 4 mm thick?
18. Benzene has a density of 0.87 g/ml. What is the volume of 200 g of benzene?

chapter 3

atomic structure, compound formation, and valence

3

The material presented in this chapter explains how and why atoms combine to form chemical compounds and assists the student in determining the valences of the various atoms and radicals and in writing chemical formulas.

A discussion of the periodic law and the periodic table is included in this chapter, although it is not essential to the continuity of this text. The chemists of the nineteenth century realized that chemistry was becoming more detailed and difficult to study as more chemical elements were discovered and more and more chemical compounds were described. Mendeleev did a great service for chemistry in developing the periodic table, making it possible to correlate much of the information of chemistry and to note the similarities in properties of certain families of elements.

The structure of atoms

SUBATOMIC PARTICLES OF MATTER

As we have noted, John Dalton believed that atoms of elements were hard, indivisible particles. A number of types of particles have now been recognized as products of the disintegration of atoms. At least three of these are present in the atom, namely the *proton, electron,* and *neutron.* These are sufficient to explain the structure of atoms and the nature of compound formation. Some facts about these subatomic particles are given in Table 3.1.

THE ATOMIC NUMBER AND LOCATION OF THE SUBATOMIC PARTICLES IN THE ATOM

An atom contains a compact dense center called the *nucleus*, in which all the protons and neutrons of the atom are located. Hydrogen is an exception, because its nucleus is composed of a single proton and contains no neutrons. The outer (extra-nuclear) structure of the atom is composed of electrons, which are not arranged in a random fashion but occur in definite orbits and energy shells. The nucleus of the atom always bears a positive charge, because it contains protons and neutrons but no electrons.

Although it has a dense nucleus, an atom has a very open structure. This was demonstrated by Lord Rutherford, who fired alpha (α)

TABLE 3.1 SOME INFORMATION CONCERNING PROTONS, ELECTRONS, AND NEUTRONS

	Proton (p)	*Electron* (e)	*Neutron* (n)
Electrical charge	+1	−1	0
"Mass"	1[a]	1/1845[a]	1[a]
Discoverer	Goldstein	Sir J. J. Thomson	Chadwick
Year of discovery	1886	c. 1900	1932

[a]As compared to a hydrogen atom with a "mass" (atomic weight) of approximately 1.

particles (charged helium atoms) into a sheet of thin gold foil. Most of the α particles passed through the gold foil undeflected; however, a few were deflected and some were completely reflected (Figure 3.1). These observations were explained by the fact that most of the α particles sped through the open part of the atom, while a few of the positively charged α particles approached the nucleus closely and were deflected by its positive charge. A few particles approached the nucleus head-on and were reflected.

Because the masses of the protons and neutrons are about unity and the electrons are of very small mass when compared to the former, the atomic weight of an atom will be equal to the sum of the number of protons and neutrons in the nucleus.

All of the atoms of the various elements have been assigned numbers. Hydrogen has been assigned the number 1 because it is the simplest of atoms, while nobelium has been assigned the number 102. These are called the *atomic numbers* of the atoms and are numerically equal to the positive charge on the nucleus; or in other words,

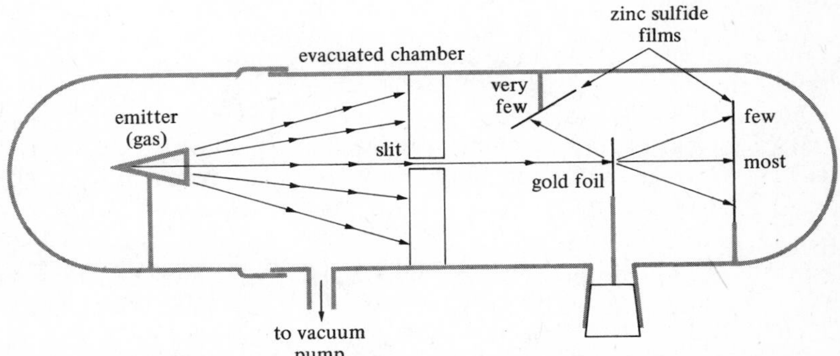

***Figure 3.1** An experiment in which gold foil is bombarded by a beam of α-particles. After passing through the gold foil, each α-particle collides with the zinc sulfide screen and emits a small flash of light.*

TABLE 3.2 NUMBER OF PROTONS, NEUTRONS, AND ELECTRONS IN SOME SIMPLE ATOMS

Name of Element	*Atomic Weight*	*Atomic Number*	*Number of Protons*	*Number of Neutrons*	*Number of Electrons*
Helium	4	2	2	2	2
Lithium	7	3	3	4	3
Carbon	12	6	6	6	6
Nitrogen	14	7	7	7	7
Fluorine	19	9	9	10	9
Sodium	23	11	11	12	11
Chlorine	35	17	17	18	17
Argon	40	18	18	22	18
Potassium	39	19	19	20	19
Calcium	40	20	20	20	20

the atomic number is equal to the number of protons in the nucleus. The electrons in an atom are equal to the number of protons, because an atom has zero charge.

The number of neutrons in an atom is equal to the difference between the atomic weight and the atomic number of the atom. In Table 3.2, the atomic weights, atomic numbers, and number of protons, neutrons, and electrons in some typical atoms are indicated.

THE STRUCTURE OF INDIVIDUAL ATOMS

The ordinary hydrogen atom, the simplest of atoms, contains a proton in the nucleus and an electron present in the first shell outside the nucleus (Figure 3.2).

Helium, with an atomic number of 2 and an atomic weight of 4, has a nucleus that contains two protons and two neutrons. The two electrons revolve about the nucleus in a single shell (Figure 3.3). The α particle is a helium atom that has lost its two electrons.

In the lithium atom, a second electron shell is formed which in this element contains only one electron (Figure 3.4). In neon, this second shell is complete with eight electrons (Figure 3.5), and in sodium, a third shell of electrons begins to fill (Figure 3.6). In potassium (Figure 3.7) and calcium (Figure 3.8), the first three shells of electrons contain two, eight, and eight electrons, respectively, and a fourth shell is beginning to fill.

In atoms that are more complex than calcium, the third and outer shells of electrons may contain 18 to 32 electrons when complete. However, the number of electrons in the *outermost* shell of an atom will never exceed eight.

The electrons present in the outer incomplete shell of an atom are called the *valence electrons.* These electrons largely determine the chemical properties of an element. With the exception of the first

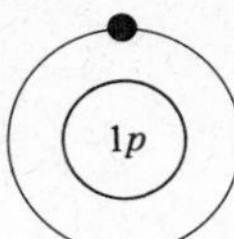

Figure 3.2 The hydrogen atom.

shell of electrons, the maximum number of electrons in the outermost shell is eight.

Two electrons in the first shell and eight in all other outermost shells represent a very stable arrangement of electrons. Prior to 1962, atoms such as helium, neon, argon, xenon, and radon that contained this arrangement of electrons were considered to be chemically unreactive, and these elements were therefore called the *inert* or *rare gases*. In that year, xenon difluoride, XeF_2, xenon tetrafluoride, XeF_4, and a fluoride of radon or undetermined composition were prepared. Because some of the members of this family have exhibited some reactivity, the inert gases are now usually called the *noble gases*. Although the preparation of these compounds of the noble gases will likely require some modification of our ideas concerning compound formation, the following discussion should still be quite satisfactory.

ISOTOPES

Because protons and neutrons have approximately unit mass and the electrons have negligible mass, all atomic weights should be nearly whole numbers. For many elements this is not the case; for example, chlorine has an atomic weight of 35.453, and lithium has an atomic weight of 6.939. It has now been demonstrated experimentally that most elements are mixtures of two or more kinds of atoms. Lithium is a mixture of atoms of atomic weights of 6 and 7, while chlorine is a mixture of atoms of atomic weights 35 and 37. Structurally, the two kinds of lithium atoms differ only in the number of neutrons in the nucleus; thus the atomic weights of the two kinds of atoms will be different. In all other respects, the two kinds of lithium atoms will be identical and are atoms of the same element. Because both atoms contain the same arrangement of electrons, and because the number of valence electrons is the same, they will have almost identical chemical properties.

Atoms that have the same atomic number but different atomic weights because of a different number of neutrons in the nucleus are

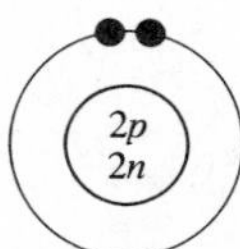

Figure 3.3 The helium atom.

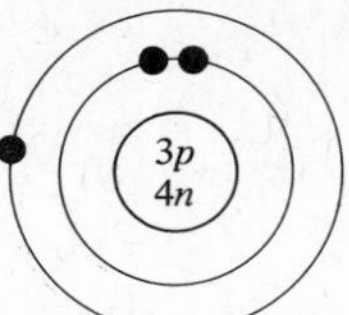

Figure 3.4 The lithium atom.

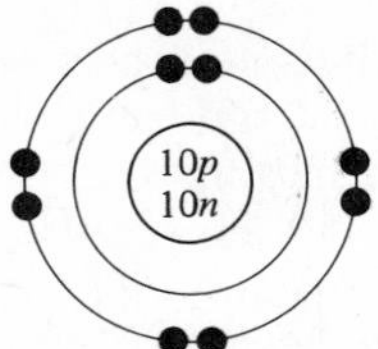

Figure 3.5 The neon atom.

called *isotopes.* The structures of the two isotopes of lithium and chlorine are illustrated in Figures 3.9 and 3.10. The isotope of hydrogen that has an atomic weight of 2 is called *deuterium* and is usually represented by the symbol D. Each atom of deuterium contains a proton and a neutron. Heavy water, D_2O, in which two deuterium atoms are combined with oxygen, was separated from ordinary water by Dr. Urey. In ordinary water, about one molecule in 6,000 is a heavy-water molecule.

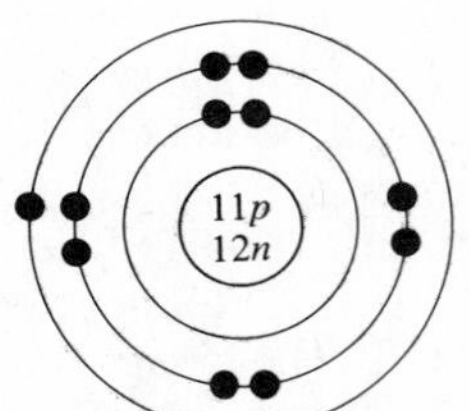

Figure 3.6 The sodium atom.

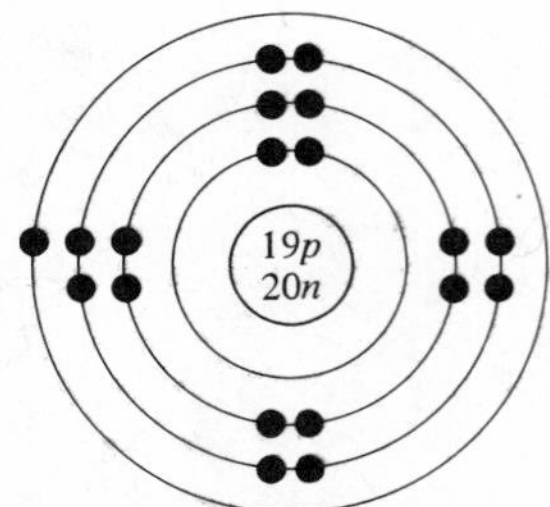

Figure 3.7 The potassium atom.

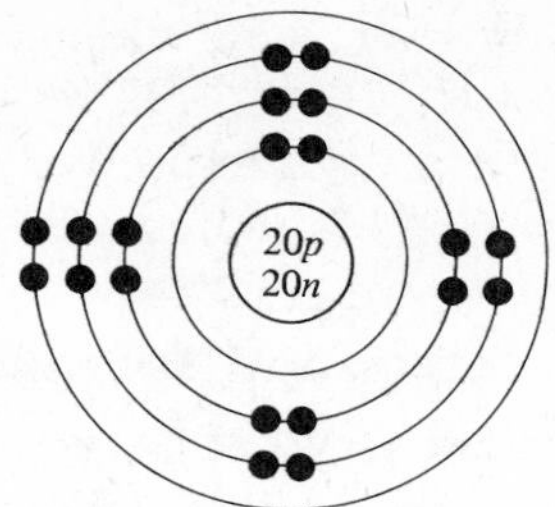

***Figure** 3.8 The calcium atom.*

The atomic weights of the elements that we have noted so far are the average atomic weights of all the atoms present. Thus the atomic weight of 80 is assigned to bromine, because the bromine that occurs in nature is composed of a nearly equal number of atoms of each of two isotopes of bromine with "masses" of 79 and 81.

The atomic weight of an individual atom is frequently called the *mass* or *mass number* of the atom and is equal to the sum of the number of protons and neutrons in the nucleus of the atom. The mass or mass number is currently represented as a superscript number to the left of the symbol of the atom, while the atomic number is placed as a subscript number also to the left of the symbol. The two common isotopes of chlorine are represented as ${}^{35}_{17}Cl$ and ${}^{37}_{17}Cl$. Protium

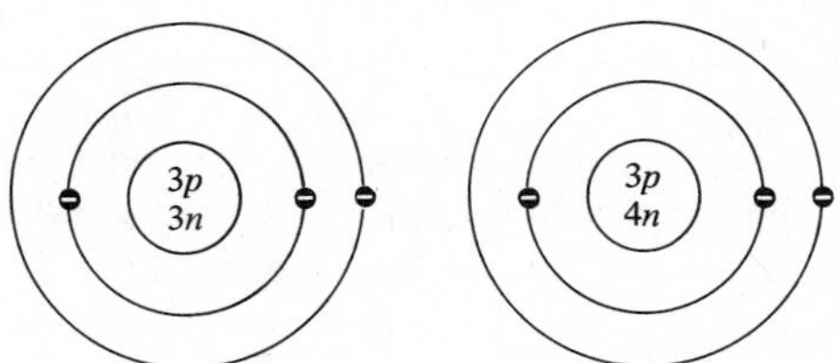

***Figure 3.9** Atoms of the two isotopes of lithium, with masses of 6 and 7, respectively.*

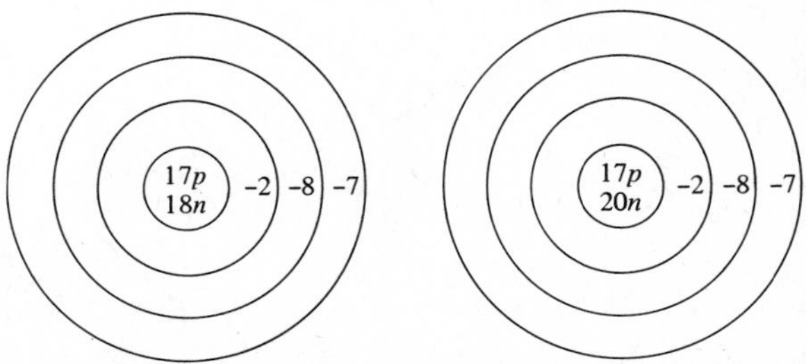

***Figure 3.10** Atoms of the two isotopes of chlorine, with masses of 35 and 37, respectively.*

(ordinary hydrogen), deuterium, and tritium (both heavy hydrogen isotopes) are represented as $^{1}_{1}H$, $^{2}_{1}D$, $^{3}_{1}T$, respectively. Ordinary oxygen and carbon are $^{16}_{8}O$ and $^{12}_{6}C$, respectively, while the radioactive isotope of carbon with mass 14 is represented as $^{14}_{6}C$.

The nature of compound formation

ELECTROVALENT OR IONIC COMPOUNDS

The valence electrons in an atom appear to be most important in compound formation. If there are less than four valence electrons in an atom, that atom will tend to lose these electrons to other atoms. If there are more than four valence electrons in an atom, such an atom will tend to gain sufficient electrons to complete the outer shell of eight electrons. It is to be noted that all atoms tend to assume a *noble gas arrangement* of electrons when they undergo chemical reaction. The gases helium, neon, argon, etc., which have a complete outer shell of electrons, are designated as the noble gases. Chemical combination takes place most readily between an atom that can release electrons and an atom that will accept them readily.

Sodium, which contains one electron in the valence shell, tends to lose this electron to other atoms, and it will assume the electron arrangement of stable neon. Chlorine, which possesses seven electrons in its outer shell, will usually capture an electron to complete this shell, thus attaining the electron structure of argon. If sodium reacts with chlorine to form sodium chloride, an electron is transferred from the sodium atom to the chlorine atom (Figure 3.11). The resulting particle of sodium, called an *ion,* is positively charged because it has one more proton than it has electrons. The chlorine atom has been converted to a chloride ion with a unit negative charge. Compounds formed by an exchange of electrons between atoms are called *electrovalent* or *ionic* compounds. In solution, ions can lead an independent existence, and a solution of an electrovalent compound will conduct the electric current.

The combination of one atom of magnesium with two atoms of fluorine to form the compound magnesium fluoride, MgF_2, is represented in Figure 3.12. The resulting compound is composed of magnesium and fluoride ions in the ratio of 1:2.

The group of elements called *metals* usually lose electrons in a chemical reaction and form positively charged ions. These are called

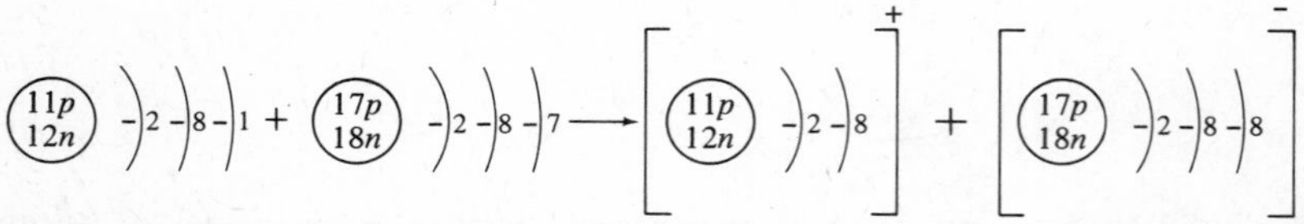

(11*p* 12*n*) 2) 8) 1 + (17*p* 18*n*) 2) 8) 7 ⟶ [(11*p* 12*n*) 2) 8]$^{+}$ + [(17*p* 18*n*) 2) 8) 8]$^{-}$

Figure 3.11 Formation of the ionic compound sodium chloride, NaCl, from sodium and chlorine atoms.

cations, because they migrate to the negative electrode, or *cathode*, when an electric current is passed through a solution that contains them. The atoms of *non metals* tend to gain electrons to form negatively charged ions. These are called *anions* because they migrate to a positively charged pole, or *anode*, during electrolysis.

In place of the detailed structures used in Figures 3.11 and 3.12, Lewis electronic formulas, in which the symbols of the elements represent all parts of the atom except the outer shell of valence electrons, can be employed to represent the structure of ionic compounds. The formation and structure of NaCl and MgF_2 would then be represented as follows:

$$\mathrm{Na}\cdot + :\ddot{\underset{\cdot\cdot}{\mathrm{Cl}}}\cdot \longrightarrow \mathrm{Na}^{+} + :\ddot{\underset{\cdot\cdot}{\mathrm{Cl}}}:^{-}$$

$$\mathrm{Mg}: + 2:\ddot{\underset{\cdot\cdot}{\mathrm{F}}}\cdot \longrightarrow \mathrm{Mg}^{2+} + 2:\ddot{\underset{\cdot\cdot}{\mathrm{F}}}:^{-}$$

COVALENT COMPOUNDS

Some elements form compounds by sharing electrons in order to complete the outer electron shell of the atoms. Carbon, which has four valence electrons, forms compounds of this type exclusively. In methane, CH_4, carbon shares its four valence electrons with the valence electrons of four hydrogen atoms. By this means, the carbon atom will have a full complement of eight electrons in its outer shell, while each hydrogen will have a complete outer shell of two electrons (Figure 3.13). A compound that has been formed by a sharing of electrons between atoms cannot ionize in solution and is called a *covalent compound.* Some other examples of covalent compounds, diagramed below, are carbon dioxide, CO_2 (Figure 3.14), ammonia, NH_3 (Figure 3.15), and phosphorous trichloride, PCl_3 (Figure 3.16).

The diagrams of ammonia and phosphorus trichloride use Lewis electronic formulas (see Figure 3.16). In the diagrams of ammonia and phosphorous trichloride, dots represent electrons from the hydrogen and chlorine atoms, respectively, while the x's represent electrons from the nitrogen and phosphorous atoms, respectively. Note, however, that electrons from whatever source are identical.

Molecules of diatomic elements, such as oxygen, hydrogen, nitrogen, chlorine, and bromine are formed by the sharing of electrons between two atoms of the element.

$\mathrm{H}:\mathrm{H}$	$:\ddot{\underset{\cdot\cdot}{\mathrm{Cl}}}:\ddot{\underset{\cdot\cdot}{\mathrm{Cl}}}:$
hydrogen	chlorine

In the covalent compounds described above, the electrons that form the "bonds' between the atoms are equally furnished by the atoms sharing them. Such a covalent bond is called a *normal covalent bond.* In some covalent bonds, one of the two atoms sharing electrons has furnished all the electrons for the formation of the

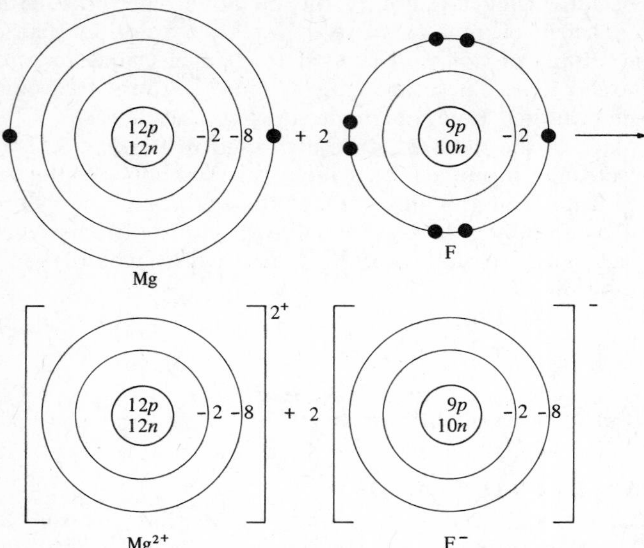

Figure 3.12 Formation of the ionic compound magnesium fluoride, MgF_2, from magnesium and fluorine atoms.

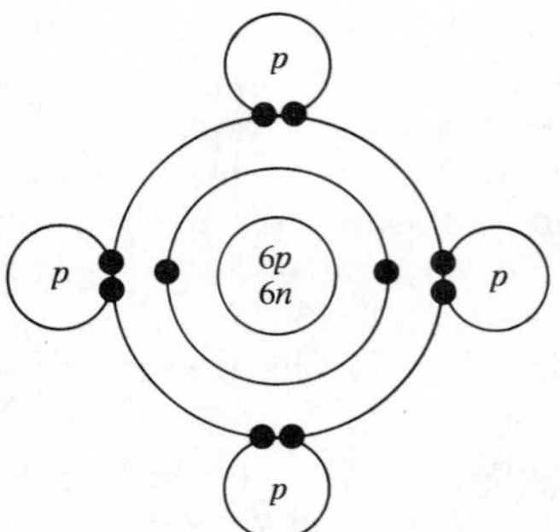

Figure 3.13 Diagram of the structure of the molecule of methane, CH_4.

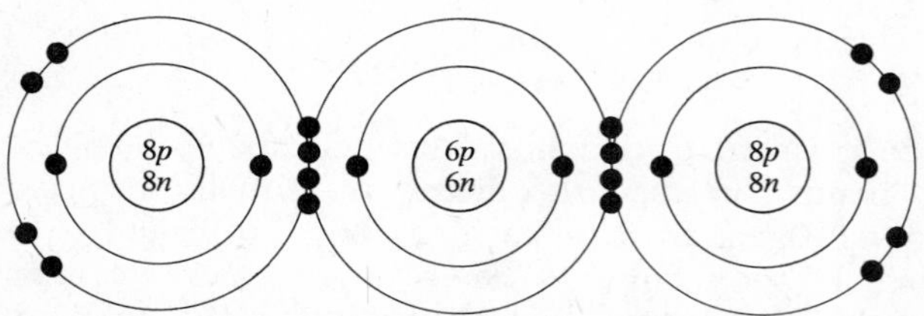

Figure 3.14 Diagram of the structure of a molecule of carbon dioxide, CO_2.

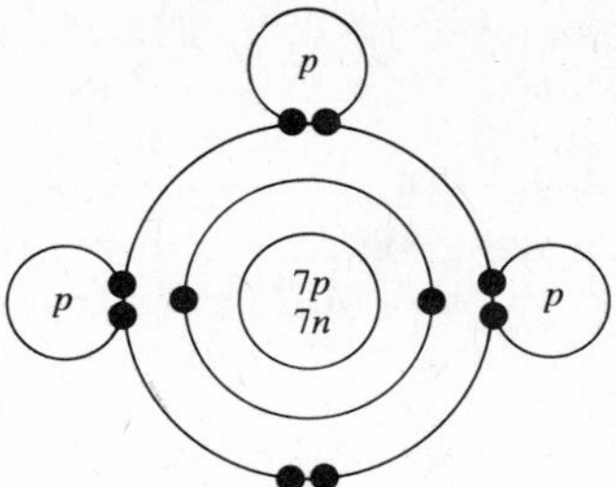

Figure 3.15 Diagram of the structure of a molecule of ammonia, NH_3.

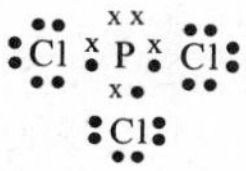

$$\begin{array}{ccccc} & & \times\times & & \\ :\overset{\bullet\bullet}{\underset{\bullet\bullet}{Cl}} & \overset{\times}{\bullet} & P & \overset{\times}{\bullet} & \overset{\bullet\bullet}{\underset{\bullet\bullet}{Cl}}: \\ & & \times\bullet & & \\ & & :\overset{}{\underset{\bullet\bullet}{Cl}}: & & \end{array}$$

Figure 3.16 The Lewis representation of the structure of a molecule of phosphorous trichloride, PCl_3.

bond, which is called a *coordinating covalent bond.* Such a bond exists in the ammonium ion, in which the unshared electron pair on the nitrogen atom of the ammonia molecule is shared with a hydrogen ion; thus

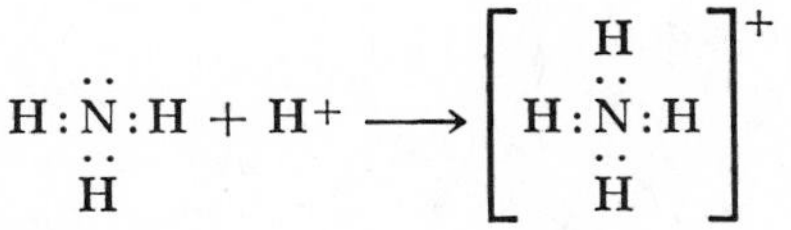

$$\begin{array}{c} \\ \mathrm{H:\overset{\cdot\cdot}{\underset{\cdot\cdot}{N}}:H} \\ \mathrm{H} \end{array} + \mathrm{H^+} \longrightarrow \left[\begin{array}{c} \mathrm{H} \\ \mathrm{H:\overset{\cdot\cdot}{\underset{\cdot\cdot}{N}}:H} \\ \mathrm{H} \end{array}\right]^+$$

ammonia ammonium ion

Hydrogen peroxide, H_2O_2, oxidizes amines of the structure R_3N to amine oxides of structure

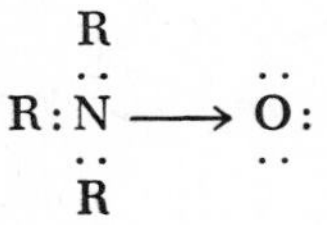

$$\begin{array}{c} \mathrm{R} \\ \mathrm{R:\overset{\cdot\cdot}{\underset{\cdot\cdot}{N}}} \\ \mathrm{R} \end{array} \longrightarrow \mathrm{\overset{\cdot\cdot}{\underset{\cdot\cdot}{O}}:}$$

in which the nitrogen atom has donated electrons to the oxygen atom. It is usually the practice to denote the coordinating covalent bond, also called the *semipolar bond,* by $\longrightarrow$ · R in the above formulas represents an organic radical (see page 228).

Valence

Before the development of the atomic theory of compound formation, *valence* was defined as a number which for a given atom of an element is the number of hydrogen atoms that this atom will combine with or displace in a chemical reaction. In the formulas HCl and H_2O, chlorine has a valence of 1 (univalent) because one atom of chlorine is combined with one atom of hydrogen; oxygen has a valence of 2 (divalent) because it combines with two hydrogen atoms. Nitrogen has a valence of 3 (trivalent) and carbon a valence of 4 (tetravalent) in NH_3 and CH_4. Because in the compounds NaCl, $CaCl_2$, $AlCl_3$, and $SiCl_4$ chlorine has a valence of 1, then sodium, calcium, aluminum, and silicon must have the valences of 1, 2, 3, and 4, respectively.

Since the development of the atomic theory of compound formation, valence has an added significance. In electrovalent compounds, the valence will be equal to the *number of electrons lost or gained by the atom.* If an atom loses electrons, it will form a positively charged ion and its valence can be called a positive valence. In sodium chloride, sodium has a valence of +1. If an atom gains electrons, it becomes a negatively charged ion and can be assigned a negative valence. Thus chlorine in sodium chloride has a valence of −1.

In covalent compounds, the valence will be equal numerically to the *number of pairs of electrons shared.* In methane, CH_4, carbon is tetravalent and hydrogen is univalent, because a carbon atom is sharing four pairs of electrons and each hydrogen atom is sharing one pair. In ammonia, NH_3, the nitrogen is trivalent, because it shares three pairs of electrons. In carbon dioxide, CO_2, each oxygen is divalent while the carbon atom is tetravalent. Positive or negative valence cannot be assigned to covalent compounds.

An element exhibits valence only when it is combined with some other element, that is, when it is part of a compound. Therefore, an element in the free state is assigned a valence of 0.

A group of elements will frequently act as a unit through many chemical reactions. Such units of atoms are called *radicals.* Radicals can be assigned valences, and some exist as charged ions. The sulfate radical, SO_4^{2-}, has a valence of −2, while the ammonium radical, NH_4^+, has a valence of +1. The nitrate radical, NO_3^-, has a valence of −1, while the phosphate radical, PO_4^{3-}, has a valence of −3.

THE USE OF VALENCE IN WRITING FORMULAS

In writing the formula of a compound, the combining capacities of the positive and negative groups must be balanced in such a way that the molecule is electrically neutral. In aluminum chloride, the valence of aluminum is +3, while that of the chlorine is −1. To balance the molecule electrically, it is necessary that three atoms of chlorine combine with one atom of aluminum.

For the formula for aluminum sulfate, write down the symbol for each element or radical and above it place its valence; thus

$$Al^{3+} \qquad SO_4^{2-}$$

To balance the positive and negative charges in this compound, a ratio of two aluminum atoms to three sulfate groups must be employed. The two aluminum atoms contribute a total of six positive charges, which just balance the six negative charges contributed by the three sulfate groups. The formula for aluminum sulfate is then

$$Al_2(SO_4)_3$$

OXIDATION NUMBERS

In covalent bonds uniting like atoms, such as the two hydrogen atoms in the hydrogen molecule, the pair of shared electrons will be centered between the two atoms. However, if the two atoms that are joined are different, the shared pair will be closer to one than to the other. Such a covalent bond is said to have *partial ionic character.* If the electron pair is completely displaced to one of the atoms or groups of atoms, then an electrovalent or ionic bond will exist. In any covalent bond in which a fluorine atom is joined to any other atom or group, the pair of electrons will be closer to the fluorine atom. If an oxygen atom is joined by a covalent bond and the other atom is not a fluorine atom, the electron pair will take a position closer to the oxygen atom.

Both the valence of elements and radicals of ionic compounds and a valence that can be assigned to elements joined by covalent bonds with partial ionic character, based on the assumption that the electron pair is completely transferred to one of the groups, are now commonly called *valence numbers* or *oxidation numbers.* These oxidation numbers are of great value in quantitative determinations involving oxidation-reduction reactions (see Chapter 12).

In most compounds, oxygen will then have an oxidation number of -2, while in all its compounds, fluorine has an oxidation number of -1. The oxidation number of manganese in potassium permanganate ($KMnO_4$) is $+7$, because the potassium ion and the oxygen atoms have oxidation numbers of $+1$ and -2, respectively, and the sum of all the oxidation numbers of all the atoms in the molecule must be zero, because a molecule is neutral. In sodium chlorate, one can determine the oxidation number of chlorine by first calling it x. From

$$\begin{array}{lll} \text{Na} & \text{Cl} & \text{O}_3 \\ +1 & x & -6 = 0 \end{array}$$

it is readily decided that its oxidation number is $+5$. In $Mg_2P_2O_7$, assign $2x$ as the sum of the oxidation number of the two P atoms. From

$$\begin{array}{lll} \text{Mg}_2 & \text{P}_2 & \text{O}_7 \\ +4 & 2x & -14 = 0 \end{array}$$

it is readily discerned that $2x$ is $+10$ and that the oxidation number of a single phosphorus atom is $+5$.

THE PERIODIC LAW AND THE PERIODIC TABLE

In 1869, the Russian chemist Mendeleev and the German chemist Lothar Meyer independently suggested a table of elements arranged in periods according to the increasing atomic weights of the elements. Because the suggestions of Meyer were not well publicized, Mendeleev has received the credit for preparing the first successful periodic table.

Assembling the periodic table

See Table 3.3 and the table inside front cover. Hydrogen is anomalous and is placed in period 0. In a second horizontal column, the elements from atomic numbers 2 to 9, that is, from helium to fluorine, are arranged according to increasing atomic weights. This is called the first period.

Period 1. He Li Be B C N O F

The next element, neon, resembles helium and is placed under it in period 2 of the table. Following neon is sodium, which resembles lithium and is placed under it. This process is continued until we reach chlorine, which resembles fluorine and is placed under it.

Period 2. Ne Na Mg Al Si P S Cl

The next element, argon, resembles helium and neon and is placed under these two elements. Argon begins the third period. Potassium is placed in this period below lithium and sodium, two elements that it resembles closely.

Periods 1 and 2 are *short periods* and contain eight elements each. Periods 3 and 4 are called *long periods* because each contains 18 elements. Period 3 begins with argon and ends with bromine, while krypton is the first and iodine the last element of period 4. Period 5 is also a long period, but it contans 32 elements because a group of elements called the *lanthanides* (atomic numbers 57 to 71) are located in a single place following barium in the table. The lanthanides were formerly known as the *rare earth elements.*

Period 6 is an incomplete period and is presumed to be a long period. The *actinides* (atomic numbers 89 to 103) fall into a single space following the element radium.

The noble gases of Group 0 were unknown when the table was first arranged by Mendeleev, and this group was added later.

The periodic law

Mendeleev attempted to arrange the elements in the periodic table in the order of increasing atomic weights. In a number of instances, this order had to be violated so as to place elements that

TABLE 3.4 ELECTRON ARRANGEMENT OF THE NOBLE GASES

	1	2	3	4	5	6
Helium (He)	2					
Neon (Ne)	2	8				
Argon (Ar)	2	8	8			
Krypton (Kr)	2	8	18	8		
Xenon (Xe)	2	8	18	18	8	
Radon (Rn)	2	8	18	32	18	8

resembled each other in the same group. He then claimed that in these cases the atomic weights of the elements were not accurate and should be redetermined. In all but three cases, the redetermination of these weights removed the difficulties.

Three sets of two elements, argon and potassium, cobalt and nickel, and tellurium and iodine, still remained in reverse order to their atomic weights. It was then decided that some property other than the atomic weights determined the position of the elements in the periodic table. This property proved to be the number of positive charges on the nucleus of the atom, that is, its atomic number. In the table, there are no exceptions to the fact that the elements are arranged in the order of increasing atomic numbers. Mendeleev's statement of the periodic law, that the properties of the elements are periodic functions of their atomic weights, must be restated as follows: *The properties of the elements are periodic functions of their atomic numbers.*

Relationship of atomic structure and valence to the periodic table

In Table 3.3, the noble gases are placed first in a period. The number of electrons in each of the shells of the noble gases is indicated in Table 3.4. From this table it is to be noted that in periods 1 and 2, the second and third shells are being filled to eight. In periods 3 and 4, although the fourth and fifth shells are being filled to eight, the third and fourth shells are being filled to 18. In period 5, the fourth shell is being filled from 18 to 32 and the fifth shell from 8 to 18, while the sixth shell is being filled to eight. Period 6 is incomplete. In radium, two electrons are present in the seventh shell. Additional electrons enter the fifth shell of the actinides.

The elements of group 0 (vertical column) have an outer shell of eight electrons, except for helium, whose outer shell is complete with two electrons. In Group I, one electron is present in the outermost shell; in Group II, two electrons are present in the valence shell; etc. Thus the group number usually indicates the number of electrons in the outer shell.

TABLE 3.3 A MODERN PERIODIC TABLE BASED ON MENDELEEV'S ORIGINAL ARRANGEMENT. ELEMENTS IN THE DARKER BOXES WERE NOT KNOWN IN 1869

Period	hydrides oxides group 0	RH R_2O group I	RH_2 RO group II	RH_3 R_2O_3 group III	RH_4 RO_2 group IV	RH_3 R_2O_5 group V	RH_2 RO_3 group VI	RH R_2O_7 group VII	RO_4 group VIII		
0		1 **H** 1.0080									
		A B	A B	A B	A B	A B	A B	A B			
1	2 **He** 4.0026	3 **Li** 6.941	4 **Be** 9.012	5 **B** 10.81	6 **C** 12.011	7 **N** 14.007	8 **O** 15.999	9 **F** 18.998			
2	10 **Ne** 20.179	11 **Na** 22.99	12 **Mg** 24.31	13 **Al** 26.98	14 **Si** 28.09	15 **P** 30.974	16 **S** 32.06	17 **Cl** 35.453			
3	18 **Ar** 39.948	19 **K** 39.102	20 **Ca** 40.08	21 **Sc** 44.96	22 **Ti** 47.90	23 **V** 50.94	24 **Cr** 52.00	25 **Mn** 54.94	26 **Fe** 55.85	27 **Co** 58.93	28 **Ni** 58.71
		29 **Cu** 63.55	30 **Zn** 65.37	31 **Ga** 69.72	32 **Ge** 72.59	33 **As** 74.92	34 **Se** 78.96	35 **Br** 79.904			

Period											
4	36 **Kr** 83.80	37 **Rb** 85.47	38 **Sr** 87.62	39 **Y** 88.91	40 **Zr** 91.22	41 **Nb** 92.91	42 **Mo** 95.94	43 **Tc** (99)	44 **Ru** 101.07	45 **Rh** 102.91	46 **Pd** 106.4
		47 **Ag** 107.87	48 **Cd** 112.40	49 **In** 114.82	50 **Sn** 118.69	51 **Sb** 121.75	52 **Te** 127.60	53 **I** 126.90			
5	54 **Xe** 131.30	55 **Cs** 132.91	56 **Ba** 137.34	57 ***La** 138.91	72 **Hf** 178.49	73 **Ta** 180.95	74 **W** 183.85	75 **Re** 186.2	76 **Os** 190.2	77 **Ir** 192.2	78 **Pt** 195.09
		79 **Au** 196.97	80 **Hg** 200.59	81 **Tl** 204.37	82 **Pb** 207.2	83 **Bi** 208.98	84 **Po** (210)	85 **At** (210)			
6	86 **Rn** (222)	87 **Fr** (223)	88 **Ra** (226)	89 **†Ac** (227)	104 ?	105 **(Ha)** (260)					

* lanthanide series

58	59	60	61	62	63	64	65	66	67	68	69	70	71
Ce	**Pr**	**Nd**	**Pm**	**Sm**	**Eu**	**Gd**	**Tb**	**Dy**	**Ho**	**Er**	**Tm**	**Yb**	**Lu**
140.12	140.91	144.24	(147)	150.4	151.96	157.25	158.93	162.50	164.93	167.26	168.93	173.04	174.97

† actinide series

90	91	92	93	94	95	96	97	98	99	100	101	102	103
Th	**Pa**	**U**	**Np**	**Pu**	**Am**	**Cm**	**Bk**	**Cf**	**Es**	**Fm**	**Md**	**No**	**Lr**
232.04	(231)	238.03	(237)	(244)	(243)	(245)	(247)	(249)	(254)	(256)	(256)	(254)	(257)

The elements of Group 0 are generally inert chemically and have a valence of 0. The most common electrovalencies of elements of Groups I, II, III, and IV are +1, +2, +3, and +4, respectively. The elements of Groups V to VII will usually exhibit electrovalencies of −3 to −1, respectively, when combined with hydrogen (the group number minus 8). The maximum valencies or oxidation numbers of the elements of Groups V to VII (when combined with oxygen) will be equal to the group number. Thus the oxidation number of nitrogen in N_2O_5 and HNO_3 is +5, of sulfur in SO_3 and H_2SO_4 is +6, and of chlorine in Cl_2O_7 and $KClO_4$ is +7.

Uses of the periodic table

The properties of an element can usually be predicted from the position of the element in the periodic table. Thus the properties of potassium resemble closely the known properties of sodium, because potassium occurs just below sodium in the table and both are members of the alkali family of metals. The properties of several elements unknown at the time that Mendeleev arranged the table were very accurately predicted by him from the location of the blank positions in the table.

Groups of closely related elements can easily be located in the table. These groups of elements are members of distinct chemical families, with the members of a certain family exhibiting very similar chemical properties. Thus the chlorine or halogen family, including fluorine, chlorine, bromine, iodine, and astatine, occurs in Group VII. The alkali metals lithium, sodium, potassium, rubidium, and cesium compose one subgroup or family in Group I. Copper, silver, and gold form the other family in Group I. Other important chemical families are the alkaline earths (Ba, Ca, Sr) of Group II, the phosporus family (P, As, Sb, Bi) of Group V, and the sulfur family (S, Se, Te) of Group VI.

The periodic table is also helpful in determining the common valences and the formulas of certain compounds of an element. As an example, because arsenic occurs in the fifth group and closely resembles phosphorus, it should exhibit a valence (oxidation number) of −3 to hydrogen and a maximum oxidation number of +5 to oxygen. Its hydride, arsine, has the formula AsH_3; one of its oxides, As_4O_{10}; and its common acid, the formula H_3AsO_4 (arsenic acid).

ATOMIC ORBITALS

It has previously been noted that in an atom, a number of electrons have about the same energy and constitute an energy shell of electrons. It was also intimated that the electrons in an energy shell occupy certain characteristic areas in space called *atomic orbitals.*

Only two electrons can occupy a certain atomic orbital; hence the lowest shell (the K shell or preferably, shell #1) contains but one orbital, while the second shell (L, or 2), which is complete with eight electrons, is composed of a maximum of four orbitals.

Electrons spin on their axes. Such a spinning charge produces a magnetic field about the electron. A specific orbital can contain a maximum of two electrons, if their magnetic fields become stabilized by opposite spins of the two electrons; that is, one must spin clockwise while the other is spinning counterclockwise.

Niels Bohr was one of the first scientists to speculate on how electrons were arranged about the nucleus. He suggested that the electrons rotated in a circular path about the nucleus, that they were arranged in shells, and that all the electrons in a shell have identical energies. He believed that electrons in a specific shell rotated at a fixed radii from the nucleus. Thus Bohr calculated that the lone electron of the hydrogen atom rotated at a distance of 0.529 Å (1 Å = 10^{-8} cm) from its nucleus.

If it is in the ground state, that is, at its lowest energy level, an electron will not radiate or absorb energy. If the atom should absorb energy, the electrons would be activated and jump to orbits in outer shells. Spectra are produced when these activated electrons return to their ground state.

When atomic spectra were examined, besides the sharp lines predicted on the basis that all the electrons were in identical orbits with identical energies in any one energy shell, there were many other fine lines. It is now accepted that although electrons of any one energy shell have about the same energy, there are slight differences in their energies. Those in the orbit of lowest energy in a shell are rotating in a spherical orbital and are designated as *s* electrons. (They are responsible for the sharp lines in atomic spectra.) Other electrons in certain shells are rotating in a number of elliptical orbitals of various and sometimes complicated geometric shapes. In increasing order of energy content in any one shell, these are designated as *p* (principal), *d* (diffuse), and *f* (fundamental) orbitals, these designations being based on the nature of the spectral lines they produce.

For an electron to move from one orbital to one of higher energy, a discrete amount of energy, usually furnished by light or some other form of electromagnetic wave, must be absorbed by it. If an insufficient amount of energy should be available, the electron will remain in its ground state. It will not shift into a higher energy level until the discrete amount of energy is furnished to it. This is the basis of the quantum theory of atomic structure.

Heisenberg noted that because the electron is a minutely small particle, it posesses about the same amount of energy that light or other forms of wave motion possesses. If the latter is then used in an attempt to measure the position and velocity of the electron at any one instant, it is impossible to determine the momentum (velocity) of the electron if one locates it in space. If one determines the momentum of an electron, then its position in space cannot be determined at any one instant. This is called the *Heisenberg uncertainty principle.*

However, Heisenberg did realize that it should be possible to calculate the probability of finding the electron in any certain space about the nucleus.

Then Schrödinger developed equations to calculate the probability of finding the electron in any specified space in the atom. These are wave functions based on an established proposal made by Louis de Broglie in 1924, that the rapidly moving electrons, besides being particles of matter, have certain characteristics similar to wave motions, such as are produced by light. Although these wave function equations are quite complex, solutions of these have been made for the hydrogen atom, and approximate solutions have been made for some other simpler atoms. Although any one electron can be present in any position about the nucleus, the Schrödinger equations can be solved to indicate the area that one can expect to find an electron 90 percent of the time. The graph in Figure 3.17 indicates the probability of finding the lone electron of hydrogen at any distance r from the center of the atom.

The electron structure of hydrogen can be designated as $1s^1$, where s indicates that the electron spends most of its time in a spherical orbital. The 1 before s indicates that this electron in its ground state is at the lowest energy level in an atom; that is, n, the *principal quantum number*, is 1. The superscript indicates that this orbital contains but one electron. It is interesting to note that the average distance of the electron from the nucleus in a hydrogen atom is 0.529 Å, which corresponds to the fixed radius of the electron of hydrogen as assigned by Bohr.

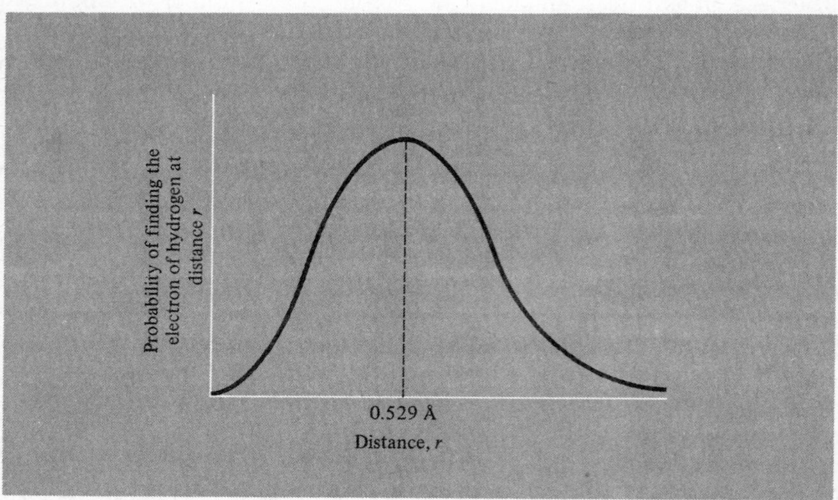

Figure 3.17 *The probability of finding the electron of the hydrogen atom at distance* r *from the center of the atom.*

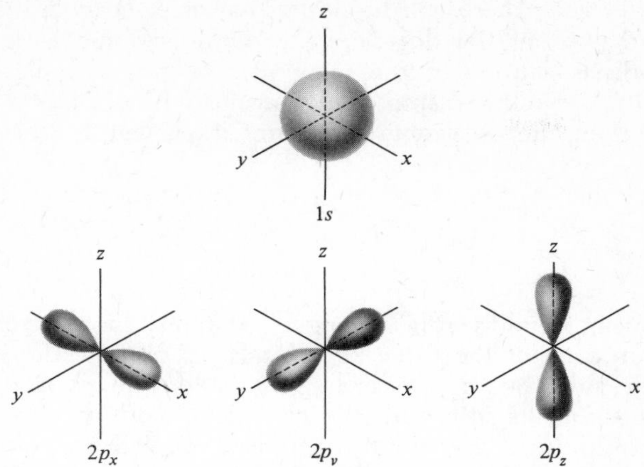

Figure 3.18 The shape of the 1s and the 2p atomic orbitals.

In building up the electron structure of succeeding elements of the periodic table, two important principles have to be recognized. One, called the *Pauli exclusion principle,* states that there may be zero, one, or a maximum of two electrons in any specific atomic orbital. If two are present in an orbital, they cannot have the same spin but must be of opposite spin. In the *Aufbau principle,* each added electron will seek the lowest energy level available to it if it is in the ground state.

In hydrogen, the electron is unpaired; that is, no electron of opposite spin is present to stabilize it. In helium, a second electron of opposite spin enters the orbital to stabilize the spin of the first electron. Because there is only one orbital in the first shell or first quantum state ($n = 1$), this completes the first shell and gives a very stable unreactive atom. The electron structure of helium can be designated as $1s^2$.

Because the first energy shell is complete, the third electron in lithium enters a spherical or *s* orbital at a greater average radius than the 1*s* electrons in the second energy level. The lithium atom has the electronic configuration $1s^22s^1$.

In carbon, the first two electrons enter the 1*s* orbital and the next two enter the 2*s* orbital, leaving two electrons to enter one or two of the 2*p* orbitals. There are three 2*p* orbitals available. These are dumbbell-shaped orbitals consisting of two lobes, as illustrated in Figure 3.18. There is a zero probability of finding *p* electrons in the center of the atom. The three *p* orbitals are perpendicular (90°) to each other and are arranged symmetrically about the *x*, *y*, and *z*, axes, and the electrons occupying these *p* orbitals can be designated as p_x, p_y, and p_z electrons, respectively. Although of slightly higher energy than the 2*s* electrons, electrons in any one of these three *p* orbitals are of equal energy and are said to be degenerate. According to Hund's rule, there will be no pairing of electrons in an atomic orbital until each of the degenerate orbitals contains a single electron, and these will be of the same spin. If we can designate an electron with a clockwise spin as ↓ and one with a counterclockwise spin as ↑, then the electronic structure of carbon is illustrated as follows:

$1s$	$2s$	$2p_x$	$2p_y$	$2p_z$
⇃↾	⇃↾	↓	↓	○

Following Hund's rule, nitrogen contains an electron of the same spin in each of the three $2p$ orbitals. In oxygen, the next electron, which must be of opposite spin, enters one of the three $2p$ orbitals to couple its spin with the electron already present. In fluorine, two of the three $2p$ orbitals contain a coupled pair of electrons, while the third contains an unpaired electron. In neon, the $2s$ and all of the $2p$ orbitals are filled. This represents a very stable noble gas structure. The electronic structure of nitrogen, oxygen, fluorine, and neon are represented as follows:

nitrogen	$1s^2 2s^2 2p_x{}^1 2p_y{}^1 2p_z{}^1$
oxygen	$1s^2 2s^2 2p_x{}^2 2p_y{}^1 2p_z{}^1$
fluorine	$1s^2 2s^2 2p_x{}^2 2p_y{}^2 2p_z{}^1$
neon	$1s^2 2s^2 2p_x{}^2 2p_y{}^2 2p_z{}^2$

In the second period of the periodic table, the $1s$, $2s$, and $2p$ orbitals are filled, and the $3s$ and $3p$ orbitals are filling as indicated below for the elements of this period:

sodium	$1s^2 2s^2 2p^2 3s^1$
magnesium	$1s^2 2s^2 2p^2 3s^2$
aluminum	$1s^2 2s^2 2p^2 3s^2 3p_x{}^1$
silicon	$1s^2 2s^2 2p^2 3s^2 3p_x{}^1 3p_y{}^1$
phosphorus	$1s^2 2s^2 2p^2 3s^2 3p_x{}^1 3p_y{}^1 3p_z{}^1$
sulfur	$1s^2 2s^2 2p^2 3s^2 3p_x{}^2 3p_y{}^1 3p_z{}^1$
chlorine	$1s^2 2s^2 2p^2 3s^2 3p_x{}^2 3p_y{}^2 3p_z{}^1$
argon	$1s^2 2s^2 2p^2 3s^2 3p_x{}^2 3p_y{}^2 3p_z{}^2$

In potassium and calcium, one and two electrons, respectively, enter the $4s$ orbital. In more complex atoms, the d and f orbitals, as well as the s and p orbitals are filled. Because we are interested in biological products, and because elements more complex than calcium rarely occur in these, we are fortunate in not having to consider the complex and sometimes confusing Aufbau principle of these more complex atoms.

In covalent bonds, the atomic orbitals of the two atoms being bonded overlap to give new characteristic probability distributions of the bonding electrons. These are designated as molecular orbitals, and, using organic compounds as characteristic examples, these will be illustrated in Chapter 16.

STUDY QUESTIONS

1. What are the three fundamental units of the atom? Indicate mass, charge, and location of these units in the atom.
2. What is the difference between an atom and a molecule?
3. Define atomic number. If the atomic number and the atomic weight of an element are known, how are the number of protons, electrons, and neutrons in an atom of the element determined?
4. What are α particles? What did the bombardment of a thin gold foil by α particles prove?
5. How are electrons arranged about the nucleus of an atom? What is the maximum number of electrons present in the first shell? In the second shell?
6. What are the electrons in the outermost shell of an atom called? Why are they so important?
7. What part of the atom has the greatest influence on the chemical properties of the atom?
8. What are isotopes? Which of the following are identical in the isotopes of any one element: atomic number, atomic weight, number of protons, number of electrons, number of neutrons, number of valence electrons, chemical properties?
9. The following atoms have the atomic numbers and the approximate atomic weights indicated. Diagram these atoms.

Symbol	Atomic number	Atomic weight
K	19	39
P	15	31
Mg	12	24
N	7	14
O	8	16
F	9	19
Ar	18	40
Al	13	27
Ca	20	40

10. Which of the atoms in Question 9 will readily lose electrons? Which will gain them easily? Which are metals, nonmetals, noble gases?
11. Is the mass of an atom due principally to the electrons or to protons and neutrons?
12. Which two of the following are isotopes of the same element?

(a)	$17p$ $18n$	−2	−8	−7	
(b)	$3p$ $4n$	−2	−1		
(c)	$20p$ $20n$	−2	−8	−8	−2
(d)	$18p$ $18n$	−2	−8	−8	
(e)	$3p$ $3n$	−2	−1		

13. Diagram the atoms of sodium, potassium, oxygen, sulfur, fluorine, and chlorine.
14. Describe the similarity in the structures of fluorine and chlorine, sodium and potassium, oxygen and sulfur. Are there also similarities in chemical properties?
15. What are the stable electron configurations? What elements have these stable configurations?
16. Describe the changes that occur when sulfur, sodium, fluorine, aluminum, and magnesium enter into chemical reactions.
17. Explain the difference between 4H, H_2, and $3H_2$.
18. The following are electrovalent or ionic compounds: KF, MgO, $CaCl_2$, Na_2S, and Li_2O. By means of appropriate diagrams, indicate the formation of these compounds from their constituent atoms.
19. If an atom loses or gains electrons, what is the resulting particle called? What atoms produce positively charged particles? Which produce negatively charged particles?
20. Explain the fact that the atomic weight of chlorine is 35.5, not a whole number.
21. Diagram the isotopes of hydrogen with masses 1, 2, and 3. Also diagram the isotopes of oxygen with masses 16 and 17. Indicate the total number of different molecules of water than can be formed by the different combinations of these atoms.
22. Distinguish between electrovalence and covalence.
23. The following are covalent compounds: CO_2, CH_4, NH_3, CCl_4, SiO_2 and $SiCl_4$. By means of Lewis structures, indicate their structure.
24. By means of Lewis structures, predict the formulas of compounds formed by sulfur with potassium, magnesium, and aluminum.
25. By means of Lewis structures indicate the structure of the molecules H_2, Cl_2, and F_2.
26. Which types of elements exhibit negative valence? Which exhibit positive valence?
27. Indicate the valence of the underlined elements in the following:

(a) $H\underline{Cl}$ ______	(b) $\underline{Ba}Cl_2$ ______	(c) $\underline{Mg}O$ ______
(d) $\underline{Al}_2O_3$ ______	(e) $\underline{Zn}Cl_2$ ______	(f) $H_2\underline{S}$ ______
(g) $\underline{Cu}O$ ______	(h) $\underline{Cu}_2O$ ______	(i) $\underline{Fe}_2O_3$ ______
(j) $\underline{Na}Cl$ ______	(k) $Na\underline{I}$ ______	(l) $\underline{Mn}O_2$ ______
(m) $\underline{Ca}Cl_2$ ______	(n) $\underline{As}Cl_3$ ______	(o) $\underline{P}Cl_3$ ______

28. Use information established in Question 27, as well as information derived within this question, to determine the valences of the following underlined elements and radicals.

(a) $Na\underline{NO}_3$ ______	(b) $Ca_3(\underline{PO_4})_2$ ______
(c) $Ca\underline{SO_4}$ ______	(d) $\underline{Cd}SO_4$ ______
(e) $\underline{Ag}Cl$ ______	(f) $Al(\underline{C_2H_3O_2})_3$ ______
(g) $Na_3\underline{AsO}_4$ ______	(h) $Na_3\underline{BO_3}$ ______
(i) $\underline{Li}Cl$ ______	(j) $Ba(\underline{ClO_2})_2$ ______
(k) $\underline{Sr}Cl_2$ ______	(l) $Sr\underline{CO_3}$ ______
(m) $Li\underline{ClO_4}$ ______	(n) $Na\underline{NO_2}$ ______
(o) $Mg\underline{SiO_3}$ ______	(p) $Na_4\underline{P_2O_7}$ ______

29. Assuming that the oxidation number of oxygen is −2, what are the oxidation numbers of the other elements in the following ions? (a) MnO_3^-,

(b) ClO_4^-, (c) NO_3^-, (d) PO_4^{3-}, (e) SbO_3^{3-}, (f) BrO_3^-, (g) SO_3^{2-}, (h) SiO_4^{4-}

30. Determine the oxidation number of the underlined element:

(a) $K_2\underline{Cr}O_4$ ______	(b) $Mg\underline{C}O_3$ ______
(c) $H_2\underline{Se}$ ______	(d) $Ca_3(\underline{V}O_4)_2$ ______
(e) $Ag_4\underline{P}_2O_7$ ______	(f) $Ba(\underline{Cl}O_2)_2$ ______
(g) $K_2\underline{Mo}O_4$ ______	(h) $Ba\underline{S}O_3$ ______

31. What kinds of elements are usually written first in the formula of an inorganic compound?
32. Place the appropriate subscripts in the following formulas, if needed.

(a) Al Br	(b) Cd I
(c) Al (PO_4)	(d) Sr (NO_3)
(e) Ag CO_3	(f) Mg (SO_3)
(g) Zn (PO_4)	(h) Ba (ClO)
(i) Ca (NO_2)	

33. The electron shell arrangements of three atoms are

A	B	C
2-8-8-3	2-8	2-8-6

______________ is a metal.
C has______________valence electrons.
______________is chemically inert.
The formula of a compound formed between A and C is A______C______.

34. Who at the same time as Mendeleev developed a table of the elements arranged in periods?
35. State the periodic law. How has the statement of this law changed since it was first enunciated? Why was this change necessary?
36. Explain how the elements are arranged in the periodic table.
37. List three instances in which the atomic weight of the elements does not increase with increasing atomic number.
38. What is the significance of the atomic number?
39. Explain the relationship of the structure of an atom of an element to its position in the periodic table.
40. How can the periodic table be utilized in predicting the valences of elements?
41. List several uses that can be made of the periodic table.
42. Name several families of chemical elements and the members of each of these families. How do the members of a family resemble each other? Locate these families in the periodic table.
43. Compare the chemical properties of a typical family of chemical elements.
44. Describe the Bohr concept of the electronic structure of an atom.
45. What is an atomic orbital?
46. How many orbitals are present in the first energy shell of an atom? In the second energy level of electrons?
47. Why must an atomic orbital contain no more than two electrons? What is the Pauli exclusion principle?
48. What information is obtained from the solution of the Schrödinger equations? What was Heisenberg's contributions to our knowledge of

the electron structure of atoms?

49. What is the shape of an *s* atomic orbital? Of a *p* atomic orbital? What are the coordinates of *p* atomic orbitals?
50. How are spectra related to the positions of electrons in atoms?
51. What does the Aufbau principle teach us?
52. On the basis of electronic structure, why is hydrogen more reactive than helium? Fluorine than neon?
53. What is the average distance of the lone electron of the hydrogen atom from the nucleus?
54. What is Hund's rule, and what does it tell us about the electronic structure of the carbon, nitrogen, and oxygen atoms?
55. Using the same style of representation as was used for the carbon atom on page 48, indicate the electronic structure of lithium, beryllium, boron, argon, potassium, and calcium.
56. Why is it not necessary for us to concern ourselves with the electronic structures of atoms more complex than calcium?

chapter 4

---

oxygen and combustion

4

Oxygen is not only one of the most important of the elements, but it is also the most abundant. It is essential for respiration, a process necessary for life itself, in which air is introduced into the lungs. Oxygen diffuses through the walls of the lungs and combines with the red pigment of the blood, called *hemoglobin.* On uniting with oxygen, the hemoglobin is converted into *oxyhemoglobin,* and the oxygenated blood is then carried to the tissues of the body. In the tissues, the oxygen released by the blood combines with the digested particles of food and produces the heat and mechanical energy required by the body for warmth and motion. The demand for oxygen by plants is much less than that needed by animals because the former are stationary and meet some of their needs for this element by a process called *photosynthesis.* In this process, plants use sunlight to convert carbon dioxide and water to sugars and oxygen. Certain microorganisms, called *anaerobes,* cannot exist in the presence of free oxygen and meet their needs for this element by extracting it from some of its compounds.

Combustions supported by oxygen heat our homes and cook our food. Oxygen is also used in the treatment of certain diseases, for the purification of water, and for the disposal of sewage. Many important commercial processes depend on the oxygen of the air. Large quantities of oxygen, in the form of air, are consumed in the manufacture of iron, steel, and other metals from their ores.

The discovery of oxygen

The Swedish chemist Scheele was possibly the first to prepare oxygen. Because his claim is somewhat obscure and little publicity was given to his discovery, the Englishman Joseph Priestley has been given credit. Priestley decomposed mercuric oxide into the liquid metal mercury and the gas oxygen. This oxide was decomposed by focusing the sun's rays on it by means of a burning glass (magnifying glass). Priestley found that combustion proceeded much more rapidly in this gas than in air. He firmly believed in the phlogiston theory and did not recognize the true nature of burning.

A short time after Priestley's discovery of oxygen, the French chemist Antoine Laurent Lavoisier demonstrated that burning occurred by the union of this gas with a combustible substance. Lavoisier assigned to this new gas the name oxygen or "acid former," which is unfortunate, because many acids contain no oxygen.

Occurrence of oxygen

In the air, oxygen in the elementary state is mixed with nitrogen and small amounts of other gases and represents about 20 percent of the air. It is combined chemically with hydrogen in water, and the latter is about eight-ninths oxygen. Almost all organic substances, such as starch, alcohol, proteins, and fats, contain appreciable amounts of oxygen, and about 60 percent of our body is composed of it. Inorganic substances, such as limestone, and the rocks and soil of the earth contain this element.

Preparation of oxygen

By Heating Certain Metallic Oxides or Peroxides. Oxygen may be obtained in the laboratory by heating red mercuric oxide, although this compound is too expensive to serve as a commercial source of oxygen. The following is a statement of this reaction:

Mercuric oxide gives mercury and oxygen.

Replacing the names of the compound and the elements by their proper formulas and symbols, and separating the reactants and products by an arrow, one obtains an unbalanced equation:

$$HgO \longrightarrow Hg + O_2$$

This equation is not balanced, because only one atom of oxygen is represented on the left, whereas two are present on the right of the arrow. By doubling the number of molecules of mercuric oxide and the atoms of mercury, the total number of atoms on the left of the arrow will equal those on the right, and this equation will be balanced:

$$2HgO \longrightarrow 2Hg + O_2$$

Heating lead dioxide, PbO_2 will also produce oxygen.

Metallic oxides may be considered as derived from water, H_2O, by replacing the two hydrogen atoms by metallic atoms. The peroxides are derived in a similar fashion from hydrogen peroxide, H_2O_2. Upon heating, most metallic peroxides are decomposed into the oxide and oxygen. Thus the heating of barium peroxide, BaO_2, produces barium oxide, BaO, and oxygen.

$$BaO_2 \longrightarrow BaO + O_2 \qquad \text{(unbalanced equation)}$$
$$2BaO_2 \longrightarrow 2BaO + O_2 \qquad \text{(balanced equation)}$$

Figure 4.1 The laboratory preparation of oxygen.

Also of biological importance is the decomposition of hydrogen peroxide into water and oxygen, a reaction that is accelerated by manganese dioxide or catalase, a substance present in most living tissue.

Preparation by Heating Certain Salts. If potassium chlorate, $KClO_3$, is heated above its melting point, it will slowly decompose into oxygen and solid potassium chloride, a reaction represented by the following equation:

$$KClO_3 \longrightarrow KCl + O_2 \quad \text{(unbalanced equation)}$$
$$2KClO_3 \longrightarrow 2KCl + 3O_2 \quad \text{(balanced equation)}$$

This is the usual method employed in the laboratory to prepare oxygen.

If a very small amount of manganese dioxide, MnO_2, is added to molten potassium chloride, the latter decomposes rapidly into potassium chloride and oxygen (Figure 4.1). At the end of the reaction, the manganese dioxide can be recovered unchanged. There are many other examples of a substance changing the speed of a chemical reaction without apparently entering into the reaction. Such substances are called *catalytic agents,* or *catalysts.* A catalyst may then be defined as any substance that speeds up the rate of a chemical reaction and can be recovered unchanged at the end of the reaction.

Preparation by the Electrolysis of Water. Small quantities of oxygen are prepared commercially by the electrolysis of water. Oxygen may be prepared by this method in the laboratory if a suitable apparatus is available (Figure 4.2). For each volume of oxygen that is liberated at the positive electrode (anode), two volumes of hydrogen will be liberated at the negative (cathode). Because water is a very poor conductor of electricity, some substance, such as sodium hydroxide or sulfuric acid, must be added. The chemical equation for the electrolysis of water is

$$2H_2O \longrightarrow 2H_2 + O_2$$

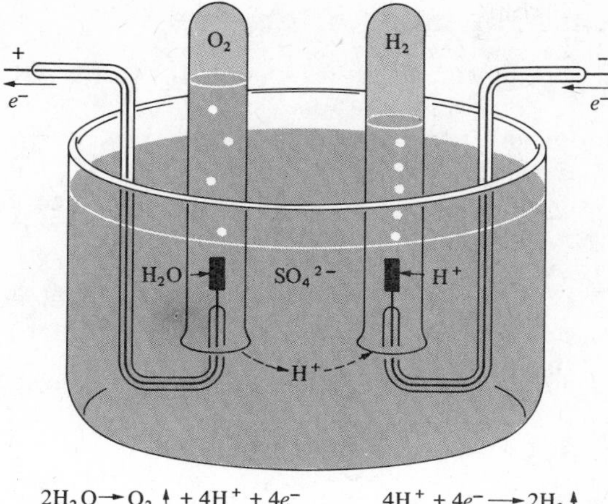

$2H_2O \rightarrow O_2 \uparrow + 4H^+ + 4e^-$ $\qquad$ $4H^+ + 4e^- \rightarrow 2H_2 \uparrow$

Figure 4.2 *Diagram of an apparatus used in the electrolysis of water. The addition of sulfuric acid makes the water a conductor of the electric current. At the positive electrode (or anode) water gives up electrons and releases oxygen. At the negative electrode (or cathode) the hydrogen ions present in the water combine with electrons and hydrogen gas is formed.*

Preparation from Air. Most commercial oxygen is prepared by the distillation of liquid air (Table 4.1). In this process, air is cooled and compressed and is then permitted to expand rapidly. Upon expanding, the air is further cooled, and part of it will liquefy. Because oxygen boils at −183°C while nitrogen boils at a lower temperature (−195.8°C), if the liquid air is permitted to warm, the nitrogen will distill and a residue of nearly pure oxygen will remain.

TABLE 4.1 THE CHEMICAL COMPOSITION OF THE AIR

Substance	*Vol/100 Vol of Air*	*Boiling Point, °C, at 760 mm Pressure*
Nitrogen	78	−195.8
Oxygen	21	−183
Argon	0.94	−185.8
Carbon dioxide	0.03–0.04	− 78.5[a]
Water vapor	Variable	100
Helium	Trace	−268.9
Neon	Trace	−246.3
Krypton	Trace	−152.9
Xenon	Trace	−107.1

[a]Sublimes.

Physical properties of oxygen

Physical properties of substances are those associated with physical changes. The following are some physical properties of oxygen: (1) Under ordinary conditions oxygen is colorless, odorless, and tasteless; (2) liquid oxygen boils at −183°C; (3) upon sufficient cooling, oxygen is obtained as a solid; (4) as a gas it is slightly heavier than air, and a liter at standard conditions (0°C and 760 mm pressure) weighs 1.429 g. About 4 ml of oxygen dissolves in 100 ml of water at 0°C. This small amount of oxygen dissolved in water is important to marine vegetation and animal life, which depend upon it for respiration.

Chemical properties of oxygen

Chemical properties are involved when matter undergoes chemical change.

Substances that burn in air will burn much more vigorously in pure oxygen, where there is about five times the concentration of oxygen. It should be recognized that the therapeutic use of oxygen in the hospital presents a serious fire hazard and that caution is demanded in its use. Besides the combustion hazard of pure oxygen, there are certain other dangers in its use, especially when its source is a high pressure tank. These should always be strapped to a desk or firmly secured.

A glowing splint will burst into flame when thrust into pure oxygen; this is used as a test for oxygen in the laboratory.

Both groups of elements, metals and nonmetals, combine with oxygen to form compounds called *oxides*. The nonmetals sulfur, carbon, and phosphorus combine with oxygen to form sulfur dioxide (SO_2), carbon dioxide (CO_2), and phosphoric oxide (P_4O_{10}), respectively.

$$C + O_2 \longrightarrow CO_2$$
$$S + O_2 \longrightarrow SO_2$$
$$P_4 + 5O_2 \longrightarrow P_4O_{10}$$

The following equations represent the union of oxygen with the metals iron and magnesium:

$$2Mg + O_2 \longrightarrow 2MgO$$
$$3Fe + 2O_2 \longrightarrow Fe_3O_4$$

Oxygen will also combine with many compounds, including organic compounds. The latter all contain carbon. The following equation represents the union of the organic compound octane (C_8H_{18}) with oxygen to form carbon dioxide and water:

$$2C_8H_{18} + 25O_2 \longrightarrow 16CO_2 + 18H_2O$$

The power necessary to operate automobile engines is produced by the union of oxygen with hydrocarbons present in gasoline. The hydrocarbons, like octane above, are composed of carbon and hydrogen exclusively.

OXIDATION AND COMBUSTION

Oxidation, in a more restricted sense, is the union of some element or compound with oxygen. (Later a more general definition of oxidation will be considered. See Chapter 12.) Some oxidations proceed very slowly without the production of light, such as the rusting of iron or the "burning" of the products of digestion in the living tissues. Such oxidations are called *slow oxidations.* The product of the slow oxidation of iron is called *rust* and has the formula Fe_2O_3. The formation of rust requires the presence of moisture as well as oxygen.

A rapid oxidation that is accompanied by the evolution of heat and light is called a *combustion.* The burning of coal and kerosene are examples of combustion.

REACTION OF OXIDES WITH WATER

The oxides of nonmetals combine with water to form a group of compounds called *acids.* These oxides are also called *acid anhydrides.* Sulfur dioxide combines with water to form sulfurous acid, H_2SO_3; carbon dioxide produces carbonic acid, H_2CO_3; while phosphoric oxide forms phosphoric acid, H_3PO_4.

$$SO_2 + H_2O \longrightarrow H_2SO_3$$
$$CO_2 + H_2O \longrightarrow H_2CO_3$$
$$P_4O_{10} + 6H_2O \longrightarrow 4H_3PO_4$$

Acids always contain hydrogen, have a sour taste, and change blue litmus to a red color.

Although most oxides of metals are insoluble in water, a few dissolve to form compounds called *bases.* These metallic oxides are also called *basic anhydrides.* Calcium oxide, CaO, unites with water to form the base calcium hydroxide, $Ca(OH)_2$.

$$CaO + H_2O \longrightarrow Ca(OH)_2$$

Bases contain the hydroxyl (OH) group, usually have a bitter taste, are soapy to the touch, and change red litmus to a blue color.

KINDLING TEMPERATURE

If the temperature of a combustible substance is raised to a certain point, it will burst into flame. The lowest temperature at which this combustion takes place is called the *kindling temperature* of the substance and varies with the nature of the substance. The following substances are listed in order of increasing kindling temperatures: white phosphorous, sulfur, wood, coal.

FACTORS INFLUENCING THE RATE OR SPEED OF OXIDATION

The rate or speed of a chemical reaction is expressed as the quantity of a chemical substance that undergoes change in a unit period of time. The following are factors that influence the rate of oxidation:

1. nature of the oxidizable substance
2. temperature
3. presence of a catalyst
4. concentration or pressure
5. contact or abundance of surface

It will be noted in Chapter 8 that these same factors influence the rate of all types of chemical reactions.

SPONTANEOUS COMBUSTION

Spontaneous combustion occurs when a substance has been heated to its kindling temperature by slow oxidation or some other exothermic reaction, such as fermentation. The material is usually well insulated, and the heat generated by the exothermic reaction accumulates. Spontaneous combustion frequently occurs when oily paint rags are permitted to accumulate in a heap, or when fine coal is piled in the hold of a ship. Barn fires are frequently produced by the fermentation of damp, improperly cured hay.

METHODS OF EXTINGUISHING FLAMES

To extinguish a flame, the source of oxygen is removed or the temperature of the burning substance is lowered below its kindling temperature. In extinguishing flames, liquid carbon dioxide extinguishers or dry extinguishers have generally replaced the formerly used soda-acid and the carbon tetrachloride extinguisher. The latter has been abandoned because of the toxic properties of carbon tetrachloride (page 257). The dry extinguisher contains solid sodium bicarbonate. The heat of the flame converts this substance to carbon dioxide according to the following equation:

$$2NaHCO_3 \longrightarrow Na_2CO_3 + H_2O + CO_2$$

Carbon dioxide gas smothers a fire by preventing the oxygen of the air from contacting the combustible substance.

Oil and gasoline fires cannot be extinguished with water because they are lighter than water and will float on its surface. The layer of water underneath will be unable to lower the temperature of the oil or gasoline below the kindling temperature. An effective fire extinguisher for oil or gasoline fires contains foamite, a mixture of chemicals and licorice, which forms a thick foam (containing carbon dioxide bubbles) on the surface of the burning oil and thus removes the oxygen. A fire-fighting foam is also produced by me-

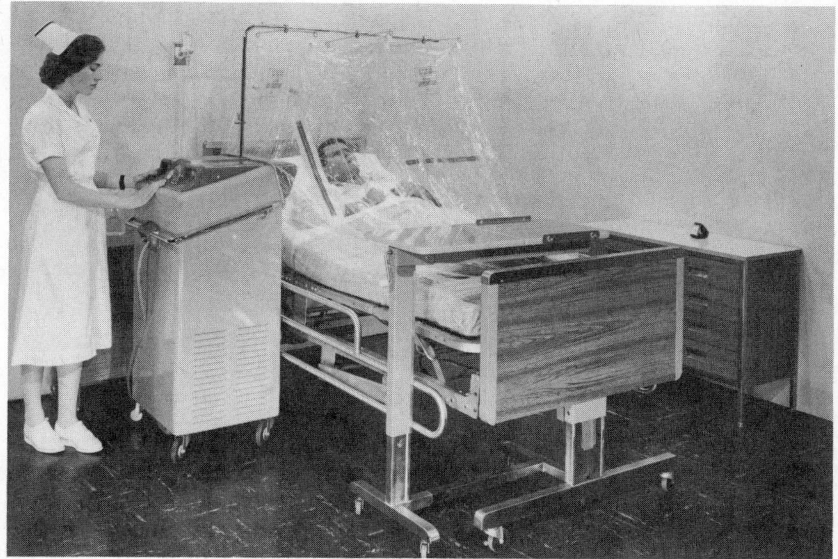

Figure 4.3 A patient in an oxygen tent. (Courtesy Linde Company, a Division of Union Carbide Corp.)

chanically foaming protein hydrolysates from soybeans or fish scraps. Sand may be employed to smother small oil fires.

Uses of oxygen

Anoxia is any pathological condition in which there is an inability to furnish the tissues of the body with sufficient oxygen to carry out their natural functions. Anoxia may be due to such causes as an oxygen deficient atmosphere in the lungs, inability of the oxygen in the lungs to be absorbed into the blood, or inability of the blood to transport oxygen. Oxygen is commonly administered in anoxic conditions (Figure 4.3).

Thus oxygen is used in asphyxiations due to smoke, carbon monoxide, oxygen-low atmospheres, drowning, and electric shock. In general anesthesia, it must be mixed with the anesthetic agent because the pure form of the anesthetic agent will not support respiration. It is also used in the treatment of pneumonia patients, whose lung capacity is reduced by the presence of mucus and who can not obtain sufficient amounts of oxygen when breathing air. It is used for newborn babies who experience difficulty in breathing or whose lungs do not readily expand, and for cyanosed surgical patients and patients suffering from cardiac or respiratory failure. Carbon dioxide is frequently administered in small amounts with oxygen, because this is a very effective respiratory stimulant. Oxygen is administered by an inhalator or a nasal catheter or by placing the patient in an oxygen tent.

Oxygen is used for respiration by aviators at high altitudes, by sailors in submarines, and in mine and fire rescue. In industry it is used in oxyacetylene and in oxyhydrogen torches used in the cutting and welding of metals.

STUDY QUESTIONS

1. Describe a method of preparing oxygen from (a) a metallic oxide, (b) a suitable salt, (c) water, (d) air, and (e) a peroxide.
2. Write balanced equations for the preparation of oxygen from each of the following substances: mercuric oxide, potassium chlorate, and water.
3. How does one test for oxygen in the laboratory?
4. What is a catalyst? Give an example of a catalyst.
5. What effect does a catalyst have on the rate of a chemical reaction?
6. Why is oxygen the first element studied in chemistry?
7. Discuss the occurrence of oxygen.
8. In what form is oxygen present in the air? In water?
9. Who besides Priestley has claim to the discovery of oxygen?
10. What does the word "oxygen" mean, and who named it?
11. What is now the most important commercial method of preparing oxygen?
12. Discuss the importance of oxygen to animal and vegetable organisms.
13. What is hemoglobin, and what role does it play in respirations?
14. How do anaerobes differ from aerobes?
15. Since animals are using the oxygen of the air for respiration, why does the percentage of oxygen in the air remain nearly constant?
16. Why do substances burn more rapidly in oxygen than in air?
17. Why are dampers placed on stoves and furnaces?
18. Define oxide, combustion, kindling temperature, slow oxidation.
19. Which oxide of iron forms when iron rusts? What type of oxidation occurs when iron rusts?
20. Complete and balance the equation for the complete combustion of pentane, C_5H_{12}, to carbon and water.
21. Write balanced equations for the burning of magnesium, iron, sulfur, carbon, and phosphorus.
22. Write balanced equations for the action of water on CO_2, SO_2, P_4O_{10}, CaO, K_2O, MgO, and SO_3.
23. Give an alternate name for the nonmetallic oxides. Why are these oxides so named?
24. What are some of the common properties of acids and bases?
25. What five factors influence the speed of an oxidation?
26. What two conditions are necessary for a spontaneous combustion?
27. Give several examples of spontaneous combustion.
28. For what purposes is oxygen used in the hospital and in medical practice?
29. What precaution would you expect to be exercised when oxygen is used in the hospital?
30. What are some of the uses of oxygen in industry?
31. What two conditions are necessary if a fire is to be extinguished?
32. How do the following extinguish a flame: carbon dioxide, a wet blanket, water, foamite foam?
33. Why are oil fires usually so difficult to extinguish? Explain how they may be extinguished.

34. What is meant by oxidation?
35. Balance the following chemical equations:

(a) $Zn + KOH \longrightarrow K_2ZnO_2 + H_2$
(b) $NaHCO_3 + H_2SO_4 \longrightarrow Na_2SO_4 + H_2O + CO_2$
(c) $Zn(OH)_2 + H_3PO_4 \longrightarrow Zn_3(PO_4)_2 + H_2O$
(d) $PI_3 + H_2O \longrightarrow HI + H_3PO_3$
(e) $H_2S + SO_2 \longrightarrow S + H_2O$
(f) $Fe_3O_4 + C \longrightarrow Fe + CO$
(g) $Hg(NO_3)_2 + Al \longrightarrow Al(NO_3)_3 + Hg$
(h) $Fe + HC_2H_3O_2 \longrightarrow Fe(C_2H_3O_2)_2 + H_2$
(i) $C_2H_4 + O_2 \longrightarrow CO_2 + H_2O$
(j) $CaC_2 + H_2O \longrightarrow Ca(OH)_2 + C_2H_2$
(k) $CaCO_3 + C \longrightarrow CaC_2 + CO$
(l) $Na + H_2O \longrightarrow NaOH + H_2$
(m) $Na_2O_2 + H_2O \longrightarrow NaOH + O_2$

chapter 5

chemical arithmetic, gas laws, and the kinetic-molecular theory of gases

5

The recent rapid development of the field of bioenergetics in the study of biochemistry requires that the student of biology become acquainted with the mathematics of chemistry. This study of chemical mathematics is no less necessary for those in paramedical programs, such as nurses and laboratory technicians, who may have a need to appreciate the interpretation of the data obtained in the hospital laboratory.

Some chemical arithmetic was introduced in Chapter 2 in the study of the metric system, density, specific gravity, and units of heat measurement. The mathematics concerned with concentration of solutions and acidity will be introduced in Chapters 7 and 10, respectively.

This chapter includes a study of molecular weights and calculations from chemical equations, concepts necessary for a basic knowledge of chemistry, as well as a study of the gas laws and a discussion of the kinetic-molecular theory. The latter is an excellent illustration of the application of the scientific method.

A knowledge of the gas laws should be of value in performing certain experiments in physiology, and a knowledge of these will be needed in a study of calorimetry and in the use of much of the specialized equipment now being introduced into the hospital. However, omission of the gas laws and the kinetic-molecular theory of gases will not destroy the continuity of this text.

Before solving problems involving calculations from equations, the student should review the discussion on page 55 concerned with the technique of balancing chemical equations.

Molecular weights

The weight of a molecule is the sum of the weights of its constituent atoms and is called the *molecular weight.* The molecular weight of sulfur dioxide (SO_2) will be the sum of 32 (at. wt. of sulfur) and 2 times 16 (at. wt. of oxygen), or 64. Molecular weights, like atomic weights, are relative weights. A molecule of sulfur dioxide is twice as heavy as a sulfur atom.

The molecular weight of a substance expressed in grams is called a *gram-molecular weight,* or a *mole.* Note that the gram-molecular weight or mole will be equal to the sum of all the atomic weights expressed in grams. Thus the mole of sulfuric acid will be 2×1 (at. wt. of hydrogen) + 32 (at. wt. of sulfur) + 4×15 (at. wt. of oxygen), or 98 g.

There are 6.023×10^{23} (or 6023 with 20 zeros following it) molecules in a mole of a substance. This is called the *Avogadro number* and is frequently represented by the symbol *N*.

The quantitative interpretation of formulas and equations

A chemical formula is not only a shorthand means of representing a chemical compound but also represents the composition of the compound. The formula H_2SO_4 indicates that each molecule of sulfuric acid contains two atoms of hydrogen, an atom of sulfur, and four atoms of oxygen. Because the molecular weight of a compound is equal to the sum of all its atomic weights, the molecular weight of sulfuric acid is 2 + 32 + 64 or 98; a mole of sulfuric acid is then 98 g.

A chemical equation, like a chemical formula, has a quantitative significance as well as a qualitative one. The reaction

$$Fe + S \longrightarrow FeS$$

may be interpreted as follows:

Qualitatively — Iron and sulfur combine to form iron sulfide.

Quantitatively

1. One atom of iron combines with one atom of sulfur to form one molecule of iron sulfide.
2. One gram-atom of iron combines with one gram-atom of sulfur to form one mole of iron sulfide.
3. Fifty-six grams of iron combine with 32 g of sulfur to form 88 g of iron sulfide.

Note that the sum of the weights of reactants will be equal to the sum of the weights of the products in a chemical reaction. This is a requirement of the law of the conservation of matter (see page 13).

Calculation of the percentage composition of a compound from its formula

The molecular weight of water is 18 ($2 \times 1 + 16$), and thus water is composed of 2 parts by weight of hydrogen and 16 parts of oxygen. The fraction of water that is hydrogen is two-eighteenths, and the fraction that is oxygen is sixteen-eighteenths. These fractions are converted to percentages by multiplying by 100. The percentage of hydrogen in water is $\frac{2}{18} \times 100$, or 11 percent, while the percentage of oxygen is $\frac{16}{18} \times 100$ or 89 percent.

ILLUSTRATIVE PROBLEM 1

What is the percentage of zinc in zinc sulfide, ZnS?

The molecular weight of ZnS is 65 + 32, or 97. Because in every 97 g of ZnS there is 65 g of zinc, the percentage of zinc in this compound is

$$\% \text{ Zn} = 65 \text{ g}/97 \text{ g} \times 100 = 67\%$$

ILLUSTRATIVE PROBLEM 2

What are the percentages of potassium, chlorine, and oxygen in potassium chlorate, $KClO_3$?

The molecular weight of $KClO_3$ is 39 + 35.5 + (3 × 16) = 122.5.

$$\% \text{ K} = \frac{\text{g of K}}{\text{mole of KClO}_3} \times 100 = 39 \text{ g}/122.5 \text{ g} \times 100 = 32\%$$

$$\% \text{ Cl} = \frac{\text{g of Cl}}{\text{mole of KClO}_3} \times 100 = 35.5 \text{ g}/122.5 \text{ g} \times 100 = 29\%$$

$$\% \text{ O} = \frac{\text{g of O}}{\text{mole of KClO}_3} \times 100 = 48 \text{ g}/122.5 \text{ g} \times 100 = 39\%$$

Determination of the empirical and molecular formula of a compound from its percentage composition

The empirical formula of a compound indicates the simplest ratio of the various atoms in the molecule. If the molecular weight of the compound is known, the *molecular formula*, a formula in which the total number of the various kinds of atoms in the molecule are indicated, can be established.

Ethylene glycol is composed of 38.7 percent carbon, 9.7 percent hydrogen, and 51.6 percent oxygen. If these percentages are now interpreted as grams of these elements in 100 g of the compound, the ratio of gram-atoms and the empirical formula may be established as follows: Dividing grams of the element per 100 g of the compound by the atomic weight of the element gives the gram-atoms per 100 g of the compound.

$$38.7 \text{ g of C} \div \frac{12 \text{ g of C}}{1 \text{ g-atom of C}} = 38.7 \text{ g} \times \frac{1 \text{ g-atom}}{12 \text{ g}}$$

$$= \frac{3.23 \text{ g-atoms of carbon}}{100 \text{ g of the compound}}$$

$$9.7 \text{ g of H} \div \frac{1 \text{ g of H}}{1 \text{ g-atom of H}} = 9.7 \text{ g} \times \frac{1 \text{ g-atom}}{1 \text{ g}}$$

$$= \frac{9.7 \text{ g-atoms of hydrogen}}{100 \text{ g of the compound}}$$

$$51.6 \text{ g of O} \div \frac{16 \text{ g of O}}{1 \text{ g-atom of O}} = 51.6 \text{ g} \times \frac{1 \text{ g-atom}}{16 \text{ g}}$$

$$= \frac{3.23 \text{ g-atoms of oxygen}}{100 \text{ g of the compound}}$$

This indicates that the gram-atoms of the elements are in the ratio 3.23:9.7:3.23, or, more simply, 1:3:1. Because a gram-atom of any element is equal to 6.02×10^{23} atoms, then the simplest ratio of the individual atoms in this compound is 1:3:1, and its empirical formula is CH_3O.

If the empirical formula is also the true molecular formula, then the molecular weight of ethylene glycol should be 31. Because the determined molecular weight of this substance is 62, the true molecular formula is double the empirical formula, or $C_2H_6O_2$.

THE GAS LAWS

Matter may exist in the gaseous, liquid, or solid state. A gas has no definite size and shape and will fill the container in which it is enclosed. A liquid has a fixed size but no definite shape; that is, it will take the shape of the container. Solids are rigid, have a fixed size and shape, and most of them have some mechanical strength (see page 19).

Unlike solids and liquids, gases can be compressed and expanded. If cooled sufficiently, all gases are converted to liquids, helium and hydrogen being the most difficult to liquefy. Gases exert pressure if they are confined in a definite space. This pressure is exerted equally on the sides, top, and bottom of the enclosing vessel (Figure 5.1).

The barometer: measurement of the pressure of a gas

To prepare a mercury barometer, a glass tube more than 30 in. long and sealed at one end is filled with mercury and then inverted in a pool of mercury. The mercury inside will drop when the tube is inverted, and the top of the tube will contain a nearly perfect vacuum (Figure 5.2). The mercury column remaining will be equal in weight to a column of air of the same cross-sectional area and which extends upward to the top of the atmosphere. The barometer is sensitive to changes in such a column of air, or in other words, to changes in atmospheric pressure (Figure 5.3).

The atmospheric pressure is determined by finding the difference between the level of mercury in the tube and in the pool of mercury. In the metric system, this pressure is expressed in millimeters of mercury; in the English system, in inches. Standard atmospheric pressure is the average pressure of the atmosphere at sea level, that is, 760 mm or 29.9 in. of mercury. The standard pressure is often designated as *one atmosphere* of pressure.

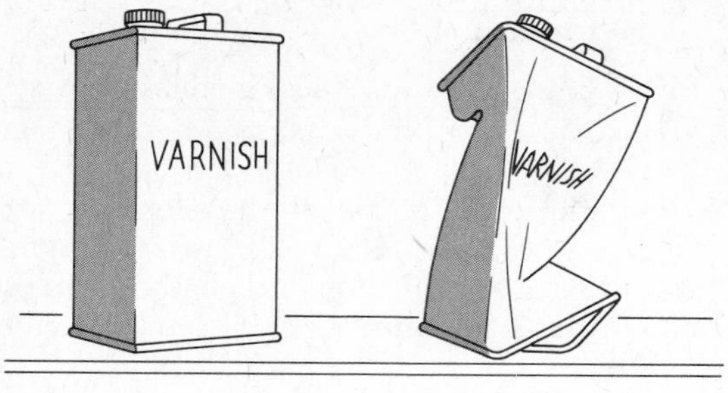

***Figure 5.1** The collapse of a can of steam when it is cooled is evidence for the existence of pressure.*

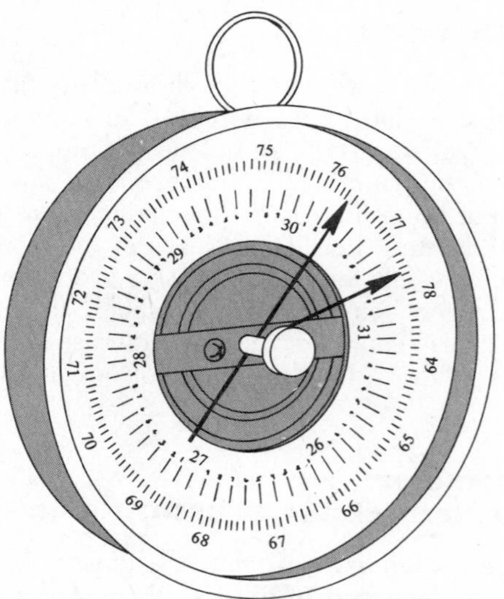

***Figure 5.2** An aneroid barometer. (Courtesy VWR Scientific)*

Boyle's law: the effect of pressure on the volume of a gas

One frequently observes that if the pressure exerted on a confined volume of gas is increased, the volume of the gas will decrease. If the pressure exerted on the gas is doubled, the volume of the gas decreases to one-half its original volume. If the pressure is tripled, the volume will be one-third the original, and if pressure is halved, the volume will be doubled. In 1662, Robert Boyle made similar observations and enunciated the law that now bears his name. He stated: *If the temperature of a gas remains constant, the volume of the gas will be inversely proportional to the pressure exerted on the gas.*

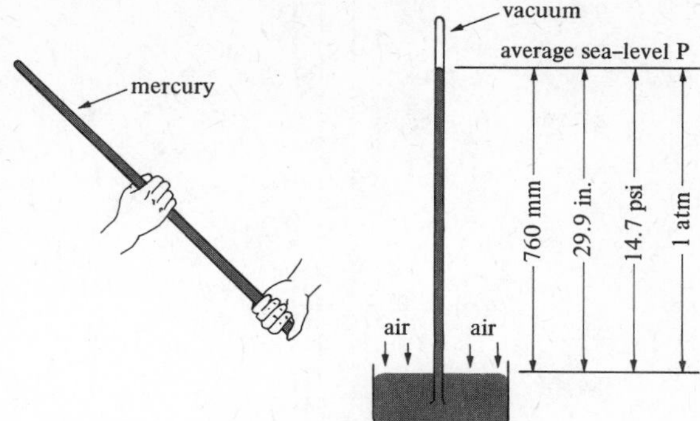

Figure 5.3 A Torricelli barometer is made by completely filling a long glass tube with mercury, inverting it while taking care not to spill any mercury, and then releasing the open end under the surface of a pool of mercury. (From C. W. Keenan and J. H. Wood, General College Chemistry, *4th ed., Harper & Row, N.Y., 1971.)*

Although Boyle's law may be expressed in mathematical terms, it will be unnecessary to apply such an expression to solve gas volume–pressure problems. The following illustrative problems will serve as examples of the method of reasoning that is to be employed in solving this type of problem.

ILLUSTRATIVE PROBLEM 3

Two hundred milliliters of a gas are confined in a vessel at standard pressure. What is the volume of this gas under a pressure of 1140 mm of mercury?

Because the pressure is increasing, the volume will decrease. The volume will decrease by the ratio or the fraction 760/1140 of its original volume. The new volume will be 200 ml $\times$ 760 mm/1140 mm $=$ 133 ml. Note that in the fraction, the smaller pressure is placed above the larger pressure, because the fraction must be less than unity.

ILLUSTRATIVE PROBLEM 4

At a pressure of 760 mm, a gas occupies a volume of 22.4 liters. What will be its volume if the pressure decreases to 720 mm?

Because the pressure decreases, the volume will increase and the original volume must be multiplied by a fraction greater than unity, namely 760/720. The solution to this problem is then

$$22.4 \text{ liters} \times 760 \text{ mm}/720 \text{ mm} = 23.7 \text{ liters}$$

The effect of temperature on the volume of a gas, absolute temperature, and Charles' law

If the pressure is maintained constant, a gas will increase in volume when heated and decrease in volume when cooled.

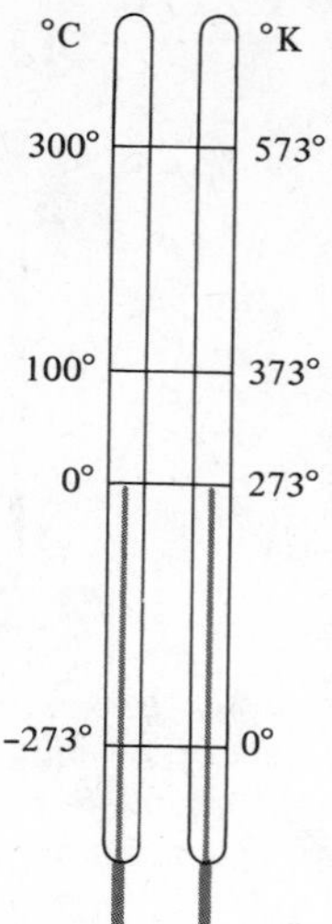

Figure 5.4 A comparison of the Celsius (centigrade) and absolute scales, as shown by two identical thermometers. °K equals °C plus 273.

The following quantitative relationships between the temperature and volume of a gas have been found when the temperature of a gas is changed. If a gas is at a temperature of 0°C, and if the pressure is maintained constant, its volume will increase by $\frac{1}{273}$ of the original volume when it is heated to 1°C. If it is cooled to −1°C, it will contract in volume by $\frac{1}{273}$ and will occupy $\frac{272}{273}$ of its original volume. If cooled to −10°C, it will occupy $\frac{263}{273}$ of its original volume, and at −100°C, $\frac{173}{273}$ of its original volume. If a gas is cooled to −273°C, it should have zero volume. This temperature is called the *absolute zero of temperature.* At this temperature, all molecular motion should cease. It is a physical impossibility for matter to occupy no space, and in fact all gases are converted to liquids before this temperature is reached. The absolute temperature of zero has been approached but never attained.

The absolute, or Kelvin, scale of temperature is used to measure changes of volumes or pressures of gases due to changes in temperature. It is only necessary to add 273 to the Celsius temperature to convert to the Kelvin scale of temperature (Figure 5.4).

$$\text{Kelvin temperature} = \text{Celsius temperature} + 273°$$

or

$$°K = °C + 273°$$

Charles and Gay-Lussac made quantitative observations of the effect of a change in temperature on the volume of a gas. Charles concluded that *if the pressure of a gas is maintained constant, the*

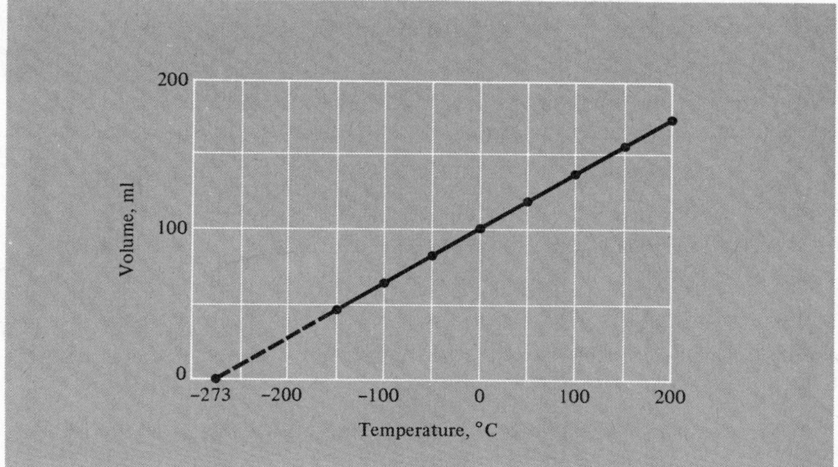

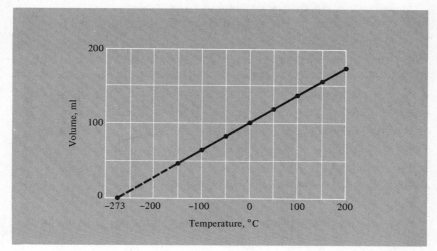

Figure 5.5 *Charles' law, which states that the volume of a gas is directly proportional to its (absolute) temperature, is readily inferred from this graph.*

volume of the gas will be directly proportional to the absolute temperature. This is known as *Charles' law* (Figure 5.5).

The standard temperature used in measuring gases is the freezing point of water, that is, 0°C or 273°K.

The solution of problems that involve Charles' law are illustrated in the following solved problems:

ILLUSTRATIVE PROBLEM 5

A certain gas occupies a volume of 500 ml at a temperature of 30°C. What will be its volume at 60°C, if the pressure is maintained constant?

Because the temperature increases, the volume should increase and the original volume should be multiplied by a fraction greater than unity to obtain the new volume. Because the volume is proportional to the absolute temperature and not to the Celsius temperature, the given temperature must be changed by adding 273 before proceeding with the solution of this problem. The initial temperature in this problem is 30°C + 273°, or 303°K, and the final temperature is 60°C + 273°, or 333°K. The solution of the problem is performed as follows:

$$500 \text{ ml} \times 333°\text{K}/303°\text{K} = 548 \text{ ml}$$

ILLUSTRATIVE PROBLEM 6

The volume of a gas at 10°C is 255 ml. What is its volume at −10° C, if the pressure is maintained constant?

Changing the Celsius temperature to the absolute scale, the initial temperature is 10°C + 273° = 283°K, and the final temperature is −10°C + 273° = 263°K. Because the temperature of the gas is decreasing, the volume of the gas will decrease and the following is the solution of this problem. Note that the fraction used is less than unity.

$$225 \text{ ml} \times 263°\text{K}/283°\text{K} = 237 \text{ ml}$$

Solution of gas problems that involve changes in temperature and pressure

In solving gas problems that involve changes in both pressure and temperature, first maintain the temperature constant and determine the change in volume with the change in pressure. Using the value for the new volume, a correction for change in temperature can then be made while maintaining the pressure constant. In the following solved problems, the correction of the volume due to change of both temperature and pressure is performed in one calculation.

ILLUSTRATIVE PROBLEM 7

A gas occupies a volume of 10 liters at a temperature of 27°C and a pressure of 500 mm of mercury. Calculate the volume of the gas if the temperature is changed to 60°C and the pressure is changed to 700 mm.

	INITIAL CONDITIONS	FINAL CONDITIONS
Volume =	10 liters	? liters
Temperature =	27°C or 300°K	60°C or 333°K
Pressure =	500 mm	700 mm

Because the temperature and pressure both increase, the volume will increase by a factor of 333/300 due to the temperature change, and will decrease by a factor of 500/700 due to the pressure change. The new volume will be

10 liters × 333°K/300°K × 500 mm/700 mm = 7.9 liters

ILLUSTRATIVE PROBLEM 8

A gas occupies a volume of 725 ml at 20°C and a pressure of 745 mm. What is its volume at standard conditions?

	INITIAL CONDITIONS	FINAL CONDITIONS
Volume =	725 ml	? ml
Temperature =	20°C or 293°K	0°C or 273°K
Pressure =	745 mm	760 mm

Because the temperature is decreasing, the volume will decrease by the factor 273/293, while the increase in pressure will decrease the volume by the factor 745/760.

725 ml × 745 mm/760 mm × 273°K/293°K = 662 ml

Dalton's law of partial pressures

When various gases are mixed, a true solution is formed, and *the total pressure of the resulting solution will be equal to the sum of the partial pressure of each constituent gas* (Figure 5.6 and 5.7). This latter statement is known as *Dalton's law of partial pressures.* The partial pressure of a gas constituent will be equal to the pressure that would be exerted by the constituent if it alone occupied the volume of the gas mixture. Thus at atmospheric pressure the partial

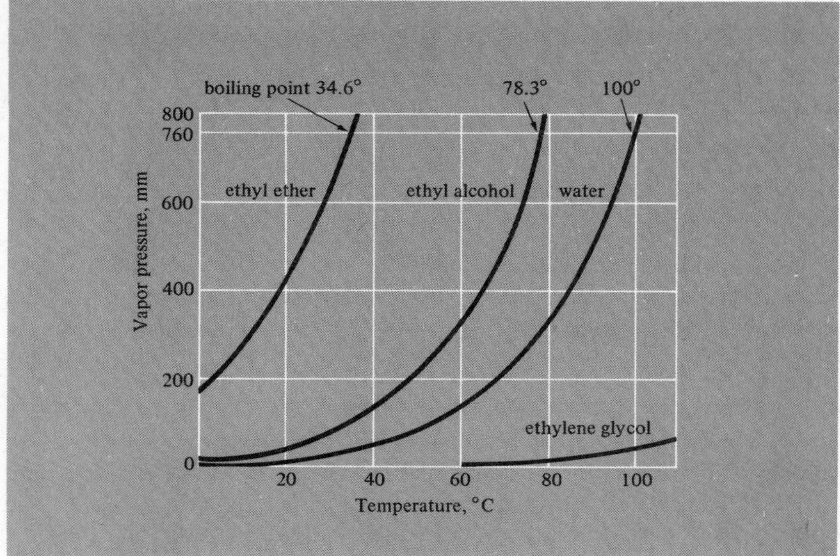

Figure 5.6 A graph in which the vapor pressure of four liquids is plotted against temperature (in degrees Celsius). The temperature at which each liquid attains a vapor pressure of 760 mm will be the boiling point of the liquid.

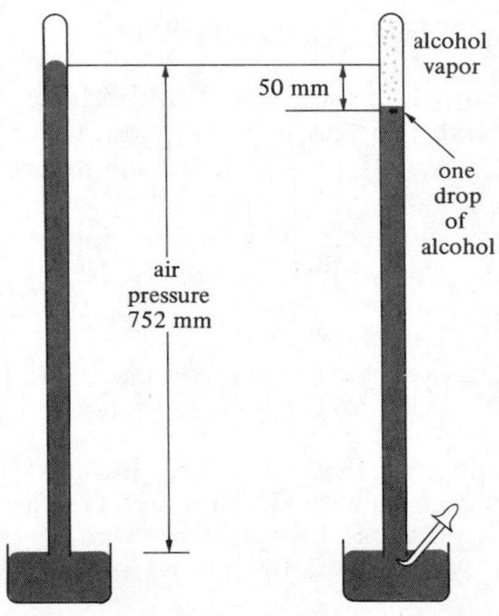

Figure 5.7 The vapor pressure of a liquid can be determined by introducing a drop of the liquid into a barometer. The drop of liquid will quickly rise to the top of the column of mercury in the barometer and evaporate. The fall that occurs in the height of the mercury column will equal the vapor pressure of the liquid.

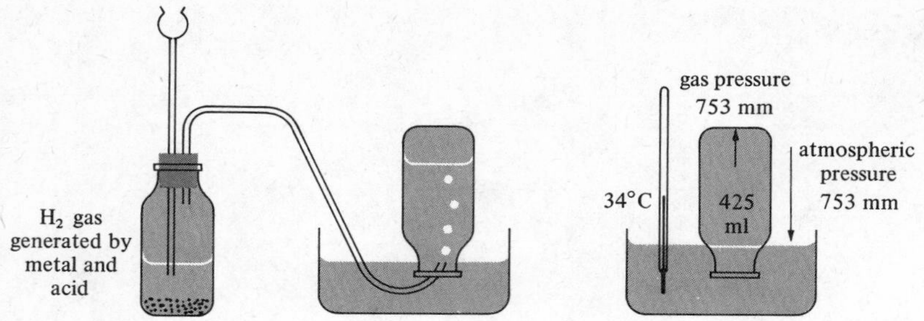

***Figure 5.8** In collecting gases it is usual to note their pressure, volume, and temperature. Most frequently, gases are collected in the laboratory at atmospheric pressure. In such cases the pressure of the gas is obtained by reading the pressure recorded on a barometer placed in the laboratory.*

pressure of oxygen in the atmosphere will be about $\frac{20}{100} \times 760$ mm, or 152 mm, and of nitrogen about $\frac{80}{100} \times 760$ mm, or 608 mm. (The sum of these partial pressures will equal the total pressure of 760 mm.)

Because the solubility of a gas in a liquid is proportional to its partial pressure, it is quite evident that respiration will have to be increased in a rarefied atmosphere in order to supply the cells with sufficient oxygen for their normal function.

Because problems involving gas laws usually require using the value of the pressure of the dry gas, a correction for the partial pressure exerted by water vapor, if the gas is collected over water, must be subtracted from the total observed pressure (Figure 5.8). The vapor pressure of water is dependent upon the temperature. Thus at 20°C the vapor pressure of water is 17.54 mm (of Hg), while at 30°C it has risen to 31.82 mm. See Table 5.1 for the relationship of vapor pressure of water to temperature. Note that at 100°C the vapor pressure of water is equal to the atmospheric pressure (760 mm of Hg).

A typical gas law calculation involving a correction for the presence of water vapor is illustrated in the following solved problem.

ILLUSTRATIVE PROBLEM 9

A gas collected over water has a volume of 550 ml at 22°C and a pressure of 740 mm. What is the volume of the dry gas under standard conditions?

Note from Table 5.1 that the vapor pressure of water at 22°C is 19.83 mm. Because according to Dalton's law (see above) the total pressure measured is the sum of the vapor pressure of the dry gas and the partial pressure of water vapor, the partial pressure exerted by the dry gas will be

$$740 \text{ mm} - 19.83 \text{ mm} = 720.17 \text{ mm}$$

From this point on, this problem is solved similarly to Illustrative Problem 8 above, thus:

TABLE 5.1 VAPOR PRESSURE OF WATER AT VARIOUS TEMPERATURES

Temperature, °C	*Vapor pressure, mm of Hg*	*Temperature, °C*	*Vapor pressure, mm of Hg*
0	4.58	29	30.04
5	6.54	30	31.82
10	9.21	31	33.70
11	9.84	32	35.66
12	10.52	33	37.73
13	11.23	34	39.90
14	11.99	35	42.18
15	12.79	40	55.32
16	13.63	45	71.88
17	14.53	50	92.51
18	15.48	55	118.04
19	16.48	60	149.38
20	17.54	65	187.54
21	18.65	70	233.7
22	19.83	75	289.1
23	21.07	80	355.1
24	22.38	85	433.6
25	23.76	90	525.8
26	25.21	95	633.9
27	26.74	100	760.0
28	28.35	150	3570.5

	INITIAL CONDITIONS	FINAL CONDITIONS
Volume	= 550 ml	? ml
Temperature	= 22°C or 295°K	0°C or 273°K
Pressure	= 720.17 mm	760 mm

The new volume will be

$$550 \text{ ml} \times \frac{720.17 \text{ mm}}{760 \text{ mm}} \times \frac{273°\text{K}}{295°\text{K}} = 483 \text{ ml}$$

Avogadro's law and the gram-molecular volume

The law first stated by the Italian chemist Amedeo Avogadro, that is, that *equal volumes of gases at the same temperature and pressure will contain equal number of molecules,* and now known as Avogadro's law, has been affirmed and is now accepted as an important law of chemistry.

As noted previously in this chapter, a mole of a chemical substance consists of 6.02×10^{23} molecules, this number being noted as Avogadro's number and designated often as N. All that is necessary to establish a molecular weight experimentally is to determine the weight of this number of molecules (Figure 5.9). Because a liter

Figure 5.9 Arranged from left to right around 1 mole of sugar (342 g) are one mole of aluminum (27 g), one mole of water (18 g), one mole of copper (64 g), and one mole of iron (56 g). Each contains 6 × 10^{23} atoms of an element, or that number of molecules of a compound. (From C. W. Keenan and J. H. Wood, General College Chemistry, *4th ed., Harper & Row, N.Y., 1971.)*

of oxygen weighs 1.429 g at standard conditions, it can be calculated that a mole of oxygen gas (32 g) will occupy

$$\frac{32\text{ g}}{1.429\text{ g/1 liter}} = 32\text{ g} \times \frac{1\text{ liter}}{1.429\text{ g}} = 22.4\text{ liters}$$

under standard conditions. Then according to Avogadro's law, a mole of all gases occupies a volume of 22.4 liters at standard conditions, and this numerical value is now known as the gram-molecular volume, or GMV.

ILLUSTRATIVE PROBLEM 10

What is the weight of a mole of methane, CH_4, if a liter of this gas at standard conditions weighs 0.713 g?

A mole will be the weight of 22.4 liters of this gas at standard conditions, or

$$\frac{0.713\text{ g}}{1\text{ liter}} \times 22.4\text{ liters} = 16.0\text{ g}$$

ILLUSTRATIVE PROBLEM 11

What is the weight of a liter of ethylene, C_2H_4, at standard conditions?

The gram-molecular weight of ethylene is 24 plus 4, or 28 g. This weight of ethylene will occupy a volume of 22.4 liters at standard conditions. Then under these conditions, a liter of ethylene will weigh

$$\frac{28\text{ g}}{1\text{ mole}} \times \frac{1\text{ mole}}{22.4\text{ liters}} = 1.25\text{ g/liter}$$

The kinetic-molecular theory of gases

It is to be noted that all gases behave alike under similar conditions. The volume of all gases is affected in the same manner by changes in temperature and pressure, regardless of the chemical nature of the gases, a behavior that finds explanation in the *kinetic-molecular theory of gases*. Some of the assumptions of this theory are listed below.

1. All gases are composed of molecules whose diameters are small compared to the distances between the molecules.
2. The molecules of gases are in a constant state of motion and travel in a straight line untii they collide with another molecule of the gas or with the walls of the container.
3. When the gas molecules collide with each other or with the walls of the container, they rebound without loss of velocity; that is, they are perfectly elastic.
4. The velocities of the gas molecules increase with an increase in temperature and decrease when the temperature is reduced.

HOW DOES THE KINETIC-MOLECULAR THEORY EXPLAIN THE BEHAVIOR OF GASES?

Because gas molecules are relatively far apart, gases may be readily compressed, or crowded closer together. Pressure is due to the bombardment of the walls of the container by the molecules of the gas. An equal number of gas molecules with equal average velocities are traveling in all directions. For this reason, the same pressure will be exerted on all the sides and the top and bottom of the container. The pressure of a gas on a given surface of the containing vessel will be equal to the total number of collisions of the gas molecules on that surface at any one instant. The pressure of a gas is a manifestation of the bombardment of a surface by the fast-moving molecules of the gas.

If the volume of a given quantity of gas at constant temperature is decreased, the pressure of the gas is increased because the collisions of the gas molecules with the surface of the container will become more frequent. If a given quantity of a gas is expanded (temperature remaining constant), the pressure exerted by this gas will decrease, because there will be fewer collisions of the gas molecules on a certain area of the surface of the container. These facts are in harmony with Boyle's law.

An increase in temperature of a gas is manifested by an increased velocity of the gas molecules. The increase of pressure that accompanies an increase in temperature at constant gas volume is explained by an increased number of collisions of the gas molecules on a unit area of surface because of their greater velocity. If the pressure of the gas is to be maintained constant while the temperature of the gas is increased, the gas must be permitted to expand.

That is, its volume must increase, because upon increase in temperature, fewer molecules per unit volume are needed to maintain the same pressure, that is, number of collisions on a unit surface of the container. One must agree that this is a very realistic explanation of Charles' law.

Calculations based on chemical equations

WEIGHT–WEIGHT CALCULATIONS FROM EQUATIONS

It is frequently necessary to know the amount of a chemical that is formed in a chemical reaction or that must enter into a chemical reaction to form a specified amount of material. *In solving such a problem, the first requirement is a chemical equation to represent the reaction, and such an equation must be balanced before any calculations can be made.* After the equation is balanced, the relationship of the weights of the reactants and products indicated by the chemical equation is determined. The weight of the material to be used or produced, as well as the unknown quantity, should be indicated. The solution of this type of problem is now illustrated.

ILLUSTRATIVE PROBLEM 12

What weight of barium chloride is needed to produce 150 g of barium sulfate?

Step 1. Write an equation for the reaction.

$$BaCl_2 + H_2SO_4 \longrightarrow BaSO_4 + HCl$$

Step 2. Balance the equation.

$$BaCl_2 + H_2SO_4 \longrightarrow BaSO_4 + 2HCl$$

Step 3. Indicate the theoretical quantity of reactant and product.

$$\overset{208\text{ g}}{BaCl_2} + H_2SO_4 \longrightarrow \overset{233\text{ g}}{BaSO_4} + 2HCl$$

Step 4. Indicate known and unknown quantities.

$$\underset{x\text{ g}}{\overset{208\text{ g}}{BaCl_2}} + H_2SO_4 \longrightarrow \underset{150\text{ g}}{\overset{233\text{ g}}{BaSO_4}} + 2HCl$$

Empirically this problem may be solved by proportion, because the ratio of the theoretical quantity of barium chloride to x will be equal to the ratio of the theoretical quantity of barium sulfate to the quantity of barium sulfate to be produced. Thus

$$\frac{208\text{ g}}{x\text{ g}} = \frac{233\text{ g}}{150\text{ g}}$$

$$x\text{ g} = \frac{208\text{ g} \times 150\text{ g}}{233\text{ g}} = 134\text{ g of } BaCl_2$$

The following is a clearer method of solving this problem: According to the above equation, a mole of $BaCl_2$ produces a mole of $BaSO_4$. Therefore, because 150/233 mole of $BaSO_4$ is to be prepared, 150/233 mole of $BaCl_2$ will be required, or on changing moles to grams,

$$\frac{150}{233}\text{ mole} \times \frac{208\text{ g}}{1\text{ mole}} = 0.64\text{ mole} \times \frac{208\text{ g}}{1\text{ mole}} = 134\text{ g of BaCl}_2$$

ILLUSTRATIVE PROBLEM 13

Twenty grams of aluminum are reacted with sulfuric acid to produce aluminum sulfate. How many grams of the latter will be produced?

Writing and balancing the equation and indicating theoretical, unknown, and given amounts of reactants and products gives the following:

$$\underset{20\text{ g}}{\overset{54\text{ g}}{2\text{Al}}} + 3\text{H}_2\text{SO}_4 \longrightarrow \underset{x\text{ g}}{\overset{342\text{ g}}{\text{Al}_2(\text{SO}_4)_3}} + 3\text{H}_2$$

Because 20/27 g-atom of aluminum is used and each g-atom of this metal produces ½ mole of aluminum sulfate, then 20/(2 × 27) mole or 20/(2 × 27) × 342 g = 127 g of aluminum sulfate is produced.

WEIGHT–VOLUME CALCULATIONS FROM EQUATIONS

If a gas is produced during a chemical reaction, or if a gas is utilized in forming a product and it is preferred to determine the volume of the gas produced or utilized at standard conditions, it is only necessary to substitute the GMV, or 22.4 liters for each mole of the gas indicated in the balanced equation.

ILLUSTRATIVE PROBLEM 14

What volume of oxygen at standard conditions can be produced by heating 5 g of potassium chlorate?

The following is the balanced equation for this reaction with the theoretical, known, and unknown quantities indicated.

$$\underset{5\text{ g}}{\overset{2\times 122.5 = 254\text{ g}}{2\text{KClO}_3}} \longrightarrow 2\text{KCl} + \underset{x\text{ liters}}{\overset{3\times 22.4,\text{ or }67.2\text{ liters}}{3\text{O}_2}}$$

The solution is

$$\frac{254\text{ g}}{5\text{ g}} = \frac{67.2\text{ liters}}{x\text{ liters}}$$

$$x\text{ liters} = \frac{67.2\text{ liters} \times 5\text{ g}}{254\text{ g}} = 1.32\text{ liters at standard conditions}$$

VOLUME–VOLUME CALCULATIONS FROM EQUATIONS

Because a mole of any gas occupies a volume of 22.4 liters at standard conditions, then the ratio of the volumes of gases undergoing reaction and/or produced during chemical reaction will be the ratio of moles of the gases that appear in the balanced equation for that reaction. This statement constitutes Gay-Lussac's law of volumes.

ILLUSTRATIVE PROBLEM 15

In the formation of hydrogen chloride gas, what volume of chlorine gas will combine with 12 liters of hydrogen at standard conditions, and what volume of HCl gas will be formed?

From the balanced equation

$$H_2 + Cl_2 \longrightarrow 2HCl(g)$$

it will be noted that each volume of hydrogen will combine with an equal volume of chlorine and will produce two volumes of hydrogen chloride gas. Thus 12 liters of hydrogen will require 12 liters of chlorine for complete reaction and will produce 24 liters of hydrogen chloride gas, all measured at standard conditions.

STUDY QUESTIONS

1. What is the standard of atomic weights?
2. A sulfur trioxide molecule is 2.5 times as heavy as an oxygen molecule. What is the molecular weight of sulfur trioxide?
3. If the atomic weights of the constitutent elements of a compound are known, how can the molecular weight of the compound be determined?
4. What is the qualitative and quantitative interpretation of the formulas NaOH and $Ca(NO_3)_2$?
5. Give a quantitative interpretation of the formulas $Zn(NO_3)_2$ and $C_3H_5(OH)_3$.
6. Give several quantitative interpretations of the equation $2SO_2 + O_2 \longrightarrow 2SO_3$
7. How do gases in general differ in properties from liquids and solids?
8. How is the pressure of a gas determined? What is standard pressure?
9. Describe the construction of a mercury or Torricelli barometer.
10. If a gas is heated from 0°C to 1°C, by what fraction of its original volume will the gas increase? Also for a change from 0°C to 10°C?
11. What is the significance of absolute temperature? Of the absolute zero? How is a Celsius temperature converted to the Kelvin temperature?
12. What temperature has been chosen as the standard temperature?
13. State Boyle's and Charles' laws.
14. If a gallon tin can is filled with steam and is then sealed and cooled, it will collapse. Why?
15. Why are automobile tires more prone to burst in hot weather?
16. What are some of the assumptions of the kinetic-molecular theory?
17. How does the above theory explain pressure?
18. What happens to gas molecules when they are heated?
19. How does the kinetic-molecular theory explain the effect of an increase in pressure on the volume of a gas? An increase in temperature?
20. What is the significance and numerical value of the GMV?
21. State and indicate a practical application of Avogadro's law.

PROBLEMS

1. Calculate the weight of iron in 22 tons of ore which contains 72 percent iron.
2. Eighteen grams of magnesium will liberate 1.5 g of hydrogen. How much magnesium will be required to liberate 4.5 g of hydrogen?
3. Calculate the molecular weight of the following:

(a) barium nitrate, $Ba(NO_3)_2$
(b) acetylene, C_2H_2
(c) glucose, $C_6H_{12}O_6$
(d) propyl alcohol, C_3H_7OH
(e) ammonium chloride, NH_4Cl
(f) thiourea $(NH_2)_2CS$

4. What is the percentage of carbon in the following compounds?

(a) ethane, C_2H_6
(b) ethyl alcohol, C_2H_5OH
(c) sodium carbonate, Na_2CO_3
(d) benzene, C_6H_6
(e) glucose, $C_6H_{12}O_6$
(f) acetone, CH_3COCH_3

5. Calculate the percentage of the element requested:

(a) calcium in $CaCO_3$
(b) iron in $K_4Fe(CN)_6$
(c) cadmium in $CdSO_4$
(d) manganese in $KMnO_4$
(e) copper in $Cu(H_2AsO_3)_2$
(f) sodium in $NaC_7H_5O_2$
(g) phosphorus in $CaHPO_4$
(h) lead in $Pb(C_2H_5)_4$

6. Five grams of magnesium are burned. What weight of ash (MgO) will remain?
7. What weight of calcium oxide, CaO, will be required to produce a kilogram of calcium hydroxide, $Ca(OH)_2$?
8. How much mercury is produced when 2.3 g of its oxide is heated?
9. What weights of carbon dioxide and water are produced when 10 g of octane, C_8H_{18}, is completely burned? What volume of carbon dioxide will be produced?
10. Hydrogen sulfate is produced by the action of water on sulfur trioxide:

$$SO_3 + H_2O \longrightarrow H_2SO_4$$

How much sulfur trioxide will be needed to make a kilogram of sulfuric acid that contains 90 percent hydrogen sulfate?
11. What weight of hydrogen phosphate, H_3PO_4, is produced by burning 34 g of phosphorous and treating the resulting oxide with water?
12. A volume of hydrogen measures 235 ml at 715 mm pressure. If the temperature remains constant, what is the volume of the gas at 840 mm pressure?
13. A gas has a volume of 1155 ml at a pressure of 775 mm. What is the volume of the gas at standard pressure if the temperature remains constant?
14. Five hundred milliliters of oxygen at a temperature of 20°C are heated to 40°C. What is the new volume of oxygen if this experiment is performed under constant pressure?
15. A gas has a volume of 15 liters at room temperature (20°C). What will be its volume at 100°C assuming that no change in pressure occurs?
16. What will be the volume of 225 ml of a gas if its temperature is changed from 35°C to the standard temperature? Assume that there is no change in pressure.
17. A gas measures 415 ml in volume at 10°C and 735 mm pressure. What will be its volume at −10°C and 745 mm pressure?
18. A quantity of carbon dioxide gas occupies 405 ml volume at 710 mm pressure and 35°C. What volume will this gas occupy at 20°C and 775 mm pressure?
19. A gas has a volume of 22.4 liters at standard conditions. What will be its volume at 22°C and 755 mm pressure?

20. A gas has a volume of 335 ml at 15°C and 780 mm pressure. What is its volume at standard conditions?
21. A gas at 1350 mm pressure and 100°C is permitted to expand. If the original volume of the gas was 5.5 liters and the new temperature is 75°C, what is the new volume of the gas if the new pressure is 800 mm?
22. From the indicated data, establish the empirical and molecular formulas for the following:

Compound (a) carbon = 92.5 percent; hydrogen = 7.5 percent; molecular weight = 78.

Compound (b) carbon = 52.2 percent; hydrogen = 13.0 percent; oxygen = 34.8 percent; molecular weight = 46.

23. What volume of carbon dioxide measured at standard conditions will be formed when 6 g of carbon burns? What volume of oxygen will be required for its combustion?
24. What weight of mercuric oxide must be heated to produce a liter of oxygen measured at standard conditions?
25. Upon electrolysis of water, 300 ml of hydrogen measured at standard conditions was formed. What volume of oxygen was simultaneously formed?
26. Ten liters each of carbon monoxide, CO, and of oxygen are mixed and ignited. What is the amount and nature of the products that remain?
27. Fill in the following blanks:

	WEIGHT PER LITER AT STANDARD CONDITIONS	MOLECULAR WEIGHT
(a) ethane	1.34 g	________
(b) carbon dioxide	________	44
(c) carbon monoxide	1.25 g	________
(d) dimethyl ether	________	46
(e) sulfur dioxide	________	64

28. A gas was collected over water at 27°C and 755 mm pressure. Under these conditions, it has a volume of 72 ml. What will be the volume of the dry gas at standard conditions?
29. A gas collected over water has a volume of 740 ml at 30°C and a pressure of 760 mm. What will be the volume of the dry gas at 20°C and at standard pressure?

chapter 6

hydrogen and its oxides: water and hydrogen peroxide

6

Hydrogen

The discovery of the element hydrogen is credited to the Englishman Sir Henry Cavendish, who prepared it in 1776. It must have been prepared many times before, but no one had recognized it as a distinct element. Lavoisier assigned it the name hydrogen, which means "water former."

The atmosphere of the sun and many stars must contain this element in the elementary state in appreciable amounts. Hydrogen occurs in only minute amounts in the free state in the atmosphere of the earth, but in combination with other elements its occurrence is quite widespread. It represents about one-ninth of the weight of water and is present in compound form in natural gas, gasoline, sugar, fats, proteins, and other food constituents. It is present in all acids and bases.

PREPARATION OF HYDROGEN FROM WATER

Hydrogen is one of the products formed when water is electrolyzed (page 56). Sodium, potassium, and calcium will react with water at room temperature to produce hydrogen and a base. Thus sodium reacts vigorously with water to form hydrogen and sodium hydroxide.

$$2Na + 2H_2O \longrightarrow 2NaOH + H_2$$

Note that sodium replaces only one of the hydrogen atoms of water. Potassium reacts so vigorously with water that the hydrogen formed frequently ignites. Great caution must be exercised when these metals are reacted with water, and only very small quantities of these should be employed in this reaction at any one time.

The metals magnesium, aluminum, and iron are less active than the metals mentioned above. These will release hydrogen from water which is in the form of steam but will not react with water at room temperature. In this reaction, a metallic oxide and hydrogen are produced.

$$Mg + H_2O \longrightarrow MgO + H_2$$
$$3Fe + 4H_2O \longrightarrow Fe_3O_4 + 4H_2$$

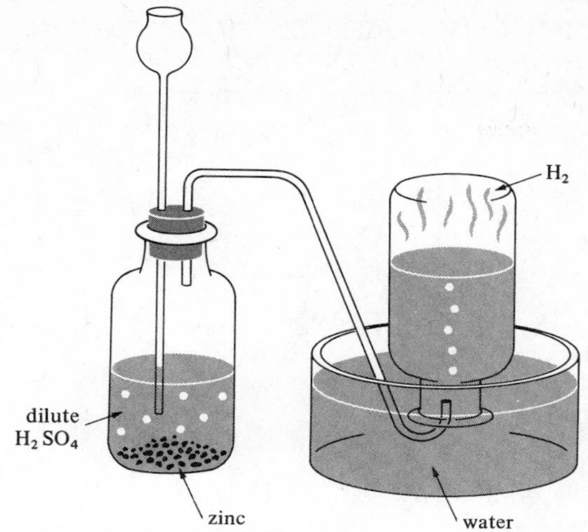

Figure 6.1 The laboratory preparation of hydrogen.

Some metals, such as copper and silver, are so inert that they will not release hydrogen from water under any conditions.

PREPARATION OF HYDROGEN FROM ACIDS

All acids contain hydrogen. Certain active metals will react with many acids to liberate hydrogen. The other product of this reaction is a salt. Iron and zinc are the metals most frequently used, and hydrochloric and dilute sulfuric acids are the acids usually employed in the laboratory preparation of hydrogen (Figure 6.1). The following equations represent the preparation of hydrogen from these metals and acids:

$$Fe + 2HCl \longrightarrow \underset{\text{ferrous chloride}}{FeCl_2} + H_2$$
$$Fe + H_2SO_4 \longrightarrow \underset{\text{ferrous sulfate}}{FeSO_4} + H_2$$
$$Zn + 2HCl \longrightarrow \underset{\text{zinc chloride}}{ZnCl_2} + H_2$$
$$Zn + H_2SO_4 \longrightarrow \underset{\text{zinc sulfate}}{ZnSO_4} + H_2$$

Iron and tin react less vigorously with acids than the more active magnesium, aluminum, and zinc. Copper, silver, gold, and platinum will not liberate hydrogen from an acid. The action of nitric acid, HNO_3, is somewhat different from that of most acids, because it is a strong oxidizing agent as well as an acid. It produces water rather than hydrogen when it reacts with metals.

THE ACTIVITY SERIES OF THE METALS

The common metals can be arranged in a series according to the rate with which the metals displace hydrogen from an acid. Such

TABLE 6.1 THE ACTIVITY, OR ELECTROMOTIVE, SERIES OF THE METALS

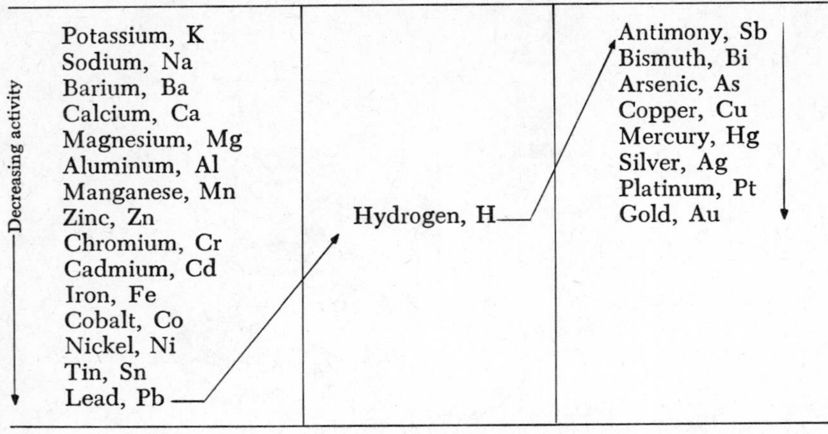

Decreasing activity		
Potassium, K		Antimony, Sb
Sodium, Na		Bismuth, Bi
Barium, Ba		Arsenic, As
Calcium, Ca		Copper, Cu
Magnesium, Mg		Mercury, Hg
Aluminum, Al		Silver, Ag
Manganese, Mn		Platinum, Pt
Zinc, Zn	Hydrogen, H	Gold, Au
Chromium, Cr		
Cadmium, Cd		
Iron, Fe		
Cobalt, Co		
Nickel, Ni		
Tin, Sn		
Lead, Pb		

an arrangement of the metals is called the *activity*, or *electromotive, series of the metals.* (See Table 6.1.) Potassium and sodium are the most active, and tin and lead are the least active, in displacing hydrogen from an acid. Those metals, like copper, mercury, silver, and gold, that do not liberate hydrogen from acids are arranged in the series by other reactions. Hydrogen is placed in the table to separate the metals that displace it from acids from those that do not. The metals to the left of hydrogen in the table belong to the former group, while those to the right of it belong to the latter.

The activity series also indicates the relative activity of metals with water. The most active metals, such as potassium, sodium, and calcium, will liberate hydrogen from water at room temperature. Most of the metals from magnesium to nickel do not react with water at room temperature but liberate hydrogen from steam. Metals below nickel in this series will not react with steam.

The activity series of the metals has a further significance. Any metal will displace a metal lying below it in the series from the latter's aqueous salt solution. Thus copper displaces mercury from mercury salts:

$$Cu + HgSO_4 \longrightarrow CuSO_4 + Hg$$

while an iron nail when placed in a copper sulfate solution will become coated with copper:

$$Fe + CuSO_4 \longrightarrow FeSO_4 + Cu$$

The reverse of this reaction will not occur; that is, copper will not displace iron from a solution of ferrous sulfate. Surgical instruments should not be sterilized with corrosive sublimate (mercuric chloride) solutions, because mercury may be deposited on them.

REVERSIBLE REACTIONS

In the study of the preparation of hydrogen (page 86), it was indicated that hydrogen could be prepared by the action of steam on iron:

$$3Fe + 4H_2O \longrightarrow Fe_3O_4 + 4H_2$$

However, Fe_3O_4 can be converted to metallic iron by heating it in the presence of hydrogen.

$$Fe_3O_4 + 4H_2 \longrightarrow 3Fe + 4H_2O$$

Note that these two reactions are the reverse of each other (page 122). Many such *reversible reactions* are known in chemistry in which, under certain conditions, an equilibrium mixture of products and reactants will form. These reversible reactions can be made to go to completion by using an excess of one of the reactants and/or by removing one of the products as fast as it is formed. If iron is to be converted to the tetroxide, then large volumes of steam are used and the hydrogen must be swept from the apparatus as fast as it is formed. If the oxide is to be reduced to iron, a large volume of hydrogen must be employed and the water vapor removed as it is formed.

PHYSICAL PROPERTIES OF HYDROGEN

Hydrogen is a colorless, odorless, tasteless gas. It is very slightly soluble in water and is the lightest of all gases. A liter weighs 0.08987 g at 0°C and 760 mm pressure. At a very low temperature, hydrogen can be liquefied (−240°C) and solidified (−260°C). Because it is the lightest of the gases, it diffuses the most rapidly. Large quantities of hydrogen are adsorbed by the metal palladium.

CHEMICAL PROPERTIES OF HYDROGEN

Hydrogen reacts quite slowly with oxygen at room temperature. At higher temperatures or in the presence of a flame, these two gases combine to form water with the production of much heat.

$$2H_2 + O_2 \longrightarrow 2H_2O$$

A mixture of hydrogen and oxygen is a dangerous explosive. Be sure that no air is mixed with hydrogen if the latter is to be ignited.

Hydrogen has the ability to remove oxygen from certain oxygen-containing compounds. If a stream of hydrogen gas is passed over heated copper oxide, CuO, the latter is converted to free copper and water will condense in the cooler portions of the apparatus (Figure 6.2).

$$CuO + H_2 \longrightarrow Cu + H_2O$$

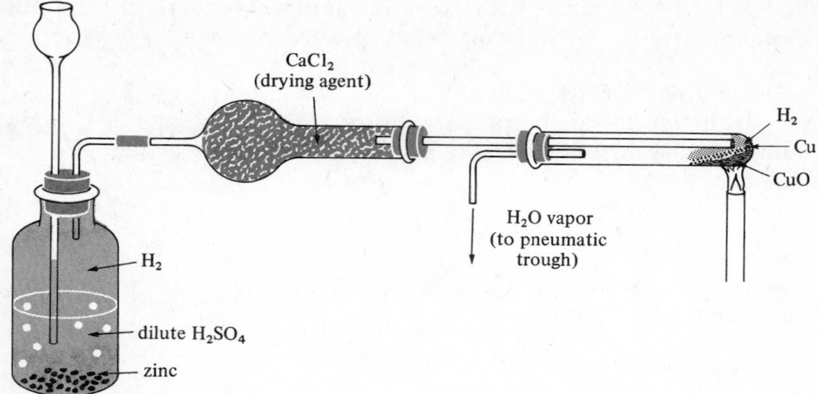

Figure 6.2 When hydrogen is passed over hot copper oxide, copper and water are formed.

In a restricted sense, a reaction such as this in which oxygen is removed from a compound is called a *reduction* and hydrogen is called the *reducing agent.* Because the copper oxide furnishes oxygen to the hydrogen, it is an *oxidizing agent.* A more general definition for oxidation and reduction will be given in Chapter 12.

Hydrogen combines readily with chlorine when a mixture of these gases is heated or exposed to ultraviolet light, and the gas hydrogen chloride is formed.

$$H_2 + Cl_2 \longrightarrow 2HCl$$

At higher temperatures and pressures, and in the presence of a catalyst, hydrogen combines with nitrogen gas to form ammonia, NH_3. It combines incompletely with molten sulfur to form hydrogen sulfide, H_2S.

USES OF HYDROGEN

Small amounts of hydrogen are used in the preparation of some of the more expensive metals, such as tungsten (wolfram) and molybdenum. Hydrogen has been used to inflate lighter-than-air craft, but because it is so combustible, it is being replaced by noncombustible helium. Hydrogen is used for the *hydrogenation* of liquid fats, such as cottonseed oil, peanut oil, and soybean oil. The solid products obtained in this process are used as substitutes for butter and lard. In hydrogenation, hydrogen is added to the reducible substance under pressure and at higher temperatures in the presence of a suitable catalyst. Nickel and platinum are commonly employed as catalysts for hydrogenations. Hydrogen is combined industrially with nitrogen and carbon monoxide to produce synthetic ammonia and methanol, respectively.

Water

Water, with the chemical formula H_2O, must be considered the simple oxide of hydrogen.

Vast quantities of water are present in the seas, as ice and snow on mountains and in the polar regions, in all fertile soils, and as vapor in the atmosphere. It is abundant in animal and vegetable tissues and composes about 70 percent of the human body. It is also a constituent of many rocks and minerals, where it is present in chemical combination with other substances.

Water, our most important and most abundant chemical compound, is a common solvent in many chemical reactions that occur in solutions. Most reactions occurring in living tissue take place in an aqueous media. In the body, water is also used to control the temperature of the body and to carry nutrients to the cells and waste products to the excretory organs.

COMPOSITION OF WATER

There are two methods used in the determination of the composition of chemical substances: *analysis* and *synthesis*. In analysis, the substance is decomposed into its constituent elements in order to determine its composition. In synthesis, the composition of the substance is determined by combining the proper elements to form the compound.

When water is decomposed by electrolysis (page 56), approximately two volumes of hydrogen are produced for each volume of oxygen liberated. This method is unsuitable for the accurate determination of the composition of water because it is subject to several sources of error. The most accurate results for the composition of water have been obtained by the direct union of hydrogen and oxygen, a synthetic method.

The most exact determination of the composition of water by a synthetic method was performed by Dr. Edward Morley of Western Reserve University. The weight of pure hydrogen and oxygen employed was determined, as well as the weight of water resulting from their combination. Dr. Morley found that 1 part by weight of hydrogen combined with 7.94 parts by weight of oxygen to form water.

THE PHYSICAL PROPERTIES OF WATER

Water has a flat taste, is nearly colorless in thin layers, and in a deep section appears pale blue in color.

Water has a *maximum density* (page 16) at 4°C. At this temperature 1 ml of water weighs 1 g. Above and below this temperature, 1 ml of water will weigh less than 1 g.

The *freezing point* of a substance is defined as the temperature at which both the solid and liquid forms of the substance will remain together in equilibrium. The statement that the freezing point of a substance is the temperature at which the liquid changes to the solid state is unsatisfactory, because many liquids can be supercooled;

that is, they can be cooled below their true freezing point without solidifying. The freezing point of water, an important constant of nature, is 0°C, or 32°F.

The *specific heat* of a substance can be defined as the amount of heat in calories that will be required to raise the temperature of 1 g of the substance 1° on the Celsius scale. The specific heat of water is 1 cal per degree. Water has a very high specific heat, exceeded only by that of liquid ammonia. Because of its high specific heat, water is frequently used in nursing procedures for its warming effect. The specific heat of ice is about one-half that of liquid water.

About 80 cal of heat are required to convert 1 g of ice at 0°C to 1 g of water at this temperature. This is known as the *heat of fusion.* When 1 g of water at 0°C is converted to ice at the same temperature, 80 cal of heat are liberated. This is known as the *heat of solidification* of water. The heats of solidification and fusion of a substance will always be equal.

About 540 cal of heat are required to convert 1 g of water at 100°C to 1 g of steam at this temperature. This is the *heat of vaporization* of water. The *heat of condensation* is the amount of heat that is liberated when 1 g of steam at 100°C condenses to form 1 g of water at this temperature, and is also 540 cal. This high heat of condensation makes water an excellent heat transfer agent in heating systems. The high heat of vaporization of water plays an important role in controlling body temperature in the presence of a high environmental temperature.

ILLUSTRATIVE PROBLEM 1

How many calories of heat will be required to convert 100 g of ice to steam?

TO CONVERT	REQUIRES
ice at 0°C to water at 0°C	80 cal/g
water at 0°C to water at 100°C	100 cal/g
water at 100°C to steam at 100°C	540 cal/g
Total =	720 cal/g

Thus for 100 g of water,

(100 g)(720 cal/g) = 72,000 cal

If a liquid, such as water, is confined in a closed vessel, molecules of the liquid will escape from its surface into the atmosphere above. In time, this atmosphere will become saturated with the vapors of the liquid, and no more of the latter will appear to evaporate. However, this does not represent a state of rest but an equilibrium between two changes. The apparent state of rest is due to the fact that the number of molecules condensing to the liquid state is equal to the number of liquid molecules that are evaporating; that is, the rate of condensation is equal to the rate of evaporation of the liquid. In physical as well as in chemical equilibriums, we will find that the rate of two opposed reactions is equal. The pressure exerted by the water vapor in an atmosphere saturated with it is called the *vapor pressure* of water (see Table 5.1). The vapor pressure of a liquid depends on and increases with the temperature. Comfort is related to relative humidity as well as to room temperature.

If the vapor pressure of a liquid becomes equal to the atmospheric pressure, the liquid will boil. The vapor pressure of water is equal to the standard atmospheric pressure at 100°C, which is the boiling point of water. If, however, the atmospheric pressure is less than 760 mm, the boiling point of water will be below 100°C, as is the situation on high mountains. In a boiler or a sterilizing autoclave in which the pressure is above atmospheric pressure, water will boil above 100°C. The *boiling point* of a liquid may be defined as the temperature at which the vapor pressure of the liquid is just equal to the external pressure.

THE CHEMICAL PROPERTIES OF WATER

Stability. Water is a very stable substance and is decomposed into hydrogen and oxygen to the extent of only 1 percent at 2000°C and 11 percent at 2700°C. On cooling, the liberated hydrogen and oxygen spontaneously recombine to form water.

Reaction with Metals. Some of the more active metals displace hydrogen from water. The student should review the reaction of sodium, iron, potassium, and magnesium on water or steam (see page 86).

Reaction with Oxides. Recall that water reacts with metallic oxides to form bases and with nonmetallic oxides to form acids page 59).

Reaction with Salts. Water enters into chemical reaction with certain substances, particularly salts, to form compounds called *hydrates.* Sodium sulfate combines with 10 molecules of water to form sodium sulfate decahydrate, $Na_2SO_4 \cdot 10H_2O$ (Glauber's salt). The centered dot in this formula indicates that the sodium sulfate and water are chemically combined. Most hydrates are crystalline but not all crystalline substances are hydrates, which, like all true chemical compounds, have a definite composition. Most hydrates are not stable when heated and decompose to form the anhydrous substance. The chemically combined water in a hydrate is called *water of hydration* or *water of crystallization.* Other hydrates are copper sulfate pentahydrate, $CuSO_4 \cdot 5H2O$ (blue vitriol), sodium carbonate decahydrate, $Na_2CO_3 \cdot 10H_2O$ (washing soda), and potassium aluminum sulfate dodecahydrate, $KAlSO_4 \cdot 12H_2O$ (alum). Calcium chloride is a good drying agent because it forms calcium chloride hexahydrate, $CaCl_2 \cdot 6H_2O$.

Hydrates such as Glauber's salt, which on standing in the air readily lose water to form the anhydrous salts, are said to be *efflorescent.* A chemical substance that readily combines with the moisture of the atmosphere is said to be *deliquescent.* Most of these latter substances, like calcium chloride, are salts that readily form hydrates. Because it is deliquescent, calcium chloride is used to settle dust on roads and in mines.

Plaster of Paris, $(CaSO_4)_2 \cdot H_2O$, the hemihydrate (or "half" hydrate) of calcium sulfate, is prepared by carefully heating gypsum, $CaSO_4 \cdot 2H_2O$, which is the dihydrate of calcium sulfate.

$$2CaSO_4 \cdot 2H_2O \longrightarrow (CaSO_4)_2 \cdot H_2O + 3H_2O$$

Plaster of Paris is employed in the preparation of molds, casts, and plaster board. When treated with water it generates gypsum, a process called *setting*, whereupon it expands slightly and gives a faithful reproduction of the mold by filling all the cracks and crevices. It is used to make plaster casts of the body, especially in the setting of fractured bones.

NATURAL WATER SUPPLIES AND THEIR PURIFICATION

Water, as it occurs in nature, may contain the following impurities:

1. dissolved gases
2. bacteria and decaying organic matter obtained from decomposed refuse
3. suspended particles such as sand, clay, and silt
4. salts of sodium, potassium, calcium, magnesium, and iron

These latter salts are present in well water and surface water that has percolated through soil and rocks containing these salts or other readily modified compounds of these metals.

For use in the chemical laboratory, water must be freed from all impurities. This purification may be accomplished by *distillation*. In this process, water is converted into steam in a boiler, then passed into a cooled jacket called a *condenser*, where it is reconverted to water. All of the impurities are retained in the boiler except the dissolved gases, and most of these will appear in the first portion of the distillate, which may be discarded.

Water for drinking purposes must be freed from suspended material, bacteria, and organic matter. Suspended material is usually removed by filtration through sand, gravel, or charcoal or by adding aluminum sulfate or alum and a mild base, such as sodium carbonate, to the water. A gelatinous suspension of aluminum hydroxide, $Al(OH)_3$, forms, and this is permitted to settle. The water from this treatment is usually passed through sand filters, removing nearly all the suspended material and about half the bacteria. The bacteria and organic matter that still remain are destroyed by the addition of a small amount of chlorine gas. *Aeration* of water, that is, the mixing of water with air, will also destroy bacteria and organic matter. Inorganic salts and gases are not usually removed from drinking water, because they give the water a palatable taste, and the salts aid in the nutrition of the body. If one is suspicious of the quality of his drinking water, it should be boiled, or a small quantity of iodine or bleaching powder should be added to make it safe.

The salts of calcium and magnesium must be removed from water which is to be used by industries and laundries. Iron and manganese salts in water can produce objectionable stains in laundered material. The salts of calcium and magnesium form precipitates with soap, and an additional quantity of the latter will be required before a lather and proper cleansing action can be obtained. These salts are also objectionable, because they form scale in boilers. The bicarbonates of calcium and magnesium may be removed from water by boiling, during which these salts are decomposed and carbon dioxide escapes.

$$Ca(HCO_3)_2 \longrightarrow \underline{CaCO_3} + H_2O + CO_2$$

The normal carbonate salts of calcium and magnesium are insoluble in water and precipitate. Water that contains the bicarbonates of calcium and magnesium is said to have *temporary hardness* or *bicarbonate hardness.* The chloride and sulfate of calcium and magnesium occur in hard water, and these salts cannot be removed by boiling. Water that contains these latter salts is said to have *permanent hardness* or *nonbicarbonate hardness.*

Hard water may be softened by chemical means. Agents frequently employed to soften permanently hard water are soda ash, Na_2CO_3,

$$CaSO_4 + Na_2CO_3 \longrightarrow \underline{CaCO_3} + Na_2SO_4$$

calgon (sodium hexametaphosphate, $Na_6P_6O_{18}$),

$$2CaSO_4 + Na_6P_6O_{18} \longrightarrow Ca_2Na_2P_6O_{18} + 2NaSO_4$$

and the natural and synthetic zeolites,

$$CaSO_4 + 2NaAlSiO_4 \longrightarrow Ca(AlSiO_4)_2 + Na_2SO_4$$

In the above equations, $CaCO_3$ is insoluble and precipitates. The $Ca_2Na_2P_6O_{18}$ is soluble but does not dissociate into Ca^{2+} ions. The zeolites have a complicated chemical structure and, for the sake of simplicity, are represented as $NaAlSiO_4$. Permutit is a synthetic zeolite. The zeolites are regenerated by washing with a strong brine solution.

De-ionized water is produced by passing water through two resinous materials, one of which replaces the cations, such as Na^+, Ca^{2+}, Mg^{2+}, by hydrogen ions; the other replaces anions, such as Cl^-, SO_4^{2-}, by OH^-. For many uses, de-ionized water can replace distilled water and is much cheaper to prepare. Occassionally, organic impurities that produce fevers appear in water and are not removed by this treatment. Therefore caution should be exercised in using de-ionized water in subcutaneous or intravenous injections.

Hydrogen peroxide

Hydrogen peroxide, H_2O_2 is a heavy, oily, colorless liquid with a characteristic odor. It has the structure

$$\begin{array}{l} H:\ddot{O}: \\ \quad :\ddot{\underset{..}{O}}:H \end{array}$$

It is interesting to compare its physical properties with those of water. It has a boiling point of 150.2°C, a freezing point of −0.4°C, and a density of 1.45 g/ml.

Commercially, hydrogen peroxide is prepared by the action of water on peroxydisulfuric acid, $H_2S_2O_8$:

$$H_2S_2O_8 + 2H_2O \longrightarrow 2H_2SO_4 + H_2O_2$$

which is prepared by the electrolysis of a cold dilute solution of sulfuric acid.

Hydrogen peroxide decomposes into water and oxygen:

$$2H_2O_2 \longrightarrow 2H_2O + O_2$$

This decomposition is catalyzed by metallic oxides, especially manganese dioxide, and by the enzyme catalase that occurs in most living tissue.

A 3 percent solution of hydrogen peroxide, stabilized by suitable preservatives to prevent its decomposition by light, is used as an antiseptic and disinfectant and as a bleach for hair, wool, and silk.

Concentration of dilute solutions of hydrogen peroxide by distillation must be carried out in a vacuum because of its instability. A 90 percent solution is available. This solution is an excellent oxidizing agent but must be handled with extreme caution, because heat and certain impurities can induce an explosive decomposition.

STUDY QUESTIONS

1. Discuss the occurrence and physical properties of hydrogen.
2. What is reduction? Give an example of a reduction in which hydrogen acts as a reducing agent.
3. What precautions must be observed in the storage of sodium metal?
4. Write an equation for a reaction in which water acts as an oxidizing agent.
5. How could one distinguish the following gases: hydrogen, nitrogen, oxygen?
6. Which of the following metals will displace lead from a solution of lead chloride, $PbCl_2$: Sn, As, Cu, Zn, Mg, Fe, Cd, and Hg?
7. Name and give formulas for five compounds in which hydrogen is combined with some other element.
8. Complete and balance the following reactions. If no reaction occurs, write N.R.

(a) $CuO + H_2 \longrightarrow$
(b) $Na + H_2O \longrightarrow$
(c) $Cu + HCl \longrightarrow$
(d) $AgNO_3 + Fe \longrightarrow$
(e) $H_2 + O_2 \longrightarrow$
(f) $CdSO_4 + Sn \longrightarrow$
(g) $Al + H_2SO_4 \longrightarrow$
(h) H_2O (electrolysis) $\longrightarrow$
(i) $Ca + H_2O \longrightarrow$
(j) $Cd + HCl \longrightarrow$
(k) $Pb + HgCl_2 \longrightarrow$
(l) $Fe + H_2O$ (steam) $\longrightarrow$
(m) $FeSO_4 + Zn \longrightarrow$
(n) $Fe_3O_4 + H_2 \longrightarrow$

9. Who discovered hydrogen? Why is it so named?
10. How can hydrogen be obtained from water?
11. What metals will displace hydrogen from water at room temperature? From steam?
12. List several generalizations that are deduced from the activity series of the metals.
13. The alchemists believed that they could transmute iron to copper by placing a piece of iron in a solution of blue vitriol (copper sulfate). Explain their fallacy.
14. What is meant by a reversible reaction? Are these common in chemistry? How can the reaction of steam with iron be made to go to completion?
15. How can hydrogen be made to combine with the following: oxygen, chlorine nitrogen, and sulfur?
16. What precautions should be exercised when handling hydrogen gas?
17. Why is hydrogen generally no longer used in filling balloons? What gas is used in its place? Why?
18. Why are some of the liquid fats combined with hydrogen? What is this process called? What metallic catalysts are usually employed in this process?
19. What synthetic products require hydrogen in their manufacture.
20. Define synthesis, analysis, efflorescence, deliquescence, boiling point, freezing point, hydrate, anhydrous substance, water of crystallization.
21. What evidence can be cited to prove that water is a compound?
22. What is the composition of water by weight and volume?
23. Why has the composition of water been determined by a synthetic method?
24. At what temperature does a certain weight of water have its least volume?
25. If tubs of water are placed in a cellar in cold weather, freezing of food products stored in the cellar may be prevented. Why?
26. Why does it take longer to boil potatoes at the top of a mountain than at the seashore? How may the time required to cook potatoes at the top of a mountain be shortened?
27. Why does spring come later to an area adjacent to a large lake than to an area far removed from a body of water?
28. Of what importance is the fact that ice is less dense than water?
29. Why is sterilization more efficient in an autoclave than in boiling water?

30. Why does freshly boiled water have a flat taste?
31. Define heat of vaporization, heat of fusion, specific heat, vapor pressure of a liquid.
32. What are the heat of vaporization, heat of fusion, specific heat, boiling point, and freezing point of water?
33. Upon what factor does the vapor pressure of a certain liquid depend? What is the relationship between the vapor pressure of a liquid and its boiling point?
34. How stable is water to heat?
35. Review the action of water on certain metals, on metallic oxides, and on nonmetallic oxides.
36. What is a hydrate?
37. How does a hydrate change when it is heated strongly?
38. What evidence can be cited to prove that hydrates are true chemical compounds?
39. Which could be the better drying agent, an efflorescent or a deliquescent substance?
40. In hot dusty weather, calcium chloride sprinkled on a road will settle the dust. Why?
41. What is the chemical composition of the following substances: blue vitriol, Glauber's salt, washing soda, gypsum, and plaster of Paris?
42. What is the chemical reaction involved in the setting of plaster of Paris?
43. Distinguish between soft and hard water; temporary and permanent hardness. Give alternate names for these types of hardness.
44. How is water purified by distillation? When is it necessary to use distilled water?
45. How may water containing the following impurities be purified?

    (a) suspended dirt
    (b) algae
    (c) copper sulfate
    (d) volatile substance
    (e) dead organic matter

46. What type of hard water can be softened by boiling?
47. Describe some of the methods of softening hard water. What chemical substances are present in hard water? For what uses is water softened?
48. List several methods that could be used, if you were uncertain of the purity of water, to make it safe for drinking.
49. What is the structure of hydrogen peroxide? How is it prepared commercially?
50. Compare the physical properties of water and hydrogen peroxide.
51. Which enzyme in living tissue efficiently decomposes hydrogen peroxide? Suggest a reason for the presence of this enzyme in living tissue. What other type of substance catalyzes the decomposition of hydrogen peroxide to water and oxygen?
52. Because hydrogen peroxide is rich in oxygen, what type of chemical substance is it?
53. What are some of the uses of a 3 percent hydrogen peroxide solution? How are dilute hydrogen peroxide solutions concentrated?
54. What precautions are necessary in the storage and use of a 90 percent hydrogen peroxide solution?

PROBLEMS

1. How much sulfuric acid will be required to react with 12.5 g of zinc?
2. Twenty grams of magnesium are reacted with 50 g of sulfuric acid. Which reactant and how much of it remains when the reaction is complete? How much magnesium sulfate will be formed?
3. How much copper will be formed when 125 g of copper oxide is reduced by hydrogen?
4. An iron nail is added to a solution of silver nitrate. If the nail weighs 2.7 g how much silver will be deposited when all of the iron has reacted?
5. What is the percentage of water in blue vitriol, $CuSO_4 \cdot 5H_2O$?
6. What weight of anhydrous sodium sulfate will be formed by heating a kilogram of Glauber's salt, $Na_2SO_4 \cdot 10H_2O$?
7. In the electrolysis of water, what volume of oxygen is formed when 38 ml of hydrogen is liberated?
8. If 550 ml of hydrogen gas and 325 ml of oxygen are mixed and ignited, what volume of gas will remain? Which gas remains?
9. How many calories of heat are required to melt 250 g of ice?
10. How many calories of heat are required to convert 15 liters of water at 30°C to steam?
11. How much heat is required to change 20 g of ice at −10°C to steam at 100°C?
12. When 1 g of fat is burned, 9 kcal of heat is generated. How many grams of fat must be burned to heat the water in a 25-gal can from 20°C to 100°C? A gallon is equal to approximately 4 liters.

chapter 7

solutions and methods of expressing concentrations in solutions

7

Solutions are homogeneous combinations of two or more chemical substances and, like mixtures, have a variable composition. They are homogeneous because the constituents are molecularily dispersed. Dispersions of aggregates or clusters of molecules are called *colloidal dispersions* (see Chapter 13).

The dissolved material in a solution is called the solute, while the medium in which the solute is dissolved is called the *solvent*. The terms "solute" and "solvent" are arbitrary, and in certain situations they may lose their significance. In a solution of ethyl alcohol and water, it may be difficult to say which is the solute and which the solvent. In such a case, it is the usual practice to designate the substance in smaller amount as the solute and that in larger amount as the solvent.

Types of solutions

Although there are nine types of solutions, only solutions of gases in liquids, liquids in liquids, solids in liquids, and gases in solids are important.

Types of Solutions	*Example*
Gases in liquids	Soda water (CO_2 in water)
Liquids in liquids	Alcohol in water Gasoline in ether
Solids in liquids	Sugar in water Benzoic acid in ether
Gases in solids	Hydrogen adsorbed on platinum or palladium

Saturated, undersaturated, and supersaturated solutions

If a crystal of urea is placed in water, molecules of urea will escape from the crystal and pass into solution and, in a short time, collide with and be retained by the urea crystal. However, as long as

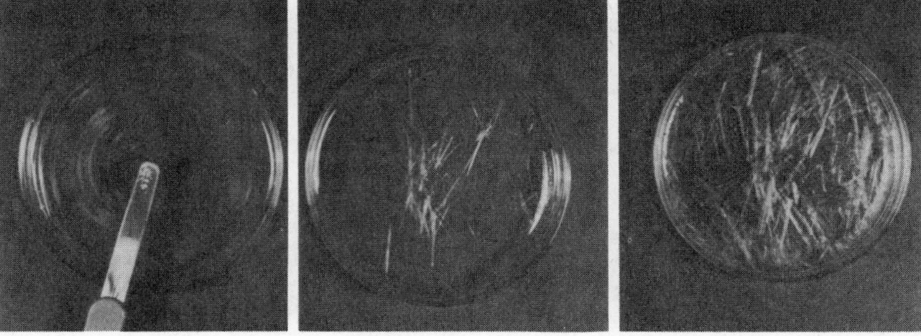

Figure 7.1 A supersaturated solution of sodium acetate is seeded (left); excess solute precipitates as needlelike crystals (middle); final result is a saturated solution in equilibrium with excess solid. (From C. W. Keenan and J. H. Wood, General College Chemistry, *4th ed., Harper & Row, N.Y., 1971.)*

molecules of the crystalline urea are dissolving more rapidly than the urea molecules in solution are crystallizing, the solution will be *undersaturated.*

If sufficient solid urea is present, the rate of solution and of crystallization will become equal, and no more urea will appear to dissolve. Such a solution is called a *saturated solution. A saturated solution is defined as a solution in which the solid solute is in equilibrium with the solute in solution.* At saturation, the rate of crystallization will be equal to the rate of solution of the solid solute.

If a saturated solution is cooled, the excess solute will crystallize from solution. However, if no solid solute is present and the solution is cooled very carefully, the solution will contain more of the solute than will normally be present in a saturated solution at that temperature. Such a solution, called a *supersaturated solution,* is unstable, and if a crystal of the solute is added to it, the excess solute will immediately crystallize and the remaining solution will be saturated (Figure 7.1).

In order to determine whether a solution is saturated, undersaturated, or supersaturated, a crystal of the solute may be added. If part of the added crystal dissolves, the solution is undersaturated; if it remains unchanged in weight, the solution is saturated; if more of the substance crystallizes from the solution, it was supersaturated and the solute will crystallize until saturation is obtained.

Effect of temperature on solubility of solids in liquids

The solubility of most solids, such as potassium nitrate, increases as the temperature rises. Sodium chloride shows little increase in solubility in water with an increase in temperature. A few compounds, in particular calcium compounds, become less soluble in water at higher temperatures. Calcium sulfate, calcium hydroxide, and calcium chromate are more soluble in a cold solution than in a hot one.

Methods of expressing concentrations of solutions

The terms "dilute" and "concentrated" used in describing solutions are only relative terms and are of little value in quantitative work. A few methods of expressing concentration of solutions will now be described.

1. The concentration of a solute in a solution may be expressed as a percentage. The percentage (weight–volume) will be equal to the number of grams of the solute in 100 ml of the solution.
2. The concentration of a solute may be expressed in grams dissolved in 100 ml of the solvent.
3. Concentration may also be expressed in grams of solute dissolved in a specified weight of the solvent.
4. Chemists also use the terms "molarity" and "normality" to express concentrations of solutions. Normality and molarity will be discussed below.

If the concentration of the solute present in a solution is known, such a solution is called a *standard solution.*

Percent solutions (weight–volume)

In weight–volume percent solutions, as indicated above, the percentage will be equal to the number of grams of solute dissolved in 100 ml of solution. Thus for intravenous injections, a 0.9 percent salt or 5 percent glucose solution is usually employed. The former will contain 0.9 g of salt, while the latter will contain 5 g of glucose per 100 ml of solution.

Weight–volume percent solutions are more frequently employed and are easier to prepare than weight–weight percent solutions, in which the percentage is equal to the number of grams of solute in 100 g of solution.

ILLUSTRATIVE PROBLEM 1

What weight of sodium chloride will be required to prepare 5 liters of 0.9 percent (weight–volume) solution?

This solution will contain 0.9 g of NaCl per 100 ml, or 9 g per liter, that is, 9×5 g, or 45 g of salt in 5 liters of this solution.

ILLUSTRATIVE PROBLEM 2

Compare the preparation of 500 ml of 5 percent (weight–volume) and 5 percent (weight–weight) glucose solutions.

Because in the former solution, 5 g of glucose will be present in each 100 ml of solution, dissolve 25 g of glucose in a small volume of water, add this to a 500-ml graduate or volumetric flask (Figure 7.2), and add

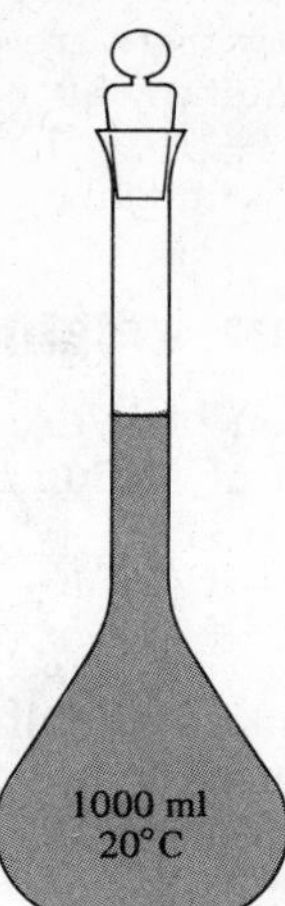

Figure 7.2 When filled to the mark on the neck, a volumetric flask will hold a specified volume at the temperature indicated on the flask.

water to the mark, and mix well. To prepare 500 ml of 5 percent (weight–weight) solution, add 25 g of glucose to 475 ml (475 g) of water and mix well.

Molar solutions

The *molarity* of a solution is equal to the number of moles of solute that are dissolved in a liter of solution. A two-molar solution can be expressed as 2 M and contains 2 moles of solute per liter of solution, while a tenth molar, or 0.1 M solution contains $\frac{1}{10}$ mole of the solute per liter.

ILLUSTRATIVE PROBLEM 3

What is the molarity of a sodium nitrate solution that contains 6 g of the salt in 100 ml of the solution?

Such a solution will contain $6 \times 1000/100$ g, or 60 g, of sodium nitrate in a liter of solution. A molar solution of sodium nitrate will contain $23 + 14 + 48 = 85$ g of the salt in a liter of solution.

Because

$$\text{molarity} = \frac{\text{moles}}{\text{liter}} = \frac{\text{grams per liter}}{\text{gram-molecular weight}}$$

the molarity of this solution will be

$$\frac{60 \text{ g/liter}}{85 \text{ g}} = 0.71\ M$$

ILLUSTRATIVE PROBLEM 4

How many grams of sodium hydroxide are present in 5 liters of a 0.4 M solution?

In 5 liters of 0.4 *M* solution there are 5 × 0.4, or 2, moles of solute. Changing moles to grams by multiplying number of moles by the gram-molecular weight of NaOH (23 + 16 + 1 = 40), one finds that there are 80 g of NaOH in the 5 liters of solution.

The gram-equivalent weight

A liter of a molar solution of hydrochloric acid will just neutralize a liter of a molar solution of sodium hydroxide.

$$\underset{\text{1 liter of 1 }M}{HCl} + \underset{\text{1 liter of 1 }M}{NaOH} \longrightarrow NaCl + H_2O$$

However, a liter of molar solution of sulfuric acid will require 2 liters of a molar solution of sodium hydroxide to neutralize it:

$$\underset{\text{1 liter of 1 }M}{H_2SO_4} + \underset{\text{2 liters of 1 }M}{2NaOH} \longrightarrow Na_2SO_4 + 2H_2O$$

while a liter of a molar solution of phosphoric acid will require 3 liters of a molar solution of sodium hydroxide to completely neutralize it:

$$\underset{\text{1 liter of 1 }M}{H_3PO_4} + \underset{\text{3 liters of 1 }M}{3NaOH} \longrightarrow Na_3PO_4 + 3H_2O$$

In order to establish a unit of concentration in which a certain volume of one chemical will just combine with the same volume of an equivalent concentration of another substance (because, as shown above, molar solutions are unsuitable for this purpose), chemists have introduced the concept of the normal solution. Before studying normal solutions, another new concept, the *gram-equivalent weight,* also called the *equivalent,* must be introduced.

The gram-equivalent weight of an element is equal to the weight of that element that will combine with or displace one gram-atom, or 1.008 g, of hydrogen.

Because 8 g of oxygen, 3 g of carbon, and 35.453 g of chlorine are combined with 1.008 g of hydrogen in water, methane, and hydrogen chloride, respectively, these are the gram-equivalent weights of these elements. Twenty-three grams of sodium, 9 g of aluminum, and 12 g of magnesium liberate 1.008 g of hydrogen from an acid, so these weights represent the gram-equivalent weights of these metals.

If the atomic weight and the valence of an element are known, the gram-equivalent weight can also be determined by dividing the gram-atomic weight by the valence:

$$\text{gram-equivalent weight} = \frac{\text{gram-atomic weight of element}}{\text{valence}}$$

The gram-equivalent weight of potassium is then $^{39}/_{1}$, or 39 g, that of calcium is $^{40}/_{2}$, or 20 g, and aluminum is $^{27}/_{3}$, or 9 g. Carbon in carbon monoxide has a gram-equivalent weight of $^{12}/_{2}$, or 6 g, while the gram-equivalent weight of carbon in carbon dioxide is $^{12}/_{4}$, or 3 g.

In summary, three methods may be used in computing the gram-equivalent weight of an element, and the choice of the method employed must depend upon which of the following facts is available:

1. the amount of the element required to combine with or displace 1.008 g of hydrogen
2. the amount of the element that will combine with a gram-equivalent weight of another element, such as oxygen (g-eq. wt. of 8) or chlorine (g-eq. wt. of 35.5)
3. the gram-atomic weight divided by the valence

Normal solutions

A normal solution of a chemical substance contains one gram-equivalent weight of that substance in a liter of solution and is often indicated as 1 *N*. A normal solution of an acid contains 1.008 g of replaceable hydrogen in a liter of solution. Thus a normal solution of hydrochloric acid contains a mole, or 35.5 + 1, or 36.5 g of hydrogen chloride. A normal solution of sulfuric acid contains a half-mole, $^{98}/_{2}$, or 49 g, per liter of hydrogen sulfate. A normal solution of phosphoric acid contains $\frac{1}{3}$ mole, $^{98}/_{3}$, or 32.7 g, of hydrogen phosphate per liter of solution.

A normal solution of a base contains 17 g of replaceable hydroxyl ions per liter of solution. Thus a normal solution of sodium hydroxide contains 23 + 16 + 1, or 40 g, of NaOH per liter; a 2 *N* sodium hydroxide solution contains 80 g of the base; a 0.1 *N* NaOH solution contains 4 g of the base in a liter of solution.

A normal solution of a salt contains a gram-equivalent weight of the metallic ion or radical and a gram-equivalent weight of the acid ion or radical. A normal solution of sodium chloride contains

$$\frac{23 + 35.5}{1}$$

or 58.5 g of the salt per liter, while a 1 *N* barium chloride, $BaCl_2$, solution contains

$$\frac{137 + (2 \times 35.5)}{2}$$

or 104 g per liter. A 1 *N* silver nitrate, $AgNO_3$, solution contains

$$\frac{108 + 14 + 48}{1}$$

or 170 g of this salt per liter.

The normality of a solution will be equal to the number of gram-equivalent weights of the solute in a liter of the solution.

ILLUSTRATIVE PROBLEM 5

Calculate the normality of a sulfuric acid solution that contains 9.8 g of hydrogen sulfate in 4 liters of the solution.

Because sulfuric acid contains two replaceable hydrogen atoms in a molecule of the acid, a normal solution will contain ½ mole, or 49 g, per liter. In this problem, the solution will contain 9.4 g/4 liters, or 2.45 g per liter. Because

$$\text{normality} = \frac{\text{gram-equivalent weights}}{\text{liters}} = \frac{\text{grams per liter}}{\text{gram-equivalent weight}}$$

the normality of this solution is

$$\frac{2.45\ \text{g/liter}}{49\ \text{g}} = 0.05\ N$$

ILLUSTRATIVE PROBLEM 6

How many grams of calcium hydroxide are present in 6 liters of a 0.1 *N* solution of this base?

Six liters of a 0.1 *N* solution contain 0.6 g-equivalent wt. of the solute. Because the gram-equivalent weight of $Ca(OH)_2$ is

$$\frac{40 + 32 + 2}{2} = 37\ \text{g}$$

there are 0.6 g-eq. wt. × 37 g, or 22 g of the solute in 6 liters of this solution.

Titration

The quantity of acid in an unknown sample may be determined by the quantitative neutralization (page 137) of the unknown acid with a standard solution of a base or vice versa, a process called *titration* (Figure 7.3). If not in solution, the sample of the unknown acid or base is dissolved in a proper solvent, usually water. An indicator, a complex organic molecule that exhibits different colors in acid and base solution, is now added. The standard acid (or base) is added from a *buret* until the neutralization of the unknown is complete, at which point the indicator will change color. This is called the *end-point* of the titration. The volume of the standard acid or base is readily read from the scale on the buret. Because the volume and concentration of the standard solution are known, if the volume of the unknown solution used in the titration is also known, it is necessary only to substitute the known values in the equation

$$V_A \times N_A = V_B \times N_B$$

which will now be discussed, and then solve for the concentration of the unknown acid or base solution.

Because a liter of a normal solution of a base will just neutralize a liter of a normal solution of any acid, a liter of a 0.1 *N* base will require a liter of a 0.1 *N* or 500 ml of a 0.2 *N* acid to neutralize it. To solve problems involving quantitative neutralizations of acids and bases, the formula

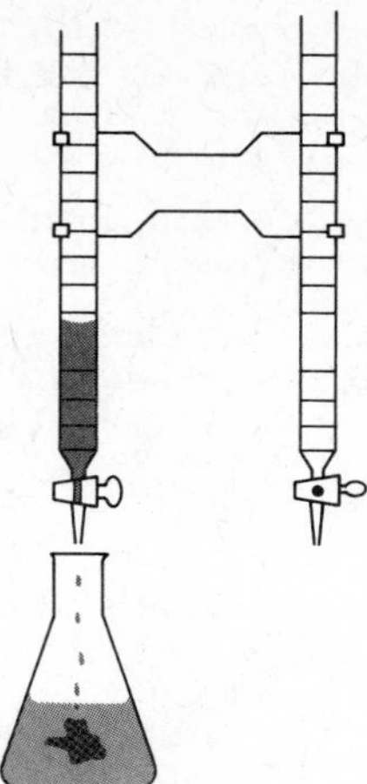

Figure 7.3 Apparatus for titration.

$$V_A \times N_A = V_B \times N_B$$

is used. This formula is derived from the fact that the volumes of acids and bases that react together are inversely proportional to their normalities. That is, the more concentrated the solution of the acid or base, the less is required. In the formula, V_A and N_A represent, respectively, the volume and normality of the acid, and V_B and N_B the volume and normality of the base.

ILLUSTRATIVE PROBLEM 7

How many milliliters of 0.2 *N* NaOH will be required to neutralize 300 ml of a 0.5 *N* phosphoric acid solution?

Substituting in the formula,

$$V_A \times N_A = V_B \times N_B$$
$$300 \times 0.5 = 0.2 \times V_B$$
$$0.2\ V_B = 150$$

then

$$V_B = 150/0.2 = 1500/2 = 750 \text{ ml}$$

Titration of stomach contents with a standard solution of a base, using dimethylaminoazobenzene as an indicator, is of considerable diagnostic value. The free acidity of the gastric contents is expressed in volume of 0.1 *N* NaOH that is required to neutralize 100 ml of the contents of the stomach. Normally about 27 ml of 0.1 *N* base is required for this titration; however, in hyperacidity 44 ml or more may be required, while in hypoacidity the amount needed may be 10 ml or less.

The formula $V_A \times N_A = V_B \times N_B$ is also used to compute the amount of a concentrated acid or base that must be used to prepare a dilute solution. This is illustrated by the following problem.

ILLUSTRATIVE PROBLEM 8

Concentrated hydrochloric acid is about 12 *N*. How much concentrated hydrochloric acid must be used to prepare 250 ml of 3 *N* acid?

Let V_A and N_A represent the volume and normality of the concentrated acid and V_B and N_B represent these for the dilute acid. Substituting in the formula,

$$12 \times V_A = 250 \times 3$$

$V_A = 750/12 = 62.5$ ml of the concentrated acid that must be diluted to 250 ml with water to make a 3 *N* acid solution.

Milliequivalents

Certain instruments, such as flame photometers (Figure 7.4), used for analytical purposes in the hospital laboratories are now calibrated to give results in terms of milliequivalents of the element. Quite frequently the concentrations of certain ionic constituents in a blood sample are being reported in terms of this unit.

A *milliequivalent* (meq) is $\frac{1}{1000}$ of an equivalent weight of a chemical substance. The number of milliequivalents per milliliter equals numerically the number of equivalents per liter.

ILLUSTRATIVE PROBLEM 9

How many milliequivalents are present in 50 ml of a 0.24 *N* sulfuric acid solution?

The number of milliequivalents will equal the product of the normality and the number of milliliters of solution. Therefore,

$$\text{meq} = 0.24 \times 50 = 12.00 \text{ meq}$$

ILLUSTRATIVE PROBLEM 10

A sample of blood contains 13.7 meq of sodium ions in a 100 ml sample. Calculate the number of milligrams of sodium in this 100 ml sample of blood:

$$1 \text{ meq of } Na^+ = 23 \text{ mg}$$

(because 1 eq of Na = 23 g). Thus the number of milligrams of $Na^+ = 13.7 \times 23 = 315$ mg.

ILLUSTRATIVE PROBLEM 11

Normal blood contains 17 mg of K^+ per 100 ml sample. Express concentration of K^+ in 100 ml of blood in terms of milliequivalents.

$$1 \text{ meq of } K^+ = 39 \text{ mg}$$

Number of meq of K^+ in 100 ml of blood = 17/39, or 0.43 meq

Figure 7.4 A flame spectrophotometer used in the rapid determination of many metallic elements. (Courtesy Beckman Instruments, Inc.)

The effect of a solute on the freezing and boiling points of a solvent

A nonvolatile solute decreases the vapor pressure (see p. 92) of a solvent because the solute molecules interfere with the escape of molecules of the solvent from the surface of the liquid. Because the vapor pressure of the solvent is decreased, it will be necessary to heat the solution to a higher temperature so that the vapor pressure of the solvent will become equal to the atmospheric pressure. The boiling point of the solution will, therefore, be higher than that of the pure solvent.

The rise in the boiling point of the solution is not directly related to the weight of the solute present, but is related to the relative number of molecules of the solute. A gram-molecular weight of an un-ionized solute, such as sugar, which does not dissociate into ions in solution, will raise the boiling point of 1000 g of water by 0.52°C. The boiling point of 1000 g of water that contains 120 g of urea (2 moles) in solution will be 101.04°C. The elevation of the boiling point of a liquid that contains 1 mole of an un-ionized solute dissolved in 1000 g of the solvent is a constant for that solvent and is called the *molal* boiling point elevation constant.* This constant for water is 0.52°C.

*A solution that contains a mole of solute in *1000 g of solvent* is known as a *molal* solution.

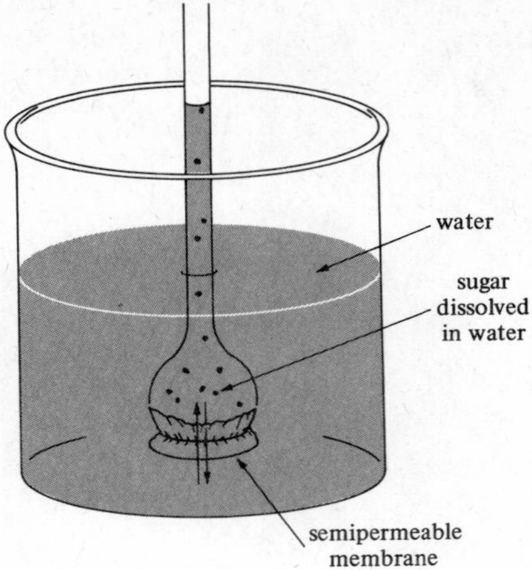

Figure 7.5 A demonstration of osmosis.

The addition of a solute to a solvent lowers the freezing point of the solvent. This lowering of the freezing point of a solvent is proportional to the number of solute particles present in the solution. A mole of an un-ionized solute will lower the freezing point of 1000 g of water by 1.86°C. This is called the *molal freezing point depression constant of water.*

Because un-ionized substances, such as methyl alcohol, ethyl alcohol, and glycol (Prestone), are quite effective in lowering the freezing point of water, they are added to the radiators of automobiles to prevent freezing during cold weather.

ILLUSTRATIVE PROBLEM 12

How many grams of ethyl alcohol (C_2H_5OH) will be needed to lower the freezing point of 1 kg of water to −10°C?

One gram-molecular weight of alcohol will lower the freezing point of 1000 g of water to −1.86°C. Therefore, it will require 10/1.86, or 5.37 gram-molecular weights of ethyl alcohol, to lower this 1000 g of water to −10°C. Since one gram-molecular weight of alcohol is 24 + 5 + 16 + 1 or 46 g, then 46 × 5.37 or 247 g of alcohol must be added to the 1000 g of water to produce a lowering of the freezing point of 10°C.

Osmotic pressure

Certain membranes will permit the passage of water through the membrane but will not permit large molecules, such as sugar, to pass. Such membranes exhibit *selective permeability* and are called *semipermeable membranes.* If such a membrane is put on the enlarged end of a thistle tube in which a sugar solution is placed and the whole apparatus is immersed in water (as is illustrated in Figure 7.5), then the solution will rise in the tube. Because sugar cannot diffuse through the membrane, water will diffuse into the tube faster than it diffuses out of the tube in an attempt to dilute the sugar solution. There will be a tendency to equalize the concentrations inside and out the tube, a situation which can never be completely attained. Eventually the pressure of the column of solution will be great enough to equilibrate the water diffusing into and out of the tube. This pressure is called the *osmotic pressure* of the sugar solution.

It is now known that the osmotic pressure is not directly dependent on the mass of the solute but upon the relative number of solute molecules in solution.

The phenomenon in which a solvent passes through a semipermeable membrane from a solution of lesser concentration to one of greater concentration is called *osmosis.* Many animal and vegetable membranes are semipermeable, and osmosis is necessary in the proper functioning of living matter.

The influence of osmosis on the erythrocytes, or red blood cells

A solution of 0.85 to 0.90 percent sodium chloride or 5 percent glucose has the same osmotic pressure as many body fluids including the blood plasma. This 0.85 to 0.90 percent solution of sodium chloride is called a *physiological salt solution,* and in it the living tissues of animals will not undergo change due to osmosis. A physiological salt solution is also called an *isotonic salt solution,* or a *normal saline.* Ringer's solution is not only isotonic with blood plasma, but also has an inorganic composition that closely approximates that of the blood fluid. It contains 0.86 percent sodium chloride, 0.03 percent potassium chloride, and 0.033 percent calcium chloride.

A solution is said to be isotonic when it has the same osmotic pressure as some body fluid. A physiological salt solution is isotonic to blood plasma. If placed in a solution that is isotonic with the blood plasma, the erythrocytes, or red blood cells, will undergo no change in size or shape.

If the red blood cells are placed in a solution that has a lower osmotic pressure than blood plasma, the cell will expand and finally burst. This process is known as *hemolysis* and is due to the fact that the fluid in the cells has a higher osmotic pressure than the solution in which they are suspended. Water flows into these cells faster than it flows out of them. Because such a solution has a lower osmotic pressure than blood plasma, it is said to be *hypotonic* to the blood fluid. A solution that has a higher osmotic pressure than the blood plasma is said to be *hypertonic.* If red blood cells are placed in such a solution, water will pass out of them faster than it enters, and the red blood cells will shrink or *crenate.*

Care and good judgment should be exercised when fluids are introduced intravenously into the bloodstream, especially if they are not isotonic with the blood plasma. When it is necessary to give intravenous injections of fluids such as salt or glucose solutions, isotonic solutions are usually employed. Drugs that are to be injected into the bloodstream should be made of such a concentration that the drug solution will be nearly isotonic with the blood plasma, or they should be dissolved in a physiological salt or glucose solution.

These phenomena that have just been described for blood plasma and blood cells are also characteristic of plant fluids and plant cells.

Saline laxatives, such as Epsom salts, act by forming a hypertonic solution, drawing water from the intestinal wall and moistening the intestinal contents for easier evacuation.

STUDY QUESTIONS

1. Define solute, solvent, solution, saturated solution, supersaturated solution.
2. How could you determine whether a solution is saturated, supersaturated, or undersaturated with a certain solid solute?
3. Give an example of a solution in which a liquid is dissolved in a liquid; a gas is dissolved in a liquid; a solid is dissolved in a liquid; a gas is dissolved in a gas.
4. Are all solids more soluble in the hot solvent than in the cold solvent?
5. Define molar solution, normal solution, gram-equivalent weight of an element, milliequivalent.
6. How would you prepare a molar solution of a chemical substance? A normal solution?
7. What is the relationship between the gram-atomic weight, the gram-equivalent weight, and the valence of an element?
8. What is meant by titration? How is a titration performed?
9. Why is it more difficult to freeze a saltwater bay than a freshwater lake?
10. What effect does an added solute have on the freezing point and boiling point of a solvent? If 2 moles of glycerol, $C_3H_5(OH)_3$, are added to 1000 g of water, what will be the freezing and boiling points of the resulting solution?
11. Which will be more effective in lowering the freezing point of a certain weight of water, a mole of ethyl alcohol, C_2H_5OH, of glycerol, $C_3H_5(OH)_3$, or of cane sugar, $C_{21}H_{22}O_{11}$?
12. A physiological salt solution is isotonic with blood plasma. Explain.
13. Define osmosis, osmotic pressure, isotonic solution, hypotonic solution, hypertonic solution, semipermeable membrane, physiological salt solution, crenation, and hemolysis of red blood cells.
14. Why are plants so dependent on osmotic pressure?
15. Why must solutions that are to be used for intravenous injections be isotonic with the blood?
16. What changes would occur in red blood cells if they were placed in a hypotonic salt solution? A hypertonic salt solution?

PROBLEMS

1. What weight of sugar is required to make 5 liters of a 3 percent solution?
2. How many grams of sodium chloride are needed to make 300 ml of a physiological salt solution (0.9 percent solution of sodium chloride)?
3. About 9 percent of the weight of the body is blood. Normal blood

contains about 0.1 percent of glucose. How many grams of glucose are present in the blood of an individual who weighs 70 kg?

4. Calculate the amount of solute that will be required to make the following solutions:

   (a) 2 liters of 0.1 *M* NaOH solution
   (b) 500 ml of a 1.5 *M* H_3PO_4 solution
   (c) 2250 ml of a 3.6 *M* H_2SO_4 solution
   (d) 150 ml of a 2 *M* $MgCl_2$ solution
   (e) 1.5 liters of a 5 *M* K_2CO_3 solution

5. How many grams of calcium chloride are present in 250 ml of a 7 percent solution?
6. Determine the gram-equivalent weight of the following ions: Ca^{2+}, $PO_4{}^{3-}$, Al^{3+}, Cu^{+}, Cu^{2+}, K^{+}, Fe^{2+}, and Sn^{4+}.
7. Calculate the gram-equivalent weight of the following metals and nonmetals from the amount of hydrogen that the given weight of the substance combines with or displaces:

   (a) 18 g of aluminum liberates 2 g of hydrogen from an acid
   (b) 7.5 g of nitrogen combines with 1.5 g of hydrogen to form ammonia
   (c) 8 g of zinc liberates 0.26 g of hydrogen from an acid
   (d) 3.55 g of chlorine combines with 0.1 g of hydrogen to form HCl
   (e) 1 g of calcium liberates 0.05 g of hydrogen from water

8. What is the gram-equivalent weight of the following elements?

   (a) 10 g of element A combines with 4 g of oxygen
   (b) 2 g of element B combines with 3 g of oxygen
   (c) 3.5 g of element C combines with 1 g of oxygen
   (d) 56 g of element D combines with 24 g of oxygen

9. If reacted with water, 0.35 g of an active metal liberates 560 ml of hydrogen measured at standard conditions. A liter of hydrogen weighs 0.09 g at standard conditions. What is the gram-equivalent weight of the metal?
10. Calculate the gram-equivalent weight of the following compounds: NaOH, H_3PO_4, H_2SO_4, KBr, $Ca(NO_3)_2$, HNO_3, $Ba(OH)_2$, $(NH_4)_2$-CO_3, and $MgSO_4$.
11. Calculate the amount of solute required to make the following solutions:

    (a) 2 liters of 0.1 *N* KOH
    (b) 750 ml of 3 *N* $KClO_4$
    (c) 350 ml of 0.3 *N* H_3PO_4
    (d) 1.5 liters of 1.5 *N* $CaBr_2$
    (e) 100 ml of 0.5 *N* $ZnSO_4$

12. What is the normality of the following solutions?

    (a) 1 liter contains 120 g of NaOH
    (b) 1 liter contains 1.4 kg of H_2SO_4
    (c) 200 ml contains 48 g of HCl

(d) 1.5 liters contain 255 g of Na_2SO_4
(e) 750 ml contains 3 g of $Ca(OH)_2$

13. What volume of 4 *N* HCl will be required to neutralize 5 liters of 0.8 *N* NH_4OH?
14. In the neutralization of 48.4 ml of an unknown acid, 40 ml of 0.35 *N* base is required. What is the normality of the acid?
15. What is the normality of a base if 37 ml of the base requires 42 ml of a 0.12 *N* acid for its neutralization?
16. Calculate the depression of the freezing point and the elevation of the boiling point of 1000 g of water if 200 g of the following substances are dissolved in it: sugar, $C_{12}H_{22}O_{11}$; urea, N_2H_4CO; glycerol, $C_3H_5(OH)_3$; glycol, $C_2H_6O_2$; ethyl alcohol, C_2H_5OH; methyl alcohol, CH_3OH.
17. What will be the freezing point of the water in a radiator if 10 liters of water and 1 kg of glycol, $C_2H_6O_2$, are added to it? Express the freezing point of this solution in both the Celsius and Fahrenheit scale.
18. What volume of 36 *N* sulfuric acid would be needed to prepare 2 liters of 6 *N* sulfuric acid?
19. How much 0.5 *N* sodium hydroxide solution can be prepared from 75 ml of 2.5 *N* sodium hydroxide?
20. Convert 3 mg of Mg^{2+} to milliequivalents; convert 10 meq of chloride ions to milligrams.

chapter 8

---

chemical equilibrium

8

Because one will always note that in a chemical equilibrium two opposed chemical reactions are proceeding at the same rate or speed, a study of the latter will be our first concern in this chapter.

The rate or speed of a chemical reaction and some factors that affect it

The rate or speed of a chemical reaction is expressed as the amount of chemical substance that undergoes change in a unit period of time, such as an hour or a minute. A discussion of the five factors that affect the rate of a chemical reaction follows.

NATURE OF THE REACTING SUBSTANCE

White phosphorus burns readily and usually ignites spontaneously in air. Carbon must be heated to a high temperature before it will burn, while sand, SiO_2, is noncombustible. Sodium reacts quite vigorously with water at room temperature. Magnesium and iron undergo change very slowly when in contact with water at room temperature but readily liberate hydrogen from steam. Platinum will not decompose water. It is evident that the rate of a chemical reaction depends upon the nature of the reacting substances.

TEMPERATURE

The rate of a chemical reaction is dependent upon and is quite sensitive to a rise in temperature. Although no simple quantitative relationship has been established between the change of the rate of a chemical reaction with change in temperature, the rate will usually be doubled or tripled with a rise of 10°C in the temperature of the reacting medium. This is called the *law of Ostwald.*

For two chemical substances to react, the molecules of the two substances must collide. However, not all collisions between molecules of the reacting substances will lead to a reaction. For reaction to occur, the molecules must possess an energy equal to or greater than a certain minimum, because if they possess less than this amount, they will rebound from each other unchanged.

At any certain temperature, molecules will have a certain characteristic average velocity and an average energy. However, there

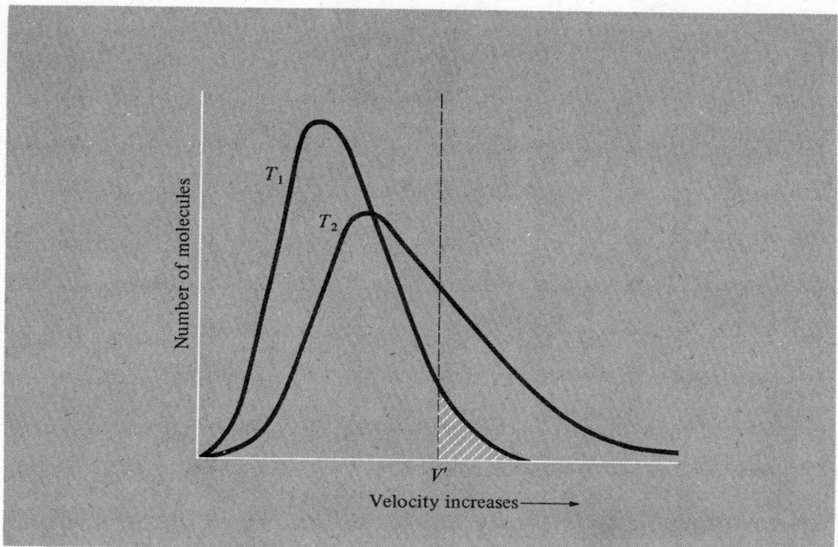

***Figure 8.1** Only molecules with a velocity equal to or greater than* V′ *will collide with sufficient energy to undergo chemical reaction. Thus, because a greater fraction of the molecules at temperature* T_2 *than at* T_1 *have an energy equal to or greater than* V′, *the reaction will be much faster at* T_2. *Note that at both* T_1 *and* T_2, *a large fraction of the molecules have energies below that required for reaction.*

will be a wide distribution of actual energies of the various molecules; a large portion of these will possess an energy that is near the average. The two curves labeled T_1 and T_2 in Figure 8.1 are a typical representation of the usual distribution of energies, as well as the velocities of the molecules, of a chemical substance at the lower temperature T_1 and the higher temperature T_2.

Under the curve labeled T_1, the shaded area represents the fraction of the molecules that have a velocity equal to or greater than V', the minimum velocity for reaction. Note that the fraction of molecules that have a velocity equal to or greater than V' at the higher temperature T_2 has greatly increased over the fraction of such molecules at the lower temperature T_1, in harmony with Ostwald's rule.

As noted earlier, in an exothermic reaction, heat is generated and lost during the reaction, while in an endothermic reaction, heat must be furnished in order that the reaction proceed. If the heating of such a reaction is stopped, the reaction ceases.

In many exothermic reactions, as well as endothermic ones, a quantity of heat, called the *heat of activation* and labeled E_a in the graph in Figure 8.2, must be furnished before any reaction will occur.

It is generally assumed that some high-energy activated complex must be formed as an intermediate in these reactions. In Figure

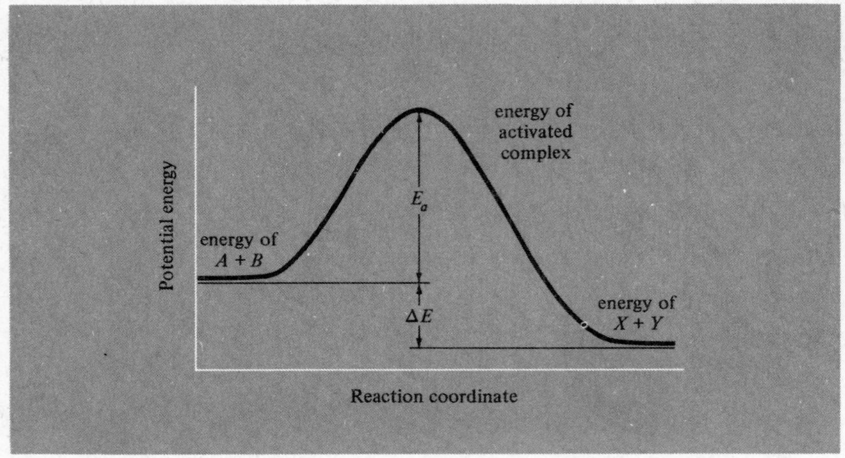

***Figure 8.2** Energy and progress of a reaction.*

8.2, this complex is represented as decomposing to give off energy equal to the sum of E_a and ΔE. Thus one can represent the stages in the reaction indicated in Figure 8.2, this way:

$$A + B \longrightarrow \text{activated complex} \longrightarrow X + Y$$

According to Figure 8.2, the conversion of A and B to X and Y is an exothermic reaction in which ΔE will be the heat of reaction. In an endothermic reaction, the heat obtained when the activated complex decomposes will be less than the heat of activation E_a. The difference between these represents the "endothermic" heat of reaction, that is, the net amount of heat absorbed in converting A and B to X and Y.

CATALYSTS

A catalyst increases the rate of a chemical reaction without apparently undergoing any change. Substances that decrease the rate of a chemical reaction are called *inhibitors*. The increased rate of decomposition of potassium chlorate in the presence of manganese dioxide is disscussed in Chapter 4.

In no case will a catalyst initiate a reaction that is incapable of occurring in the absence of the catalyst.

A catalyst apparently changes the mechanism of a chemical reaction from that of the uncatalyzed reaction. In the presence of the catalyst, a lower-energy activated complex is formed, which for its formation requires a smaller heat of activation E_a than the activated complex formed in the uncatalyzed reaction.* This effect of a catalyst on the activation energy required to induce a chemical reaction is illustrated in Figure 8.3. The reduction in the required activation

*If it is an enzyme-catalyzed reaction, this could be identical to the enzyme-substrate complex (page 376).

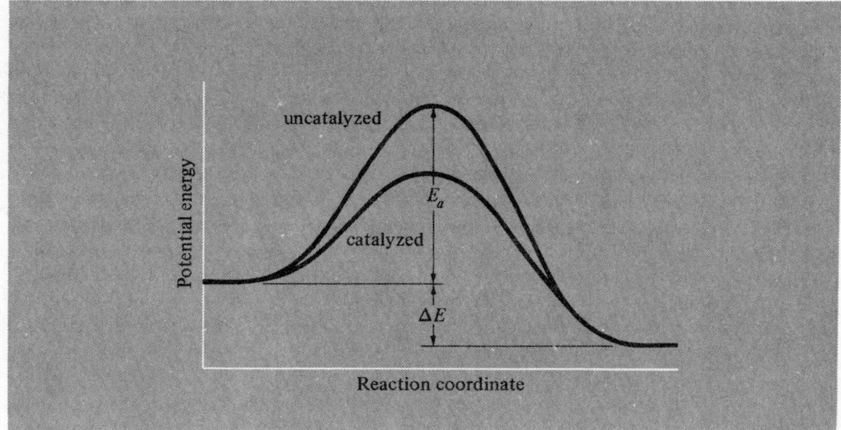

Figure 8.3 A larger fraction of the molecules in a catalyzed reaction have an energy equal to or greater than the minimum required for reaction than in the uncatalyzed reaction.

energy for the reaction induced by the catalyst results in a much lower minimum energy and minimum velocity for the molecules to react on collision than in the uncatalyzed reaction. This is clearly illustrated in Figure 8.4, where the fraction of molecules that can react in the uncatalyzed reaction is represented by the darker shaded area, whereas the fraction in the catalyzed reaction that has sufficient velocity and energy to react is indicated by a combination of both shaded areas.

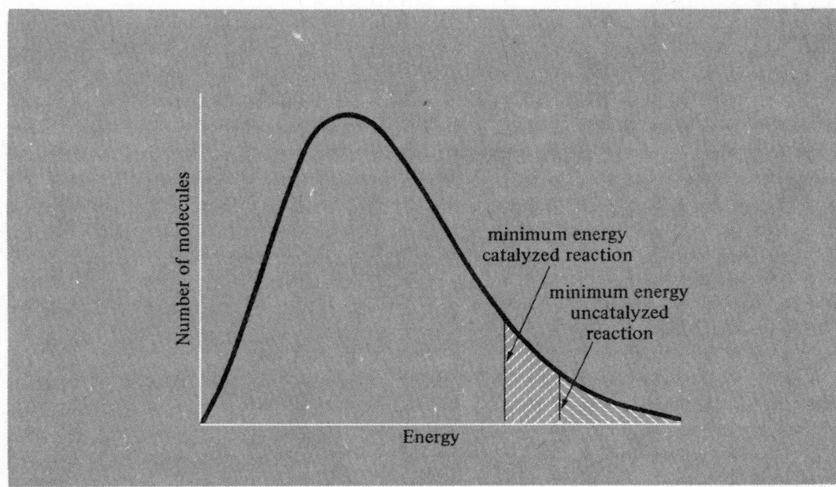

Figure 8.4 Comparison of the fraction of molecules that can react in catalyzed and uncatalyzed reactions.

CONCENTRATION

The effect of concentration can be illustrated by the difference in rate of combustion of substances in air and in pure oxygen (where the concentration of oxygen is five times as great). The rate of a chemical reaction is proportional to the concentration of the reacting substances. This is called the *law of mass action* and is a very important law of chemistry. See also pages 125–128.

STATE OF SUBDIVISION

The finer a solid is divided and the greater the surface exposed by a certain mass of the substance, the more rapid its reaction will be. Kindling wood burns much more rapidly than logs; sawdust, when suspended in air, will burn explosively; and grain dust has been responsible for many explosions in grain elevators and flour mills.

Physical equilibrium

A saturated solution of a solid solute in a liquid is an example of a physical equilibrium (see page 103). In such a solution, the rate of solution of the solid solute is equal to the rate of crystallization of some of the solute from the solution. It can then be stated that the solute in solution is in equilibrium with the solid solute. When the solute is placed in the pure solvent, solute molecules will escape into solution at a rapid rate. After a short interval of time, there will be sufficient solute molecules in solution so that some of them will crystallize on the solid solute. At this point, the rate of solution will be much greater than the rate of crystallization. As more of the solute dissolves, the rate of crystallization will continually increase. Eventually the two rates will become equal, no more of the solute will apparently dissolve, and the solution is said to be saturated. This is only an apparent state of rest, because at equilibrium the solute will be dissolving while the solute in solution will be crystallizing. But because the rate of solution and crystallization will be equal, no change in the system will be apparent.

Another example of a physical equilibrium is that of a gas which is saturated with the vapors of some liquid. The rate of evaporation of molecules from the surface of the liquid will be equal to the rate of condensation of the molecules of the liquid from the vapor state when the gas is saturated with the vapors of the liquid.

Reversible and irreversible chemical changes

The action of steam on iron is a true reversible chemical change, because the products iron tetroxide and hydrogen will react together to regenerate steam and metallic iron. Many chemical changes are reversible. Some chemical changes are irreversible. Examples of the latter are the burning of sugar, the decomposition of potassium chlorate, and the burning of calcium to form lime.

Chemical equilibrium

A chemical equilibrium differs from a physical equilibrium only in the fact that the former involves chemical changes, while the latter involves physical changes.

If steam is passed over red-hot iron, the reaction will go to completion, because hydrogen is removed as fast as it is formed, and no reversal of the reaction can occur.

$$3Fe + 4H_2O \longrightarrow Fe_3O_4 + 4H_2$$

If hydrogen is passed over heated iron oxide, the reaction will also go to completion, because the steam will escape as fast as it is formed

$$Fe_3O_4 + 4H_2 \longrightarrow 3Fe + 4H_2O$$

If steam is placed in contact with iron in a closed container or bomb, the hydrogen that is formed cannot escape but will combine with the iron tetroxide that has been formed to re-form steam and iron. At first the reverse reaction will be slow, but as the quantity of the iron tetroxide and hydrogen increases, the rate of this reaction will increase, while the rate of the initial reaction will decrease as the amount of steam and iron decreases. Eventually the two rates will become equal, and from this time on, no change in the composition of the system will occur. Such an apparent state of rest is called a *chemical equilibrium.* In a chemical equilibrium, *the rates of two opposing chemical reactions are equal.*

A chemical equilibrium may be represented by the following general chemical equation:

$$A + B \rightleftharpoons C + D$$

(*Note:* The double arrows represent a chemical system in equilibrium.) If such a hypothetical system is in equilibrium, the reaction can be made to go to completion by removing one of the products of the reaction, that is, either C or D. If some B is removed from the reaction at equilibrium, more C and D will combine to form A and B. If now B is removed as it is formed, C and D will eventually disappear from the system and only A will remain.

In a chemical equilibrium, the equilibrium state will not usually occur when 50 percent of the reactants have been converted to products. The actual percentage of reactants and products at equilibrium will depend upon the nature of both the reactants and products, and it may vary between wide limits.

In the hypothetical reaction

$$A + B \rightleftharpoons C + D$$

if at equilibrium more D is added to the system, the system will no longer be in equilibrium. It will then change in such a manner as to come to a new equilibrium state. In this second equilibrium state, the quantity of *A* and *B* will be greater but *C* will be less than in the original equilibrium.

Why certain ionic reactions in solution go to completion

Because a reversible chemical change can be made to go to completion by removing one of the products of the reaction, it is possible to make a reaction of ions in solution go to completion under the three conditions discussed below.

Formation of an Insoluble Gas. If a gas that is insoluble in the solution is formed in a chemical reaction, it will escape as it is formed, and no reversal of the reaction can then occur. Such a reaction will readily go to completion. For example, if a metallic carbonate is treated with an acid, carbon dioxide will escape from solution because it is only slightly soluble in water.

$$Na_2CO_3 + H_2SO_4 \longrightarrow Na_2SO_4 + H_2O + \overline{CO_2}$$

(A line over a formula indicates a volatile substance.) In the reaction

$$NaCl + H_2SO_4 \longrightarrow NaHSO_4 + HCl$$

if the NaCl is dissolved in water before the sulfuric acid is added, an equilibrium will exist at room temperature, because the hydrogen chloride is quite soluble in water at this temperature and will not escape. Upon heating, hydrogen chloride gas will escape and the reaction will go to completion.

Formation of an Un-ionized Substance. Neutralization reactions go to completion because almost completely un-ionized water molecules are formed and most of the H^+ and OH^- ions are removed from solution.

Formation of an Insoluble Solid. The following three equations are illustrative of such situations:

$$AgNO_3 + HCl \longrightarrow \underline{AgCl} + HNO_3$$
$$Ba(C_2H_3O_2)_2 + K_2SO_4 \longrightarrow \underline{BaSO_4} + 2KC_2H_3O_2$$
$$CuSO_4 + 2NH_4OH \longrightarrow \underline{Cu(OH)_2} + (NH_4)_2SO_4$$

A line under the above formulas indicates a water insoluble solid.

If dilute solutions of potassium sulfate and sodium nitrate are mixed, an equilibrium between four salts, namely K_2SO_4, $NaNO_3$, KNO_3, and Na_2SO_4, will result, because none of the products can easily be removed from the equilibrium.

$$K_2SO_4 + 2NaNO_3 \rightleftharpoons Na_2SO_4 + 2KNO_3$$
$$2K^+,SO_4^{2-} + 2Na^+,2NO_3^- \rightleftharpoons 2Na^+,SO_4^{2-} + 2K^+,2NO_3^-$$

The influence of a catalyst on a chemical equilibrium

A catalyst shortens the time required for a system to reach equilibrium. The time required for a reaction to reach equilibrium is an important factor in industry, because a reaction that requires a long period of time will not usually be economically feasible.

Catalysts cannot change the proportions of reactants and products at equilibrium, nor can they initiate a reaction that cannot occur in their absence. Certan catalysts produced by living tissues are called *enzymes* and are of great importance in the proper functioning of such tissues. Enzymes will be discussed in Chapter 25.

The quantitative aspects of a chemical equilibrium and the equilibrium constant, K_{eq}

Because it has now been accepted that molecules cannot react together unless they collide, the rate of collision and the rate of reaction must be proportional to the concentration of the reacting molecules, a most important principle of chemistry previously noted as the law of mass action (page 122).

In a reaction where single molecules of two individual substances are reacting to form a product, such as in the hypothetical reaction

$$A + B \longrightarrow C + D$$

the rate of the chemical reaction in which A and B combine to form products will be *directly proportional* to the concentration of both of the reactants A and B. This statement may be expressed mathematically as follows:

$$R_{\rightarrow} \propto cA \times cB$$

in which $\propto$ is a symbol indicating "is proportional to," and where the symbols cA and cB indicate the concentration of A and B either in moles per reacting vessel, in moles per liter, or in any other suitable concentration units. $R_{\rightarrow}$ is the measured rate of the reaction occurring to the right.

The above statement can be made an equality if $\propto$ is replaced by $= k$. k is called the *velocity constant* and depends on the nature of A and B and the temperature. Restating the above expression as an equality gives

$$R_{\rightarrow} = k \times cA \times cB$$

In a reversible reaction, C and D will also be reacting to re-form A and B. The rate of this reverse reaction is expressed as

$$R_{\leftarrow} = k' \times cC \times cD$$

where k' is the velocity constant for this reaction and will not be the same as k above, because its value will depend on the nature of C and D rather than A and B.

The fact that at equilibrium the forward and reverse reactions are occurring at the same rate can be indicated by the following expression

$$R_{\rightarrow} = R_{\leftarrow}$$

that is, the rates of the forward and reverse reactions are equal. If these two rates are equal, then the following equality is also true:

$$k \times cA \times cB = k' \times cC \times cD$$

Moving constants to one side of the equality and concentrations to the other gives us the expression

$$\frac{k}{k'} = \frac{cC \times cD}{cA \times cB}$$

Because a constant divided by a constant gives a new constant, then k/k' is made equal to K_{eq}, called the *equilibrium constant.* Then

$$K_{eq} = \frac{cC \times cD}{cA \times cB}$$

In the use of K_{eq}, it is to be noted that

1. It is common practice to place the concentration of products in the numerator and reactants in the denominator.
2. K_{eq} has a constant numerical value at any one temperature irrespective of the actual concentration of reactants and products. It does vary in numerical value with changes in the temperature.
3. If K_{eq} is greater than 1, the concentration of products will be greater than the concentration of reactants. If it is much smaller than 1, equilibrium will have been established when only a small amount of the reactants have been converted to products.

The formation of an ester (page 275) from an alcohol and an organic acid is a reversible reaction and can be represented by the following general equation:

$$\text{alcohol} + \text{organic acid} \rightleftharpoons \text{ester} + \text{water}$$

In many of these esterification reactions, equilibrium is established when a mole each of alcohol and organic acid are present with two moles each of ester and water. Then the numerical value of K_{eq} for this equilibrium is established as follows:

$$K_{eq} = \frac{\text{moles of ester} \times \text{moles of water}}{\text{moles of alcohol} \times \text{moles of acid}}$$

$$= \frac{2 \text{ moles} \times 2 \text{ moles}}{1 \text{ mole} \times 1 \text{ mole}} = 4$$

If two molecules of a substance must react to give products, the concentration of this substance will be squared in determining the numerical value for K_{eq}. The K_{eq} expression for the equilibrium

$$2HI \rightleftharpoons H_2 + I_2$$

is

$$K_{eq} = \frac{cH_2 \times cI_2}{c^2HI}$$

For the generalized equilibrium

$$nA + mB \rightleftharpoons pC + qD$$

then

$$K_{eq} = \frac{c^pC \times c^qD}{c^nA \times c^mB}$$

and K_{eq} will be established for a reversible reaction by employing the concentration of reactants and products at equilibrium, each raised to a power equal to the coefficient for each in the chemical equation which represents the reaction.

ILLUSTRATIVE PROBLEM 1

Calculate the numerical value of the equilibrium constant for the system

$$NO_2 + SO_2 \rightleftharpoons NO + SO_3$$

if the following concentrations of reactants and products are found to be present at equilibrium: 0.6 mole of NO_2; 1.2 moles of SO_2, 2.2 moles of SO_3, and 1.64 moles of NO.

Then

$$K_{eq} = \frac{\text{moles of NO} \times \text{moles of } SO_3}{\text{moles of } NO_2 \times \text{moles of } SO_2}$$

$$= \frac{1.64 \text{ moles} \times 2.2 \text{ moles}}{0.6 \text{ mole} \times 1.2 \text{ moles}} = \frac{3.61}{0.72} = 5$$

ILLUSTRATIVE PROBLEM 2

For the reaction

$$NO_2 + SO_2 \rightleftharpoons NO + SO_3$$

how many moles of NO will be in equilibrium with the following concentration of these reactants and products: 1.65 moles of SO_2; 3.3 moles of SO_3 and 1 mole of NO_2?

As established in Illustrative Problem 1 above, the numerical value for the equilibrium constant for this reaction is 5.

Then

$$K_{eq} = \frac{\text{moles of NO} \times \text{moles of } SO_3}{\text{moles of } NO_2 \times \text{moles of } SO_2}$$

$$= \frac{\text{moles of NO} \times 3.3 \text{ moles}}{1.65 \text{ moles} \times 1 \text{ mole}} = 5$$

and the moles of NO at equilibrium will be

$$\text{moles of NO} = \frac{5 \times 1.65 \text{ moles} \times 1 \text{ mole}}{3.3 \text{ moles}} = 2.5 \text{ moles}$$

STUDY QUESTIONS

1. What effect do changes in the following have on the rate of a chemical reaction: nature of the substance, temperature, concentration, catalyst, state of subdivision.
2. What is the law of Ostwald?
3. State the law of mass action.
4. Why do violent explosions sometime occur in flour mills?
5. Give some examples of physical equilibriums, a reversible chemical reaction, an irreversible chemical reaction, a chemical equilibrium.
6. How do physical and chemical equilibriums differ? In what respects are they alike?
7. In a chemical equilibrium, what two factors are equal?
8. At equilibrium, will the concentration of the reactants be equal to the concentration of the products?
9. Find the dictionary meaning of the words "static" and "dynamic," and then tell which describes a chemical equilibrium.
10. By means of a suitable equation and a suitable explanation, tell what happens to a chemical equilibrium if an additional amount of one of the products of the reaction is added to it.
11. Under what three conditions do ionic reactions in solution go to completion?
12. Why do the following reactions go to completion?

    (a) a neutralization
    (b) the action of hydrochloric acid on calcium carbonate
    (c) the treatment of solid sodium acetate, $NaC_2H_3O_2$, with hot concentrated sulfuric acid
    (d) addition of a solution of sodium chloride to a solution of silver nitrate, $AgNO_3$
    (e) the action of a solution of sodium hydroxide on a ferric chloride, $FeCl_3$, solution

13. What effect does a catalyst have on a chemical reaction? On a chemical equilibrium?

14. For a certain chemical reaction, how are the following related: rate of reaction; concentration of the reactants; rate of collisions of molecules?
15. Interpret the expression

$$R_{\rightarrow} = R_{\leftarrow}$$

16. Is the numerical value for the equilibrium constant (K_{eq}) of a certain reaction dependent upon the concentration of reactants?
17. In the K_{eq} expression, does the concentration of the reactants or the products appear in the numerator?
18. In the reaction

$$N_2 + O_2 \rightleftharpoons 2NO$$

$R_{\rightarrow} \propto$ ________ and $R_{\rightarrow} = k \times$ ________, where $R_{\rightarrow}$ is the rate of the reaction occurring from left to right. What is the significance of $\propto$ in the expression above? Write a mathematical expression to indicate the rate of the reaction that is occurring from right to left ($R_{\leftarrow}$).
19. Using the expressions for $R_{\rightarrow}$ and $R_{\leftarrow}$ established in Question 18, derive the K_{eq} expression for the reversible reaction

$$N_2 + O_2 \rightleftharpoons 2NO$$

20. For the following reversible reactions

$$H \cdot CHO_2 + CH_3OH \rightleftharpoons CH_3 \cdot CHO_2 + H_2O$$
$$H_2 + Br_2 \rightleftharpoons 2HBr$$
$$2NO_2 \rightleftharpoons 2NO + O_2$$
$$N_2 + 3H_2 \rightleftharpoons 2NH_3$$
$$2SO_2 + O_2 \rightleftharpoons 2SO_3$$

write expressions for

(a) $R_{\rightarrow}$
(b) $R_{\leftarrow}$
(c) K_{eq}

PROBLEMS

1. Calculate K_{eq} for the reversible reaction

$$2HI \rightleftharpoons H_2 + I_2$$

if 7.5 moles of HI are in equilibrium with 2.5 moles each of hydrogen and iodine.

2. Calculate K_{eq} for the reversible reaction

$$PCl_3 + Cl_2 \rightleftharpoons PCl_5$$

if at equilibrium

$[PCl_3] = 0.2$ mole/liter
$[Cl_2] = 0.1$ mole/liter

and

$[PCl_5] = 0.8$ mole/liter

3. How many moles of NH_3 will be in equilibrium with 0.4 mole of N_2 and 0.2 mole of H_2 if the K_{eq} for the reversible reaction

$$N_2 + 3H_2 \rightleftharpoons 2NH_2$$

has the numerical value 3.13?

4. The numerical value for K_{eq} for the reaction

$$2SO_2 + O_2 \rightleftharpoons 2SO_3$$

is 30. A liter reaction vessel in which the above reactants and product have reached equilibrium contains 0.6 mole of SO_3 and 0.2 mole of SO_2. Calculate the number of moles of O_2 present in this vessel.

chapter 9

ionization; acids, bases, and salts; and nomenclature of inorganic compounds

9

Ionization of electrolytes

In the discussion of atomic structure and compound formation (Chapter 3), it was noted that certain atoms form compounds by the loss or gain of electrons. The resulting particles are electrically charged and are called *ions*. Those that are positively charged are called *cations*, because they migrate to the cathode, or negative electrode of an electrolytic cell. The negatively charged ions migrate to the anode, or positive pole, and are called *anions*.

In 1832, the Englishman Michael Faraday suggested the existence of ions. Faraday believed that ions did not form in solution in the absence of an electric current but were produced by the latter. In 1887, the Swedish chemist Svante Arrhenius suggested a new theory of ionization. He recognized that ions exist in solutions of electrovalent compounds and are responsible for the passage of the electric current through these solutions. For highly ionized substances, the Arrhenius theory is rather unsatisfactory, and Debye and Hückel have modified this theory for these strong electrolytes.

Substances whose aqueous solutions conduct an electric current are called *electrolytes*, while nonconductors are called *nonelectrolytes*. Nearly all salts, acids, and bases that are soluble in water are electrolytes. Most organic compounds are nonelectrolytes. Thus electrolytes must exist in solution as charged particles.

EVIDENCE FOR THE IONIZATION OF ELECTROVALENT SUBSTANCES IN SOLUTION

Electrolytes produce abnormal freezing point lowerings and boiling point elevations of solvents. If a chemical substance is not dissociated into ions, 1 mole of a nonvolatile substance will elevate the boiling point of 1000 g of water 0.52°C and lower the freezing point of this amount of water 1.86°C. A solution that contains two moles of such a substance in 1000 g of water will depress the freezing point 3.72°C and elevate the boiling point 1.04°C (see page 111).

If a mole of sodium chloride is dissolved in 1000 g of water, the boiling point elevation and the freezing point depression of the solvent will be not quite equal to twice the value that would be expected if sodium chloride was not dissociated into ions. From this information it can be concluded that the molecules of sodium chloride are nearly all dissociated into two ions apiece. One mole of calcium chloride, $CaCl_2$, which dissociates into three ions, produces a lowering of the freezing point and an elevation of the boiling point of nearly three times the normal values. This is to be expected if calcium chloride is extensively dissociated into ions in an aqueous solution.

Instantaneous Reaction of Ions in Solution. If powdered dry silver nitrate and sodium chloride are intimately mixed, no reaction will occur. If, however, aqueous solutions of these two reagents are mixed, the reaction that forms silver chloride is instantaneous. This reaction is so rapid that its rate has never been measured. On the other hand, organic substances, which are primarily covalent in nature and are nonelectrolytes, react together very slowly and may require hours or days for the completion of the reaction. To summarize, the reaction of silver nitrate and sodium chloride in aqueous solution is an ionic reaction and is so rapid that the time of reaction is not measurable.

SOME OF THE ASSUMPTIONS OF THE ARRHENIUS THEORY OF IONIZATION; THE DEGREE OF IONIZATION

Arrhenius recognized that certain chemical substances, called nonelectrolytes, do not dissociate into ions in solution and their solutions do not conduct the electric current. He also noted that many inorganic substances, such as acids, bases, and salts, which he called electrolytes, dissociate into ions in solution and the resulting solutions conduct the electric current.

The main assumptions of Arrhenius were that all pure substances exist in a molecular form and that the molecules of an electrolyte, if dissolved in an ionizing solvent, such as water, will partially dissociate into ions. Therefore, his theory is frequently called the *partial ionization theory*. He suggested that an equilibrium is established in a solution of electrolytes between the undissociated molecules in solution and the ions:

$$AB \rightleftharpoons A^+ + B^-$$

At this equilibrium point, the rate of dissociation of the compound AB into ions will be equal to the rate at which the ions A^+ and B^- associate to re-form the molecule AB.

Arrhenius also introduced the concept of the *degree of ionization.* This he defined as *the fraction of the molecules in solution that dissociate into ions.* Thus in the case of acetic acid, $H \cdot C_2H_3O_2$, if three molecules of each hundred present in solution dissociate into hydrogen and acetate ions, then the degree of ionization is 0.03. This may also be expressed as 3 percent. If one of each four molecules in solution dissociates into ions, then the degree of ionization is $\frac{1}{4}$, 0.25, or 25 percent. As may be expected, the degree of ionization of an electrolyte increases with increasing dilution of the electrolyte in solution. An acetic acid solution 0.1 N is more highly ionized than a 1 N solution of this acid.

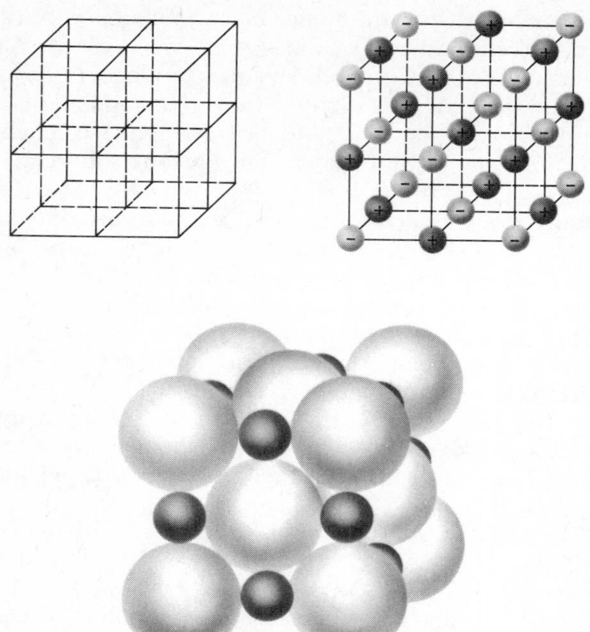

Figure 9.1 Three ways of representing the structure of sodium chloride. The positive sodium ions are smaller than the negative chloride ions.

THE DEBYE–HÜCKEL THEORY OF IONIZATION OF STRONG ELECTROLYTES

It was suggested by Debye and Hückel in 1922 that many strong electrolytes are 100 percent ionized. By means of X rays it has been demonstrated beyond doubt that sodium ions and chloride ions (Figure 9.1) exist in the crystal lattice of the salt crystal and that no sodium chloride molecules are present. Similar studies have shown that many other salts exist exclusively as ions in the crystalline state.

If sodium chloride is 100 percent ionized in the solid state, why does it not exhibit complete dissociation in an aqueous solution? The experimentally determined degree of ionization of a dilute solution of sodium chloride will be between 90 and 95 percent. Instead of lowering the freezing point of 1000 g of water to −3.72°C, a mole of salt produces a freezing point lowering of 3.37°C. These anomalous results have been explained by Debye and Hückel in the following manner: In a solution, some sodium ions will attract a number of chloride ions, while a chloride ion will attract a number of sodium ions. The ions in these clusters of ions will no longer act as independent ions or particles, but the whole cluster will act somewhat as a single particle. This is one of the main tenets of the Debye-Hückel theory of ionization, also called the *interionic attraction theory*.

The Debye-Hückel theory does not supplant the Arrhenius

theory of ionization, because weak electrolytes behave according to the pattern described by Arrhenius. However, it does appear to explain the ionization of strong electrolytes better than the Arrhenius theory.

Acids, bases, salts, and neutralization

Chemists have known for some time that all acids contain hydrogen, that they have a sour taste, that they change the color of certain indicators (blue litmus to red, red phenolphthalein to a colorless solution, amber-yellow methyl orange to an orange-red color, etc.), and that they neutralize the properties of bases. Chemists have recognized as bases those substances that contain hydroxyl (OH) radicals, that reverse the color changes induced by acids on the indicators noted above, that "neutralize" the properties of acids, and that usually have a bitter taste and a soapy feel.

Arrhenius, following his development of the partial ionization theory, formulated ionic definitions for acids and bases. He defined an acid as *a substance that would furnish hydrogen ions,* and a base as *a substance that would furnish hydroxyl ions* in solution. I have frequently labeled this as the Arrhenius, or classical, theory of acids and bases.

Recently, several new theories, which are much more general in their scope, have been suggested for acids and bases. In part, these have been developed because the Arrhenius theory of acids and bases is quite inadequate when one tries to extend it to nonaqueous solutions. One of these theories (the Brönsted-Lowry theory of acids and bases) is discussed later in this chapter.

The strength of acids and bases is related to their degree of ionization. An acid such as hydrochloric acid, which is highly ionized, will be a strong acid; one such as acetic acid, which is only slightly ionized, will be a weak acid. The highly ionized sodium hydroxide is a strong base, while the slightly ionized ammonium hydroxide is a weak base.

Other strong acids are nitric, sulfuric, hydriodic, and hydrobromic acids. Potassium hydroxide is another strong base, while the hydroxides of calcium, barium, and strontium are moderately strong.

The hydrogen ion is hydrated in an aqueous solution of an acid. The monohydrated hydrogen ion has the formula $H^+ \cdot H_2O$, or H_3O^+, and is called the *oxonium* or *hydronium ion.* However, in future discussions in this text, its hydration will usually by ignored and the hydrogen ion will be represented simply as H^+.

GENERAL REACTIONS OF ACIDS

The following are some of the general reactions of acids:

1. Certain metals can displace hydrogen from an acid. This reaction may be represented by the "molecular" equation

$$2Al + 6HCl \longrightarrow 2AlCl_3 + 3H_2$$

or by the "ionic" equation

$$2Al + 6H^+,6Cl^- \longrightarrow 2Al^{3+},6Cl^- + 3H_2$$

This serves as an important method for the preparation of hydrogen in the laboratory. Remember that only those metals above hydrogen in the activity series of the metals will displace hydrogen from an acid (see Chapter 6).

2. Acids react with bases to form salts and water (neutralization).

$$2NaOH + H_2SO_4 \longrightarrow Na_2SO_4 + 2H_2O$$

or

$$2Na^+,2OH^- + 2H^+,SO_4^{2-} \longrightarrow 2Na^+,SO_4^{2-} + 2H_2O$$

3. They react with metallic oxides to form salts and water.

$$2HCl + CaO \longrightarrow CaCl_2 + H_2O$$

or

$$2H^+,2Cl^- + CaO \longrightarrow Ca^{2+},2Cl^- + H_2O$$

4. They react with carbonates to form salts, carbon dioxide, and water

$$Na_2CO_3 + 2HCl \longrightarrow 2NaCl + CO_2 + H_2O$$

or

$$2Na^+,CO_3^{2+} + 2H^+Cl^- \longrightarrow 2Na^+,2Cl^- + CO_2 + H_2O$$

5. An acid may react with a salt to form a new acid and a new salt.

$$2NaNO_3 + H_2SO_4 \longrightarrow Na_2SO_4 + 2HNO_3$$

or

$$2Na^+,2NO_3^- + 2H^+,SO_4^{2-} \longrightarrow 2Na^+,SO_4^{2-} + 2HNO_3$$

If this reaction is to go to completion, the new acid must be either volatile or insoluble in the reacting medium.

GENERAL REACTIONS OF BASES

The following are general reactions of bases:

1. They react with acids to form salts and water (neutralization).

$$KOH + HNO_3 \longrightarrow KNO_3 + H_2O$$

or

$$K^+,OH^- + H^+,NO_3^- \longrightarrow K^+,NO_3^- + H_2O$$

2. Bases react with some salts to form new bases and new salts.

$$CuSO_4 + 2NaOH \longrightarrow \underline{Cu(OH)_2} + Na_2SO_4$$

or

$$Cu^{2+},SO_4^{2-} + 2Na^+,2OH^- \longrightarrow \underline{Cu(OH)_2} + 2Na^+,SO_4^{2-}$$

3. Salts are also formed by the reaction of bases with nonmetallic oxides.

$$2NaOH + SO_2 \longrightarrow Na_2SO_3 + H_2O$$

or

$$2Na^+,2OH^- + SO_2 \longrightarrow 2Na^+,SO_3^{2-} + H_2O$$

SALTS AND NEUTRALIZATION

Neutralization is the union of an acid and a base to form a salt and water. From the ionic viewpoint, neutralization is essentially *the union of the hydrogen ion of an acid with the hydroxyl ion of a base to form un-ionized water;* thus

$$H^+ + OH^- \longrightarrow H_2O$$

Salts may be defined ionically as *the product of the union of a cation of a base with the anion of an acid.*

With few exceptions, salts are highly ionized in an aqueous solution. Besides the neutralization reaction, salts may be formed by the following reactions:

1. reaction of an acid and a metallic oxide
2. reaction of a base and a nonmetallic oxide
3. action of an acid on a metal
4. union of a metallic and a nonmetallic oxide; thus

$$CaO + CO_2 \longrightarrow CaCO_3$$

The salts are the most numerous of all inorganic compounds and many are of great industrial and biological importance.

Some general reactions of salts are as follows:

1. Action of a base and a salt to form a new base and a new salt:

$$FeCl_3 + 3KOH \longrightarrow \underline{Fe(OH)_3} + 3KCl$$

or

$$Fe^{3+},3Cl^- + 3K^+,3OH^- \longrightarrow \underline{Fe(OH)_3} + 3K^+,3Cl^-$$

2. The interaction of two salts will form two new salts if one is insoluble in the reacting medium.

$$NaCl + AgNO_3 \longrightarrow \underline{AgCl} + NaNO_3$$

or

$$Na^+,Cl^- + Ag^+,NO_3^- \longrightarrow \underline{AgCl} + Na^+,NO_3^-$$

or the "net" ionic equation

$$Cl^- + Ag^+ \longrightarrow \underline{AgCl}$$

3. The action of an acid and a salt to form a new acid and a new salt:

$$2NaCl + H_2SO_4 \longrightarrow \overline{2HCl} + Na_2SO_4$$
$$2Na^+,2Cl^- + 2H^+,SO_4^{2-} \longrightarrow \overline{2HCl} + 2Na^+,SO_4^{2-}$$

The acid produced must be volatile or insoluble.

AMPHOTERISM

The hydroxides of some of the elements with weak metallic properties react with both acids and bases to form salts. This phenomenon is called *amphoterism,* and the hydroxides are called *amphoteric hydroxides.*

Lead hydroxide, $Pb(OH)_2$, is amphoteric, because it reacts with hydrochloric acid to form lead chloride, $PbCl_2$, and with sodium hydroxide to form the salt sodium plumbite, Na_2PbO_2. In the former reaction, the lead hydroxide is acting as a weak base, whereas in the latter reaction, it is acting as a weak acid.

$$Pb(OH)_2 + 2HCl \longrightarrow PbCl_2 + 2H_2O$$
$$Pb(OH)_2 + 2NaOH \longrightarrow Na_2PbO_2 + 2H_2O$$

Other amphoteric hydroxides are those of the metals tin, arsenic, antimony, aluminum, zinc, and chromium.

Amphoteric hydroxides always exhibit the properties of a weak base and a weak acid.

Amino acids and proteins are organic substances that possess amphoteric properties (see Chapters 22 and 24).

IONIZATION EQUILIBRIUMS OF ACETIC ACID, OF AMMONIUM HYDROXIDE, AND OF THE POLYBASIC ACIDS, SULFURIC ACID AND PHOSPHORIC ACID

Acids that ionize to produce one hydrogen ion per molecule, such as acetic acid, $HC_2H_3O_2$, are called *monobasic acids.* Acids that produce two or more hydrogen ions per molecule, such as sulfuric acid and phosphoric acid, are called *polybasic acids.*

When acetic acid ionizes in an aqueous solution, an equilibrium is rapidly established between acetic acid molecules and the hydrogen and acetate ions (see page 133). This equilibrium is represented by the following equation:

$$H \cdot C_2H_2O_3 \rightleftharpoons H^+ + C_2H_3O_2^-$$

In a similar fashion, a molecule of a weak base, such as ammonium hydroxide, will exist in an aqueous solution in equilibrium with its constituent ions.

$$NH_4OH \rightleftharpoons NH_4^+ + OH^-$$

Sulfuric acid dissociates first into a hydrogen ion and a bisulfate ion, and these ions will be in equilibrium with undissociated sulfuric acid molecules.

$$H_2SO_4 \rightleftharpoons H^+ + HSO_4^-$$

This is called the *primary ionization* of sulfuric acid. Because the bisulfate ion is also an acid (because it dissociates in solution to give hydrogen ions), it dissociates into hydrogen and sulfate ions, and this ionization step is called the *secondary ionization* of sulfuric acid.

$$HSO_4^- \rightleftharpoons H^+ + SO_4^{2-}$$

In a polybasic acid, the secondary ionization is not as complete as the primary ionization. For this reason, bisulfate ions are weaker as an acid than sulfuric acid. Actually the latter is five to ten times as strong an acid as the bisulfate ion.

The primary ionization of phosphoric acid involves an ionic equilibrium between phosphoric acid molecules and hydrogen and dihydrogen phosphate, $H_2PO_4^-$, ions.

$$H_3PO_4 \rightleftharpoons H^+ + H_2PO_4^-$$

The dihydrogen phosphate ions further dissociate into hydrogen and monohydrogen phosphate, HPO_4^{2-}, ions.

$$H_2PO_4^- \rightleftharpoons H^+ + HPO_4^{2-}$$

This represents the secondary ionization of phosphoric acid. In the third ionization step, the monohydrogen phosphate ions are in equilibrium with hydrogen and phosphate ions.

$$HPO_4^{2-} \rightleftharpoons H^+ + PO_4^{3-}$$

Each of these stages of ionization occurs to a lesser extent than the previous one.

THE GENERALIZED THEORY OF ACIDS AND BASES

Although the definitions of an acid and a base given above will generally be satisfactory for the nonchemist, chemists have accepted a more generalized definition. This new definition is called the *generalized theory of acids and bases* of Brönsted and Lowry. Because paramedical and biological science students may occasionally meet discussions in which this newer concept of acids and bases is employed, it is considered here.

According to this generalized theory, acids are substances that can furnish protons, H^+; that is, *they are proton donors.* Bases are defined as those substances that combine with protons; that is, *they are proton acceptors.* Acetic acid is an acid because it furnishes a proton to a hydroxide ion. The latter is a base because it combines with a proton to form a molecule of water. Ammonia combines with protons to form the ammonium ion, NH_4^+; hence ammonia is a base. Under certain conditions, the ammonium ion can act as an acid by releasing a proton to form the base NH_3. Acids and bases can then be defined by the reversible general equation

$$H^+ + A^- \rightleftharpoons HA$$

in which A^- represents a base and HA represents an acid. In the reaction proceeding to the right, a base is combining with a proton to form an acid, whereas in the reverse reaction the acid is losing a proton to form a base. Such a base as A^-, derived from the acid HA by the loss of a proton, is called the *conjugate base* of the acid. The acetate ion, $C_2H_3O_2^-$, is the conjugate base of acetic acid, $HC_2H_3O_2$, while the chloride ion is the conjugate base of hydrochloric acid.

A number of chemical substances may act both as acids and bases. Water is such a substance. When hydrogen chloride is dissolved in water, protons are transferred to the water to form hydronium ions, and because water has accepted a proton, it has acted as a base.

$$HCl + H_2O \rightleftharpoons H_3O^+ + Cl^-$$

If acetate ions are added to water, the water will lose protons to the acetate ion, forming acetic acid, and the water will have acted as an acid.

$$C_2H_3O_2^- + H_2O \rightleftharpoons HC_2H_3O_2 + OH^-$$

The bisulfate ion, HSO_4^-, is both an acid and a base. As a base, albeit a very weak one, it accepts protons to form sulfuric acid.

$$HSO_4^- + H^+ \rightleftharpoons H_2SO_4$$

As an acid it furnishes protons to a base and forms sulfate ions.

$$HSO_4^- + OH^- \rightleftharpoons H_2O + SO_4^{2-}$$

The relative strength of acids, from the standpoint of this generalized theory, will depend upon the ease with which they lose protons. Hydrochloric acid and hydronium ions, H_3O^+, are strong acids, because they release protons readily. Acetic acid is a fairly weak acid, because it is reluctant to part with protons. Water is a very weak acid, because protons are not easily removed from water to form the hydroxyl ion. Both acetate and hydroxyl ions are strong bases, because they combine readily with protons to form acetic acid and water, respectively. Because it does not hold the proton strongly in the hydronium ion, water is a weak base. The chloride ion and the sulfate ion, which are the conjugate bases of hydrochloric and sulfuric acid, are extremely weak bases, because they exhibit little or no tendency to combine with protons in aqueous solutions to form molecules of the undissociated acid.

In many reactions, in solution two bases are in competition for protons and the stronger base will capture most of the protons. Thus hydrogen chloride loses protons to water to form the hydronium ion, because the chloride ion is a much weaker base than water.

$$HCl + H_2O \rightleftharpoons H_3O^+ + Cl^-$$

Ionization constants of acids, K_a, and bases, K_b

Because in the ionization of weak acids and weak bases a chemical equilibrium is rapidly established between the ions formed and the undissociated acid or base (see page 133), equilibrium constants for these ionizations have been experimentally determined. Equilibrium constants for ionization reactions are called *ionization constants* and are generally represented as K_a for weak acids and K_b for weak bases.

Because the ionization of acetic acid can be represented by the equation

$$HC_2H_3O_2 + H_2O \rightleftharpoons H_3O^+ + C_2H_3O_2^-$$

then

$$K = \frac{[H_3O^+] \times [C_2H_3O_2^-]}{[HC_2H_3O_2] \times [H_2O]}$$

The molar concentration of water is a large number (about 55 moles per liter) and does not change significantly during the ionization of acetic acid, so it can be considered a constant and combined with K to form K_a. Then

$$K_a = \frac{[H_3O^+] \times [C_2H_3O_2^-]}{[HC_2H_3O_2]}$$

The numerical value for this constant for acetic acid is about $1.8 \times 1/100{,}000$, or 1.8×10^{-5}. (See pages 555–558 for the significance of exponential numbers.)

ILLUSTRATIVE PROBLEM 1

What is the ionization constant (K_a) of isovaleric acid, $H \cdot C_5H_9O_2$, if 0.01 *M* solution of this acid ionizes to give 4×10^{-4} mole of H^+ per liter?

Because each molecule of this acid that ionizes gives one hydrogen ion and one isovalerate ion, then $[H^+] = [C_5H_9O_2^-] = 4 \times 10^{-4}$ mole/liter. This indicates that only a very few molecules of the acid ionize (4×10^{-4} mole/liter). Then no serious error is made if one assumes that the concentration of the undissociated isovaleric acid after ionization is equal to the initial concentration of this acid, or $[HC_5H_9O_2] = 0.01$, or 10^{-2} mole/liter in the expression for K_a, namely,

$$K_a = \frac{[H^+] \times [C_5H_9O_2^-]}{[HC_5H_9O_2]}$$

Substituting known values in this equation,

$$K_a = \frac{4 \times 10^{-4}\ \text{mole/liter} \times 4 \times 10^{-4}\ \text{mole/liter}}{10^{-2}\ \text{mole/liter}} = \frac{16 \times 10^{-8}}{10^{-2}}$$

$$= 16 \times 10^{-6} = 1.6 \times 10^{-5}$$

A weak base solution is formed when ammonia is passed into water, the formation of this basic solution may be represented by the following equation:

$$NH_3 + H_2O \rightleftharpoons NH_4^+ + OH^-$$

$$K = \frac{[NH_4^+] \times [OH^-]}{[NH_3] \times [H_2O]}$$

and because $[H_2O]$ can be considered to have a constant value, this expression becomes

$$K_b = \frac{[NH_4^+] \times [OH^-]}{[NH_3]}$$

and has a numerical value of about 1.8×10^{-5}.

Because a polybasic acid, such as phosphoric acid, ionizes in several stages (see page 139), ionization constants for the primary, secondary, and tertiary ionizations, represented by K_1, K_2, and K_3, respectively, can be established. Thus for phosphoric acid,

$$K_1 = \frac{[H_2PO_4^-] \times [H_3O^+]}{[H_3PO_4]} = 1.1 \times 10^{-2}$$

$$K_2 = \frac{[HPO_4^{-2}] \times [H_3O^+]}{[H_2PO_4^-]} = 7.5 \times 10^{-8}$$

$$K_3 = \frac{[PO_4^{-3}] \times [H_3O^+]}{[HPO_4^{-2}]} = 4.8 \times 10^{-12}$$

The following are ionization constants for some other weak acids:

hydrocyanic, HCN	$K_a = 4.9 \times 10^{-10}$
hydrofluoric, HF	$K_a = 3.5 \times 10^{-4}$
nitrous, HNO_2	$K_a = 4 \times 10^{-4}$

sulfurous, H_2SO_3	$K_1 = 1.7 \times 10^{-2}$
	$K_2 = 5 \times 10^{-6}$
carbonic, H_2CO_3	$K_1 = 4.3 \times 10^{-7}$
	$K_2 = 5.6 \times 10^{-11}$
hydrosulfuric, H_2S	$K_1 = 9.1 \times 10^{-8}$
	$K_2 = 1.1 \times 10^{-12}$
formic, HCOOH	$K_a = 1.8 \times 10^{-4}$

By comparing their ionization constants, one can establish the relative strength of weak acids. The weaker the acid, the smaller the ionization constant. Thus HF ($K_a = 3.5 \times 10^{-4}$) and HNO_2 ($K_a = 4 \times 10^{-4}$) are of about the same strength and are stronger acids than HCN ($K_a = 4.9 \times 10^{-10}$). By comparing the numerical values of the various ionization constants of di- and polybasic acids, as stated in a previous discussion (page 139), one can note that the secondary ionization occurs to a lesser extent than the primary, and the tertiary to a lesser extent than the secondary.

If the K_a of a weak acid or the K_b of a weak base can be established experimentally at one concentration, this numerical value can be used to calculate the concentration of ions and undissociated molecules of the acid or base in equilibrium at some other concentration of the acid or base.

ILLUSTRATIVE PROBLEM 2

The numerical value for the ionization constant (K_a) of hypochlorous acid, HClO, is 4×10^{-8}. Calculate the concentration of $[H^+]$ and $[ClO^-]$ in 0.01 *M* solution of this acid.

Because $[H^+]$ and $[ClO^-]$ will be equal, let
$[H^+] = [ClO^-] = x$ moles/liter

Few molecules ionize, so the concentration of undissociated hypochlorous acid in the solution can be taken equal to its initial concentration, namely $[HClO] = 0.01$ mole/liter $= 10^{-2}$ mole/liter.

Substituting into the ionization constant expression

$$K_a = \frac{[H^+] \times [ClO^-]}{[HClO]} = 4 \times 10^{-8}$$

the known and unknown quantities gives

$$\frac{x \text{ moles/liter} \times x \text{ moles/liter}}{10^{-2} \text{ mole/liter}} = \frac{x^2}{10^{-2}} = 4 \times 10^{-8}$$

$$x^2 = 4 \times 10^{-8} \times 10^{-2} = 4 \times 10^{-10}$$

$$[H^+] = [ClO^-] = x = 2 \times 10^{-5} \text{ mole/liter}$$

Note: The calculated pH of this solution is 4.7.

The hydrolysis of salts

The ions of a salt in an aqueous solution may combine with hydrogen or hydroxyl ions derived from the water. If hydrogen ions are removed by the ions of the salt, an excess of hydroxyl ions will remain and the solution will give a basic reaction to indicators. If hydroxyl ions are removed from water, the solution will be acidic due to the excess of hydrogen ions present. This reaction of the ions of a salt with water is called *hydrolysis*. Hydrolysis may be considered a reversal of the neutralization reaction. The reaction produced by a salt in solution will depend upon the strength of the acid and base from which the salt is derived.

A salt derived from a strong base and a weak acid will give a basic reaction in an aqueous solution. Sodium acetate and sodium carbonate are examples of such salts. Acetate ions combine with hydrogen ions derived from water and liberate hydroxyl ions.

$$C_2H_3O_2^- + H_2O \rightleftharpoons HC_2H_3O_2 + OH^-$$

Because the sodium ions will have little affinity for the hydroxyl ions, and because the latter will be in excess, the solution will give a basic test. Carbonate ions, CO_3^{2-}, combine with water to form bicarbonate ions and hydroxyl ions,

$$CO_3^{2-} + H_2O \rightleftharpoons HCO_3^- + OH^-$$

Therefore, its sodium salt will give a basic reaction in solution. Even though sodium bicarbonate, $NaHCO_3$, is an "acid" salt, its solution will be slightly basic due to hydrolysis.

A salt derived from a weak base and a strong acid will give an acidic reaction in an aqueous solution. In a solution of ammonium chloride, ammonium ions will combine with the hydroxyl ions of water,

$$NH_4^+ + H_2O \rightleftharpoons H^+ + NH_4OH$$

while the chloride ions have no tendency to abstract hydrogen ions from the water. Salts of aluminum, copper, iron, cobalt, nickel, etc., usually produce an acid reaction to litmus in aqueous solution. An aluminum ion evidently combines with six molecules of water to form a "hydrated" ion.

$$Al^{3+} + 6H_2O \rightleftharpoons \left[\begin{array}{ccc} H_2O & & H_2O \\ & \diagdown \quad \diagup & \\ H_2O & —Al— & H_2O \\ & \diagup \quad \diagdown & \\ H_2O & & H_2O \end{array}\right]^{3+}$$

One of the water molecules in this hydrate readily loses a hydrogen ion, and this liberated ion produces an acid reaction in the solution.

$$\begin{bmatrix} H_2O & & H_2O \\ & \diagdown \quad \diagup & \\ H_2O\text{—} & Al & \text{—}H_2O \\ & \diagup \quad \diagdown & \\ H_2O & & H_2O \end{bmatrix}^{3+} \longrightarrow \begin{bmatrix} H_2O & & OH \\ & \diagdown \quad \diagup & \\ H_2O\text{—} & Al & \text{—}H_2O \\ & \diagup \quad \diagdown & \\ H_2O & & H_2O \end{bmatrix}^{2+} + H^+$$

The aqueous solutions of salts derived from weak acids and weak bases will be nearly neutral in reaction. However, salts of this class are usually appreciably hydrolyzed in solution. The reaction of such a solution will depend upon the relative strength of the acid and the base from which the salt was formed. An example of a salt of this class is ammonium acetate. The ammonium ions and the acetate ions will remove about equal numbers of hydrogen and hydroxyl ions from the solution, and after the hydrolysis has occurred, the solution will remain very nearly neutral.

The aqueous solution of a salt derived from a strong acid and a strong base will be neutral. An example of such a salt is sodium chloride. Neither the sodium nor the chloride ions will have any tendency to combine with hydroxyl or hydrogen ions, respectively.

In this discussion, hydrolysis has been explained primarily on the basis of the classical theory of acids and bases. For an interpretation of hydrolysis in the terms of the generalized theory of acids and bases, the student is referred to a text in inorganic chemistry.

Nomenclature of inorganic compounds

THE NAMING OF ACIDS

Compounds of any kind that contain two different elements are called *binary* compounds. Binary acids are named by employing the prefix *hydro-*, the name or a shortened form of the name of the characteristic nonmetal, and the suffix *-ic* followed by the word "acid." Some examples follow:

HCl hydrochloric acid
hydro — chlor — ic acid
↑ ↑ ↑
prefix stem suffix
HI hydriodic acid
H_2S hydrosulfuric acid

Compounds that consist of three different elements are called *ternary* compounds. Most ternary acids contain hydrogen, a characteristic nonmetal, and oxygen. The suffix *-ic*, used to denote the most common ternary acid of a nonmetal, is added to the name or a shortened form of the name of the characteristic nonmetal followed by the word "acid." In the acid with one fewer oxygen atom than the common acid, the ending *-ic* is replaced by *-ous*. An acid that contains two fewer oxygen atoms than the common acid is given the prefix *hypo-* and the suffix *-ous* followed by the word "acid." An acid that contains one more oxygen atom than the common acid is given the prefix *per-* and the suffix *-ic* followed by the word "acid." Some examples of ternary acids and their names follow:

$HClO$	hypochlorous acid
$HClO_2$	chlorous acid
$HClO_3$	chloric acid
$HClO_4$	perchloric acid
H_2SO_3	sulfurous acid
H_2SO_4	sulfuric acid
H_2SO_5	persulfuric acid
H_3PO_2	hypophosphorous acid
H_3PO_3	phosphorous acid
H_3PO_4	phosphoric acid

The following are the common ternary acids of some of the important nonmetals: $HClO_3$, H_2SO_4, HNO_3, H_3PO_4, H_2CO_3, HIO_3, H_3AsO_4, and $HBrO_3$.

THE NAMING OF SALTS

Salts of binary acids are named by using the name of the metal plus a shortened form of the name of the nonmetal with the suffix *-ide*. The name of these salts always consists of two separate words.

$NaCl$	sodium chloride
KI	potassium iodide
$(NH_4)_2S$	ammonium sulfide

The salts derived from ternary acids are named by changing the *-ic* ending of acids to *-ate* and the *-ous* endings to *-ite*. The prefixes *hypo-* and *per-* are retained in the name of the *salts*. Some examples follow:

$NaClO$	sodium hypochlorite
$NaClO_2$	sodium chlorite
$NaClO_3$	sodium chlorate
$NaClO_4$	sodium perchlorate

THE NAMING OF BASES

Bases in which the valence of the metallic portion is not variable are named by using the name of the metal followed by the word "hydroxide." If the metallic ion in the base has a variable valence, a shortened form of the name of the metal with the suffix *-ic* is used for the higher valence, whereas in the lower valence the *-ous* ending is used.

$Ba(OH)_2$	barium hydroxide
$Sn(OH)_2$	stannous hydroxide
$Fe(OH)_2$	ferrous hydroxide
$LiOH$	lithium hydroxide
$Sn(OH)_4$	stannic hydroxide
$Fe(OH)_3$	ferric hydroxide

This distinction in valence is retained in the naming of the salts of metals with variable valence; thus

$CuCl$	cuprous chloride
$FeCl_2$	ferrous chloride
$CuCl_2$	cupric chloride
$FeCl_3$	ferric chloride

In inorganic compounds, where metals exhibit variable valence, the name of the metal followed by the Roman numeral to indicate its valence may be used in place of the usual *-ous* and *-ic* endings. The following examples are illustrative:

$FeCl_2$	iron(II) chloride
$SnBr_4$	tin(IV) bromide
$FeCl_3$	iron(III) chloride
CuS	copper(II) sulfide

THE NAMING OF OXIDES

The nomenclature of metallic oxides is similar to that of bases. The oxide in which the metal exhibits the higher valence is an *-ic* oxide, whereas the oxide in which the metal has a lower valence is the *-ous* oxide; for example,

FeO	ferrous oxide
Fe_2O_3	ferric oxide

To name nonmetallic oxides, one usually uses a prefix to denote the number of oxygen atoms in the molecule.

SO_2	sulfur dioxide
SO_3	sulfur trioxide
NO_2	nitrogen dioxide
CO	carbon monoxide

ACID AND BASIC SALTS

There is a group of inorganic compounds that have been formed by the partial neutralization of the hydrogens of an acid. Such substances still retain acid properties as well as the properties of salts. These are called *acid salts.* If only one of the hydrogens of sulfuric acid is neutralized by sodium hydroxide, an acid salt of the formula $NaHSO_4$ is formed. It may be called either sodium acid sulfate, sodium hydrogen sulfate, or sodium bisulfate. The *normal salt* of this acid is Na_2SO_4.

Because phosphoric acid, H_3PO_4, contains three hydrogens per molecule, it can be converted into two series of acid salts and a normal salt. The primary acid salt has the formula NaH_2PO_4 and is called sodium dihydrogen phosphate, while Na_2HPO_4 is the secondary acid salt and is called sodium monohydrogen phosphate. Both of these acid salts are important in maintaining the nearly constant acidity of the blood (page 161). Because of this property of maintaining a nearly constant acidity in solution, the combination of these two salts in solution is called a *buffer solution,* a term that will be explained more fully in a later chapter (page 160). Normal sodium phosphate, also called trisodium phosphate, has the formula Na_3PO_4.

If a base contains two or more hydroxyl groups per molecule, it may be neutralized in steps and a basic salt may be formed. These salts still retain the properties of a base as well as those of a salt. If one molecule of lead hydroxide, $Pb(OH)_2$, is neutralized by a molecule of nitric acid, a basic salt, $Pb(OH)NO_3$, is formed which is called basic lead nitrate.

$$Pb(OH)_2 + HNO_3 \longrightarrow Pb(OH)NO_3 + H_2O$$

Other examples of basic salts are $Bi(OH)Cl_2$ (monohydroxy bismuth chloride), $Bi(OH)_2Cl$ (dihydroxy bismuth chloride), $Mg(OH)Cl$ (basic magnesium chloride), and $Ca(OH)Cl$ (basic calcium chloride). The dehydration product of dihydroxy bismuth chloride is bismuthyl chloride ($BiOCl$).

$$Bi(OH)_2Cl \longrightarrow BiOCl + H_2O$$

STUDY QUESTIONS

1. Give examples of some electrolytes and some nonelectrolytes.
2. How did Faraday explain the ionization of electrolytes in solution?
3. Give evidences that electrolytes ionize in aqueous solutions.
4. Why does magnesium chloride depress the freezing point of water nearly three times as much as a solution of sugar that contains the same number of moles of the solute? Why is the freezing point depression of such a magnesium chloride solution not exactly three times that of the sugar solution?
5. How does a solution of sodium chloride conduct the electric current when placed between two electrodes?
6. Write an equilibrium equation to represent the ionization in an aqueous solution of (a) nitric acid, (b) calcium bromide, (c) potassium iodide, (d) magnesium sulfate, and (e) ammonium carbonate.
7. What three classes of chemical substances are usually ionized in their aqueous solutions?
8. Why is the aqueous solution of an electrolyte electrically neutral?
9. What is a cation? An anion? A cathode? An anode?
10. Distinguish between a potassium atom and a potassium ion; between a sulfur atom and a sulfide ion.
11. How rapid are ionic reactions? Why?
12. What are the main assumptions of the Arrhenius theory of ionization?
13. Define and explain the concept of the degree of ionization.
14. Metallic zinc will displace hydrogen from a dilute sulfuric acid solution more rapidly than from the concentrated acid (98 percent). Why?
15. Give the ionic definition of the following: an acid, a base, a salt, neutralization.
16. What is the essential reaction in neutralization? Write a simplified ionic equation for the neutralization reaction.

17. Why are some acids strong whereas others are weak? Give examples of both classes of acids.
18. How does the degree of ionization of a weak acid compare to that of a strong acid?
19. Why, and in what cases, have Debye and Hückel modified the ionization theory of Arrhenius?
20. How does the degree of ionization change with the dilution of a solution?
21. What are some of the other names for the Debye-Hückel theory of ionization?
22. The ionization of what type of substances may best be explained by the Arrhenius theory of ionization? By the Debye-Hückel theory of ionization?
23. Are the units present in the crystal lattice of sodium chloride molecules or ions?
24. In the Debye-Hückel theory of ionization, how is the apparent degree of ionization explained?
25. Briefly summarize the general properties of acids and bases.
26. What are the two classes of oxides, and how do they differ in their reaction with water? Give examples of these two classes of oxides.
27. Complete the following statements:

    (a) An acid and a base give ________________.
    (b) A metal and an acid give ________________.
    (c) A metal and a salt give ________________.
    (d) An acid and a carbonate give ________________.
    (e) A salt and a salt give ________________.
    (f) A metallic oxide and water give ________________.
    (g) An acid and a metallic oxide give ________________.
    (h) A salt and a base give ________________.
    (i) A metal and a nonmetal give ________________.
    (j) A nonmetallic oxide and water give ________________.

28. Which two of the reactions mentioned in Question 27 are somewhat limited in their general nature? What are the limitations in these two general reactions?
29. Complete and balance the following equations. Where no reaction occurs, indicate by the abbreviation N.R.

    (a) $K_2O + H_2O \longrightarrow$
    (b) $Zn + Cu(NO_3)_2 \longrightarrow$
    (c) $Ag + SnCl_2 \longrightarrow$
    (d) $BaO + HNO_3 \longrightarrow$
    (e) $FeCl_3 + NH_4OH \longrightarrow$
    (f) $NaNO_3 + H_2SO_4 \longrightarrow$
    (g) $MgCO_3 + HC_2H_3O_2 \longrightarrow$
    (h) $Sr(OH)_2 + H_3PO_4 \longrightarrow$
    (i) $AgNO_3 + NH_4I \longrightarrow$
    (j) $Cu + Cl_2 \longrightarrow$
    (k) $Hg + HCl \longrightarrow$
    (l) $As_2O_3 + H_2O \longrightarrow$
    (m) $Al + H_2SO_4 \longrightarrow$
    (n) $MgO + SO_2 \longrightarrow$
    (o) $KClO_3 + \text{heat} \longrightarrow$

30. Suggest first-aid treatment for an acid burn, for an alkali burn.
31. Suppose that *W*, *X*, *Y*, and *Z* are hypothetical elements. Complete the following equations that illustrate the reaction of these elements:

$W(OH)_3 + H_2X \longrightarrow$
$YCl + Ag_3Z \longrightarrow$
$BaX + Y_2SO_4 \longrightarrow$
$W_2(CO_3)_3 + H_3Z \longrightarrow$
$Y_2O + H_2X \longrightarrow$
$WCl_3 + YOH \longrightarrow$

32. What is meant by hydrolysis?
33. A solid substance is not an acid. When dissolved in water, however, it changes the color of blue litmus to red. Explain.
34. Divide salts into four classes depending upon hydrolysis and the reaction of their aqueous solutions with litmus.
35. Make a list of the strong acids and another list of strong bases. Use these two lists in answering Question 36.
36. What reaction to litmus will the aqueous solutions of the following salts produce?

(a) potassium carbonate
(b) ammonium sulfate
(c) calcium chloride
(d) sodium dihydrogen phosphate
(e) sodium bicarbonate
(f) copper sulfate
(g) sodium acetate
(h) aluminum nitrate
(i) zinc chloride

37. What are indicators? Give some examples of indicators and state their color when present in solutions of acids and bases.
38. What are amphoteric substances? Give examples.
39. Are amphoteric hydroxides ever strong bases? Strong acids?
40. Write equations for the reaction of HCl and NaOH with the amphoteric hydroxide $Zn(OH)_2$.
41. Indicate the different stages in the ionization of sulfuric acid; of phosphoric acid. What is meant by the primary and secondary ionization of sulfuric acid?
42. Explain why we consider a bisulfate ion an acid. How does it compare in strength as an acid with an equal concentration of sulfuric acid?
43. Give an example of a monobasic acid; a polybasic acid.
44. The common ternary acids of Cl, S, P, N, C, Si, As, Br, I Mn, and Cr are $HClO_3$, H_2SO_4 H_3PO_4, HNO_3, H_2CO_3, H_2SiO_3 H_3AsO_4, $HBrO_3$, HIO_3, $HMnO_3$, and H_2CrO_4. Give suitable names to each of these acids.
45. Write formulas for the following compounds:

hypochlorous acid	ferrous sulfate
sulfurous acid	silver nitrite
hydrobromic acid	barium sulfite
periodic acid	magnesium nitride

persulfuric acid	ferric nitrate
chlorous acid	mercurous chloride
arsenious acid	calcium hypochlorite
sodium bromide	zinc hydroxide
potassium bromite	ammonium fluoride

46. Name the following:

HIO_3	HI
H_3AsO_3	$HMnO_4$
$Zn(NO_2)_2$	HgO
SnO_2	$Mg(HCO_3)_2$
$(NH_4)_2HPO_4$	$AlBr_3$
Li_2CO_3	SrS
SO_3	$FeCl_2$
$Pb(OH)Cl$	NaH_2PO_4

47. Name the following ions:

IO_3^-	Cu^{2+}
P^{3-}	ClO_2^-
SO_3^{2-}	NO_3^-
HCO_3^-	$H_2PO_4^-$
BrO^-	AsO_3^{3-}
SiO_3^{2-}	Fe^{2+}
Sn^{4+}	Hg^+

48. For the following compounds, insert the proper subscript numbers:

(NH_4^+)____(CO_3^{2-})____	Ba^{2+}____(OH^-)____
Al^{3+}____(SO_4^{2-})____	Sn^{4+}____(PO_4^{3-})____
Fe^{3+}____(ClO_3^-)____	Zn^{2+}____(NO_3^-)____
K^+____S^{2-}____	Ag^+____(BO_3^{3-})____
Na^+____(PO_4^{3-})____	Sr^{2+}____(SO_4^{2-})____
Ca^{2+}____(SO_4^{2-})____	Mg^{2+}____(BO_3^{3-})____
Cr^{3+}____(OH^-)____	Fe^{3+}____(CO_3^{2-})____

49. What are acidic salts? Give an example of each.
50. Write an ionization constant (K_a) expression for each of the following weak monobasic acids: propionic acid ($H \cdot C_3H_5O_2$); nitrous acid (HNO_2); benzoic acid ($H \cdot C_7H_5O_2$); hydrocyanic acid (HCN); and hypochlorous acid ($HClO$).
51. The equation

$$CH_3NH_2 + H_2O \rightleftharpoons CH_3\overset{+}{N}H_3 + OH^-$$

represents the ionization of the weak base methylamine. Write an expression for the K_b of this weak base. Recall that the concentration of water (55 moles/liter) will be considered as remaining constant during ionization.
52. Write an expression for the secondary ionization constant (K_2) for the following weak polybasic acids: (a) sulfurous acid (H_2SO_3); (b) arsenic acid (H_3AsO_4); (c) oxalic acid ($H_2C_2O_4$); (d) carbonic acid (H_2CO_3); (e) phosphorous acid (H_3PO_3).

53. Based on the numerical value of their ionization constants, arrange the following acids in order of *decreasing* acidity:

ACID	K_a
chloroacetic acid	1.6×10^{-3}
hydrocyanic acid	4.9×10^{-10}
hypochlorous acid	4×10^{-8}
acetic acid	1.8×10^{-5}
formic acid	1.8×10^{-4}
picolinic acid	3×10^{-6}

54. Based on the numerical value of their ionization constants, arrange the following bases in order of *increasing* basicity:

BASE	K_b
diethylamine	1.3×10^{-3}
methylamine	5×10^{-4}
aniline	5×10^{-10}
ammonia	1.8×10^{-5}
cinchonine	1.6×10^{-7}

PROBLEMS

1. Determine the numerical value of the ionization constant (K_a) for lactic acid, $H \cdot C_3H_5O_3$, if in a 0.1 N solution $[H_3O^+] = [C_3H_5O_3^-] = 4 \times 10^{-3}$ mole/liter. Assume that the concentration of the undissociated lactic acid present after ionization is equal to the original concentration of this acid placed in solution (because only a minute amount of the lactic acid will have ionized).
2. In an aqueous solution 1 M with trimethylamine, $(CH_3)_3N$, the trimethylammonium, $(CH_3)_3\overset{+}{N}H$, and hydroxyl ion OH^-, concentrations will each equal 9×10^{-3} mole/liter. Determine the numerical value of the ionization constant (K_b) for this weak base.
3. Determine the concentration of dimethylammonium ion, $(CH_3)_2\overset{+}{N}H_2$, and hydroxyl ion, OH^-, in a 0.01 M dimethylamine solution. The numerical value for the ionization constant for this weak base is 8×10^{-4}. *Suggestion:* Because the two ionic concentrations will be equal, let $[(CH_3)_2\overset{+}{N}H_2] = [OH^-] = x$ moles/liter.
4. Calculate the concentration of H_3O^+ and NO_2^- in a 1 M nitrous acid, HNO_2, solution. The K_a for HNO_2 is 4×10^{-4}. *Suggestion:* $[H_3O^+] = [NO_2^-] = x$ moles/liter.

chapter 10

hydrogen ion concentration, pH, and buffer solutions

10

The dissociation of water into hydrogen and hydroxyl ions

Although water is usually considered to be undissociated and not a conductor of electricity, it actually ionizes to a very slight extent. In each liter of pure water, 1/10,000,000 mole of water dissociates into 1/10,000,000 mole*, or $1/10^7$ (10^{-7}), mole of hydrogen ions and 1/10,000,000 or 10^{-7} mole of hydroxyl ions; thus

$$H_2O \rightleftharpoons H^+ + OH^-$$

(At this time, the student should review the discussion of exponents, powers, and exponential numbers in Appendix A.)

The expression for the equilibrium constant for this ionization of water will be (see page 125)

$$K_{eq} = \frac{[H^+] \times [OH^-]}{[H_2O]}$$

Because $[H_2O]$ will equal about 55 moles per liter and will not change appreciably when water ionizes, it can be considered a constant. Then

$$[H^+] \times [OH^-] = 55 \times K_{eq} = K_w$$

Thus in water, the product of the hydrogen and hydroxyl ion concentration is equal to a constant K_w, which is called the *ion-product constant of water*. Substituting 10^{-7} mole/liter for the hydrogen and hydroxyl ion concentration in the K_w expression above gives

$$K_w = 10^{-7} \times 10^{-7} = 10^{-14}$$

because it is necessary to add exponents only when multiplying exponential numbers.

The relationship that the product of the hydrogen and hydroxyl ion concentrations expressed in moles per liter is equal numerically

*I am using mole to designate not only gram-molecular weights of compounds but also gram-ionic weight of ions.

to the constant 10^{-14} holds not only for pure water but also for aqueous solutions of acids and bases. If an acid solution contains 10^{-3} mole of hydrogen ions, then it will contain 10^{-11} mole of hydroxyl ions, because

$$10^{-3} \times [OH^-] = K_w = 10^{-14}$$

and

$$[OH^-] = \frac{10^{-14}}{10^{-3}} = 10^{-11}$$

To divide 10^{-14} by 10^{-3} it is necessary only to subtract the exponent -3 from -14, or to add $+3$ to -14.

It should now be evident that an acid solution does not contain hydrogen ions exclusively, but also hydroxyl ions. In an acid solution, however, there is a greater number of hydrogen than hydroxyl ions. A basic solution that contains 10^{-2} mole of hydroxyl ions will contain 10^{-12} mole of hydrogen ions. All basic solutions contain hydrogen ions, but in such solutions the hydroxyl ions are in excess. A solution in which the concentrations of these two ions are equal is neutral.

Hydrogen ion concentration and pH

The relative strength of acid solutions can be indicated by means of the hydrogen ion concentration. An acid solution that contains 10^{-2} mole of hydrogen ions per liter is ten times as strong an acid as one that contains 10^{-3} mole per liter (Figure 10.1).

Sørensen, a Swedish biologist, introduced a new method of expressing the strength of an acid. He called this new notation the pH, and this is usually preferred to the term *hydrogen ion concentration* by both biologists and chemists.

The pH is the exponent of the hydrogen ion concentration with its sign reversed. Thus an acid solution with a hydrogen ion concentration of 10^{-2} has a pH of 2, while one with a concentration of 10^{-5} mole of hydrogen ions per liter has a pH of 5. A basic solution that contains 10^{-5} mole of OH^- has a hydrogen ion concentration of 10^{-9} and a pH of 9.

Because pure water contains 10^{-7} mole of hydrogen ions, it has a pH of 7. Basic solutions contain less than 10^{-7} mole of hydrogen ions, and their pH will be greater than 7. In acid solutions, the hydrogen ion concentration is greater than 10^{-7}, and the pH will be numerically less than 7.

One can easily fall into error when using the pH notation. It should be remembered that the pH decreases numerically as the acidity of the solution increases. Furthermore, a solution with a pH of 3 is actually ten times stronger an acid than a solution with a pH of 4, and a hundred times as strong an acid as one with a pH of 5.

The mathematical expression that relates the pH with the hydrogen ion concentration is $pH = \log 1/[H^+]$. If a solution contains

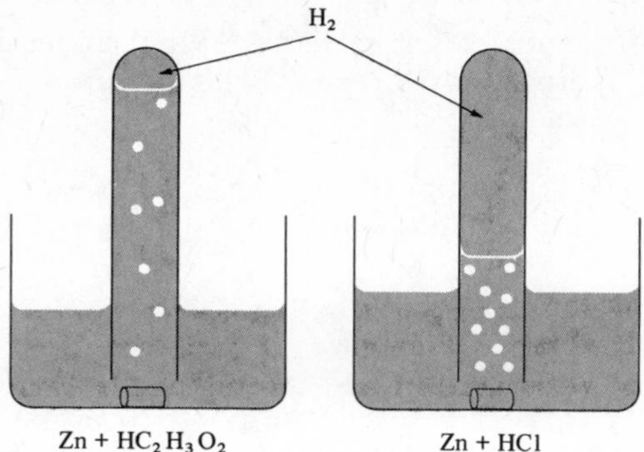

Figure 10.1 The relative amount of ionization of two acids can be approximately determined by comparing the volume of hydrogen that each liberates when they are reacted with an active metal for an equal period of time.

10^{-4} mole of hydrogen ions per liter, then

$$pH = \log 1/10^{-4} = \log 10^4$$

Because the logarithm of a number is the exponent (to base 10) of that number, then the pH of this solution is 4.

If fractional, the pH can usually be approximated to a whole-number value. The following solved problem illustrates the calculation of a fractional pH from the known hydrogen ion concentration.

ILLUSTRATIVE PROBLEM 1

The hydrogen ion concentration of a solution is 0.006 mole per liter. What is the pH?

The pH of this solution is equal to log 1/0.006, or −log 0.006. Note that 0.006 can be placed in the numerator if the sign of the logarithm is changed. Then

$$-\log 0.006 = -\log (6 \times 10^{-3}) = -\log 10^{-3} - \log 6$$

The log of $10^{-3} = -3$, and the log of 6 is 0.778 (see the log table on pages 560–561); thus

$$\begin{aligned} pH &= -(-3) - 0.778 \\ &= 3 - 0.778 = 2.222 \end{aligned}$$

Relationships between hydrogen ion concentration, hydroxyl ion concentration, and pH are illustrated in Table 10.1.

EXPERIMENTAL DETERMINATION OF pH

Two methods are used to determine pH experimentally, the *electrometric* and the *colorimetric* methods.

In the electrometric method, an instrument usually called the pH *meter* (Figure 10.2) is employed. The sample is generally placed in a

TABLE 10.1 SOME IMPORTANT RELATIONSHIPS BETWEEN HYDROGEN ION CONCENTRATION [H+], HYDROXYL ION CONCENTRATION [OH−], and pH

[H+]	pH	[OH−]	[H+] × [OH−] = K_w	Reaction
10^{-2}, or 0.01	2	10^{-12}	10^{-14}	Acid
10^{-4}, or 0.0001	4	10^{-10}	10^{-14}	Acid
10^{-7}	7	10^{-7}	10^{-14}	Neutral
10^{-9}	9	10^{-5}	10^{-14}	Basic
10^{-12}	12	10^{-2}	10^{-14}	Basic

small vessel, which is placed in the instrument. A button is pushed and a needle moves to a point on a scale from which the pH can be read directly. The operation of this instrument is based on the fact that the voltage of an electric current that passes through the test solution is dependent upon the pH of this solution.

The colorimetric method of measuring pH depends upon the fact that although indicators (page 135) show a characteristic color in certain ranges of acidity and another in certain ranges of alkalinity, over a certain pH range (usually not at the neutral point) there is a gradual transition between the two colors (see Table 10.2). An indi-

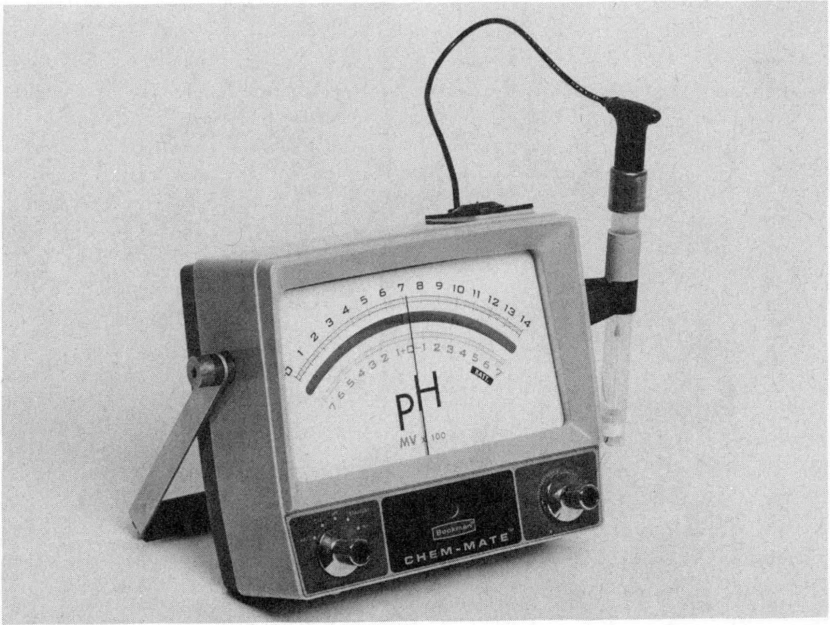

Figure 10.2 The Beckman laboratory pH meter. This instrument will accurately determine pH to 0.02 unit. (Courtesy Beckman Instruments, Inc.)

TABLE 10.2 DEPENDENCE OF COLOR OF INDICATORS ON pH RANGE

Indicator	pH *Range of Color Change*	*Change in Color in this* pH *Range*	*Color in Solutions of Lower* pH *than Transition* pH *Range*	*Color in Solutions of Higher* pH *than Transition* pH *Range*
Bromcresol purple	5.2–6.8	Yellow to purple	Yellow	Purple
Bromcresol green	3.8–5.4	Yellow to green	Yellow	Green
Litmus	4.5–8.3	Red to blue	Red	Blue
Methyl orange	3.1–4.4	Orange to yellow	Orange-red	Yellow
Methyl red	4.2–6.3	Red to yellow	Red	Yellow
Phenolphthalein	8.2–10	Colorless to red	Colorless	Red
Phenol red	6.8–8.4	Yellow to red	Yellow	Red
Thymol blue[a]	8–9.6	Yellow to blue	Yellow	Blue
Thymol blue[b]	1.2–2.8	Red to yellow	Red	Yellow

[a]Alkaline range.
[b]Acid range.

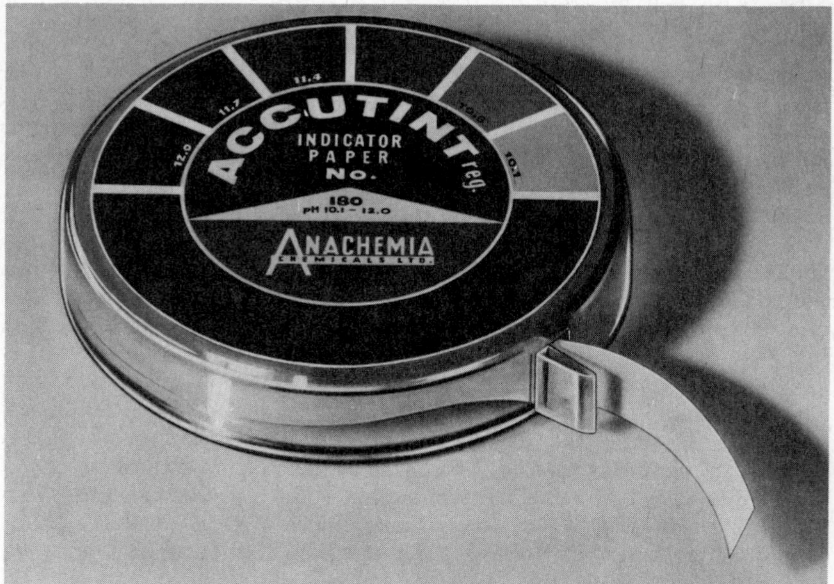

Figure 10.3 *Indicator paper used in the rapid determination of pH. (Courtesy Anachemia Chemicals Ltd., Quebec, Canada.)*

cator must be so chosen that it exhibits this color change in the range of the pH of the solution to be tested. A series of standard pH solutions in this range are prepared; these usually vary by 0.1 pH unit and the indicator is added to them. A solution of the test sample and the indicator are prepared under similar conditions, and the pH is determined by matching the color of the resulting solution with that of one of the standard pH solutions. The dependence of the color of certain indicators on pH is indicated in Table 10.2.

To simplify the colorimetric method, the set of pH standard solutions has sometimes been replaced by suitable colored plates. Also, a small strip of indicator paper can be used (Figure 10.3) by placing a drop of the test solution on it and immediately matching it with a color standard furnished with the test paper.

Although the colorimetric method of pH determination is much more economical to perform than the electrometric method, the latter is more rapid and accurate and is preferred when it is necessary to measure the pH of deeply colored solutions.

The common ion effect

In the ionization equilibrium for acetic acid,

$$H \cdot C_2H_3O_2 \rightleftharpoons H^+ + C_2H_3O_2^-$$

if the concentration of any one of the constituents of this system, such as hydrogen ions, acetate ions, or undissociated acetic acid molecules, is changed, then the concentrations of the other two constituents will adjust themselves in such a fashion that a new equilibrium will be established (see Chapter 8). If more acetate ions are added to the original equilibrium, the concentration of the hydrogen ions will decrease and the concentration of undissociated acetic acid molecules will increase until a new equilibrium is established. The acetate ion concentration may be increased by adding sodium acetate, which is highly ionized. The effect of adding this salt to an acetic acid solution will be to decrease the ionization of the acetic acid and the acidity of the solution (because hydrogen ion concentration is decreased). A change such as this, which results from the addition of a salt that furnishes an ion identical with the anion of a weak acid, is called the *common ion effect.*

A weak base will also exhibit the common ion effect if a salt with a common ion is added to the aqueous solution of the base. Because ammonium chloride, NH_4Cl, dissociates in solution into an ammonium ion, NH_4^+, which is a common ion to the base ammonium hydroxide, the addition of ammonium chloride to a solution of ammonium hydroxide will repress the ionization of the latter. The

$$NH_4OH \rightleftharpoons NH_4^+ + OH^-$$

resulting solution will be a weaker base, because the hydroxyl ion concentration will also be decreased.

Buffer solutions

Buffer solutions are solutions of combinations of chemicals that *resist change in basicity or acidity.* Solutions of a weak acid with a salt of the weak acid, or of a weak base with a salt of the base, usually serve as buffer solutions.

If concentrated hydrochloric acid is added to a buffer solution that contains acetic acid and sodium acetate, most of the added hydrogen ions from the hydrochloric acid will combine with the excess of acetate ions in solution to form slightly dissociated acetic acid. The hydrogen ion concentration will not be appreciably altered from that of the original buffer solution, and the acidity of the solution will remain almost unchanged.

$$\begin{array}{llll} NaC_2H_3O_2 \rightleftharpoons Na^+ + & C_2H_3O_2^- & \\ & + & \\ HCl \rightarrow & H^+ & + Cl^- \\ & \upharpoonleft\downharpoonright & \\ & H \cdot C_2H_3O_2 & \end{array}$$

If hydroxyl ions are added to the buffer solution of acetic acid and sodium acetate, the hydroxyl ions will combine with the free hydrogen ions produced by the ionization of the acetic acid, and water will be formed. As free hydrogen ions are removed from the solution, more molecules of acetic acid will dissociate into hydrogen and acetate ions. This dissociation of the acetic acid will tend to maintain a nearly constant hydrogen ion concentration, and the solution, at most, will show a very slight decrease in acidity.

$$\begin{array}{lcl} H \cdot C_2H_3O_2 \rightleftharpoons & H^+ & + C_2H_3O_2^- \\ & + & \\ NaOH \rightarrow Na^+ + & OH^- & \\ & \downarrow & \\ & H_2O & \end{array}$$

In similar fashion, the addition of a strong acid or a strong base will not appreciably change the basicity of a buffer solution prepared by adding ammonium chloride to a solution of ammonium hydroxide.

For a method of calculating the pH of a buffer solution, knowing the concentration of weak acid and its salt, see Appendix B.

The importance of pH control and of buffer action in biological processes

Because bacteria grow best in solutions in which there is a narrow range of acidity or basicity, the pH of culture media in which bacteria are to be grown must be carefully controlled. In sterilization, the choice of pH is also important, because bacteria are destroyed more readily at a certain pH. In staining bacteria and tissues, best results are obtained when the pH is also carefully controlled.

All enzymes show activity between certain pH ranges. The pH at which the enzyme exhibits its greatest activity is called the *optimum* pH for that enzyme. Pepsin, a protein-cleaving enzyme present in the gastric juice of the stomach, has an optimum pH of about 2 and is active in moderately strong acid solutions of pH 0.1 to 4. The enzyme trypsin, which digests proteins in the small intestine, is active in a weakly alkaline medium and exhibits maximum activity at pH 8 to 9. Pepsin will be inactive in an alkaline solution, whereas trypsin will be inactive in a strong acid solution.

The pH of most body fluids will be nearly constant or will vary between very narrow limits. Saliva has a pH of about 7, whereas gastric juice has a pH of about 2. The intestinal juices are slightly alkaline. Although urine is usually very slightly acid, it may give an alkaline reaction following a meal (alkaline tide). Refer to Figure 10.4 for pH values of other substances.

The body has mechanisms for maintaining the pH of the blood between about 7.3 and 7.5. Variations from these limits will produce coma and death if below 7.3 or tetany and death if above 7.5. A number of buffer systems present in the body assist in maintaining the nearly constant pH of the blood. Some of these buffers consist of phosphate salts, carbonic acid, bicarbonates, and the sodium salts of the proteins. Acids produced in the body are, in part, neutralized by HPO_4^{2-} ions, which are thereby converted into $H_2PO_4^-$ ions; any large excess of the latter is eliminated in the urine by the kidneys.

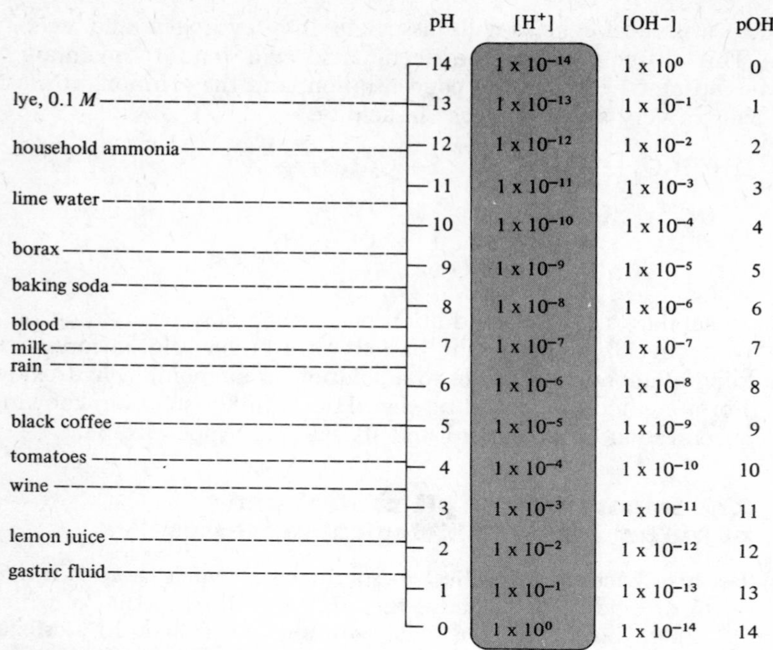

***Figure 10.4** The pH values of some common substances. (Courtesy Beckman Instruments, Inc.)*

On the other hand, $H_2PO_4^-$ ions will neutralize any base that may enter the blood. In this neutralization, HPO_4^{2-} ions will be produced, the excess of which will be eliminated in the urine (see pages 475–476).

STUDY QUESTIONS

1. What kind of ions are present in water? How many?
2. What is meant by K_w? What is its numerical value?
3. How is the pH related to the hydrogen ion concentration. To the hydroxyl ion concentration?
4. Why has the pH notation been introduced to represent the acidity and basicity of substances?
5. Why is water a neutral substance?
6. What is the pH of a neutral solution? What is the range of pH of acid solutions? Of basic solutions?
7. Would a dilute vinegar solution have a pH greater or less than 7? Is the pH of limewater, $Ca(OH)_2$, greater or less than 7?
8. What is meant by a common ion? Give an example.
9. What is the common ion effect? What effect does the common ion have on the acidity of a weak acid? On the basicity of a weak base?
10. What are buffer solutions? Give a suitable example of a buffer system, and explain how it acts as a buffer.
11. What would be the effect of adding a solution of sodium acetate to a solution of sulfuric acid?

12. Indicate which of the following can act as a buffer system:

(a) $NaCl + H \cdot C_2H_3O_2$
(b) $NaCl + HCl$
(c) $NaC_2H_3O_2 + H \cdot C_2H_3O_2$
(d) $KNO_3 + HCl$
(e) $KH_2PO_4 + K_2HPO_4$
(f) $K_2SO_4 + NaHCO_3$
(g) $Na_2CO_3 + NaHCO_3$

13. How will the addition of a strong acid or a strong base affect a buffer solution?
14. Why is the body so dependent on buffer systems? Indicate a few buffer systems that are present in the body fluids.
15. What is the normal pH of the blood plasma? What symptoms will result if there is a marked change in the pH of the blood?
16. Of what importance are pH and buffers in bacteriology? In sterilization? In digestion?
17. Explain fully the chemical changes that occur when sodium hydroxide is added to a 1 *M* NH_4OH solution in which a mole of NH_4Cl is dissolved. What changes occur when a hydrochloric acid solution is added to such a solution?
18. What chemical substance must be added to each of the following to make a satisfactory buffer solution: acetic acid, $H \cdot C_2H_3O_2$; ammonium chloride; NH_4Cl; potassium dihydrogen phosphate, KH_2PO_4?

PROBLEMS

1. Write exponential numbers for the following: 1000, 1/100, 100,000, and 1/10,000.
2. Complete the following mathematical operations:

(a) $10^3 \times 10^4 =$
(b) $10^4 \times 10^{-9} =$
(c) $10^2/10^{11} =$
(d) $10^{-3}/10^{-2} =$
(e) $10^4 \times 10^9 =$
(f) $10^{-3} \times 10^{-7} =$
(g) $10^2/10^{-11} =$
(h) $10^{-20}/10^{-13} =$

3. What is the pH of a solution that contains 10^{-6} mole of hydrogen ions per liter?
4. What is the pH of a solution that contains 10^{-5} mole of hydroxyl ions per liter?
5. What is the pH of a 0.001 *M* solution of hydrochloric acid? Assume complete ionization of the acid.
6. What is the hydroxyl ion concentration of a solution that has a pH of 5?
7. How many more H+ ions are present in a solution that has a pH of 2 than one that has a pH of 6?
8. 0.1 *N* acetic acid solution is ionized approximately 1 percent. What is the approximate pH of this solution?
9. Using the four-place logarithm table (Appendix A, pages 560 and 561), determine the pH of the following solutions:

(a) $[H^+] = 1.8 \times 10^{-5}$
(b) $[OH^-] = 1.8 \times 10^{-5}$
(c) $[H^+] = 7 \times 10^{-8}$
(d) $[OH^-] = 4 \times 10^{-11}$

chapter 11

some important nonmetals and their compounds

11

In this chapter, the important nonmetals, fluorine, chlorine, bromine, iodine, nitrogen, sulfur, carbon, and phosphorus will be briefly described. The names, formulas, and some uses of their more important compounds will also be included. The first four nonmetals listed above are called the *halide elements,* or the *halogens,* because they all form salts similar to sodium chloride, which is also called halite. The halogens, which appear in Group VII of the periodic table, resemble each other closely in chemical properties and represent an important family of elements. This family relationship is not unexpected, because the atoms of all four of these elements possess seven electrons in the outer or valence shell. They all form acids with the general formula HX (where X represents an atom of the halogen), and their sodium salts have the general formula NaX. Their most important valence is −1, although a number of other valence states are known for these elements.

For a more extensive discussion of these nonmetals and their compounds, the student is referred to the many texts in inorganic chemistry.

The halogens

FLUORINE, F_2, AND ITS COMPOUNDS

Fluorine is a very active pale yellow gas; it does not occur in the free state. It is prepared by the electrolysis of potassium acid fluoride, KHF_2, dissolved in hydrogen fluoride, HF (Figure 11.1). It combines readily with hydrogen to form the gas hydrogen fluoride. With water it combines spontaneously and explosively to form hydrogen fluoride and possibly fluorine oxide, F_2O.

Hydrogen fluoride is a colorless gas that dissolves readily in water to form the weak acid hydrofluoric acid. This acid attacks glass and is used in the etching of thermometers and laboratory glassware. Calcium fluoride, CaF_2, or *fluorite,* occurs in nature and is the most important source of fluorine and its compounds. Sodium fluoride is an important insecticide. The presence of a small amount of fluorides in water supplies appears to reduce the incidence of cavities in the teeth of children. Small amounts of soluble fluorides are now being

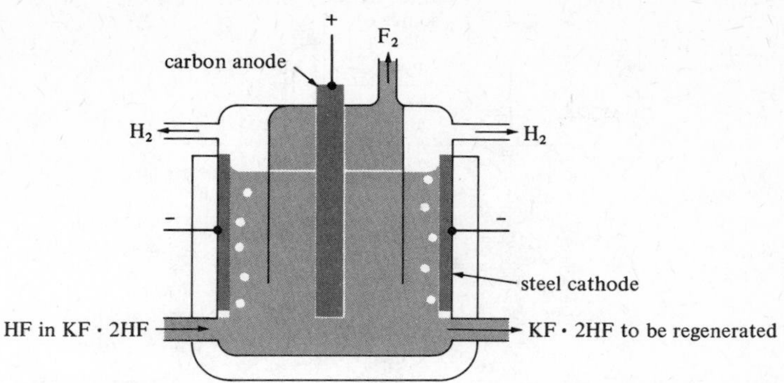

Figure 11.1 Fluorine is produced by the electrolysis of HF in molten KF·2HF. (From C. W. Keenan and J. H. Wood, General College Chemistry, *4th ed., Harper & Row, N.Y., 1971.)*

added to some water supplies deficient in fluorine. A high concentration of fluorides in water produces an undesirable mottling of the teeth. A small amount of fluorides is normally present in the inorganic matrix of the teeth (see page 484).

CHLORINE, Cl_2

Chlorine is the most important of the halogens. It is a yellowish-green gas with a sharp odor and is quite irritating to the mucous membrane of the respiratory tract. Breathing moderate amounts of this gas may be fatal; it was one of the first substances to be used as a toxic war gas. Although an active substance chemically, it is not as active as fluorine. It may be prepared by oxidizing hydrochloric acid with a suitable oxidizing agent or by the electrolysis of an aqueous solution or fused sodium chloride (Figure 11.2). Its most important uses are in bleaching, in disinfecting water supplies and sewage, and in the preparation of many chlorine-containing compounds.

COMPOUNDS OF CHLORINE

Hydrogen chloride, HCl, is a colorless gas that fumes in moist air. It is prepared by heating sodium chloride with concentrated sulfuric acid,

$$NaCl + H_2SO_4 \longrightarrow NaHSO_4 + HCl$$

or by the direct union of hydrogen and chlorine. Its aqueous solution is the strong acid called hydrochloric acid. This acid is essential for gastric digestion. The most important salt of this acid is sodium chloride, NaCl. This compound is indispensable in the proper functioning of the tissues of the body, and its solutions are frequently employed in intravenous injections. Other important salts of hydrochloric acid are the mercurous salt, Hg_2Cl_2, called *calomel,* which is

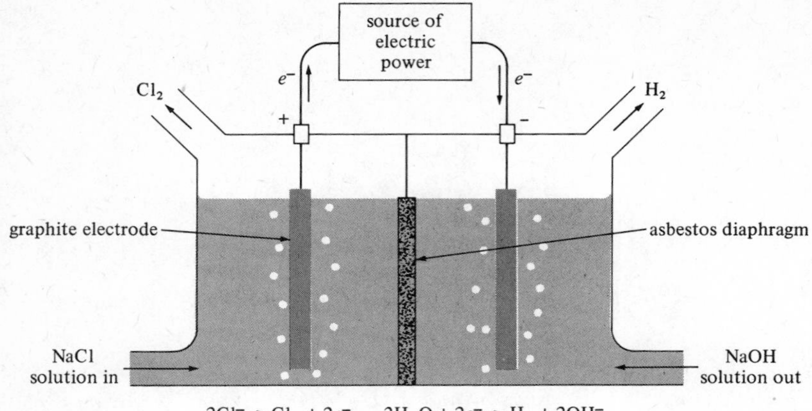

***Figure 11.2** Schematic representation of the electrolysis of concentrated aqueous sodium chloride. From C. W. Keenan and J. H. Wood,* General College Chemistry, *4th ed., Harper & Row, N.Y. 1971.)*

used as a cathartic, and the mercuric salt, $HgCl_2$, called *corrosive sublimate,* which is quite toxic and is used as an antiseptic.

Hypochlorous acid, HClO, is formed in small amounts when chlorine is passed into water. This reaction is reversible,

$$Cl_2 + H_2O \rightleftharpoons HCl + HClO$$

and if any substance that is easily oxidized is present, the hypochlorous acid loses its oxygen to this substance and the reaction proceeds to the right until all the chlorine is consumed. In bleaching with chlorine, the hypochlorous acid must be the active agent, because dry chlorine has no bleaching properties. Hypochlorous acid is also a good disinfecting agent. Both sodium and calcium hypochlorites are commercially available for bleaching and disinfecting applications. A sodium hypochlorite solution is sometimes known as *Javelle water.*

The potassium salt of chloric acid has the formula $KClO_3$ and is used in explosives, in mouthwashes and gargles, and in some toothpastes. When taken internally, it is toxic.

BROMINE, Br_2, AND SOME OF ITS COMPOUNDS

Bromine is a volatile reddish-brown liquid with a sharp irritating odor. It attacks the mucous membrane in the nose and throat and produces a copious flow of tears. Bromine occurs as bromide salts in seawater and in salt brines obtained from certain wells in Michigan.

Hydrogen bromide, HBr, is a colorless gas that fumes strongly in moist air. It dissolves readily in water, and the resulting solution is a strong acid called *hydrobromic acid.* Sodium, ammonium, and

Figure 11.3 Apparatus (left) for the purification of iodine, and a close-up view (right) of crystals of pure iodine obtained by sublimation. (Courtesy Beckwith Studios.)

potassium bromide are used in medicine as sedatives. Silver bromide, AgBr, is a light sensitive substance used on photographic plates.

IODINE, I_2, AND ITS COMPOUNDS

Iodine is a solid element and is usually obtained as shiny, black crystals. When warmed, the solid iodine is converted directly to a purple vapor (Figure 11.3), a process called *sublimation*. On cooling, the vapor condenses to form small iodine crystals; this is the usual method employed to purify iodine. Iodine is obtained from the ashes of seaweed and from the brine obtained from certain oil wells in the western United States. Another source of iodine is the saltpeter deposits of Chile, where it occurs as sodium iodate, $NaIO_3$. Iodine produces a deep blue-black color with starch, and this serves as a sensitive test for iodine.

A solution of iodine and potassium iodide in an alcohol–water solution is known as *tincture of iodine*. It is used as a mild antiseptic. Old solutions in which part of the alcohol has evaporated should not be used, because concentrated solutions of iodine may produce blisters on the skin. An aqueous solution of iodine is also used as a disinfectant. Potassium iodide is added to these solutions, because it increases the solubility of iodine in water and alcohol–water solutions.

Sodium iodide, NaI, has been used in the prevention and cure of certain types of goiters. It is sometimes introduced into the diet as iodized salt. In areas in which iodine is deficient in the soil, goiter is quite prevalent, and sometimes a small amount of iodine is added to the water supply.

Hydrogen iodide, HI, is a colorless gas that dissolves in water to form hydriodic acid, which is a strong acid. Both hydrogen iodide and hydriodic acid are very good reducing agents.

Phosphorus

THE ELEMENT

Phosphorus is the second member of the nitrogen family and appears in Group V of the periodic table. There are two common *allotropic forms*, or *allotropes*, of phosphorus. Remember that allotropes are different forms of the same element with different properties. One of these allotropic forms of phosphorus is a yellowish-white waxy solid called *white phosphorus*. When exposed to the air, white phosphorus ignites and burns to phosphoric oxide, P_4O_{10}, and it must, therefore, be preserved under water. White phosphorus is insoluble in water but soluble in carbon disulfide. It is poisonous, and contact with the skin will produce painful sores that are difficult to heal.

The other allotropic form of phosphorus is called *red phosphorus*, a reddish-violet amorphous powder which is stable in air at ordinary temperatures. It is insoluble in water and carbon disulfide and is nontoxic.

Phosphorus is an essential element for both plants and animals, and organic compounds of it are present in the cell nuclei. The most important constituent of bone is calcium phosphate, $Ca_3(PO_4)_2$. Red phosphorus and phosphorus sesquisulfide, P_4S_3, are used in the manufacture of matches. Commercial fertilizers contain compounds of phosphorus (page 178). White phosphorus is used in incendiary bombs and in some rat poisons.

COMPOUNDS OF PHOSPHORUS

The most important compound of phosphorus is phosphoric acid, H_3PO_4. The anhydride of this acid is P_4O_{10}. Because phosphoric acid has three replaceable hydrogen atoms, it forms two series of acid salts and a normal salt. Normal or tertiary sodium phosphate, Na_3PO_4, is quite basic and is used as a cleansing agent. Disodium phosphate, Na_2HPO_4, also called *sodium monohydrogen phosphate*, has been used as a mild laxative. Sodium dihydrogen phosphate, NaH_2PO_4, is given to patients when an acid urine is desired. Calcium phosphate, $Ca_3(PO_4)_2$, occurs as the main constituent of phosphate rock and is the most important natural source of phosphorus.

Sulfur

THE ELEMENT

This element is the second member of the oxygen family, which constitutes Group VI of the periodic table. The three common allotropic forms of sulfur are the prismatic or rhombic, the monoclinic, and the amorphous or plastic forms (Figure 11.4). Prismatic sulfur is the stable form and exists as pale yellow crystals that are nearly insoluble in water but quite soluble in carbon disulfide. The monoclinic form of sulfur exists as long needles above a temperature of 96°C. Below this temperature, it changes to prismatic sulfur. If molten sulfur is quickly cooled, a rubberlike amorphous sulfur is formed which slowly changes to prismatic sulfur.

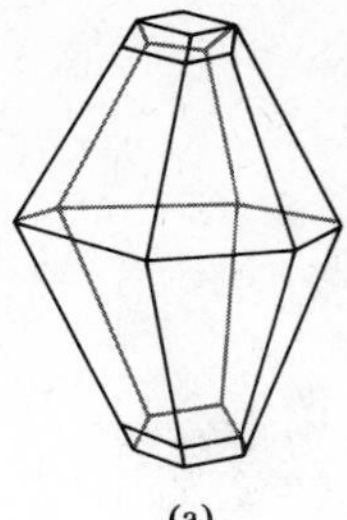

(a)

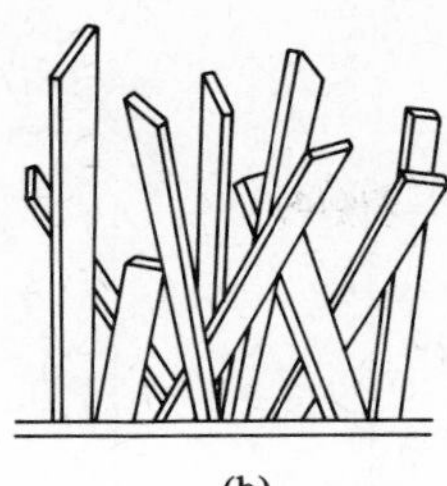

(b)

***Figure 11.4** Crystals of rhombic (a) and monoclinic (b) sulfur. (From Scott and Kanda,* The Nature of Atoms and Molecules, *Harper & Row, N.Y., 1962.)*

HYDROGEN SULFIDE, H_2S

This gas is foul-smelling, with a rotten-egg odor. It is formed when a sulfide, such as ferrous sulfide, FeS, is treated with acid:

$$FeS + 2HCl \longrightarrow H_2S + FeCl_2$$

If large quantities of hydrogen sulfide are inhaled, paralysis or death may occur.

SULFUR DIOXIDE, SO_2

Sulfur burns with a blue flame and the gas sulfur dioxide is formed.

$$S + O_2 \longrightarrow SO_2$$

This gas has a very sharp, irritating odor and is used as a fumigant. When combined with oxygen in the presence of a catalyst, such as certain oxides of nitrogen or metallic platinum, it forms sulfur trioxide. If dissolved in water, sulfur dioxide forms the weak acid, sulfurous acid.

$$H_2O + SO_2 \longrightarrow H_2SO_3$$

Sulfur dioxide and sulfurous acid are used as mild bleaches for straw, wools, and silks. Sulfur dioxide and some of the salts of sulfurous acid are used as food preservatives.

SULFURIC ACID, H_2SO_4

This is the most important compound of sulfur. It is produced by the action of water on sulfur trioxide and is a colorless, oily liquid. It has a great affinity for water, and when added to the latter, it generates a great quantity of heat. It removes the elements of water from many organic substances, such as sugar (Figure 11.5), paper, wood, and flesh and thus chars them. Concentrated sulfuric acid should not be permitted to come in contact with the skin or clothing. Sulfuric

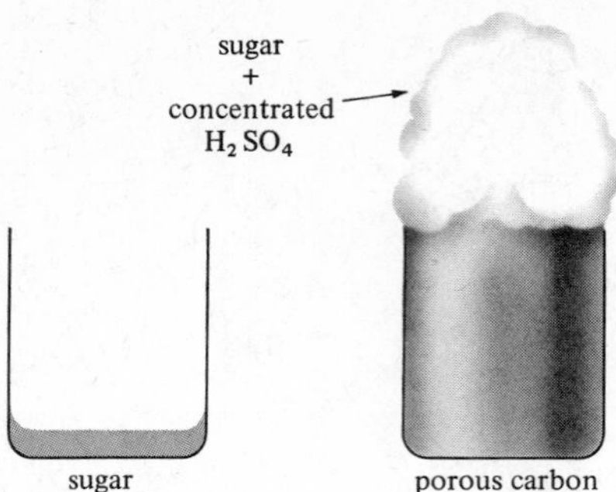

Figure 11.5 The dehydration of sugar by concentrated sulfuric acid.

acid is not only a strong acid but also a mild oxidizing agent, particularly when heated.

The following are important salts of sulfuric acid and their uses: Magnesium sulfate, or Epsom salts, $MgSO_4 \cdot 7H_2O$, and sodium sulfate, or Glauber's salt, $Na_2SO_4 \cdot 10H_2O$, are used as laxatives. Gypsum, or calcium sulfate, $CaSO_4 \cdot 2H_2O$, is used in the production of plaster of Paris, $(CaSO_4)_2 \cdot H_2O$. Because barium sulfate, $BaSO_4$, is opaque to X rays, it is used in the X raying of the intestinal tract.

Carbon

Carbon exists in several allotropic forms of which two, diamond and graphite, are crystalline. Charcoal, coke, coal, and soot are amorphous forms of carbon. On examination, these latter forms of carbon appear to be fine powders without crystalline structure. However, studies using X rays indicate that these forms may consist of microscopic crystals.

The largest deposit of diamonds occurs in South Africa. Other deposits occur in Brazil and India. The diamond, when cut, is a very hard, colorless crystalline solid that sparkles in the light. For this reason, it is much valued as an ornamental gem. Imperfect or off-color diamonds are used for cutting glass, for drill bits (for drilling holes), and for cutting other diamonds.

Graphite, unlike the diamond, is a soft, greasy, crystalline solid. Natural deposits of graphite are found in Ceylon and New York State. It is also prepared synthetically from coal. Graphite is used as a lubricant and in the manufacture of lead pencils, shoe polish, paints, and crucibles that must withstand high temperatures.

Wood charcoal is prepared by heating wood to a high temperature in the absence of air, a process called *destructive distillation.* Animal charcoal is prepared by the destructive distillation of bones and animal refuse. Specially prepared wood charcoal has the property of adsorbing gases and colored materials from solution and is used in making gas masks to adsorb poisonous gases. Animal charcoal also contains calcium carbonate and calcium phosphate, which were present in the bones. It is used in the manufacture of sugar, to remove the brown color from the sugar solution before it is evaporated to permit crystallization.

Hard coal (anthracite) and soft coal (bituminous) are impure forms of carbon containing mineral and volatile impurities. Coke is produced by the destructive distillation of soft coal; it is essentially mineral matter (ash) and carbon and contains no volatile products. Both charcoal and coke are used as reducing agents, especially in the production of metals from their oxides.

Although carbon cannot be melted or volatilized, except at very high temperatures, it burns to form carbon dioxide. If insufficient oxygen is present, carbon monoxide, CO, is produced.

CARBON MONOXIDE, CO

This is a colorless gas and is quite insidious; there is no warning of its presence because it is also odorless. The resulting anoxia (page 61) of the brain cells induces a loss of power of logical thought. Carbon monoxide is produced when any form of carbon burns in an insufficient supply of oxygen. It is formed when a coal furnace is not operating properly, from overheated gas stoves, and in the exhaust gases of the internal combustion engine. Automobile engines should not be permitted to run in a closed garage, nor should the windows of an automobile be closed when the automobile is operating on the highway. *Water gas,* which is prepared by passing steam over hot coke, is used in some cities as an artificial gas. It is composed of carbon monoxide and hydrogen, and because both of these are odorless, there would be no warning if a leak occurred. Some foul-smelling material is added to water gas as a warning agent.

Carbon monoxide burns with a hot blue flame to form carbon dioxide.

$$2CO + O_2 \longrightarrow 2CO_2$$

The blue tongues of flames above a coal fire are produced by the combustion of carbon monoxide.

Carbon monoxide is poisonous because it quickly combines with the hemoglobin of the blood to form *carboxyhemoglobin,* a very stable substance which will not combine with oxygen. When much of the hemoglobin of the body combines with carbon monoxide, the tissues of the body can no longer be supplied with oxygen, and the victim dies of asphyxiation.

CARBON DIOXIDE (CO_2) AND THE CARBONATES

Carbon dioxide is a colorless, odorless gas which is heavier than air. It is soluble in water, and the resulting solution has an acid taste. The carbon dioxide content of the atmosphere remains nearly constant at 0.04 percent by volume.

In the laboratory, carbon dioxide is usually prepared by the action of an acid such as hydrochloric acid on limestone and marble, both forms of calcium carbonate.

$$CaCO_3 + 2HCl \longrightarrow CaCl_2 + H_2O + \overline{CO_2}$$

It is produced when any form of carbon burns in a sufficient supply of oxygen and is a product of the fermentation of sugars and the decay of animal and vegetable matter.

Because carbon dioxide is heavier than air and is not combustible, it will form a blanket of gas over a flame and extinguish it. Carbon dioxide is often administered with oxygen to patients, because it stimulates respiration. It is dissolved under pressure in water to form soda water. Plants, in the presence of sunlight, convert it to sugars and oxygen, a process called *photosynthesis.* Solid carbon dioxide is used as a refrigerant. It sublimes without leaving a residue. (The trade name for this refrigerant is Dry Ice.) Because of its inertness and resistance to oxidation and reduction, carbon dioxide is used in pressure-packaging foods.

When dissolved in water, a part of the carbon dioxide combines with the water to form unstable carbonic acid, H_2CO_3,

$$H_2O + CO_2 \longrightarrow H_2CO_3$$

Carbon dioxide combines with bases to form carbonates. The resulting carbonates combine with excess carbon dioxide to form bicarbonates (acid salts).

$$2NaOH + CO_2 \longrightarrow Na_2CO_3 + H_2O$$
$$Na_2CO_3 + H_2O + CO_2 \longrightarrow 2NaHCO_3$$

Sodium carbonate, Na_2CO_3, also called *soda ash,* is used as a cleansing agent; its aqueous solution is mildly basic. Sodium bicarbonate, $NaHCO_3$, or saleratus, is a component of baking powder. When used as a leavening agent in baking, it is called *baking soda.* Because it neutralizes acids, it is used to counteract hyperacidity (see page 109).

Marble and limestone are two forms of calcium carbonate, $CaCO_3$. When heated to a high temperature, limestone loses carbon dioxide and forms quicklime, CaO.

$$CaCO_3 \longrightarrow CaO + \overline{CO_2}$$

BAKING POWDERS

Baking powder is used as a leavening agent; that is, it produces carbon dioxide gas in dough, causing the dough to become light and porous upon baking. Baking powders consist of three main ingredients: (1) sodium bicarbonate, (2) an acidic substance or some substance that will produce an acid when moistened, and (3) starch. Sodium bicarbonate is the source of the carbon dioxide. Starch acts to keep the baking powder dry so that it will not act prematurely. It is also a filler or diluent, and enough starch is usually added so that 10 percent of the weight of the baking powder will be evolved as carbon dioxide.

Three types of substances serve as the acid component of baking powders: (1) potassium hydrogen tartrate, or cream of tartar, $KHC_4H_4O_6$, (2) sodium or calcium dihydrogen phosphate, NaH_2PO_4 or $Ca(H_2PO_4)_2$, and (3) anyhydrous alum, or potassium aluminum sulfate, $KAl(SO_4)_2$. The action of water on a baking powder containing cream of tartar is represented by the following equation:

$$KHC_4H_4O_6 + NaHCO_3 \longrightarrow KNaC_4H_4O_6 + H_2O + \overline{CO_2}$$

Alum, although a normal salt, decomposes (by hydrolysis, page 144) when dissolved in water to form a small amount of sulfuric acid, which then liberates carbon dioxide from the sodium bicarbonate.

Carbon dioxide for the leavening of dough is also produced by the action of sour milk containing lactic acid on baking soda and by the fermentation of some of the sugar in the dough by the action of yeast.

Other compounds of carbon are discussed in that portion of this textbook that deals with organic compounds. In particular see Chapters 15–23.

Nitrogen

THE ELEMENT

This element, the first member of Group V of the periodic table, occurs in the free state in the air to the extent of about four-fifths of the latter. It occurs in small amounts in the combined state in all fertile soils. Deposits of sodium nitrate, $NaNO_3$, occur in Chile.

Nitrogen is a colorless, odorless, tasteless gas. It is only slightly soluble in water and can be liquefied and solidified at very low temperatures. Nitrogen is noncombustible, does not support combustion, and is not poisonous. In the air, it acts as a diluent of oxygen, for in its absence, combustions would be very vigorous and would not be easily controlled. Chemically, nitrogen is a very inert substance. It reacts with no elements at ordinary temperatures and with only a few at higher temperatures.

Nitrogen has been used to fill electric light bulbs. Commercially it is combined with hydrogen in the Haber process for the preparation of ammonia. Nitrogen bacteria, present on the root nodules of the legume plants, take nitrogen from the air and combine it with other elements to produce compounds of nitrogen. These are then used by the plant to produce proteins essential for the plant's growth.

AMMONIA, AMMONIUM HYDROXIDE, AND AMMONIUM SALTS

Ammonia, NH_3, is a colorless gas with a sharp, suffocating odor. It is quite soluble in water and is easily liquefied to form a colorless liquid that boils at $-33.5°C$.

Large amounts of ammonia are produced synthetically by combining nitrogen and hydrogen under high pressures and moderate temperatures in the presence of a suitable catalyst (Haber process).

$$N_2 + 3H_2 \rightleftharpoons 2NH_3$$

Small amounts of ammonia and ammonium salts are prepared as by-products of the coking of coal. In this process, called *destructive distillation,* soft coal is heated in the absence of air, and the small amounts of nitrogen compounds present in the coal are partially converted to ammonia, which volatilizes and is absorbed in sulfuric acid.

When ammonia dissolves in water, it combines chemically with the water to form the weak base ammonium hydroxide, NH_4OH, which when heated readily decomposes into water and ammonia; hence this is a reversible reaction.

$$NH_3 + H_2O \rightleftharpoons NH_4OH$$

Both ammonia and ammonium hydroxide react with acids to form salts.

$$NH_3 + HCl \longrightarrow NH_4Cl$$
$$2NH_4OH + H_2SO_4 \longrightarrow (NH_4)_2SO_4 + H_2O$$

The ammonium salts are usually colorless, crystalline solids. Many are used as components of fertilizer and as explosives. Ammonium carbonate, $(NH_4)_2CO_3$, is used in smelling salts as a heart stimulant.

Liquid ammonia is used in refrigeration and in the manufacture of ice. Ammonia is also used in cleansing agents and in washing (household ammonia or aqua ammonia) because it readily removes grease and oil films. Large amounts of ammonia are used in the production of its salts and of nitric acid.

NITRIC ACID AND ITS SALTS

Nitric acid, HNO_3, is a colorless, oily liquid, a strong acid, and also a good oxidizing agent. When heated strongly, it produces reddish-brown oxides of nitrogen.

Nitric acid is prepared by heating sodium nitrate with concentrated sulfuric acid.

$$NaNO_3 + H_2SO_4 \longrightarrow NaHSO_4 + HNO_3$$

It is also prepared by the action of water on certain oxides of nitrogen. Nitric acid is used in the production of explosives and fertilizers.

The salts of nitric acid are called *nitrates,* and all are soluble in water. Sodium nitrate, $NaNO_3$, occurs in the Chile saltpeter deposits. Silver nitrate, or lunar caustic, $AgNO_3$, is a colorless, crystalline solid that has a caustic action on the skin. As a 2 percent solution, it has been used as a prophylactic in the eyes of the newborn infant, but it must be quickly washed off with a dilute salt solution so that it does not cause any irritation. Nitrates occur to a slight extent in soils, and some, like sodium nitrate, are added to commercial fertilizers.

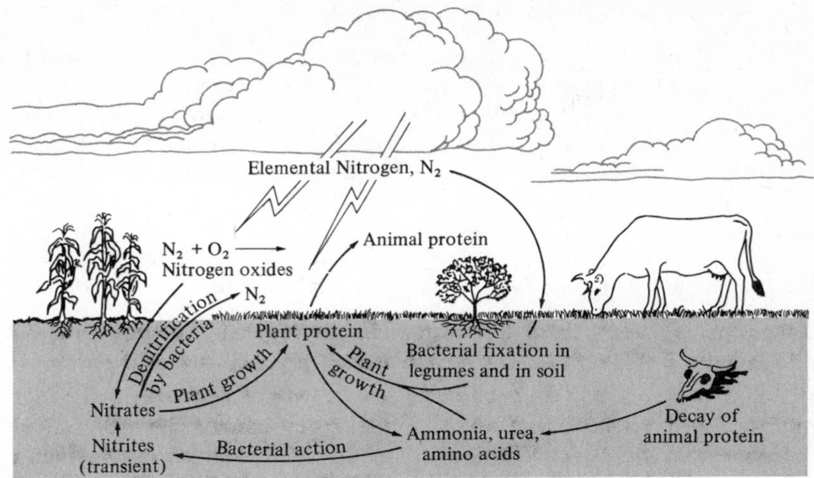

Figure 11.6 *The nitrogen cycle in living things.*

THE NITROGEN CYCLE IN NATURE

Although plants and animals require nitrogen compounds, they cannot directly synthesize these from the large reservoir of nitrogen in the atmosphere. Several types of bacteria, including those that grow in the root nodules of the legumes, convert free nitrogen into inorganic compounds. Lightning flashes cause the nitrogen and oxygen of the atmosphere to combine to form oxides of nitrogen. These oxides then combine with water to form nitric acid, which is absorbed into the soil. The Haber process for the manufacture of ammonia is the most important of the synthetic methods of converting atmospheric nitrogen to compounds that are utilizable for the synthesis of proteins by plants. The conversion of atmospheric nitrogen to nitrogen compounds, whether performed by nature or industrially, is known as the *fixation of nitrogen.*

Plants synthesize essential proteins from the nitrogen compounds in the soil. Animals cannot perform this synthesis but must depend upon vegetable proteins. Nitrogen compounds are returned to the soil either in the excretion of animals or by the decay of dead animal or plant materials. These products will consist, for the most part, of urea and ammonia. Certain bacteria convert ammonia to nitrogen gas, which is returned to the atmosphere, whereas other bacteria transform it to nitrites. This latter form of nitrogen is not assimilated and is toxic to plants. Fortunately, other bacteria are present in the soil which oxidize nitrites to nitrates; the latter are then available for plant assimilation.

The cycle in which atmospheric nitrogen is converted to inorganic nitrogen compounds, then to vegetable and animal proteins and the reconversion of the proteins to free nitrogen by decay and other means is called the *nitrogen cycle* (Figure 11.6).

Constituents of the atmosphere

Although air is essentially nitrogen and oxygen, about 1 percent of it is composed of the noble gases. Although argon accounts for the greater portion of the noble gases in the atmosphere, small amounts of helium, neon, krypton, and xenon are also present. Because these elements have very stable atomic structures, they are nearly chemically inert (see page 31).

Helium occurs in the gases issuing from certain wells in Texas and Kansas, and although present in the atmosphere in small amounts, it is obtained commercially from the former source. This gas was detected in the atmosphere of the sun and stars before its presence in the earth's atmosphere was discovered.

Helium is used as a substitute for hydrogen in filling lighter-than-air craft. Although it is slightly inferior to hydrogen in lifting power, it is noninflammable and there is a smaller loss of gas because it is not as diffusible as hydrogen.

An artificial atmosphere of helium and oxygen is used in the decompression of caisson workers to prevent *caisson sickness,* or *the bends.* Because of the lightness and ready diffusibility of helium, this artificial atmosphere is also furnished to certain patients who have difficulty in breathing. Helium is sometimes mixed with general anesthetics to reduce their inflammability and explosion hazards.

Argon is replacing nitrogen as a filler of electric light bulbs. All the naturally occurring noble gases, particularly neon and helium, are used in the discharge tubes of colored illuminating signs.

Water vapor and carbon dioxide, which occur in small amounts in the atmosphere, are important but minor constituents of the latter.

Fertilizers

As will be noted in Chapter 32, man and animal organisms require certain mineral elements for proper growth. Similarly, the growth of plants may be greatly retarded by the lack of certain elements in the soil. Continued cultivation of our land will result in the depletion of these essential elements, and unless replaced by natural or commerical fertilizers, a significant decrease in per-acre yield can be expected.

Although significant amounts of nitrogen, phosphorus, potassium, and sulfur must be present in the soil for proper plant growth, trace amounts of such elements as boron, copper, manganese, zinc, silicon, and aluminum must also be present.

In commercial fertilizers, the amount of nitrogen, phosphorus, and potassium is usually indicated on the package. Thus a 5-10-5 fertilizer will contain 5 percent nitrogen, phosphorus equivalent to 10 percent P_4O_{10}, and potassium equivalent to 5 percent K_2O.

Although the phosphorus and potassium content of a fertilizer are represented as P_4O_{10} and K_2O in its analysis, phosphorus is usually present as $Ca(H_2PO_4)_2$, called *superphosphate of calcium,* while potassium is present generally as potassium chloride, KCl. Superphosphate of calcium is produced by the action of sulfuric acid on phosphate rock, $Ca_3(PO_4)_2$, thus:

$$Ca_3(PO_4)_2 + 2H_2SO_4 \longrightarrow 2CaSO_4 + Ca(H_2PO_4)_2$$

In some commercial fertilizers, the more effective "triple phosphate," which is prepared by the action of phosphoric acid on phosphate rock,

$$Ca_3(PO_4)_2 + 4H_3PO_4 \longrightarrow 3Ca(H_2PO_4)_2$$

replaces the superphosphate of calcium (mixture of one mole of $Ca(H_2PO_4)_2$ to two moles of $CaSO_4$).

The usual sources of nitrogen in fertilizers are ammonium salts, such as ammonium sulfate and ammonium nitrate, sodium nitrate (Chile saltpeter), and urea, although liquid ammonia is frequently added directly to the soil.

The calcium sulfate that is present in the superphosphate of calcium and ammonium sulfate, a by-product of the coking industry, is the major source of sulfur in commercial fertilizers.

STUDY QUESTIONS

1. How are nonmetals distinguished from the metals?
2. Describe the physical properties of fluorine, chlorine, bromine, iodine, white phosphorus, red phosphorus, prismatic sulfur, monoclinic sulfur, amorphous sulfur, nitrogen, carbon, and helium.
3. Why are fluorine, chlorine, bromine, and iodine called halogens? Why do they represent a single chemical family?
4. What element plays an important role in the prevention of cavities in teeth? The prevention of goiter?
5. What are the medical uses of the following substances?

sodium chloride
sodium bromide
silver nitrate
ammonium carbonate
sodium iodide
tincture of iodine
mercuric chloride
sodium dihydrogen phosphate
helium
mercurous chloride
sodium monohydrogen phosphate
hydrated magnesium sulfate
barium sulfate

6. Indicate uses other than those listed in Question 5 for the following:

chlorine
hydrogen fluoride
silver bromide
potassium chlorate
normal sodium phosphate
sulfur dioxide
neon
sulfites
sodium nitrate
nitrogen gas
ammonia
argon
helium
nitric acid
phosphorus

7. (a) Compare the strength of the following acids: HCl, HI, HBr, HF, H_2SO_3, H_3PO_4, and H_2SO_4.
   (b) Which of these is a strong reducing agent?
   (c) Which are good oxidizing agents?
8. List some allotropic forms of carbon. Which of these are crystalline? Which are amorphous?
9. Name some other elements besides carbon that exist in allotropic forms.
10. What is meant by destructive distillation? Which forms of carbon are prepared by means of destructive distillation?
11. What substances besides carbon are present in animal charcoal?
12. Under what conditions is carbon monoxide formed?
13. Distinguish between oxyhemoglobin, carboxyhemoglobin, and hemoglobin.
14. Discuss the changes that occur in the body when large concentrations of carbon monoxide are inhaled.
15. In what ways are harmful concentrations of carbon monoxide frequently formed in an enclosed space?
16. How is carbon dioxide prepared in the laboratory? Write equations for its preparation.
17. Write an equation for the preparation of carbon dioxide by the action of sulfuric acid on magnesium carbonate, $MgCO_3$.
18. In what ways is carbon dioxide formed in nature?
19. What properties of carbon dioxide are utilized when it is used as a fire extinguisher?
20. What is an acid salt? A normal salt? List examples of each that have appeared in this chapter.
21. What purpose does sodium bicarbonate serve in baking powders? What is the purpose of starch in baking powders?
22. List the three types of chemicals used as the acidic constituent of baking powders.
23. Why can alum, which is not an acid itself, serve as an acid-producing constituent in baking powders?
24. By what means may dough be "leavened"?
25. Indicate important uses of the following: soda ash, baking powder, carbon dioxide, anthracite, graphite, animal charcoal, diamond, coke.
26. By means of equations, represent the reaction of (a) lactic acid, $H \cdot C_3H_5O_3$; (b) sodium dihydrogen phosphate, NaH_2PO_4, with sodium bicarbonate. How are these substances utilized in leavening bread?
27. What is meant by nitrogen fixation? Why is the fixation of nitrogen important?
28. Describe the nitrogen cycle in nature.
29. Why are animals so dependent nutritionally on plants?
30. Write balanced equations for the following:

   (a) the preparation of HCl
   (b) the combustion of phosphorus
   (c) the combustion of sulfur
   (d) the action of water on P_4O_{10}, SO_2, SO_3, and NH_3
   (e) the action of H_2SO_4 on ZnS
   (f) the neutralization of NH_4OH by HNO_3
   (g) the preparation of nitric acid
   (h) the Haber process for the manufacture of ammonia

31. What are the main constituents of the atmosphere, and in what relative proportions are they present?
32. The compounds of which of the elements described in this chapter are added to commercial fertilizers? (See page 178.)
33. What explanation can be given for the inertness of the noble gases?

chapter 12

oxidation-reduction reactions

12

Shortly after the discovery of the nature of combustion by Lavoisier, scientists began to define *oxidation* as the union of a chemical substance with oxygen. Thus when carbon burns, it unites with oxygen to form carbon dioxide and is said to be oxidized.

$$C + O_2 \longrightarrow CO_2$$

Reduction was at first defined as the removal of oxygen from or the addition of hydrogen to a chemical substance. When lead oxide is heated with coke, a form of carbon, metallic lead, is formed.

$$PbO + C \longrightarrow Pb + CO$$

In this latter reaction, carbon is the reducing agent and lead oxide is reduced to metallic lead. Note that the carbon is oxidized, because it has combined with oxygen. In the addition of hydrogen to ethylene,

$$\underset{\text{ethylene}}{C_2H_4} + H_2 \longrightarrow \underset{\text{ethane}}{C_2H_6}$$

the ethylene is reduced to ethane, because it has united with hydrogen.

The generalized concept of oxidation and reduction

The concepts of oxidation and reduction have been made more general in their scope. Oxidation is now defined as the loss of electrons, while reduction is the gain of electrons, by a chemical substance. In the burning of magnesium to magnesium oxide,

$$2Mg + O_2 \longrightarrow 2MgO$$

each magnesium atom loses two electrons, magnesium is therefore oxidized. In the reaction

$$CuO + H_2 \longrightarrow Cu + H_2O$$

copper is reduced, because it captures two electrons, and hydrogen is oxidized, because each atom loses an electron.

According to the generalized concept, the reactions

$$Hg + S \longrightarrow HgS$$

and

$$Cu + Cl_2 \longrightarrow CuCl_2$$

are considered oxidation-reduction reactions because in the first reaction, a mercury atom loses two electrons to a sulfur atom. The mercury is oxidized while the sulfur is reduced. In the second reaction, a copper atom loses two electrons to two atoms of chlorine; thus the copper is oxidized and the chlorine is reduced.

An oxidation reaction is always accompanied by a reduction reaction and vice versa, because if a chemical substance loses electrons, some other substance must gain these electrons. For this reason, all of these reactions are called *oxidation-reduction reactions* and not simply oxidation or reduction reactions.

Balancing oxidation-reduction reactions

Although various methods have been used to balance oxidation-reduction reactions, the *ion-electron method* is preferred by many chemists. Many oxidation-reduction reactions are quite complex, and their equations are difficult to balance without extensive experience with this method.

In the ion-electron method, *two partial equations* are written and balanced. One partial equation involves the oxidation reaction, while the other involves the reduction. As an example, if iron is added to a copper nitrate solution, the following reaction occurs:

$$Cu(NO_3)_2 + Fe \longrightarrow Fe(NO_3)_2 + Cu$$

In this reaction, iron atoms lose two electrons each and are oxidized to ferrous ions, as represented by the following partial equation:

$$Fe \longrightarrow Fe^{2+} + 2e$$

The electrons lost or gained are represented by e in the partial equation. The copper ions gain two electrons each and are reduced to metallic copper, as represented by the following partial equation:

$$Cu^{2+} + 2e \longrightarrow Cu$$

The ionic equation for the overall reaction is obtained by adding these two partial equations. Because the number of electrons lost and gained are the same, they cancel and do not appear in the resulting equation.

$$Fe + Cu^{2+} \longrightarrow Fe^{2+} + Cu$$

Note that the reduction of Cu^{2+} by iron is a general reaction, because the addition of iron to any copper salt (nitrate, sulfate, chloride, etc.) will produce the change shown here and will be represented by the same net ionic equation.

The partial equations for the precipitation of silver from solutions of its compounds by aluminum metal are

$$Al \longrightarrow Al^{3+} + 3e \qquad \text{(oxidation partial)}$$
$$Ag^{+} + e \longrightarrow Ag \qquad \text{(reduction partial)}$$

For an equation for the overall reaction, it is not possible to add these two equations, because the number of electrons lost in the first reaction is greater than those gained in the second. If the second equaion is multiplied by three, then addition is possible because the number of electrons lost and gained in the two reactions are now equal. The overall reaction is obtained by canceling electrons and adding the two partial equations.

$$\begin{array}{r} Al \longrightarrow Al^{3+} + 3e \\ 3Ag^{+} + 3e \longrightarrow 3Ag \\ \hline 3Ag^{+} + Al \longrightarrow 3Ag + Al^{3+} \end{array} \qquad \text{(overall ionic equation)}$$

The overall molecular equation is

$$3AgNO_3 + Al \longrightarrow 3Ag + Al(NO_3)_3$$

The laboratory preparation of chlorine by the oxidation of hydrochloric acid by potassium permanganate is an example of a more complicated type of oxidation-reduction reaction and is represented by the balanced complete ionic equation

$$10Cl^{-} + 2MnO_4^{-} + 16H^{+} \longrightarrow 5Cl_2 + 2Mn^{2+} + 8H_2O$$

Note that in an acid solution, the permanganate ions, MnO_4^{-}, are reduced to manganous ions, Mn^{2+}.

$$2Cl^{-} \longrightarrow Cl_2 + 2e$$

represents the oxidation partial equation for this reaction. When a permanganate ion is reduced to a manganous ion in an acid solution, hydrogen ions combine with the four oxygen atoms of the permanganate ion to form water. The reduction partial equation representing this reaction will be

$$MnO_4^{-} + 8H^{+} + 5e \longrightarrow Mn^{2+} + 4H_2O$$

Because the ionic partial equation must be balanced electrically, as well as chemically, and because there is a net charge of +7 on the left side and +2 on the right side of this partial equation, five electrons must be added to the left side.

Now, to obtain the overall ionic equation noted above, the oxidation partial equation must be multiplied through by 5 and the reduction partial equation by 2, giving

$$10Cl^- \longrightarrow 5Cl_2 + 10e$$
$$2MnO_4^- + 16H^+ + 10e \longrightarrow 2Mn^{2+} + 8H_2O$$

Adding and canceling electrons gives the complete ionic equation

$$10Cl^- + 2MnO_4^- + 16H^+ \longrightarrow 5Cl_2 + 2Mn^{2+} + 8H_2O$$

whereas

$$16HCl + 2KMnO_4 \longrightarrow 2MnCl_2 + 2KCl + 5Cl_2 + 8H_2O$$

represents a complete molecular equation for this reaction.

Some common oxidizing and reducing agents

The following are some common oxidizing agents:

chlorine, Cl_2
ferric chloride, $FeCl_3$
nitric acid, HNO_3
oxygen, O_2
ozone, O_3
sulfuric acid, H_2SO_4
sodium hypochlorite, NaClO
cupric oxide, CuO
hydrogen peroxide, H_2O_2
hypochlorous acid, HClO
potassium chlorate, $KClO_3$
potassium dichromate, $K_2Cr_2O_7$
potassium perchlorate, $KClO_4$
potassium permanganate, $KMnO_4$

The following is a list of some common reducing agents:

carbon, C
ferrous sulfate, $FeSO_4$
hydrogen, H_2
hydrogen sulfide, H_2S
oxalic acid, $H_2C_2O_4$
sodium sulfite, Na_2SO_3
sulfur dioxide, SO_2
carbon monoxide, CO
formaldehyde, CH_2O
hydrogen iodide, HI
mercurous chloride (calomel), Hg_2Cl_2
sodium thiosulfate (hypo), $Na_2S_2O_3$
stannous chloride, $SnCl_2$
tartaric acid, $H_2 \cdot C_4H_4O_6$

Many metals, such as sodium, magnesium, aluminum, iron, tin, and zinc, are also important reducing agents.

Importance of oxidation-reduction reactions in biological processes

Oxidation is a very important process that occurs in the tissues of the body. Oxygen and the digested products of food, brought by the blood from the lungs and the digestive organs to the cells, combine with the simultaneous production of heat and muscular energy. The latter is utilized in the various movements of the body. In the cells, the carbon and hydrogen present in the food products are converted to carbon dioxide and water, while in the liver, nitrogen is converted to ammonia and urea, NH_2CONH_2. These products of oxidation are excreted from the body by such organs as the kidneys and lungs.

In the body, each gram of carbohydrate and protein produces about 4 kcal of heat, whereas a gram of fat produces about 9 kcal. In a calorimeter (see Figure 2.3), an instrument for measuring the heat produced by the combustion of oxidizable substances, about the same amount of heat as above is generated for each gram of fat and carbohydrate burned. However, in the calorimeter, a gram of protein produces about 5.3 kcal of heat, and the nitrogen is liberated in the free state. This latter fact explains the difference between the amount of heat liberated in the oxidation of protein in the body and that in the calorimeter, because free nitrogen is a more highly oxidized form of nitrogen than is urea.

Importance of oxidizing and reducing agents in sterilization and disinfection

Because both ozone and chlorine oxidize organic matter and bacteria, they are employed in the disinfecting of water to make it potable.

Hydrogen peroxide, H_2O_2, and potassium permanganate, $KMnO_4$, are used as antiseptics in the treatment of wounds. Tincture of iodine and aqueous iodine solutions are used in the treatment of minor wounds and skin abrasions. Although hydrogen peroxide oxidizes pus and tissue material, it is not considered a good antiseptic, because the oxygen that it liberates does not usually come in close contact with bacteria.

Formaldehyde and sulfur dioxide are reducing agents frequently employed in disinfecting rooms that have been vacated by patients with contagious diseases. The formaldehyde is prepared by burning paraformaldehyde candles, while the sulfur dioxide is produced by burning sulfur candles.

Sewage is oxidized by and will thus remove the dissolved oxygen in the water in which it is suspended. When the amount of oxygen is appreciably decreased, fish and other marine life die. Sewage is frequently aerated; that is, it is mixed mechanically with oxygen by spraying it into the air. In this treatment, most of the organic material and bacteria can be destroyed before the sewage is discharged into lakes or streams.

Importance of oxidizing and reducing agents in bleaching and spot removal

Bleaching is usually accomplished by oxidizing or reducing a dye or stain to a colorless substance by means of a suitable oxidizing or reducing agent. Caution must be used when bleaching, because certain of the chemicals may weaken or destroy the fabric, paper, etc., being bleached. In other cases, the agent may be harmful only if it is not removed after the bleaching process.

Ozone and chlorine are oxidizing agents frequently employed as bleaching agents. Chlorine must be moist to bleach, because it is converted by the moisture to hypochlorous acid, HClO, which is the active oxidizing agent. Chlorine will destroy or weaken silk, wool, and other protein fibers and must not be used on these materials. It is used to bleach cotton and linen but must be removed by washing with some *anti-chlor,* such as a solution of sodium thiosulfate. Dilute solutions of salts of hypochlorous acid, such as Clorox and Javelle water, are used as mild bleaching agents in the home laundry. A 3 percent hydrogen peroxide solution in dilute ammonia is used in the bleaching of hair, silk, and feathers. Sulfurous acid and its salts, the sulfites, are employed in the bleaching of paper and straw. The latter substances are damaged by a chlorine bleach.

Because many dyes are bleached by Clorox (sodium hypochlorite solution), it is a common ink eradicator. Fresh ink can usually be removed from fabrics by washing with water. Some inks are made from colorless ferrous tannate, and a blue dye is added to this type of ink to make it visible when first placed on paper. On exposure to the air, the ferrous tannate is oxidized to black insoluble ferric tannate. Oxalic acid or ammonium oxalate will reduce the ferric tannate to the ferrous salt, which is then removed from the fabric by washing with water. Oxalic acid and oxalate salts are also effective in removing rust stains from fabrics. Oxalates reduce the rust, Fe_2O_3, to soluble ferrous oxalate.

A sodium thiosulfate solution may be used in the removal of iodine stains and stains produced by silver compounds and silver preparations, such as Argyrol. Hydrogen peroxide is used to remove blood stains, perspiration, and urine from cotton and linen fabrics but should be permitted to remain in contact with the cloth for only a short period of time.

Potassium permanganate solution may be used to remove mildew, perspiration, coffee, tea, tobacco, many fruit stains, and a diversity of other stains. It has two disadvantages: It will remove the dye from many colored fabrics and should be tested on a small portion of the fabric before use; and it leaves a dirty brown stain of manganese dioxide, MnO_2, when used on cloth. Fortunately, this is easily removed with a hydrogen peroxide or sodium thiosulfate solution. Note that hydrogen peroxide can act both as an oxidizing agent, being reduced to water, and a reducing agent, whereby it is converted to oxygen and water.

STUDY QUESTIONS

1. State the generalized definition of oxidation and reduction.
2. How can the definition of oxidation and reduction be modified for those reactions in which oxygen is involved. In which hydrogen is involved?
3. Explain why oxidation is always accompanied by reduction.
4. If an element is oxidized or reduced in a certain chemical reaction, what change in valence will it undergo?

5. In the following balanced equations, indicate (1) oxidizing agent, (2) reducing agent, (3) substance that has been oxidized, and (4) substance that has been reduced:

(a) $PbS + 4H_2O_2 \longrightarrow PbSO_4 + 4H_2O$
(b) $C + 2H_2O \longrightarrow CO_2 + 2H_2$
(c) $2CuO + \underset{\text{formaldehyde}}{CH_2O} \longrightarrow Cu_2O + \underset{\text{formic acid}}{HCOOH}$
(d) $H_2SO_4 + 2HBr \longrightarrow SO_2 + 2H_2O + Br_2$
(e) $Fe + HgSO_4 \longrightarrow FeSO_4 + Hg$
(f) $2Ca + O_2 \longrightarrow 2CaO$

6. Balance the following partial equations, and indicate the number of electrons lost or gained in each.

$Fe^{2+} \longrightarrow Fe^{3+}$
$Sn^{2+} \longrightarrow Sn^{4+}$
$I^- \longrightarrow I_2$
$Hg_2{}^{2+} \longrightarrow Hg^{2+}$
$Cl_2 \longrightarrow Cl^-$

7. By use of the ion-electron method, balance the folowing oxidation-reduction reactions.

$HgCl_2 + SnCl_2 \longrightarrow Hg_2Cl_2 + SnCl_4$
$Al_2O_3 + K \longrightarrow K_2O + Al$
$KI + Cl_2 \longrightarrow KCl + I_2$
$PbCl_2 + Cr \longrightarrow CrCl_3 + Pb$
$Sb + Cl_2 \longrightarrow SbCl_3$

8. Discuss the importance of oxidation and reduction reactions in the following.

(a) sewage disposal
(b) bleaching
(c) heating the home
(d) the production of muscular energy in the body
(e) fumigation
(f) industry
(g) spot removal
(h) the use of antiseptics
(i) sterilization

9. Indicate which of the following are good reducing or good oxidizing agents.

nitric acid
carbon
sodium thiosulfate
stannous chloride
chlorine
potassium permanganate
sulfur dioxide
oxalic acid

sodium chlorate
hydrogen iodide

10. Balance the following partial equations, and indicate the number of electrons lost or gained in each.

$ClO_3^- + H^+ \longrightarrow Cl^- + H_2O$
$SO_3^{-2} + H_2O \longrightarrow SO_4^{-2} + H^+$
$NO_3^- + H^+ \longrightarrow NO_2^- + H_2O$
$IO_4^- + H^+ \longrightarrow I_2 + H_2O$

11. Using the ion-electron method, balance the following ionic equations, which represent more complicated oxidation-reduction reactions.

$SO_4^{-2} + Br^- + H^+ \longrightarrow SO_2 + Br_2 + H_2O$
$NO_3^- + Cu + H^+ \longrightarrow Cu^{2+} + NO + H_2O$
$ClO^- + Cl^- + H^+ \longrightarrow Cl_2 + H_2O$
$Fe^{2+} + MnO_4^- + H^+ \longrightarrow Fe^{3+} + Mn^{2+} + H_2O$

chapter 13

colloidal substances

13

The difference between colloids, true solutions, and suspensions

When sugar and salt are dissolved in water, a true solution is formed. In such a solution, the solute is dispersed as single molecules or ions, and during filtration, these are not retained by a paper filter or by certain parchments and animal membranes. If clay or sand is finely ground and then shaken with water, many of the particles will remain in suspension for some time before they settle and are large enough to be retained by an ordinary filter. Such a dispersion is called a *suspension.*

If the sand or clay particles are ground finer, the particles, although not in true solution when suspended in water, will settle slowly or not at all. These fine particles, which are not retained by a filter and will not usually pass through an animal membrane, are called *colloidal dispersions*, or *colloids.* Some incorrectly refer to the resulting systems as colloidal solutions, but these are not true solutions. A colloidal particle is usually an aggregate of many thousands of molecules, whereas solutions represent a single molecular aggregate of the solute. There are some exceptions to this statement. Very large molecules, such as protein and starch molecules, exhibit the properties of colloids when they are dispersed in water. Such very large molecules are frequently called *macromolecules.*

Size of colloidal particles

Colloidal particles are so small that the millimeter is too large a unit for their measurement, and the micron (μ), one-thousandth of a millimeter, the millimicron (mμ), one-millionth of a millimeter, and the angstrom unit (Å), 0.1 mμ, or 10^{-8} cm (see Figure 13.1) are employed. Particles that exhibit the properties of colloids have diameters ranging from about 1 mμ to 200 mμ. The diameters of small molecules range in size up to 1 mμ, whereas the particles in a suspension are much larger than 200 mμ (Figure 13.1).

The properties of colloidal particles

Most colloidal particles possess either a positive or a negative charge which tends to stabilize them. If such a charged colloid par-

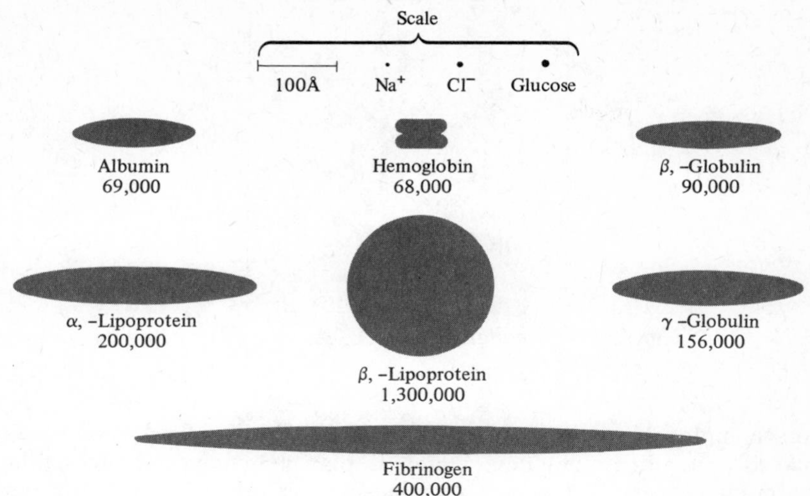

$1 \text{Å} = 0.00000001 \text{ cm} = 10^{-8} \text{ cm}$

Figure 13.1 *Relative dimensions of salt, glucose, and various proteins. (Courtesy J. L. Oncley.)*

ticle comes in contact with a colloid or an ion of opposite charge, it will be discharged and become unstable. A number of these discharged particles will then join together to form a large particle that will precipitate from solution.

Colloidal particles are so small that they cannot be seen by the naked eye or by the microscope. Many colloidal suspensions will appear cloudy because each particle will tend to reflect the light that strikes it. If a small beam of light is passed through a true solution, and if the solution is observed at right angles to the beam, the beam of light will not be observed in the solution. However, it will be easily observed in a colloidal dispersion, due to the scattering of the light by the suspended particles. This is called the *Tyndall effect.* The ultramicroscope is used to observe the Tyndall effect. It consists of an optical microscope placed at right angles to the incident beam of light. Each particle will be observed to travel in a zigzag fashion through the medium; such motion of the colloidal particles, called *Brownian movement* (Figure 13.2), was first observed by the botanist Brown. This zigzag movement of the colloidal particles must be due to the uneven bombardment of the particles by the molecules of the dispersing medium.

Dialysis

As has been mentioned above, there are certain parchments and other membranes through which solutes in true solution diffuse and through which colloidal dispersions cannot pass. Such membranes are called *dialyzing membranes* and are used to remove

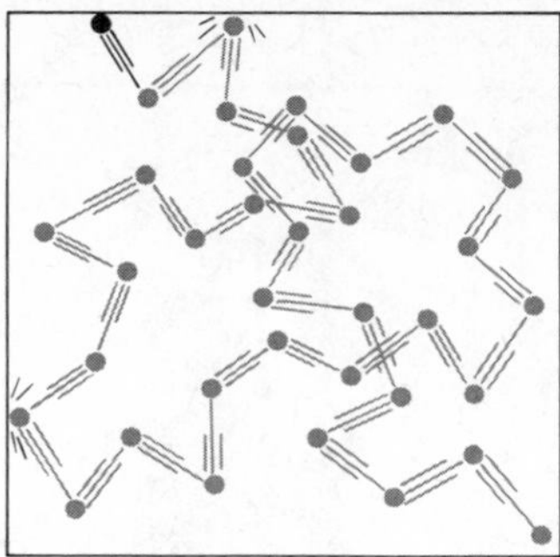

Figure 13.2 The path of a colloidal particle, an example of Brownian movement.

solute molecules and ions from a colloid. The colloidal system is placed in a parchment bag, which is placed in a vessel of distilled water. If the distilled water is replaced from time to time, only particles in true solution will have diffused out of the parchment. This method of purifying colloids is called *dialysis* (Figure 13.3).

Dialysis occurs through many membranes of living plants and animals. Most food particles are of colloidal dimension, whereas most of the digested products of the food are in true solution. Because the intestinal wall is a dialyzing membrane, digested food products will be absorbed through this wall into the bloodstream, whereas the undigested food residues will remain in the intestinal tract until they are digested or excreted.

Protective colloids and emulsions

Certain colloidal particles that tend to be unstable and easily precipitated may be stabilized by coating them with a second colloidal substance. This latter substance is called a *protective colloid.* Gelatin

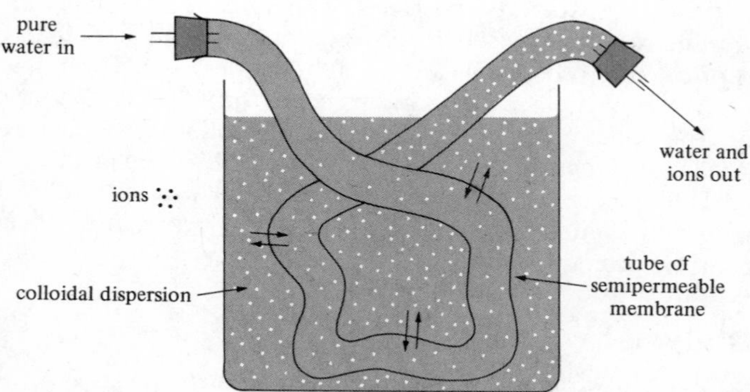

Figure 13.3 Dialysis. (From C. W. Keenan and J. H. Wood, General College Chemistry, *4th ed., Harper & Row, N.Y., 1971.)*

***Figure 13.4** An emulsion of oil and water (left) stabilized by soap. Note (right) the rapid separation of the oil–water emulsion in the absence of the stabilizing agent. (Courtesy Beckwith Studios.)*

acts as a protective colloid in ice cream; its presence prevents the formation of large ice crystals in the ice cream.

An *emulsion* (Figure 13.4) is a colloidal dispersion of one liquid in a second liquid in which it is insoluble. If olive oil or kerosene is mixed vigorously with water, the oil is broken into fine droplets of colloidal dimensions and thus forms an emulsion with water. In a short period of time, the emulsion will "break," and the two liquids will separate. These emulsions are frequently stabilized by the addition of a protective colloid, called an *emulsifying agent*. Thus the addition of soap will produce stable emulsions of olive oil or kerosene in water. Other emulsifying agents are the casein of milk, which stabilizes the emulsion of butter fat in water, and egg yolk, which in mayonnaise produces a permanent emulsion of olive oil in vinegar.

Gels

When heated in water, gelatin will be dispersed in water in a fluid state. If this fluid is cooled, it will set to a semisolid called a *gel*. If fruit juices are boiled and thickened, they will form jelly in the presence of sugar and fruit acids. A material called *pectin* is responsible for this gelling, and if it is not present in sufficient amounts in the fruit juices, they will not gel. In such cases, commercial pectin should be added to produce a satisfactory jelly.

Adsorption

Because each particle is minute in size, a small mass of colloidal particles will have a very large surface. These colloid surfaces have a great power to hold molecules and ions, a process called *adsorption*. Specially prepared collodial carbon called *activated carbon* is capable of adsorbing large amounts of certain substances. In the gas mask, it has been used to adsorb poisonous gases such as chlorine. It is used to remove the color from brown sugar in the refining of sugar, as well as the color from many solutions in the organic laboratory. Activated charcoal is also used to adsorb gases from the stomach.

The importance of a knowledge of colloids

Proteins (page 351) are macromolecules and exhibit the properties of colloids rather than true solutions in water.

Although the food that is consumed is converted, for the most part, into small water soluble particles during digestion, these substances are converted to colloidal size particles once they are absorbed in the living cell. Thus a knowledge of colloids is necessary to understand the properties of cell constituents.

Some catalytic phenomena appear to depend upon adsorption at the surface of the catalytic substance.

The Cottrell process for removing noxious impurities from the smoke of chimneys depends upon an electrical apparatus in the chimney that will discharge and cause the coalescence and precipitation of charged colloidal particles in the smoke. Air pollution is a serious public health problem, and the paramedic should aid in the education of the public concerning this problem.

STUDY QUESTIONS

1. Define the following terms: dialysis, dialyzing membrane, semipermeable membrane, emulsion, gel, emulsifying agent, protective colloid, Tyndall effect, Brownian movement.
2. Distinguish between true solutions, colloidal dispersions, and suspensions.
3. How is a suspension separated from a colloidal dispersion?
4. Distinguish between adsorption and absorption. Why are colloids such good adsorbing agents?
5. Discuss some uses of activated charcoal as an adsorbing agent.
6. How may colloids be separated from water soluble solutes?
7. Give some examples of emulsions. How are emulsions prepared, and what substances may be used to stabilize these?
8. What is the nature of many animal and plant membranes? Of what importance is this fact?
9. The addition of a second colloid or an electrolyte will frequently cause a colloid to precipitate. Why?
10. Suggest how colloids may become electrically charged.
11. Describe what one observes when a beam of light passing through a colloidal dispersion is examined with an ultramiscroscope. What is the significance of this observation?
12. Why must commercial pectin be added to some fruit juices when jellies are being prepared?
13. Discuss the size of colloidal particles.
14. What is a micron? A millimicron? An angstrom unit?

chapter 14

radioactivity and its biological applications

14

NATURAL RADIOACTIVITY

During most of the nineteenth century, it was believed that an atom of an element could not be converted into an atom of some other element; that is, atoms were indestructible.

In 1896, Henri Becquerel, a French physicist, wrapped a photographic plate with an opaque black paper and placed upon it a piece of pitchblende, an ore of uranium (U). After the photographic plate was developed, Becquerel observed that the plate had been affected by rays that had been produced by the uranium ore and had passed through the opaque paper. He also observed that these rays would discharge an electroscope (Figure 14.1). This phenomenon exhibited by uranium became known as *radioactivity*.

Two students of Becquerel, Pierre and Marie Curie (Figure 14.2) observed that pitchblende was much more radioactive than purified uranium oxide, and they suspected that the ore contained some other chemical substance which was much more radioactive than uranium. In 1898, Marie Curie isolated radium as the salt radium bromide, $RaBr_2$, and found that it was about a million times as radioactive as purified uranium oxide. It was soon recognized that these properties of uranium and radium were due to spontaneous disintegration of their atomic nuclei. Other very complex atoms are also spontaneously disintegrating, but naturally occurring atoms of simple structure are usually stable. Radioactivity is now recognized as this spontaneous decomposition of atomic nuclei.

Some properties of radium

Shortly after the discovery of radium by Madame Curie, Lord Rutherford, an English physicist, demonstrated that three different rays are emitted from radium atoms as they disintegrate. He called these alpha (α), beta (β), and gamma (γ) rays. Two of these rays, the α and the β rays, were bent out of a straight path by a magnet or electromagnet, but γ rays are not affected (Figure 14.3). Both the α and the β rays are composed of particles of matter. The α particles consist of helium nuclei, that is, helium ions composed of two protons and two neutrons. Each of these ions has a positive charge of

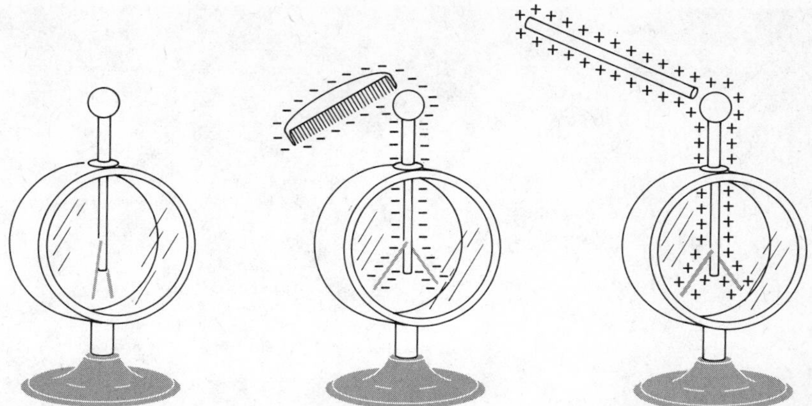

Figure 14.1 A neutral, a negatively charged, and a positively charged electroscope.

two, because they lack the two electrons present in the first shell of the natural helium atom. The β particles are electrons that are ejected from the nucleus of the radioactive atom and are probably formed from the neutrons of the nucleus. This seems reasonable, because the loss of a β particle increases the atomic number by 1. Thus the loss of a β particle from a uranium atom of atomic weight 239 and atomic number 92 produces neptunium, also of atomic weight 239 but atomic number 93. The loss of a β particle from the latter produces plutonium, of atomic weight 239 but atomic number 94. The γ rays are waves similar to light rays but of much shorter wavelength; they thus resemble X rays.

Although α particles are stopped by a piece of paper or by passage through a few centimeters of air, the higher-velocity β particles are more penetrating. The γ rays have much greater energy than either of these two, and a much greater penetrating power; they are able to pass through several centimeters of lead.

A large amount of energy as heat is liberated at all times from radium. The heat generated by the radioactive decay of a gram of radium exceeds 100 cal per hour. In 1,620 years, half of a certain quantity of radium will disappear as such, owing to radioactive disintegration. This is called the *half-life period* of radium. Thus in 1,620 years, only one-half gram of radium will remain from an initial weight of 1 g, while after the elapse of 3,240 years, one-fourth gram will remain. The half-life periods of radioactive substances are characteristic of the atoms that are disintegrating. Some radioactive substances have a half-life period of a few seconds, whereas other very stable ones may have a half-life of thousands of years.

Uranium is the parent substance from which radium is produced by radioactive decay. Radium emanations, or radon, a highly radioactive gas, is the residue remaining from the decay of radium (Table 14.1). It has a short half-life (3.82 days). Several other radio-

Figure 14.2 Marie Curie, discoverer of radium and polonium. (Courtesy New York Public Library Picture Collection.)

active atoms are produced successively from the radon atoms. This series of radioactive atoms ends with a stable isotope of lead. Thorium, another element consisting of very complex atoms, is the first member of another radioactive series, which, after the production of successive radioactive atoms, ends with the production of another stable lead isotope.

Utilization of natural radioactivity in medicine

The rays of radium produce painful sores when they come in contact with living tissue. Malignant or cancerous cells appear to be

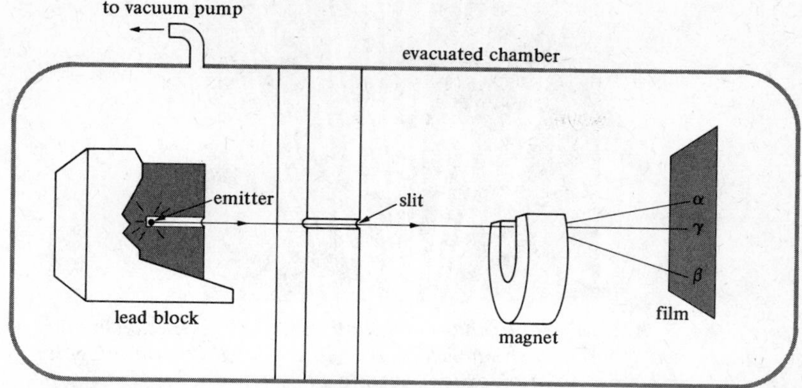

Figure 14.3 The α, β, and γ rays that are emitted by most natural radioactive substances are separated in a magnetic field. The α and β rays are particulate in nature. The α rays consist of a stream of helium atoms that are charged positively; the β rays consist of a stream of electrons. Unlike these, the γ rays are vibrations of short wavelength similar to X rays.

more easily destroyed than healthy ones. For this reason, the radium rays are used in the treatment of this dread disease. Some cures have been obtained in the early stages of cancer, and some relief has been obtained in advanced stages. In some cases, small tubes of radium emanations, or radon gas, are placed in the cancerous tissue. These radon tubes are called *radon seeds.*

In handling radium or radium emanations, the scientist and nurse should use a certain amount of protection, usually in the form of a thick lead shield, or else the sample should be placed in a lead case. Samples of radium or radon should be carried with tongs, because the greater the distance between the worker and the radium sample, the less the chance for the production of burns from the radiations.

TABLE 14.1 STEPS IN THE RADIOACTIVE DECAY OF URANIUM 238 TO RADON OF ATOMIC WEIGHT 222[a]

$$^{238}_{92}\text{U} \xrightarrow{-\alpha \text{ particle}} {}^{234}_{90}\text{Th} \xrightarrow{-\beta \text{ particle}} {}^{234}_{91}\text{Pa} \xrightarrow{-\beta \text{ particle}}$$

$$^{234}_{92}\text{U} \xrightarrow{-\alpha \text{ particle}} {}^{230}_{90}\text{Th} \xrightarrow{-\alpha \text{ particle}} {}^{226}_{88}\text{Ra} \xrightarrow{-\alpha \text{ particle}}$$

$$^{222}_{86}\text{Rn}$$

[a]This radon isotope decays, through intermediate formation of various radioactive isotopes of lead, bismuth, and polonium, to the stable lead isotope of atomic weight 206. For the significance of the letter symbols, see the table of atomic weights inside the back cover. Numerical superscripts indicate the atomic weight of the isotope, while the subscripts indicate the atomic number.

Figure 14.4 A historic cloud chamber photograph of α tracks in nitrogen gas. The forked track was shown to be due to a speeding proton (going off to the left) and an isotope of oxygen (going off to the right). It is assumed that the α particle struck the nucleus of a nitrogen atom at the point where the track forks. Blackett made 20,000 such photographs (containing about 400,000 tracks) and found 8 forked tracks. (Courtesy P. M. S. Blackett of the Cavendish Laboratory.)

ARTIFICIAL RADIOACTIVITY

Production of artificial radioactive atoms

Stable atoms may be converted to radioactive atoms by bombardment with (1) α particles (He^{2+}), (2) protons (H^+), (3) deuterium ions or deuterons (D^+), (4) neutrons, (5) electrons, or (6) γ rays (Figure 14.4). Bombardment of stable atoms by the rays from natural radioactive substances such as radon may produce a short-life radioactivity in these atoms. Protons, deuterons, and α particles are accelerated to high velocities in a special instrument called the *cyclotron,* and these fast-moving particles can be directed at the stable nuclei. Electrons can also be used for the production of artificial radioactivity if they are speeded to a high velocity in a *betatron.* The γ rays frequently knock out neutrons or protons from stable nuclei to produce radioactive nuclei. Neutrons, used in the bombardment of stable nuclei, were first produced by the action of radon on beryllium They are now generally produced as by-products of *fission,* that is, the cleavage of uranium atoms in the *atomic pile.* Many artificial radioactive atoms are now produced by bombarding suitable material with these neutrons in an atomic pile. They have been made available for scientific investigations and medical applications by the Atomic Energy Commission.

Artificially produced radioactive atoms are usually, but not always, of short-life. One of the radioactive isotopes of phosphorus has a half-life of 2½ minutes, while another isotope of this element has a half-life of 14 days. The half-life of the long-life carbon-14 isotope is 5,720 years (Figure 14.5).

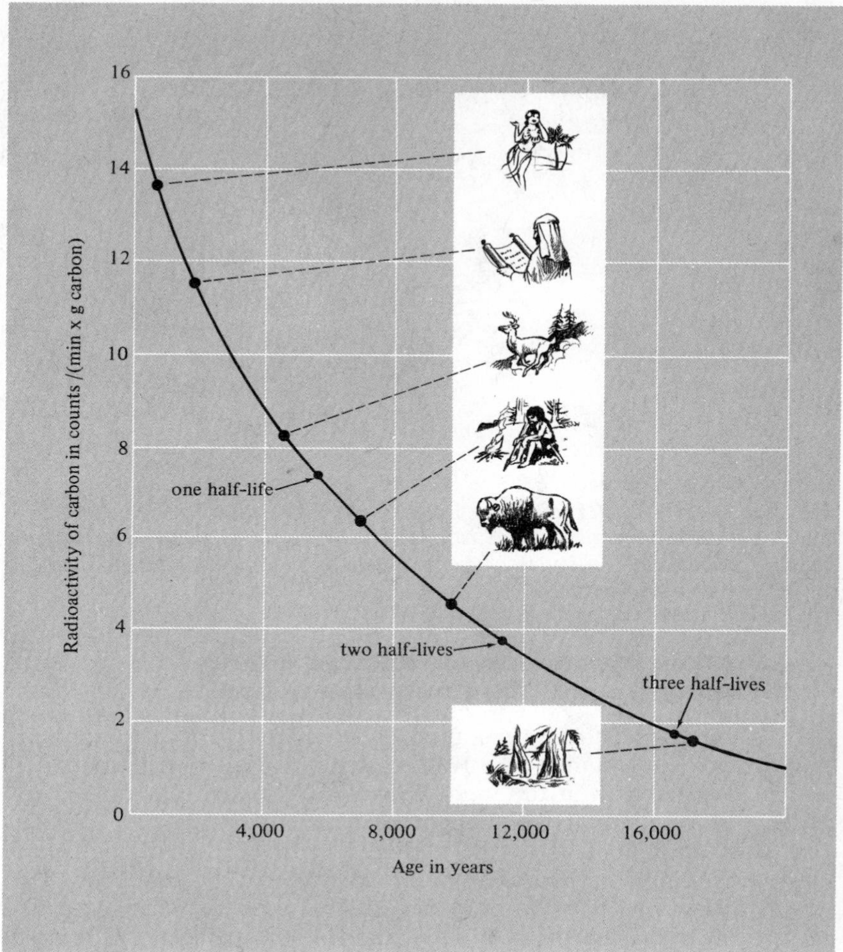

Figure 14.5. Radiocarbon dating has been of great value to archaeologists and geologists in establishing the ages of prehistoric Indian villages, glacial deposits, etc. For this work, carbon-containing materials must have been preserved, usually by being covered with earth, to the present time. Reading from the top of the graph downward along the carbon 14 decay curve are shown the ages of an early Polynesian culture established from charcoal found in Hawaii, age 950 years; the Dead Sea scrolls, age about 1,900 years; deer antler from Indian mound near Indian Knoll, Ky., age 4,900 years; charcoal from a hearth at Long Site, S. Dak., age 7,100 years; bison bone from near Lubbock, Tex., age 9,900 years; cypress wood from a buried stump near Santee, S.C., older than 17,000 years. (From Keenan, Wood, and Bull, Fundamentals of College Chemistry, *3rd ed., Harper & Row, N.Y,. 1972. Data from W. F. Libby,* Radiocarbon Dating, *courtesy of the University of Chicago Press.)*

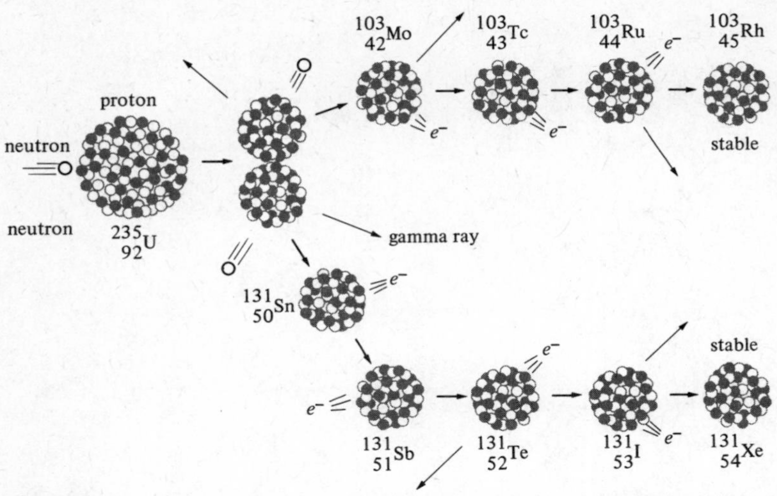

Figure 14.6 A uranium 235 nucleus after capturing a neutron (left) splits into two smaller nuclei with the emission of γ rays and two or three neutrons. The uranium 235 nuclei can split in over thirty ways, producing a total of about 200 radioactive species, generally with atomic numbers 30 to 64 and masses 72 to 161.

Fission, fusion, and the nuclear energy and hydrogen (thermonuclear) bombs

Naturally occurring uranium consists of 99.3 percent uranium-238 and 0.7 percent uranium-235. By very tedious processes, uranium-235 can be separated from the more abundant isotope. Uranium-235 can absorb "slow" neutrons, whereupon it decomposes into two elements per atom with about half the atomic weight of the original atom; thus one atom of barium and one of krypton may be produced. Two or more neutrons may be produced in this process, which is called *atomic fission* (Figures 14.6, 14.7, and 14.8). Incidentally, a small amount of mass disappears in fission and appears as a very large quantity of energy (according to the prediction of the Einstein equation on page 14).

Uranium-238 can absorb fast neutrons to form a new isotope of uranium, uranium-239. This latter isotope can lose two β particles (electrons) successively to form a short-life neptunium (neptunium-239) and the very long-life plutonium isotope, plutonium-239. Neptunium and plutonium are man-made elements: they have not been found in nature. Because they have atomic numbers greater than uranium, they are frequently called the *transuranium* or *transuranic elements.*

Like uranium-235, plutonium-239 can undergo fission when irradiated with neutrons. Because in the fission of uranium-235 and plutonium-239 more neutrons are usually produced than are used to initiate the reaction, if the quantity of these isotopes is above a certain

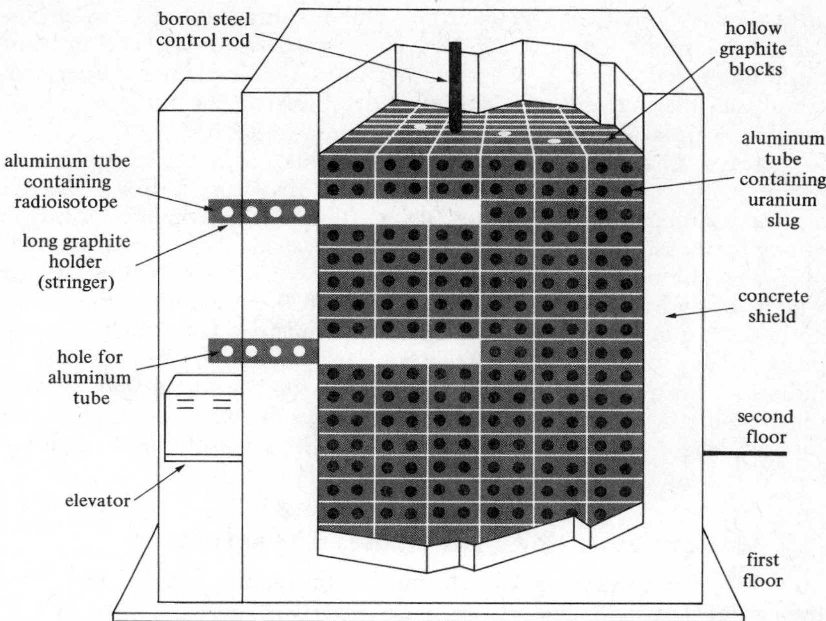

Figure 14.7 Schematic drawing of the first nuclear reactor at Oak Ridge, Tenn. (From Keenan, Wood, and Bull, Fundamentals of College Chemistry, *3rd ed., Harper & Row, N.Y., 1972.)*

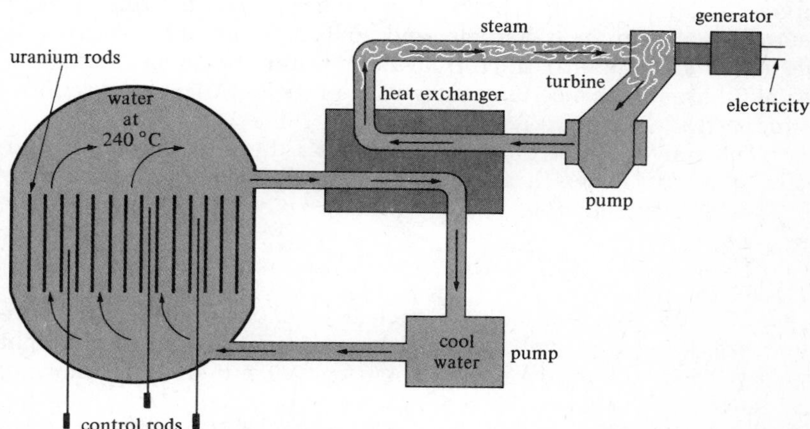

Figure 14.8 A pressurized power reactor. In this type of reactor, natural uranium is usually enriched with uranium 235 so that the reactor can be more readily brought to a critical condition. Water circulating through the core removes the heat of fission. (From Keenan, Wood, and Bull, Fundamentals of College Chemistry, *3rd ed., Harper & Row, N.Y., 1972.)*

critical mass, many of the neutrons will not escape but will be absorbed by other uranium-235 or plutonium-239 atoms, and a very rapid chain reaction will occur, with a great deal of energy produced, mainly as thermal energy. This was the basis of the nuclear energy bombs dropped on Nagasaki and Hiroshima in 1945.

It has been conjectured that at very high temperatures, such as occur on the sun, a number of light atoms may fuse to form heavier atoms, a process called *atomic fusion.* Thus four ordinary hydrogen atoms may fuse to form a helium atom. In so doing, a small amount of mass (atomic weight of helium minus four times the atomic weight of hydrogen, or 4.032 − 4.003, or 0.029 g) will disappear with an accompanying conversion to energy. It must be from such a fusion process that the sun generates the large amounts of heat that it radiates. The recently produced *hydrogen,* or *thermonuclear, bombs* must generate their great destructive power by some atomic fusion process, likely the conversion of heavy hydrogen isotopes to helium, which is triggered by an ordinary nuclear energy bomb explosion.

Modern notations for nuclear reactions

In natural radioactivity, a complex atom emits from its nucleus a lightweight particle, such as an α or β particle, and the particle that remains is usually of a complex nature. Thus the spontaneous disintegration of radium to radon is accompanied by the emission of an α particle, ${}^{4}_{2}\mathrm{He}$, and can be represented by the following equation:

$${}^{226}_{88}\mathrm{Ra} \longrightarrow {}^{222}_{86}\mathrm{Rn} + {}^{4}_{2}\mathrm{He}$$

In artificial or induced radioactivity, a heavy atom acts as a target for an accelerated light particle such as an electron, proton, neutron, deuteron, or α particle, and collision is accompanied by the ejection of some light particle from the resulting heavy nucleus. Some useful information concerning these light particles utilized or produced in induced radioactivity is summarized in Table 14.2.

The conversion of ordinary nitrogen to the radioisotope carbon-14 by neutrons, in which a proton is emitted, is now represented by the following equation:

$${}^{14}_{7}\mathrm{N} + {}^{1}_{0}n \longrightarrow {}^{14}_{6}\mathrm{C} + {}^{1}_{1}\mathrm{H}$$

A popular new notation for such nuclear reactions indicates in parentheses following the symbol for the heavy target atom first the light target particle, followed by the light particle emitted, and, following outside the parentheses, the residual heavy particle. Again using the production of carbon-14 isotopes as an example, the following illustrates this notation:

$${}^{14}_{7}\mathrm{N}\ (n,\ p)\ {}^{14}_{6}\mathrm{C}$$

Other examples of these two types of notation are:

TABLE 14.2 LIGHT PARTICLES UTILIZED OR PRODUCED IN INDUCED RADIOACTIVITY

Particle	*Approximate Mass*[a]	*Charge*	*Notation*
Neutron	1	0	${}_0^1n$ or n
Proton	1	+1	${}_1^1\mathrm{H}$ or p
Electron or β particle	1/1845	−1	${}_{-1}^{0}e$ or e
Deuteron	2	+1	${}_1^2D$ or d
α Particle or He^{2+}	4	+2	${}_2^4\mathrm{He}$ or α

[a]Compared to the hydrogen atom of mass 1.

1. the production of neutrons by the α particles of radium striking a target of beryllium:

$${}_4^9\mathrm{Be} + {}_2^4\mathrm{He} \longrightarrow {}_6^{12}\mathrm{C} + {}_0^1n$$

or

$${}_4^9\mathrm{Be}\ (\alpha, n)\ {}_6^{12}\mathrm{C}$$

2. conversion of uranium-238 to the transuranic element neptunium by bombarding the former with neutrons:

$${}_{92}^{238}\mathrm{U} + {}_0^1n \longrightarrow {}_{93}^{239}\mathrm{Np} + {}_{-1}^{0}e$$

or

$${}_{92}^{238}\mathrm{U}\ (n, e)\ {}_{93}^{239}\mathrm{Np}$$

3. the production of neptunium-238 by deuteron bombardment of uranium-238:

$${}_{92}^{238}\mathrm{U} + {}_1^2\mathrm{H} \longrightarrow {}_{93}^{238}\mathrm{Np} + 2\ {}_0^1n$$

or

$${}_{92}^{238}\mathrm{U}\ (d, 2n)\ {}_{93}^{238}\mathrm{Np}$$

Utilization of artificial radioactive atoms in biology and medicine

In the Geiger counter, the particles produced by the disintegration of radioactive nuclei produce clicks resembling those made by a telegraph key. If certain compounds that contain radioactive phos-

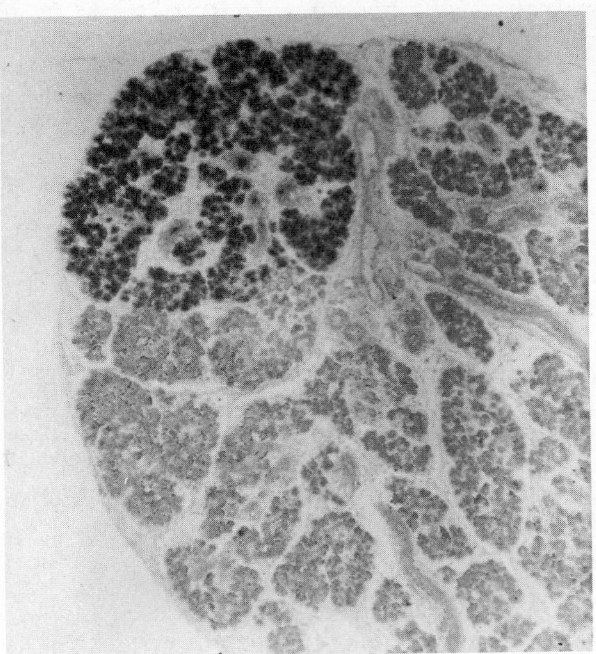

Figure 14.9 Radioautograph of sublingual (darker areas) and submaxillary (lighter areas) glands of a rat given an injection of a carbon 14-labeled bicarbonate salt immediately after birth. The dark area indicates the presence of carbon 14-labeled substances as seen in an animal sacrificed four hours after injection. The sublingual gland shows an intense reaction given by the mucous material present in the acinar cells as well as the lumen of the ducts. The elaboration and release of the sublingual mucus is extremely rapid. The submaxillary secretion is considerably slower. (From R. C. Greulich and C. P. Leblond, "Radioautographic Visualization of Radiocarbon in the Organs and Tissues of Newborn Rats Following Administration of C^{14}-Labeled Bicarbonate," Anat. Rec. ***115**, (1953), 559–585.)*

phorus and calcium are added to the diet, these two elements are rapidly deposited in bone tissue, and their concentration in the latter can be traced and determined by the Geiger counter.

Another technique, called *radioautography,* can be used to locate areas in a tissue in which radioactive isotopes have been deposited. In radioautography, a cut section of the tissue is placed adjacent to a photographic film or plate. The film or plate, when developed, will have darkened in those areas adjacent to the areas in the tissue in which the radioisotope has concentrated (Figure 14.9).

If a chemical compound is synthesized with an uncommon isotope of one of the elements present, it is said to be *tagged.* Substances tagged with radioactive isotopes are especially useful, because their concentration, location, and decomposition products in the reaction system can be established by determining their radioactivity. Thus many organic molecules are tagged by synthesizing them in such a fashion that ordinary carbon-12 atoms are replaced by the weakly radioactive carbon-14 atoms. Tagged molecules have been used to determine the amino acid composition of proteins, to assist in the elucidation of the complex mechanism of photosynthesis, to locate tumors in the body, and in many other ways.

The use of artificially produced radioactive atoms shows great promise in therapy, and much research is being carried out in this field of medical research. Radioactive iodine, iodine-131, has been employed in the detection and treatment of hyperthyroidism (overactivity of the thyroid gland) (see page 516). When taken orally, this isotope is rapidly concentrated in the thyroid gland, particularly if the latter is overactive. If large doses of iodine-131 are administered, a sufficient amount of this isotope becomes concentrated in the thyroid gland to destroy part of the tissue, thus frequently making surgery unnecessary. The isotope potassium-38 is rapidly absorbed by the erythrocytes of the blood and more slowly by the leucocytes. This isotope has been found efficacious in the treatment of polycythemia vera (an overproduction of the red blood cells) and in leukemia (an overproduction of white blood cells). Also, because this isotope is more rapidly absorbed by tumorous growths than normal tissue, it has found some use in the detection of cancer.

An isotope of cobalt, cobalt-60, can be readily prepared by the irradiation of ordinary cobalt, cobalt-59, with neutrons in a nuclear reactor. It is a very strong γ-ray emitter and can be used in place of radium in the treatment of cancer (Figure 14.10). It is much cheaper to produce than radium, has a much shorter half-life (5.3 years), and can be more easily stored. However, its high initial activity requires a great deal of shielding in transport. The use of a very short-life sodium isotope has also been suggested in cancer therapy, because it would be unnecessary to remove it after treatment of the cancerous tissue.

Radiation sickness

Gamma rays appear to be most effective in producing changes in living tissue, while α rays usually produce a lesser change. Apparently due to their very small weight, β particles are the least destructive to tissue.

Extensive radiation absorption by a human can produce such extensive chemical changes in the body that death can ensue. Prior to death, or with lesser doses of radiation, many of the symptoms of radiation sickness may appear. These may include nausea, vomiting, diarrhea, general body weakness, a significant drop in the red and white cell count, loss of hair (see Figure 14.11), extensive damage to the skin, and the appearance of unsightly ulcerative sores that are most difficult to heal.

Generally, radiation from substances absorbed into the body is more damaging than external radiation. The absorption of radio-

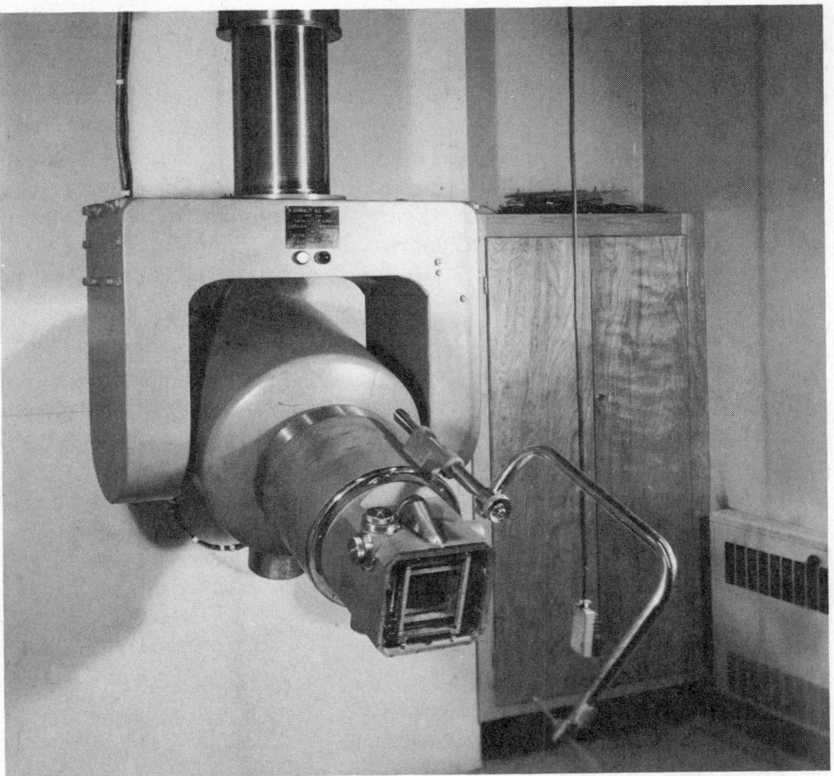

Figure 14.10 Head of a "cobalt bomb" that contains radio-cobalt and directs a beam of penetrating radiation on and into a portion of a patient's body. (Courtesy Dr. T. A. Watson of the Saskatoon Cancer Clinic, Saskatchewan, Canada.)

activity and X rays is capable of modifying the structure of chemical species and thereby disrupting certain essential reactions or trains of reactions. Although enzymes should be susceptible to modification, resulting in the breaking of some metabolic cycles essential to life, probably the most destructive action of radiation occurs when it attacks and modifies nucleoproteins. This may interfere with cell division, or if nucleoprotein material is being transferred from a mother cell to a daughter cell, mutants may be formed. Although some mutants may exhibit a new and greater adaptation than their predecessor form, many mutants either exhibit inferior functions or lack certain functions leading to their destruction, frequently through natural selection processes.

Units used to measure radiation

The curie and the roentgen are units that have been frequently employed in measuring radiation. The curie equals 37 billion nuclear

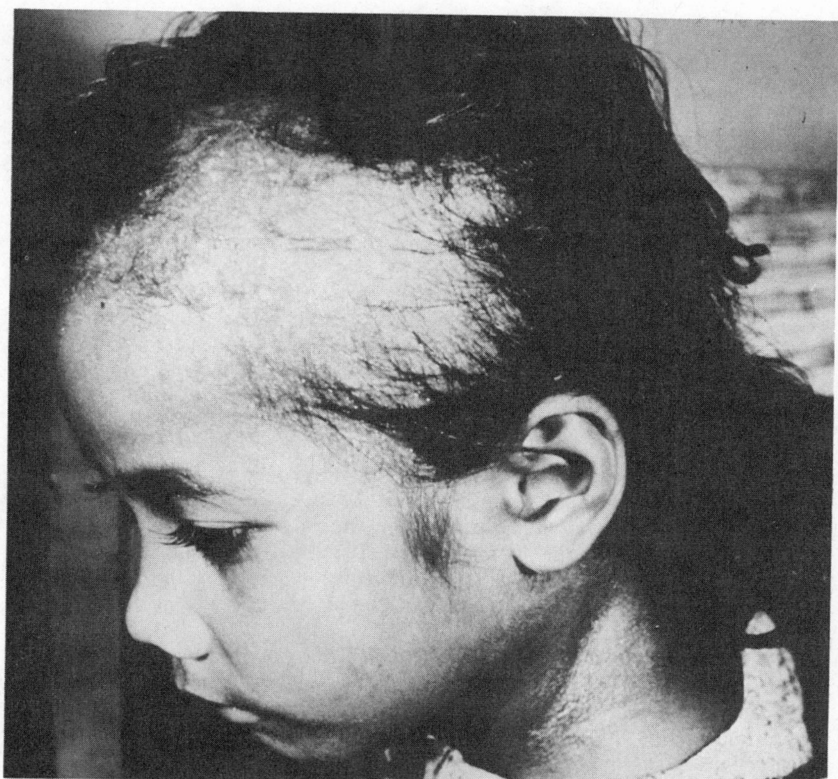

Figure 14.11 A victim of radiation sickness. (Courtesy U.S. Atomic Energy Commission.)

disintegrations. This is the number of disintegrations experienced by a gram of pure radium in 1 second. For small amounts of radiation, the millicurie (37 million disintegrations) and the microcurie (37 thousand disintegrations) are utilized. The roentgen is defined as the amount of X rays or γ rays that will produce 1 electrostatic unit (esu) of positive or negative electricity in a milliliter of air.

Neither of these units of radiation is adaptable to the measurement of radiation absorbed by living tissue. For the curie, no distinction is made in the source of the nuclear disintegrations, while the roentgen measures the absorption of radiation in air rather than in living tissue.

Recently, the rad (radiation absorbed dosage) and the rem (rad equivalent to man) have been introduced to overcome these difficulties. The rad represents the absorption of 100 ergs of energy by tissue (see page 14 for the definition of the erg). Although this may appear to be a small amount of radiation, because 1 cal is equal to about 40 million ergs, it can produce significant changes in living tissue. The rem is that amount of energy absorbed by a human from 1 roentgen of radiation.

STUDY QUESTIONS

1. What is radioactivity, and who first discovered this phenomenon?
2. What discoveries did Pierre and Marie Curie make in the field of radioactivity?
3. Describe the nature of the three rays produced by radium.
4. What is meant by the half-life of a radioactive substance?
5. What two naturally occurring elements are the parents of the radioactive series? After the disintegration of radium, what residue remains? What is the final stable substance produced by the radioactive decay of radium? What is radon?
6. How has radium been used in medical practice?
7. What precautions must be taken in handling radium and radon?
8. How are stable nuclei converted to radioactive substances?
9. How may the presence of radioactive atoms be detected?
10. What uses may be made of tagged compounds?
11. What are some of the promising uses of artificially radioactive atoms in therapy?
12. In 1917, Rutherford converted ordinary nitrogen to an isotope of oxygen with mass 17 by bombardment of the former with α particles. This conversion was accompanied by the escape of protons. Represent this conversion by the two notations discussed on page 208.
13. What are some of the characteristic symptoms of radiation sickness?
14. Discuss some of the changes that can occur in living cells following absorbtion of radiation.
15. It is known that certain types of radiation may induce the formation of mutants. What is the usual fate of mutants?
16. Name and define some of the units employed in the measurement of radioactivity, X rays and γ rays. Which of the units are best suited to measure the amount of absorption by living tissue? By man?

chapter 15

an introduction to organic chemistry

15

Early history and definition of organic chemistry

At the beginning of the nineteenth century, chemistry was divided into two fields. *Inorganic chemistry* dealt with substances obtained from mineral sources, whereas substances obtained from animal or plant sources were placed in the field of *organic chemistry.* Furthermore, the chemist believed that it was not possible to synthesize these latter substances in the laboratory, but that some special power possessed by living tissue was necessary for their formation. This was called the *vitalistic theory.*

In 1828, Wöhler, a German chemist, synthesized urea, NH_2-$CONH_2$, from ammonium cyanate, NH_4CNO. Urea was considered to be definitely organic (an important occurrence of this substance is in the urine), and the ammonium cyanate was prepared from inorganic substances. Shortly after this important contribution of Wöhler, many other organic substances were synthesized in the laboratory, and the vitalistic theory was abandoned.

Today, the division of the study of chemistry into two fields is retained, and organic chemistry is defined as the *chemistry of the compounds of carbon.* This has been made necessary because there are well over a million compounds containing carbon and because most carbon compounds have structures that are more complex than the inorganic ones.

The definition of organic chemistry just stated is not too accurate, because many chemists consider some compounds that contain carbon as inorganic. Such compounds as carbon dioxide, carbon monoxide, sodium carbonate, and calcium carbonate fall into this category (see pages 173–174).

Chemists have been able to classify organic compounds either as *hydrocarbons,* that is, compounds made up exclusively of carbon and hydrogen, or as compounds derived from the hydrocarbons. Many, therefore, prefer the following definition of organic chemistry: *Organic chemistry is the chemistry of the hydrocarbons and their derivatives.*

Sources of organic compounds

Many organic substances are still obtained from animal and vegetable sources. Some of these are glycerol, fats, proteins, sugar, starch, cellulose, alkaloids, perfumes, and flavors. Another source of organic compounds is coal tar, obtained in the destructive distillation of soft coal. Many valuable dyes, drugs, plastics, explosives, perfumes, and flavors are obtained from this source.

Natural gas and petroleum are other important sources of organic compounds. Today American chemists are preparing more and more organic compounds from this source, just as German chemists of two generations ago developed new products from coal tar.

Importance of organic compounds

Organic chemistry furnishes us many necessities and luxuries for better living, such as dyes, drugs, perfumes, flavors, and plastics. Students of nursing, medicine, home economics, and biology should realize that our foods are complex mixtures of organic compounds and that the digestion and utilization of these foods in the body involve many organic reactions. In fact, the proper functioning of the body depends upon many complex organic reactions.

Elements present in organic compounds

All organic compounds contain carbon and many also contain hydrogen, oxygen, and nitrogen. Some contain sulfur and phosphorus. Many organic compounds that contain the halogens (that is, chlorine, bromine, and iodine) have been synthesized in the laboratory, but only a few naturally occurring organic compounds contain these elements. As an example, thyroxine, obtained from the thyroid gland, is an iodine-containing compound. Other elements are of infrequent occurrence in organic compounds.

Structure of organic compounds and the tetracovalency of carbon

If the student will recall the study of atomic structure (see Chapter 3), it will be remembered that the sodium atom possesses a single electron in the valence shell of electrons, whereas chlorine has seven electrons in its outer shell. Now if sodium reacts with chlorine, it loses its single valence electron to chlorine, which builds its outer shell to the stable octet, and the compound sodium chloride is thus formed. It is an electrovalent, or ionic, compound, has a high melting and boiling point, is highly crystalline, and dissolves in water to form a solution that conducts the electric current. Electrovalent compounds such as sodium chloride are formed by loss or gain of electrons between atoms of different elements, and they all more or less show these properties that are possessed by sodium chloride.

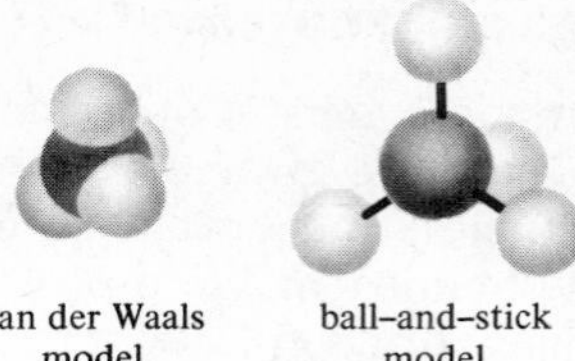

***Figure 15.1** Two models of methane.*

Carbon possesses four electrons in its outer electron shell and can form a stable arrangement of electrons either by loss or gain of four electrons, or by sharing its four electrons with other atoms capable of furnishing four additional electrons. The latter is the actual case for carbon. Compounds formed by sharing of electrons are called *covalent compounds;* unlike sodium chloride, the typical electrovalent compound, they are usually gases, liquids, or soft waxy solids. They have rather low melting and boiling points. Covalent compounds are either insoluble in water or show only a slight solubility in this solvent, and the resulting solution does not conduct an electric current.

One of the simplest covalent compounds containing carbon is the gas methane, CH_4. In methane (Figure 15.1), carbon shares its four valence electrons with electrons from four hydrogen atoms. Its structure can be represented as

```
   H
   ..
H:C:H
   ..
   H
```

The organic chemist usually simplifies this diagrammatic formula for methane by replacing a pair of shared electrons by a dash, and the resulting formula for methane, called a *structural formula,* is

```
    H
    |
H—C—H
    |
    H
```

methane

In its compounds, with few exceptions, carbon has a valence of four, and it is an important axiom of organic chemistry that carbon is *tetracovalent.* In writing structural formulas for organic compounds, the student should always check these formulas to be sure that the proper valence has been assigned to carbon.

In methane and other organic compounds in which carbon is bonded to four univalent atoms or groups, it can be considered as being in the center of a *regular tetrahedron.* The four valence bonds are directed to the corners of the tetrahedron, while the four uni-

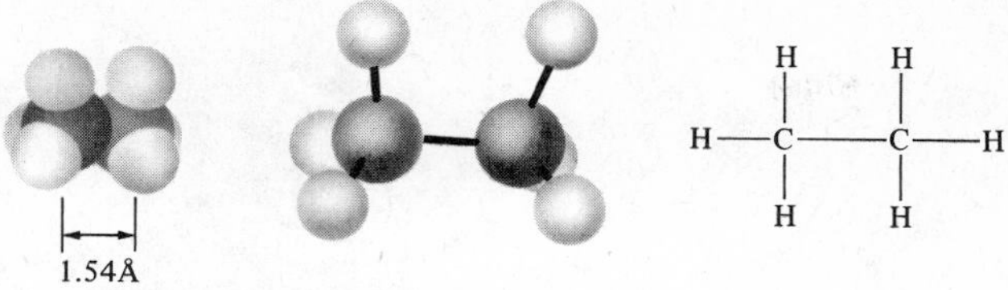

Figure 15.2 Structural representations of ethane.

valent atoms or groups are located at these corners. For this reason, carbon is said to be *tetrahedral* (Figure 15.1). The angle made by any two substituents (such as the two hydrogens in methane) and carbon will be 109° 28′ and is called the *tetrahedral angle.*

Ability of carbon atoms to share electrons with other carbon atoms

A somewhat unique property of carbon atoms is the ability of two or more of them to share electrons, thereby forming chains of carbon atoms. Ethane, C_2H_6, represents a compound in which two atoms of carbon are sharing a single pair of electrons (Figure 15.2).

$$\begin{array}{ccccc} & H & & H & \\ & \cdot\cdot & & \cdot\cdot & \\ H: & C & : & C & :H \\ & \cdot\cdot & & \cdot\cdot & \\ & H & & H & \end{array} \quad \text{or} \quad \begin{array}{ccccccc} & & H & & H & & \\ & & | & & | & & \\ H & - & C & - & C & - & H \\ & & | & & | & & \\ & & H & & H & & \end{array}$$

ethane

The resulting hydrocarbon is said to contain carbon atoms that are joined by a single bond; any organic compound in which all the carbon atoms are joined by single bonds is called a *saturated compound.* Propane (Figure 15.3) is another example of a saturated compound.

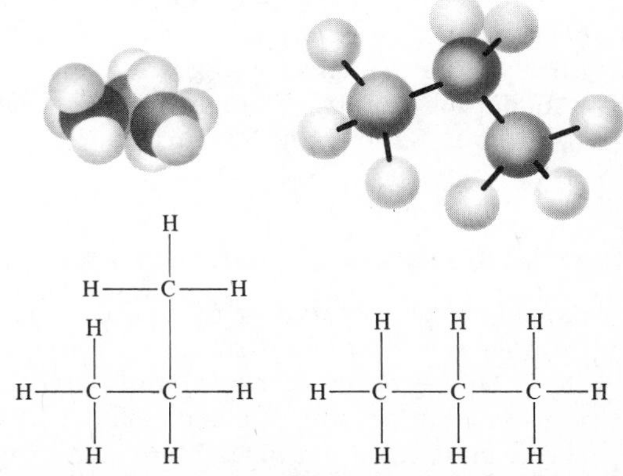

Figure 15.3 Structural representations of propane.

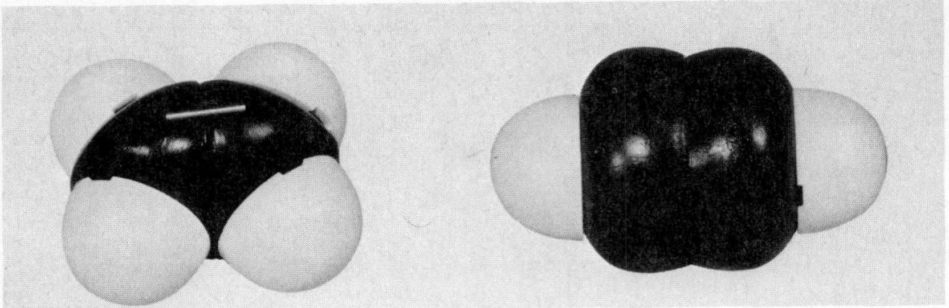

***Figure 15.4** Stuart-Briegleb models of ethylene (left) and acetylene (right) molecules. (Courtesy LaPine Scientific Company.)*

Ethylene, C_2H_4 (Figure 15.4), contains a double bond; that is, the two carbon atoms of ethylene share two pairs of electrons:

$$\begin{array}{c} \mathrm{H}\ \ \mathrm{H} \\ \cdot\cdot\ \ \cdot\cdot \\ \mathrm{H:C::C:H} \end{array} \quad \text{or} \quad \begin{array}{c} \mathrm{H}\quad\ \mathrm{H} \\ |\qquad | \\ \mathrm{H{-}C{=}C{-}H} \end{array}$$

ethylene

Acetylene, C_2H_2 (Figure 15.4), contains a triple bond; that is, the two carbon atoms share three pairs of electrons:

$$\mathrm{H:C:::C:H} \quad \text{or} \quad \mathrm{H{-}C{\equiv}C{-}H}$$

acetylene

Hydrocarbons that contain double- or triple-bonded carbon atoms are said to be *unsaturated hydrocarbons.* These unsaturated hydrocarbons are much more reactive than the saturated ones.

Homologous series and homologs

There are a large number of series of organic compounds in which one member of the series differs from the next by the group of atoms CH_2. Such a series is called a *homologous series,* and the members of such a series are called *homologs* of each other. Methane is the first and ethane the second member of such a series of compounds, called the *methane,* or *paraffin, series of hydrocarbons.* Homologs usually show similar chemical properties. Physical properties, such as melting and boiling points, either increase or decrease as the series is ascended, although the value of this change is irregular in amount (see Table 16.1, page 227).

Isomerism and isomeric compounds

There are many examples of two or more compounds having the same molecular formula but different structures. This phenomenon is called *isomerism,* and such groups of compounds are called *isomeric compounds,* or *isomers.* Incidentally, the molecular formula indicates the total number and kind of atoms present in a molecule but does not indicate the order of arrangement of these atoms in its structure.

Some simple compounds that are isomers are ethyl alcohol and dimethyl ether. Both have the molecular formula C_2H_6O but have the following structural formulas.

```
    H  H              H       H
    |  |              |       |
 H—C—C—OH         H—C—O—C—H
    |  |              |       |
    H  H              H       H
```

ethyl alcohol — dimethyl ether

Fifteen sugars have similar structures to and the same molecular formula as glucose, $C_6H_{12}O_6$, and are isomers of the latter.

The characteristic or functional group of organic compounds

Organic compounds are divided into a number of classes of compounds that possess one or more characteristic atoms or groups of atoms called *functional groups.* Alcohols contain the OH group, which is the typical functional group of this class of compounds. The reactions of an individual organic compound are usually characteristic of the functional group or groups present in it, and this fact is a great aid in simplifying the study of organic chemistry. All compounds of a certain class will show quite similar chemical properties.

Some classes of organic compounds (in addition to the alcohols) are hydrocarbons, ethers, aldehydes, ketones, organic acids, esters, amides, and amines.

The divisions of organic chemistry

Many organic compounds contain open chains of carbon atoms. These may be either straight or branched chains. Such compounds are called *aliphatic compounds,* a name derived from the long continuous-chain organic acids obtained from the fats. Normal butane (frequently abbreviated as *n*-butane) and isobutane are examples of aliphatic compounds.

```
    H  H  H  H              H      H      H
    |  |  |  |              |      |      |
 H—C—C—C—C—H          H—C————C————C—H
    |  |  |  |              |      |      |
    H  H  H  H              H      |      H
                                H—C—H
                                   |
                                   H
```

n-butane — isobutane

Organic compounds may also contain closed rings of atoms. Such compounds are called *cyclic compounds.* If the atoms in the ring are all carbon atoms, the compound is called a *homocyclic,* or *carbocyclic, compound.* Carbocyclic compounds that possess characteristic aliphatic properties are called *alicyclic compounds.* Examples are cyclopropane and cyclohexanol.

```
      H   H                H      OH
       \ /                  \    /
        C              H     C       H
       / \              \  /   \   /
  H—C———C—H            H—C       C—H
     |     |              |       |
     H     H           H—C       C—H
                         /  \   /  \
                       H     C      H
                            / \
                           H   H
```

cyclopropane

cyclohexanol

The *aromatic compounds* are the most important class of carbocyclic compounds. Benzene, C_6H_6, is the parent compound of this division of organic compounds. All the compounds of this division possess at least one ring which contains six carbon atoms and in which three double bonds alternate with three single bonds in the ring. Other aromatic compounds are nitrobenzene and benzaldehyde.

```
                                                            H
                                 NO2                       /
          H                       |                       C
          C                       C                       |\\
        /   \\                  /   \\                    C  O
  H—C        C—H          H—C        C—H          H—C  /  \\  C—H
     ||       |               ||       |              ||      |
  H—C        C—H          H—C        C—H          H—C        C—H
        \   //                  \   //                 \   //
          C                       C                       C
          |                       |                       |
          H                       H                       H
```

benzene

nitrobenzene

benzaldehyde

These have been called aromatic because the first members to be described had pleasant perfumelike odors. Many of these compounds, however, are either odorless or have unpleasant odors.

Another group of cyclic compounds is composed of the *heterocyclic compounds.* These contain other elements besides carbon in the ring. Some examples of heterocyclic compounds are thiophene, pyrrole, and nicotinic acid.

```
  H—C———C—H      H—C———C—H                H
     ||     ||        ||     ||                 C            O
  H—C      C—H    H—C      C—H             /   \\         //
       \   /            \   /           H—C        C—C
         S                N                ||       |    \
                          |             H—C        C—H   OH
                          H                  \   //
                                               N
```

thiophene

pyrrole

nicotinic acid

Nitrogen, oxygen, and sulfur atoms will be most commonly found in the ring of the heterocyclic compounds.

Our study of organic chemistry will begin in the next chapter with a study of some of the simpler aliphatic compounds. Cyclic organic compounds will be discussed in Chapter 23.

STUDY QUESTIONS

1. Why was Wöhler's synthesis of urea in 1828 of such great significance to the history of organic chemistry?
2. What was the vitalistic theory?
3. Give two definitions of organic chemistry.
4. Why are carbon compounds studied separately from the compounds of other elements?
5. List three sources of organic compounds.
6. Why should students of the biological sciences study organic chemistry?
7. What four elements occur most frequently in organic compounds?
8. How are electrovalent, or ionic, compounds formed? Summarize their general properties.
9. How are covalent compounds formed? Summarize their general properties.
10. What is meant by the following terms: isomers, double bond, functional group, homologous series, saturated hydrocarbon, unsaturated hydrocarbon, single bond, triple bond? Illustrate each with a suitable example.
11. How has the study of organic chemistry been simplified by grouping organic compounds into classes?
12. To what division of organic chemistry do the continuous- and branched-chain compounds belong?
13. What are cyclic compounds? Name and describe the three types of cyclic compounds mentioned in this chapter.
14. In the following list of organic compounds, compounds, ________ and ________ are isomers, while compounds ________ and ________ are homologs.

(a) C_2H_4 (b) $CH_3—CH_2Cl$
(c) C_3H_6 (d) $CH_3—CH_2—CH_2—OH$
(e) $CH_3—O—CH_2—CH_3$ (f) $CH_2Br—CH_2Br$

15. Study the following, then answer the questions.

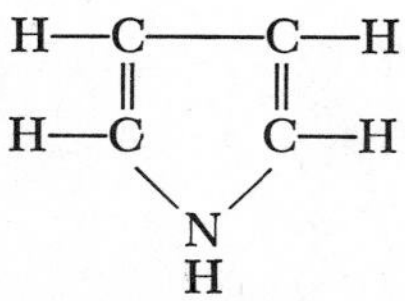

```
H—C——C—H              CH3              CH3—CH—CH2OH
  ‖    ‖               |                    |
H—C    C—H       CH3—C—OH                  CH3
   \  /                |                   (c)
    N                 CH3
    H                 (b)
   (a)
```

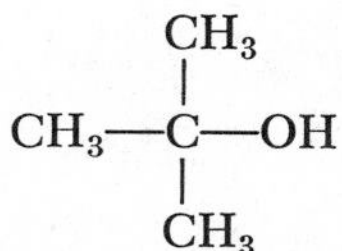

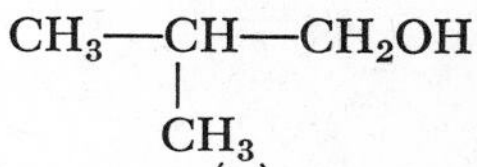

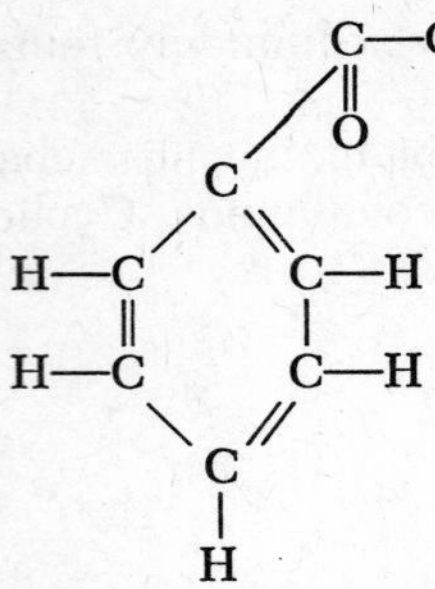

$$\begin{array}{ccc} & & C-CH_3 \\ & & \| \\ & C & O \\ H-C & & C-H \\ \| & & | \\ H-C & & C-H \\ & C & \\ & | & \\ & H & \end{array}$$

(d)

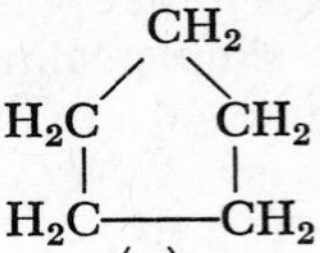

$$\begin{array}{ccc} & CH_2 & \\ H_2C & & CH_2 \\ | & & | \\ H_2C & - & CH_2 \end{array}$$

(e)

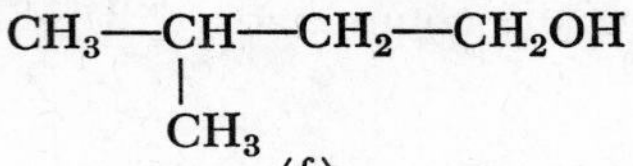

$$\begin{array}{l} CH_3-CH-CH_2-CH_2OH \\ \quad\quad\;\; | \\ \quad\quad CH_3 \end{array}$$

(f)

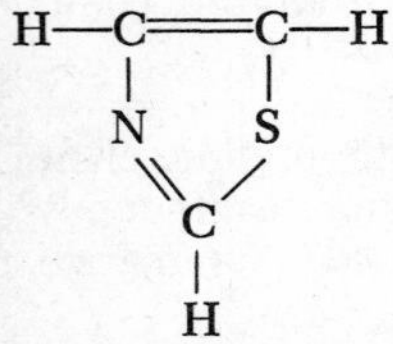

$$\begin{array}{ccc} H-C & = & C-H \\ | & & | \\ N & & S \\ & C & \\ & | & \\ & H & \end{array}$$

(g)

$$\begin{array}{ccc} & CH_3 & \\ & | & \\ & C & \\ H-C & & C-H \\ \| & & | \\ H-C & & C-CH_3 \\ & C & \\ & | & \\ & H & \end{array}$$

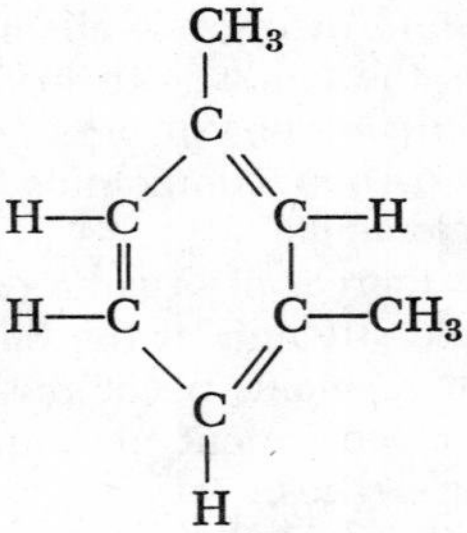

(h)

(a) Compounds ________ and ________ are homologs.
(b) Compounds ________ and ________ are isomers.
(c) Compounds ________ and ________ are aliphatic.
(d) Compound ________ is an alicyclic compound.
(e) Compounds ________ and ________ are aromatic compounds.
(f) Compounds ________ and ________ are heterocyclic compounds.

16. Indicate the division of organic chemistry to which the following compounds belong by writing the word aliphatic, aromatic, alicyclic, or heterocyclic below their formulas.

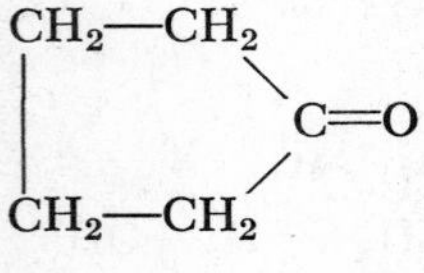

$$\begin{array}{ccc} CH_2-CH_2 & & \\ | & & C=O \\ CH_2-CH_2 & & \end{array}$$

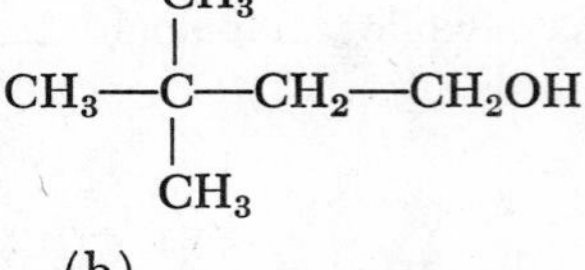

(a)____________

$$\begin{array}{l} \quad\quad\;\; CH_3 \\ \quad\quad\quad | \\ CH_3-C-CH_2-CH_2OH \\ \quad\quad\quad | \\ \quad\quad\;\; CH_3 \end{array}$$

(b)____________

$$\begin{array}{ccc} HC & - & CH \\ \| & & \| \\ HC & & CH \\ & S & \end{array}$$

(c)____________

$$\begin{array}{ccc} & CH_3 & \\ & | & \\ & C & \\ HC & & C-H \\ \| & & | \\ HC & & C-H \\ & N & \end{array}$$

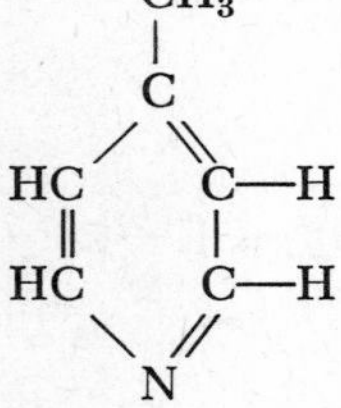

(d)____________

$$\begin{array}{ccc} & H & \\ & | & \\ & N & \\ CH_2 & & CH_2 \\ | & & | \\ CH_2 & & CH_2 \\ & N & \\ & | & \\ & H & \end{array}$$

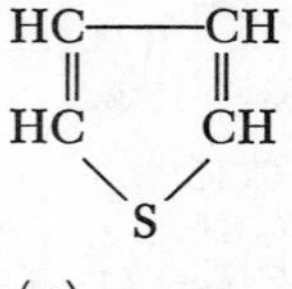

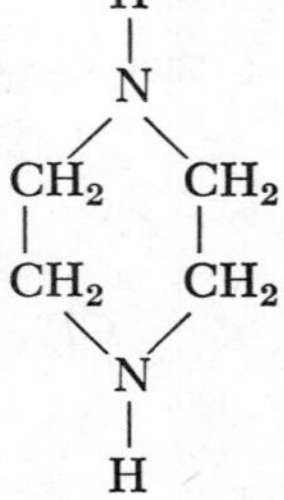

(e)____________

chapter 16

aliphatic hydrocarbons

16

The hydrocarbons are organic compounds that contain carbon and hydrogen, exclusively. There are three important series of aliphatic hydrocarbons, the most important of which is a series of saturated hydrocarbons called the *paraffin,* or *alkane, series* (Table 16.1). The other two are unsaturated hydrocarbon (see page 232) series and are called the *alkene,* or *ethylene, series* and the *alkyne,* or *acetylene, series.*

THE PARAFFIN, OR ALKANE, SERIES

Isomerism

There are only one methane (Figure 15.1), one ethane (Figure 15.2), and one propane (Figure 15.3), as represented by the following structural formulas:

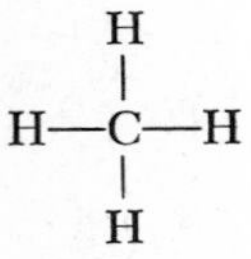

$$\begin{array}{ccccc} & & H & & \\ & & | & & \\ H & — & C & — & H \\ & & | & & \\ & & H & & \end{array}$$

methane

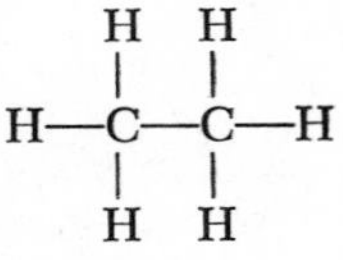

$$\begin{array}{ccccccc} & & H & & H & & \\ & & | & & | & & \\ H & — & C & — & C & — & H \\ & & | & & | & & \\ & & H & & H & & \end{array}$$

ethane

$$\begin{array}{ccccccccc} & & H & & H & & H & & \\ & & | & & | & & | & & \\ H & — & C & — & C & — & C & — & H \\ & & | & & | & & | & & \\ & & H & & H & & H & & \end{array}$$

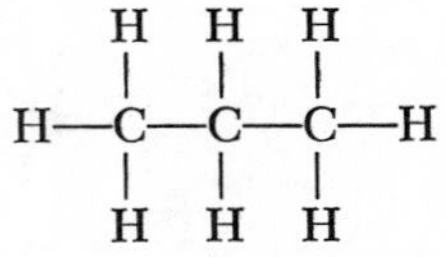

propane

There are two isomeric compounds with the molecular formula C_4H_{10}: normal, or *n*-butane (Figure 16.1*a*), and isobutane (Figure 16.1*b*).

$$CH_3—CH_2—CH_2—CH_3$$

n-butane

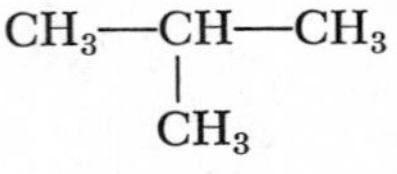

$$\begin{array}{ccccc} CH_3 & — & CH & — & CH_3 \\ & & | & & \\ & & CH_3 & & \end{array}$$

isobutane

There are three isomeric pentanes:

$$CH_3—CH_2—CH_2—CH_2—CH_3$$

n-pentane

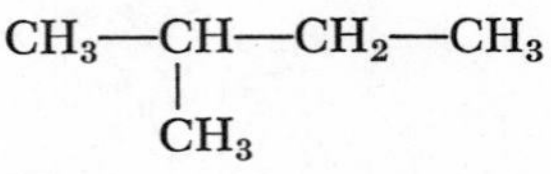

$$\begin{array}{ccccccc} CH_3 & — & CH & — & CH_2 & — & CH_3 \\ & & | & & & & \\ & & CH_3 & & & & \end{array}$$

isopentane

```
        CH3
        |
CH3—C—CH3
        |
        CH3
```

neopentane

The number of possible isomers in this series increases rapidly with the number of carbon atoms in the molecule (see Table 16.1).

TABLE 16.1 THE PARAFFIN, OR ALKANE, SERIES OF HYDROCARBONS

Name	*Molecular Formula*	*Straight Chain Isomer* b.p., °C	*Straight Chain Isomer* m.p., °C	*Number of Structural Arrangements*
Methane	CH_4	−161.5	−184	1
Ethane	C_2H_6	−88.3	−172	1
Propane	C_3H_8	−42	−189.9	1
Butane	C_4H_{10}	−0.6	−135	2
Pentane	C_5H_{12}	36.0	−132	3
Hexane	C_6H_{14}	69	−94.3	5
Heptane	C_7H_{16}	98.4	−90.5	9
Octane	C_8H_{18}	125.8	−56.5	18
Nonane	C_9H_{20}	150.8	−53.7	35
Decane	$C_{10}H_{22}$	174	−30	75

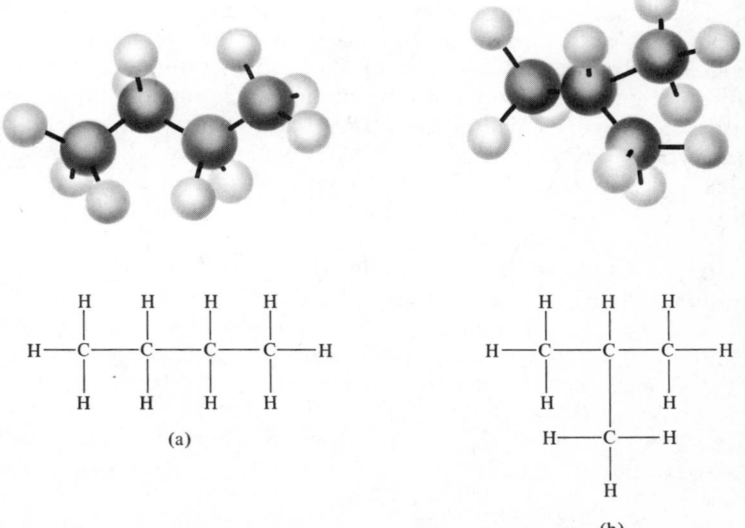

***Figure 16.1** The isomeric butanes. (a) Butane. (b) Isobutane.*

Note that the straight-chain isomer is called normal, abbreviated *n*-.

The alkyl radicals

Groups of atoms that maintain their identity through a series of reactions are called *radicals*. Sulfate, SO_4, ammonium, NH_4, and nitrate, NO_3, are some inorganic radicals.

A group of organic radicals can be derived hypothetically from the alkane hydrocarbons by removing a hydrogen atom. Such radicals are univalent and are called *alkyl radicals*. They are named by replacing the *-ane* ending of the corresponding hydrocarbon by the suffix *-yl* and are frequently represented by the general symbol R. The methyl (CH_3) radical, derived from methane, and the ethyl (C_2H_5) radical, derived from ethane, are among the best-known alkyl radicals.

It must be remembered that radicals do not usually exist as free entities but are nearly always combined with some other atom or group of atoms.

A general formula for the alkane hydrocarbons

The general formula for the members of the alkane, or paraffin, series is C_nH_{2n+2}, where n is the number of carbon atoms in the molecule. If a paraffin hydrocarbon contains 16 carbons, it will then contain 34 hydrogens ($2 \times 16 + 2$), and its formula will be $C_{16}H_{34}$.

IUPAC names of alkane hydrocarbons

The names assigned above to the paraffin hydrocarbons have been in use for a long time and are known as common, or trivial, names. The International Union of Pure and Applied Chemistry has established an international system of nomenclature for organic compounds, now usually called the *IUPAC system*. In this system, these hydrocarbons are known as alkanes. The names of all the members of this series end in *-ane*. For the continuous-chain members of this series, the accepted common names, such as methane, *n*-butane, *n*-pentane, etc., are retained.

Branched-chain alkanes are named as substitution products of the continuous-chain hydrocarbon with the same number of carbon atoms as their *longest continuous chain*. The carbon atoms of this longest continuous chain are numbered consecutively 1, 2, 3, 4, etc., starting from the end of the carbon chain nearest the substitutents. The substituents on this carbon chain are then suitably located by name and by the number of the carbon atom to which they are attached. As an example,

$$\overset{1}{CH_3}—\underset{\displaystyle CH_3}{\underset{|}{\overset{2}{CH}}}—\overset{3}{CH_2}—\overset{4}{CH_2}—\overset{5}{CH_3}$$

is 2-methylpentane, while

$$\begin{array}{c} \qquad\quad CH_3 \\ \overset{1}{C}H_3-\overset{2}{C}-\overset{3}{C}H_2-\overset{4}{C}H_3 \\ \qquad\quad | \\ \qquad\quad CH_3 \end{array}$$

is 2,2-dimethylbutane.

OCCURRENCE

Natural gas, which is used as a fuel, is composed mostly of methane but also contains small amounts of ethane, propane, butanes, pentanes, nitrogen, etc. Methane sometimes occurs in coal mines and is then known as *fire damp.* It is frequently responsible for explosions in mines. Methane is formed when decaying vegetable matter is covered with water, as in a swamp, and is then called *marsh gas.* Certain bacteria convert moist cellulosic material, such as cornstalks, to methane.

Petroleum or crude oil is another important source of saturated hydrocarbons. It is a complex mixture of these but also contains small amounts of many other organic compounds. Also see page 232.

PHYSICAL PROPERTIES

The lower members of the methane series, including butane, are colorless, nearly odorless gases. The members of the series containing 5 to 16 carbon atoms are colorless liquids that are not as dense as water and are insoluble in it. The members with more than 16 carbon atoms are waxy colorless solids. (See Table 16.1 and Figure 16.2 for melting and boiling points of the lower members of this series.)

Preparation

In the laboratory, these hydrocarbons can be prepared by heating the sodium salt of an organic acid with *soda lime,* which is a sintered mixture of sodium hydroxide, NaOH, and lime, CaO. Thus methane is prepared by heating a mixture of the sodium salt of acetic acid, $H \cdot C_2H_3O_2$, with soda lime (Figure 16.3).

$$\underset{\text{sodium acetate}}{\begin{array}{c} H \\ | \\ H-C- \\ | \\ H \end{array}\boxed{\overset{O}{\overset{/\!/}{C}}-ONa + NaO-}} \underset{\text{soda lime}}{H(CaO)} \longrightarrow \underset{\text{methane}}{\begin{array}{c} H \\ | \\ H-C-H \\ | \\ H \end{array}} + Na_2CO_3$$

In a similar fashion, ethane can be prepared by heating sodium propionate (CH_3—CH_2—COONa) with soda lime.

Normal boiling point, °C

150
100
50
0
-50
-100
-150

20 40 60 80 100 120 140

Molecular mass

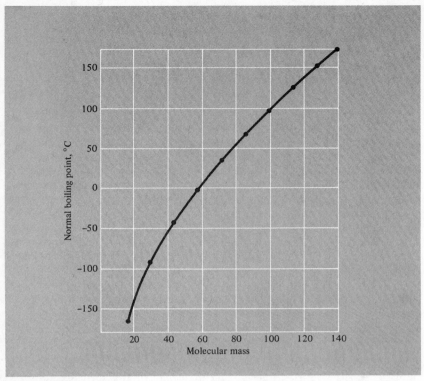

Figure 16.2 *Effect of molecular mass on the normal boiling point of alkanes.*

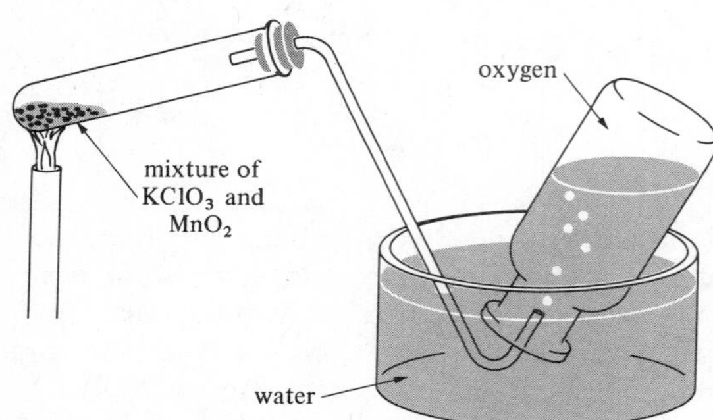

Figure 16.3 *Laboratory apparatus for the preparation of methane.*

$$\underset{\text{sodium propionate}}{H{-}\overset{\displaystyle H}{\underset{\displaystyle H}{C}}{-}\overset{\displaystyle H}{\underset{\displaystyle H}{C}}{-}\overset{\displaystyle O}{\overset{\Vert}{C}}{-}ONa} + NaO{-}H(CaO) \longrightarrow \underset{\text{ethane}}{H{-}\overset{\displaystyle H}{\underset{\displaystyle H}{C}}{-}\overset{\displaystyle H}{\underset{\displaystyle H}{C}}{-}H} + Na_2CO_3$$

If R represents an alkyl group (see page 228), then this reaction may be represented as a general reaction:

$$R\!\left|\!-\overset{\overset{\displaystyle O}{/\!/}}{C}\!-ONa + NaO\!\right|\!-H(CaO) \longrightarrow R\!-\!H + Na_2CO_3$$

Still other methods of synthesis of these hydrocarbons are discussed in organic chemistry texts.

CHEMICAL PROPERTIES

Because the members of this series are saturated, they are not very reactive. For this reason, this series is also called the *paraffin series,* the word paraffin being derived from the Latin *parum affinis,* which means *little affinity.* The two most important reactions of the members of this series are *combustion* and *halogenation.*

The members of this series are not oxidized by strong chemical oxidizing agents such as potassium permanganate, $KMnO_4$, and potassium dichromate, $K_2Cr_2O_7$. In the presence of sufficient oxygen, they burn to form carbon dioxide and water. They form explosive mixtures with air, and extreme caution should be used when the vapors of any of these hydrocarbons mixed with air are present in a confined place. The complete combustion of methane and pentane is represented by the following balanced equations:

$$CH_4 + 2O_2 \longrightarrow CO_2 + 2H_2O$$
$$C_5H_{12} + 8O_2 \longrightarrow 5CO_2 + 6H_2O$$

Incomplete combustion of these hydrocarbons produces carbon monoxide, CO, which is a dangerous poison because it is an odorless gas and the victim is not warned of its presence (page 173). This gas is always a product of the combustion of gasoline in the internal combustion engine, and many deaths are caused by breathing the exhaust gases from an automobile operating in a closed garage.

Saturated hydrocarbons can only react with other substances, such as the halogens chlorine and bromine, by replacing hydrogen atoms of the hydrocarbons with reacting atoms, a process called *substitution.* Unsaturated hydrocarbons add halogen atoms without the displacement of hydrogen atoms, a process called *addition.* Chlorine and bromine react with the saturated hydrocarbons upon heating or in the presence of sunlight or ultraviolet light, and the ensuing reaction may be so violent that an explosion may occur. This reaction, called *halogenation,* cannot usually be controlled, and a mixture of products is formed. Chlorine will combine with methane to form a mixture of methyl chloride, CH_3Cl, methylene chloride, CH_2Cl_2, chloroform, $CHCl_3$, and carbon tetrachloride, CCl_4.

$$CH_4 + Cl_2 \longrightarrow \underset{\text{methyl chloride}}{CH_3Cl} + HCl$$

$$CH_3Cl + Cl_2 \longrightarrow \underset{\text{methylene chloride}}{CH_2Cl_2} + HCl$$

$$CH_2Cl_2 + Cl_2 \longrightarrow \underset{\text{chloroform}}{CHCl_3} + HCl$$

$$CHCl_3 + Cl_2 \longrightarrow \underset{\text{carbon tetrachloride}}{CCl_4} + HCl$$

Bromine will form analogous bromine compounds when it reacts with methane.

PETROLEUM, OR CRUDE OIL

Petroleum, or crude oil, is a mixture of many saturated hydrocarbons as well as small amounts of many other organic compounds. It is separated by distillation into fractions boiling between specified ranges of temperature, a process called *fractional distillation.* Each fraction is itself a mixture of hydrocarbons.

The lowest boiling petroleum fraction, called *petroleum ether,* or *ligroin,* is used as a solvent for fats and oils. The next fraction, gasoline, is mostly a mixture of hexanes, heptanes, and octanes and is in great demand as a fuel for the internal combustion engine. Some of the higher boiling fractions are kerosene, now in use as a fuel; fuel oil, a fuel used in home furnaces, diesel engines, and ocean liners; lubricating oils; petroleum jelly (Vaseline); and mineral oils, such as Nujol. The residue from the distillation of Pennsylvania crude oils is paraffin, and such a crude oil is called a *paraffin-base oil.* Asphalt is the residue obtained from California, Mexican, and Russian oils, and such crudes are called *asphalt-base oils.* Asphaltic bitumens are used in the preparation of asphalt used in road construction and roofing.

UNSATURATED HYDROCARBONS: ETHYLENE AND ACETYLENE

These hydrocarbons contain two carbon atoms joined by a double or triple bond, and they are more reactive than the saturated hydrocarbons. Ethylene contains two double-bonded carbon atoms and is the first member of the alkene series, which is also called the *olefin series.* Acetylene contains a triple bond between two carbon atoms and is the first member of the alkyne series.

Ethylene, C_2H_4

Ethylene (Figure 15.4) is a colorless gas with a faint, pleasant odor and has the following structure:

```
H—C=C—H
   |   |
   H   H
```

Ethylene is prepared by the *dehydration* of ethyl alcohol, $CH_3—CH_2—OH$.

$$\begin{array}{c} \text{H}\ \ \text{H} \\ |\ \ \ | \\ \text{H—C—C—H} \\ |\ \ \ | \\ \boxed{\text{H}\ \ \text{OH}} \end{array} \longrightarrow CH_2{=}CH_2 + H_2O$$

This is accomplished by heating the alcohol with concentrated sulfuric acid at a temperature above 180°C, or by passing the vapors of the alcohol over heated alumina, Al_2O_3.

Ethylene burns readily to form carbon dioxide and water, and a mixture of ethylene and air is quite explosive.

$$C_2H_4 + 3O_2 \longrightarrow 2CO_2 + 2H_2O$$

Strong chemical oxidizing agents, such as potassium permanganate, convert it to ethylene glycol,

$$\begin{array}{cc} CH_2\text{—} & CH_2 \\ | & | \\ OH & OH \end{array}$$

and other oxidation products. A dilute solution of potassium permanganate is pink to purple in color. When it thus oxidizes ethylene, the solution becomes colorless. This is an important test for unsaturation and is known as the *Baeyer test.*

Because ethylene is unsaturated, it will combine directly with many reagents without the loss of hydrogen atoms, a process known as *addition.* In the presence of certain catalysts, ethylene will combine with two atoms of hydrogen to form the saturated hydrocarbon ethane.

$$CH_2{=}CH_2 + H_2 \longrightarrow \underset{\text{ethane}}{CH_3\text{—}CH_2}$$

The addition of hydrogen to a double bond is called *hydrogenation* and is the essential reaction in the conversion of vegetable oils into solid cooking fats (page 283). Ethylene will decolorize solutions of bromine by combining with the latter to form colorless ethylene bromide.

$$CH_2{=}CH_2 + Br_2 \longrightarrow \underset{\text{ethylene bromide}}{\begin{array}{c} \text{H}\ \ \text{H} \\ |\ \ \ | \\ \text{H—C—C—H} \\ |\ \ \ | \\ \text{Br}\ \ \text{Br} \end{array}}$$

This is another important test for unsaturation. Ethylene combines with hydrogen bromide to form ethyl bromide.

$$CH_2{=}CH_2 + HBr \longrightarrow \underset{\text{ethyl bromide}}{CH_3\text{—}CH_2\text{—}Br}$$

Note that the following general reaction represents any addition reaction of ethylene:

$$\underset{}{CH_2{=}CH_2} + x{-}y \longrightarrow \underset{x}{\underset{|}{CH_2}}{-}\underset{y}{\underset{|}{CH_2}}$$

Ethylene has been used as a general anesthetic, although the fact that ethylene forms explosive mixtures with air has discouraged this use. It is used to ripen fruits and vegetables that have been picked when partly ripe. Great quantities of ethylene are used in the synthesis of other organic substances, such as ethyl alcohol; ethylene glycol (Prestone), an antifreeze; ethylene bromide, used in anti-knock fuels; and the plastic polyethylene, or Polythene.

Acetylene, C_2H_2

Acetylene (Figure 15.4), the first member of the alkyne series, is a colorless gas that is insoluble in water and, when pure, has a faint, pleasant odor; however, the crude product has a disagreeable odor due to impurities. Acetylene contains two carbon atoms joined by a triple bond and has the following structure:

$$H{-}C{\equiv}C{-}H$$

It is readily prepared by adding water to calcium carbide, CaC_2.

$$CaC_2 + 2H_2O \longrightarrow Ca(OH)_2 + C_2H_2$$

Calcium carbide is prepared by heating lime, CaO, with coke, C, to a high temperature in an electric furnace.

$$CaO + 3C \longrightarrow CaC_2 + CO$$

When compressed into cylinders, acetylene is unstable and may spontaneously decompose into carbon and hydrogen. It is safely compressed into cylinders containing asbestos that has been saturated with acetone.

Acetylene burns to form carbon dioxide and water, and it forms explosive mixtures with air.

$$2C_2H_2 + 5O_2 \longrightarrow 4CO_2 + 2H_2O$$

When it burns in the presence of pure oxygen, a colorless, very hot flame is produced. This fact is utilized in the oxyacetylene torch, which is used in the cutting and welding of metals. When acetylene burns in air, a yellow luminous flame is produced; acetylene has been used as an illuminating agent, as in the miner's lamp.

Acetylene is a typical unsaturated hydrocarbon and reacts with many substances by addition. It will decolorize a permanganate (positive Baeyer test) and a bromine solution. The addition of bromine to acetylene occurs in two steps, as is indicated by the following equations:

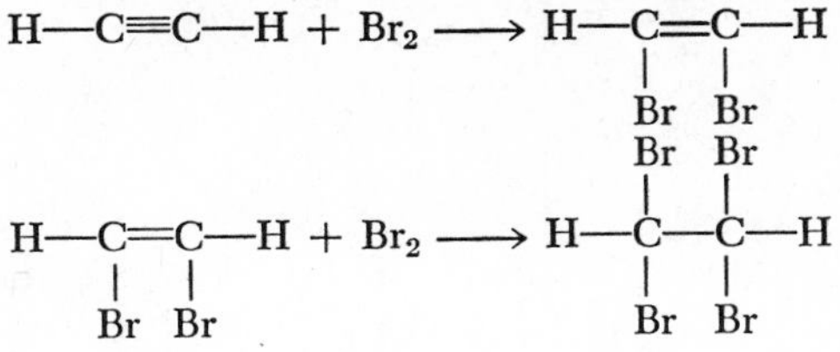

$$\mathrm{H{-}C{\equiv}C{-}H + Br_2 \longrightarrow H{-}\underset{\substack{|\\ Br}}{C}{=}\underset{\substack{|\\ Br}}{C}{-}H}$$

$$\mathrm{H{-}\underset{\substack{|\\ Br}}{C}{=}\underset{\substack{|\\ Br}}{C}{-}H + Br_2 \longrightarrow H{-}\overset{\substack{Br\\ |}}{\underset{\substack{|\\ Br}}{C}}{-}\overset{\substack{Br\\ |}}{\underset{\substack{|\\ Br}}{C}}{-}H}$$

Acetylene has been used also as a starting material for the preparation of many important industrial products, such as neoprene (a synthetic rubber), synthetic drying oils, and plastics.

General formulas for the unsaturated hydrocarbon series

The general formula for the alkene series is C_nH_{2n}, while for the alkyne series it is C_nH_{2n-2}. The hydrocarbon containing 16 carbon atoms would have the formula $C_{16}H_{32}$ in the alkene series and $C_{16}H_{30}$ in the alkyne series.

IUPAC names of the unsaturated hydrocarbons

In the scientific, or IUPAC, nomenclature, the ethylene series is called the alkene series, and the names of the members of this series are obtained by changing the *-ane* ending of the corresponding saturated hydrocarbon to *-ene*. Thus ethylene is called ethene. The acetylene series is called the alkyne series, and the names of its members are obtained by changing the *-ane* ending of the corresponding saturated hydrocarbon to *-yne*. In scientific nomenclature, acetylene is known as ethyne (see Table 16.2.)

Because in the continuous-chain pentenes and pentynes two isomeric hydrocarbons, in which the double or triple bond can be placed between different pairs of carbon atoms, are known, in such cases the multiple bonds are located by indicating the lower-numbered carbon atom to which they are attached. Thus note the following formulas and IUPAC names:

$$\mathrm{\overset{5}{C}H_3{-}\overset{4}{C}H_2{-}\overset{3}{C}H_2{-}\overset{2}{C}H{=}\overset{1}{C}H_2} \qquad \text{1-pentene}$$

$$\mathrm{\overset{5}{C}H_3{-}\overset{4}{C}H_2{-}\overset{3}{C}H{=}\overset{2}{C}H{-}\overset{1}{C}H_3} \qquad \text{2-pentene}$$

$$\mathrm{\overset{5}{C}H_3{-}\overset{4}{C}H_2{-}\overset{3}{C}H_2{-}\overset{2}{C}{\equiv}\overset{1}{C}H} \qquad \text{1-pentyne}$$

$$\mathrm{\overset{5}{C}H_3{-}\overset{4}{C}H_2{-}\overset{3}{C}{\equiv}\overset{2}{C}{-}\overset{1}{C}H_3} \qquad \text{2-pentyne}$$

Branched-chain alkenes and alkynes are named in the same

TABLE 16.2 GENERAL FORMULAS AND NAMES OF ALIPHATIC HYDROCARBON SERIES

Common Name of Series	*IUPAC Name of Series*	*Suffix of the IUPAC Name*	*IUPAC Name of First Member*	*IUPAC Name of C_5 Compound*	*General Formula*	*Formula of C_{10} Member*
Methane, or paraffin	Alkane	-ane	Methane	Pentane	C_nH_{2n+2}	$C_{10}H_{22}$
Ethylene, or olefin	Alkene	-ene	Ethene	Pentene	C_nH_{2n}	$C_{10}H_{20}$
Acetylene	Alkyne	-yne	Ethyne	Pentyne	C_nH_{2n-2}	$C_{10}H_{18}$

fashion as the alkanes (page 228). The following will illustrate the naming of these:

$$\begin{array}{ccccccc} \overset{4}{CH_3} & - & \overset{3}{CH_2} & - & \overset{2}{C} & = & \overset{1}{CH_2} \\ & & & & | & & \\ & & & & CH_3 & & \end{array} \qquad \text{2-methyl-1-butene}$$

$$\begin{array}{ccccccc} \overset{4}{CH_3} & - & \overset{3}{CH} & - & \overset{2}{CH} & = & \overset{1}{CH_2} \\ & & | & & & & \\ & & CH_3 & & & & \end{array} \qquad \text{3-methyl-1-butene}$$

$$\begin{array}{ccccccc} & & CH_3 & & & & \\ & & |{\scriptstyle 3} & & & & \\ \overset{4}{CH_3} & - & C & - & \overset{2}{C} & \equiv & \overset{1}{CH} \\ & & | & & & & \\ & & CH_3 & & & & \end{array} \qquad \text{3,3-dimethyl-1-butyne}$$

However, it is to be noted that in naming branched-chain unsaturated hydrocarbons, the longest continuous chain *that contains the multiple* bond is chosen as the parent hydrocarbon. Thus

$$\begin{array}{ccccccc} CH_2 & = & C & - & CH_2 & - & CH_3 \\ & & | & & & & \\ & & CH_2 & & & & \\ & & | & & & & \\ & & CH_3 & & & & \end{array}$$

is named 2-ethyl-1-butene.

FORMATION OF COVALENT BONDS OF ORGANIC COMPOUNDS

Single covalent bonds are formed by overlap of two *s* atomic orbitals, an *s* and a *p* atomic orbital, or two *p* atomic orbitals, whereby a σ (sigma) type molecular orbital is formed. In ethylene, one of the bonds between the two carbons is a σ bond, while the second is a π (pi) molecular orbital, which results from a sidewise overlap of *p* atomic orbitals present on the two carbons that are joined by the double bond.

SIGMA BONDS

In the hydrogen molecule, a σ molecular orbital, which is primarily centered between the two hydrogen nuclei, cements the two hydrogen atoms together. The general characteristic of a σ bond is that it resides primarily between the two nuclei that are being bonded and that it is symmetrical about the axis that joins the two nuclei.

The formation of the hydrogen molecule in which the two hydrogen atoms are bonded by a σ bond by the overlap of the 1*s* atomic orbitals of two hydrogen atoms is represented in Figure 16.4.

Because the $3p_x$, the $3p_y$, and the $3p_z$ orbitals are degenerate

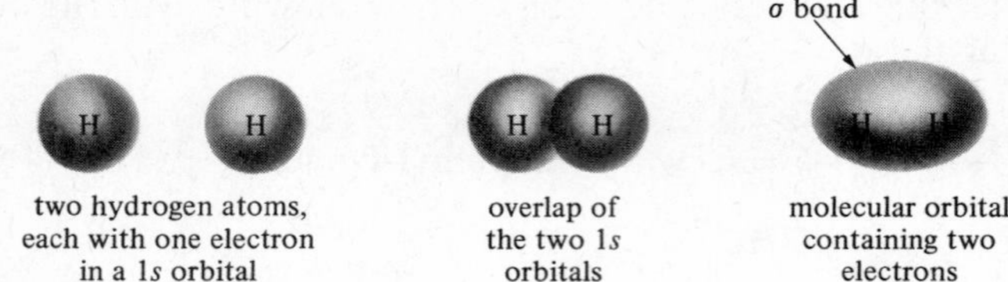

***Figure 16.4** The overlap of the 1s atomic orbitals of hydrogen to form molecular hydrogen.*

(page 47), for convenience in writing structures let us consider that the $3p_x$ orbital of chlorine contains the unpaired electron, while the $3p_y$ and the $3p_z$ orbitals each contains a pair of coupled electrons.

Hydrogen chloride gas is a covalent compound in which a σ covalent bond is formed from the overlap of the 1*s* atomic orbital of hydrogen and the incomplete $3p_x$ orbital of chlorine, as indicated in Figure 16.5.

In the chlorine molecule, a σ bond is also present, formed by the overlap of the incomplete $3p_x$ orbitals of the two chlorine atoms, as indicated in Figure 16.6.

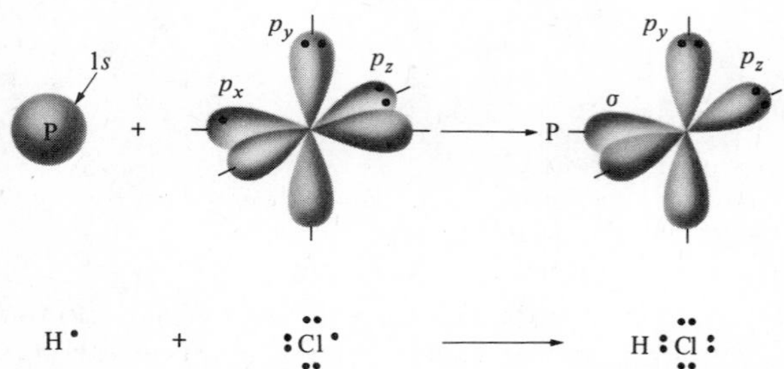

***Figure 16.5** The overlap of the 1s atomic orbital of hydrogen and the* $3p_x$ *orbital of chlorine to form covalent hydrogen chloride.*

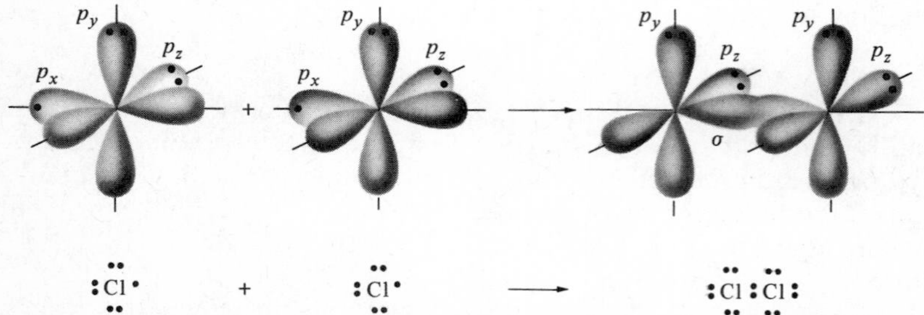

***Figure 16.6** The overlap of the two* $3p_x$ *orbitals of two chlorine atoms to form molecular chlorine.*

Sigma Bonds in Saturated Organic Compounds. As was indicated in Chapter 3 (page 48), the electronic structure of carbon can be represented as

$1s$	$2s$	$2p_x$	$2p_y$	$2p_z$
		$\downarrow$	$\downarrow$	$\bigcirc$
	$\downarrow\uparrow$			
$\downarrow\uparrow$				

in which there is an increase in energy as we proceed from the bottom to the top in this diagram and in which those that are at the same level are degenerate. In a carbon atom, the $1s$ and the $2s$ orbitals each contain a coupled pair of electrons, while the $2p_x$ and the $2p_y$ contain an uncoupled electron each. The $2p_z$ orbital is vacant.

As a first consideration, we might expect carbon to be divalent, the $2s$ orbital being complete and the $2p_x$ and the $2p_y$ orbitals overlapping with the s and the p orbital of the element joining carbon in compound formation. Actually, before compound formation, the carbon atom is activated by absorbing energy and promoting one of the $2s$ electrons to the vacant $2p_z$ orbital. In fact, all four of the electrons in the second shell of the carbon atom become degenerate. They are no longer pure s or pure p orbitals. They are said to be *hybridized*, so that each now possesses one-fourth s character and three-fourths p character. It is now conventional to represent these hybridized bond orbitals as sp^3 orbitals. The electronic structure of the carbon atom activated for union with other elements can now be represented as

$1s$	$2sp^3$	$2sp^3$	$2sp^3$	$2sp^3$
	$\downarrow$	$\downarrow$	$\downarrow$	$\downarrow$
$\downarrow\uparrow$				

and if the four sp^3 carbon atoms bond with the $1s$ orbitals of four hydrogen atoms to form methane, then the electronic structure of carbon in methane can be represented as

$1s$	$2sp^3$	$2sp^3$	$2sp^3$	$2sp^3$
	$\downarrow\uparrow$	$\downarrow\uparrow$	$\downarrow\uparrow$	$\downarrow\uparrow$
$\downarrow\uparrow$				

and four equivalent σ bonds will join the carbon with the four hydrogen atoms. Because these four σ bonds are acting like clouds of electrons, they will repel each other and take up positions as far as possible from each other. This represents a tetrahedral structure in which each of the four covalent bonds is directed from the carbon atom to the hydrogen atoms which can be considered as present at the four corners of the tetrahedron. Thus if carbon is bonded to four univalent groups, it will have a tetrahedral structure. In an organic compound of the structure CH_4, the angle formed by two of the substituents and the carbon will be 109° 28′. See Figure 16.7.

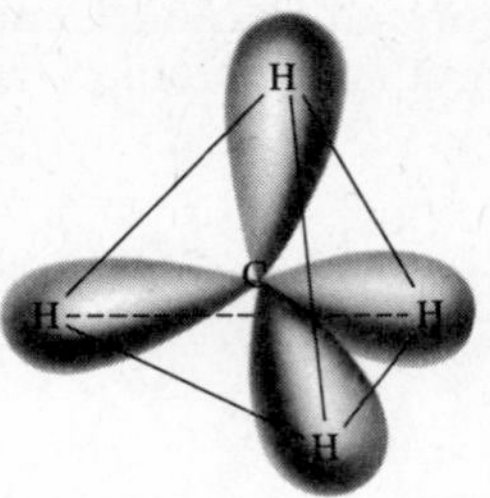

Figure 16.7 A schematic representation of the bonding orbitals of methane.

Sigma and Pi Molecular Orbitals in Unsaturated Organic Compounds. In olefinic hydrocarbons, such as ethylene, only the $2s$ and one of the $2p$ electrons become hybridized. An electron in one of the $2p$ orbitals is not hybridized and remains in a $2\,p$ orbital which is at a right angle to the hybridized sp^2 orbitals. The following is a representation of the electronic structure of double-bonded carbon atoms:

$1s$	$2sp^2$	$2sp^2$	$2sp^2$	$2p_z$
⇃↾	↓	↓	↓	↓

Each of the double-bonded carbon atoms is bonded by a σ bond to the other double-bonded carbon and by 2 σ bonds to two hydrogen atoms. The second carbon and the two hydrogens will be bonded to the carbon atom in a planar trigonal structure, and the H—C—C and the H—C—H bond angles will be very near 120°. This is typical of the bondings of atoms through sp^2 hybrids. If, and only if, the two planes of the H_2C—C set of atoms are coplanar, then the two remaining electrons in the $2p_z$ orbitals of the double-bonded carbon atoms can overlap sideways to form the second bond between these two carbon atoms.

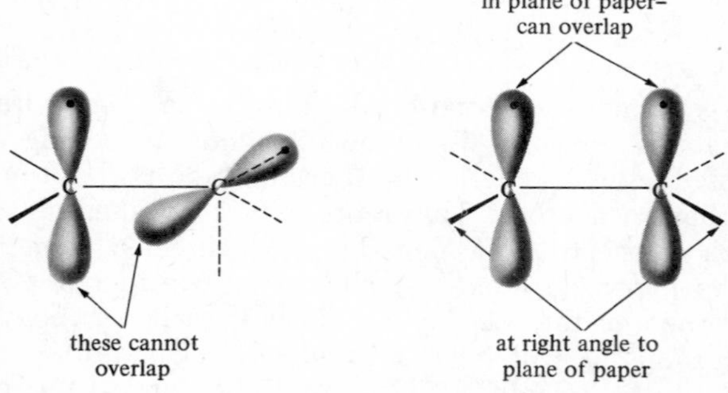

Such a bond is called a π (pi) bond, and the electrons in it reside in

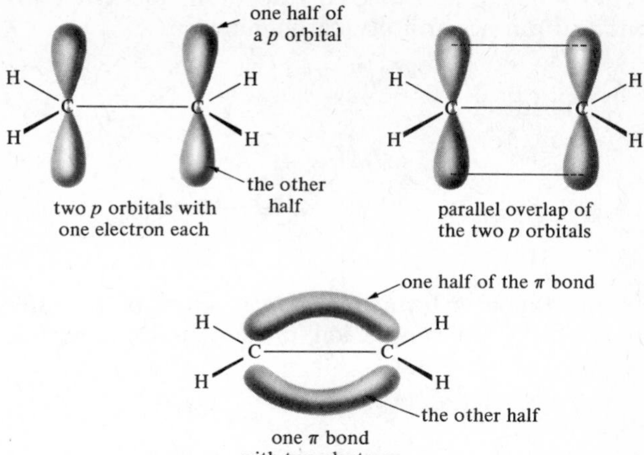

***Figure 16.8** There is a planar arrangement of the three* sp² *σ bonds about each carbon atom of ethylene. The two carbon atoms of ethylene are joined by both a σ and a π bond. Note that the π bond lies above and below the plane of the molecule.*

a lobe in part above and in part below the plane of the molecule, as represented in Figure 16.8.

The π bond represents a loose cloud of electrons that blankets the molecule. If HI adds to ethylene, the H^+ portion of the molecule is attracted by the negatively charged π cloud, and the H^+ becomes bonded to one of the carbon atoms by a σ bond by using the pair of π electrons:

```
              H+
H          H         H           H
  \       /           \ +       /
    C—C       ⟶         C—C—H
  /       \           /       \
H          H         H           H
```

One of the carbons now contains a sextet of electrons and possesses a positive charge. Such a carbon atom is called a *carbonium ion.* It is extremely unstable, has a powerful attraction for any nearby anion, and rapidly joins with the I^- of the HI, thus completing the addition reaction:

```
H          H                  H           H
  \ +     /                     \       /
    C—C—H + I⁻  ⟶    H—C—C—H
  /       \                     /       \
H          H                  I           H
```

When bromine adds to ethylene, the bromine momentarily becomes polarized into a Br^+ ion, called a *bromonium ion,* and a bromide ion (Br^-). Then, as expected, the bromonium ion first attacks the ethylene to form a transitory carbonium ion:

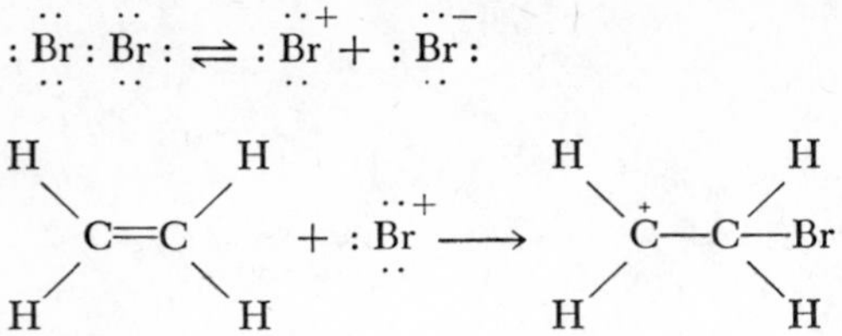

$$:\ddot{Br}:\ddot{Br}: \rightleftharpoons :\ddot{Br}^{+} + :\ddot{Br}:^{-}$$

$$\begin{matrix} H & & H \\ & C{=}C & \\ H & & H \end{matrix} \; + :\ddot{Br}^{+} \longrightarrow \begin{matrix} H & & H & \\ & \overset{+}{C}{-}C{-}Br & \\ H & & H & \end{matrix}$$

Stable dibromoethane is formed as the product of the union of the carbonium ion with a bromide ion present in the reaction mixture, thus:

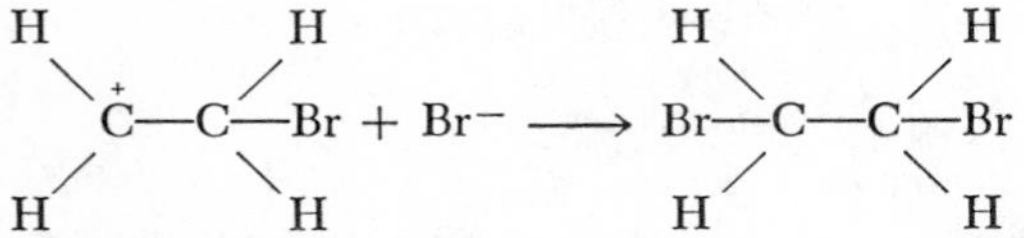

$$\begin{matrix} H & & H & \\ & \overset{+}{C}{-}C{-}Br & \\ H & & H & \end{matrix} + Br^- \longrightarrow \begin{matrix} & H & & H & \\ Br{-} & C{-}C & {-}Br \\ & H & & H & \end{matrix}$$

In aldehydes and ketones of the general formula

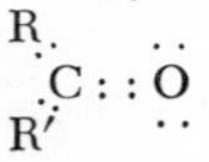

$$\begin{matrix} R \\ & C::\ddot{\underset{..}{O}} \\ R' \end{matrix}$$

in which in aldehydes one of the R groups must be a hydrogen (see page 262), the carbon of the CO group, called the *carbonyl group*, is joined to the oxygen by a double bond. The carbon of the carbonyl group is sp^2 hybridized in a planar trigonal structure to the oxygen and the two R groups. The oxygen, which contains two pairs of unshared electrons, is joined to the carbon atom by a σ bond. This leaves a lone electron in an unhybridized p orbital on both the carbon atom and the oxygen atom. These two remaining p orbitals can overlap, forming the second bond—a π bond—between the carbon and oxygen atoms of the carbonyl group, as represented in Figure 16.9.

Because oxygen is more electronegative and has a greater attraction for electrons than the carbon atom, the π bond actually will be nearer oxygen than carbon, and the carbon atom takes on a partial positive charge. Unlike an olefin, when a reagent adds to a carbonyl group, the anion, or negatively charged particle, adds to the carbon atom. In a second step, the positive addend adds to the oxygen atom.

Finally, the nature of the triply-bonded carbons in acetylene and its homologs will be discussed. In such hydrocarbons, only the 2*s* electrons become hybridized into *sp* orbitals, while the two 2*p* electrons remain unhybridized. Such a structure can be represented as

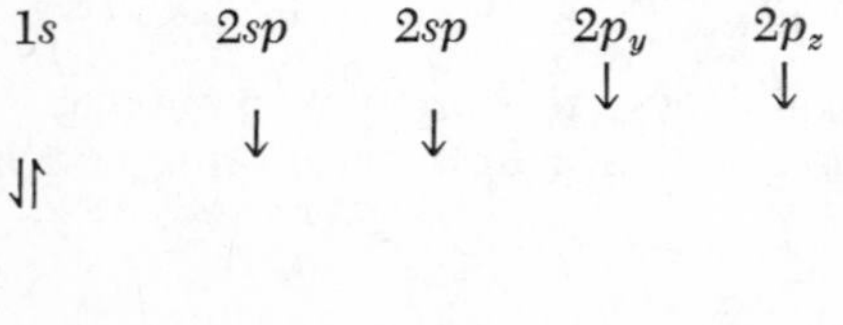

1*s*	2*sp*	2*sp*	$2p_y$	$2p_z$
			↓	↓
	↓	↓		
⇃↾				

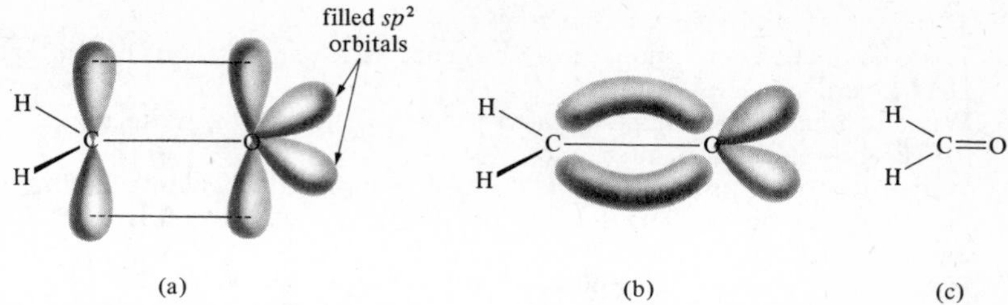

Figure 16.9 *Formaldehyde is a compound in which oxygen has a double bond. (a)* p*-orbital overlap. (b) π bond. (c) Usual representation of formaldehyde.*

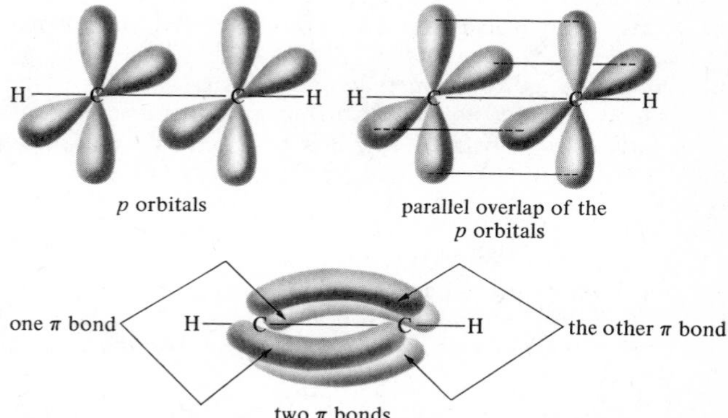

Figure 16.10 *There is a linear arrangement of the two sp σ bonds about each carbon atom of acetylene. Besides a σ bond, the two carbon atoms of acetylene are joined by two π bonds. One of these is above and below the σ bond while the other is in front and in back of the σ bond.*

In acetylene, these two *sp* hybridized electrons join with the other carbon atom and a hydrogen atom in a linear fashion. Such a geometric structure—a straight line—is typical of *sp* hybridization. The remaining unhybridized *p* electrons on the two carbon atoms can overlap to form a π bond above and below and a second π bond in front of and in back of the molecule. Note that the two carbon atoms are joined by a σ and two π bonds, as represented in Figure 16.10.

Acetylene adds reagents in the same manner as ethylene, the positive addend adding to one of the π clouds of electrons, followed by the addition of a negative addend to complete the bond. Because acetylene contains a second π cloud, another molecule of the addend can add to acetylene in the very same fashion as the first one added.

STUDY QUESTIONS

1. Write an equation for the preparation of propane, C_3H_8, from CH_3—CH_2—CH_2—COONa.
2. What is soda lime? What is the active ingredient present in soda lime?

3. What are several sources of methane?
4. Write balanced equations for the complete combustion of butane, C_4H_{10}, and hexane, C_6H_{14}.
5. What is meant by a substitution reaction? To illustrate, write equations for the bromination of methane.
6. Write structural formulas for all the dichloro- and trichlorosubstitution products of ethane. The former has the molecular formula $C_2H_4Cl_2$, and the latter $C_2H_3Cl_3$.
7. What is the chemical composition of petroleum or crude oil? How is it separated into useful products?
8. What commercial products are obtained from petroleum, or crude oil?
9. Is combustion complete in the internal combustion engine? What poisonous substance is formed in this type of engine?
10. Write equations for the preparation of ethylene and acetylene.
11. Which series of hydrocarbons are oxidized by dilute potassium permanganate solution?
12. What is meant by an addition reaction? Illustrate.
13. What two tests are used to establish that a hydrocarbon is unsaturated?
14. Write equations to illustrate the addition of (a) chlorine and (b) bromine to ethylene; the addition of HI to ethylene.
15. Write equations for the addition of two moles of chlorine to a mole of acetylene.
16. Make a list of uses of ethylene and acetylene.
17. Addition of HBr to acetylene produces a compound with the formula $CH_3—CHBr_2$. Write equations for its formation.
18. Name and give the molecular formulas for the first ten members of the paraffin series.
19. What are the general formulas for the paraffin, ethylene, and acetylene series? Give the molecular formula for the hydrocarbon of each of these series with 23 carbon atoms.
20. What are the IUPAC names for the three series of hydrocarbons studied? Write a structural formula for the three carbon-atom members of each of these series, and give the IUPAC name for each.
21. Write structural formulas for two isomeric acetylenic hydrocarbons with five carbon atoms.
22. Indicate which of the following compounds are isomers.

    (a) $CH_3—CH_2—CO—CH_3$
    (b) $CH_3—CHCl—CH_2—CH_2Cl$
    (c) $CH_3—CHCl—CHCl_2$
    (d) $CH_3—CH_2—CH_2—CO—CH_3$
    (e) $CHCl_2—CH_2—CH_2—CH_2Cl$
    (f) $CH_3—CH_2—CH_2—CH_2—CHO$
    (g) $CH_2Cl—CHCl—CHCl—CH_3$

23. Assign IUPAC names to the following branched-chain hydrocarbons:

    (a) $CH_3—\underset{\displaystyle CH_3}{\underset{|}{CH}}—CH_2—CH_3$

    (b) $CH_3—CH_2—\overset{\displaystyle CH_3}{\overset{|}{\underset{\displaystyle CH_3}{\underset{|}{C}}}}—CH_2—CH_2—CH_3$

(c) $CH_3\text{—}\underset{\overset{|}{CH_3}}{CH}\text{—}\underset{\overset{|}{CH_3}}{CH}\text{—}CH_2\text{—}CH_3$

(d) $CH_3\text{—}\underset{\overset{|}{CH_3}}{C}\text{=}CH\text{—}CH_2\text{—}CH_3$

(e) $CH_3\text{—}CH_2\text{—}\underset{\overset{|}{CH_3}}{C}\text{=}\underset{\overset{|}{CH_3}}{C}\text{—}CH_3$

(f) $CH_3\text{—}\underset{\overset{|}{CH_3}}{CH}\text{—}C\equiv C\text{—}CH_2\text{—}CH_2\text{—}CH_3$

(g) $H\text{—}C\equiv C\text{—}\underset{\overset{|}{CH_3}}{CH}\text{—}CH_2\text{—}CH_3$

24. Assign structural formulas to

    (a) 2-methylhexane
    (b) 2,2,3-trimethylpentane
    (c) 3-hexene
    (d) 3,4-dimethyl-1-pentyne
    (e) 3-ethyl-1-pentene

25. How are σ covalent bonds formed by overlap of two *s*, one *s* and one *p*, and two *p* atomic orbitals?
26. Describe the overlap of atomic orbitals to form a single bond between two carbon atoms. To form a double bond. To form a triple bond.
27. In the formation of SiF_4, the 3*s* and the 3*p* atomic orbitals of silicon are hybridized to four sp^3 orbitals. Describe the geometry of the SiF_4 molecule.
28. In BF_3, the three atomic orbitals of boron that overlap with the fluorine atoms are sp^2 hybridized. Describe the geometry of the BF_3 molecule.
29. In $BeCl_2$, the Be—Cl bond is primarily covalent in nature, and in forming this compound, the atomic orbitals of beryllium become *sp* hybridized. What is the geometry of the $BeCl_2$ molecule?
30. Write structural formulas for the following covalent compounds, and label each of the covalent bonds in these compounds as σ or π bonds. CCl_4, $CH_3CH\text{=}CH_2$, $CH_3\text{—}C\equiv CH$, C_2H_6, $CH_3\text{—}CHO$, $HC\equiv N$, $H\text{—}CHO$.
31. In molecular nitrogen (N_2), each nitrogen atom possesses an unshared pair of electrons and is bonded to the other nitrogen atom by a triple bond. Using a model like the one that appears in Figure 16.9 for acetylene, indicate the structure of N_2 gas.
32. Describe the mechanism by which (a) HBr and (b) Cl_2 adds to a molecule of 2-butene ($CH_3\text{—}CH = CH\text{—}CH_3$).
33. Show the steps by which two molecules of chlorine add to a molecule of acetylene.
34. When HCN adds to the carbonyl group (CO) of formaldehyde, CN^- adds first followed by the addition of the proton. Illustrate by two suitable equations the addition of HCN to formaldehyde.

chapter 17

alcohols, ethers, alkyl halides, and polyhalides

17

Alcohols

The *alcohols* are derived from the hydrocarbons by replacing one or more hydrogen atoms with hydroxyl groups. Their general formula is ROH, where R is an alkyl group. The hydroxyl group (OH) is the functional group of the alcohols.

Alcohols, such as ethyl alcohol ($CH_3—CH_2OH$), that contain but one hydroxyl group are called *monohydric alcohols.* Alcohols containing two hydroxyl groups in a molecule are called *dihydric,* while those that contain three such groups or more are called *polyhydric alcohols.* Ethylene glycol, $CH_2OH—CH_2OH$, is a dihydric alcohol, while glycerol, $CH_2OH—CHOH—CH_2OH$, is a polyhydric alcohol, although it is also referred to as a *trihydric alcohol.*

Alcohols may also be classed as *primary, secondary,* and *tertiary.* Primary alcohols contain only one alkyl radical attached to the carbon atom joined to the hydroxyl group. This class of alcohol has the general formula $R—CH_2OH$, and ethyl alcohol, $CH_3—CH_2OH$, belongs to this class. Although methyl alcohol, CH_3OH, does not strictly fit this definition, it is classed as a primary alcohol. A secondary alcohol has two alkyl radicals attached to the carbon atom joined to the hydroxyl group, while tertiary alcohols have three. The general formula for secondary alcohols is

```
R
 \
  CHOH
 /
R'
```

Isopropyl alcohol,

```
CH3
   \
    CHOH
   /
CH3
```

is the simplest secondary alcohol. The general formula for tertiary alcohols is

$$\begin{array}{l} R \\ \quad \diagdown \\ R'—C—OH \\ \quad \diagup \\ R'' \end{array}$$

Tertiary butyl alcohol,

$$\begin{array}{l} CH_3 \\ \quad \diagdown \\ CH_3—C—OH \\ \quad \diagup \\ CH_3 \end{array}$$

is the simplest alcohol in this class.

THE NAMING OF ALCOHOLS

Alcohols are given common names by naming the alkyl radical attached to the hydroxyl group followed by the word alcohol. Some examples follow:

Radical		*Alcohol*	
CH_3-	methyl	CH_3OH	methyl alcohol
C_2H_5-	ethyl	C_2H_5OH	ethyl alcohol
$(CH_3)_2CH-$	isopropyl	$(CH_3)_2CHOH$	isopropyl alcohol
$CH_3-CH_2-CH_2-$	*n*-propyl	$CH_3-CH_2-CH_2OH$	*n*-propyl alcohol
$CH_3-(CH_2)_2-CH_2-$	*n*-butyl	$CH_3-(CH_2)_2-CH_2OH$	*n*-butyl alcohol
$(CH_3)_2CH-CH_2-$	isobutyl	$(CH_3)_2CH-CH_2OH$	isobutyl alcohol
$CH_3-CH_2-\underset{\displaystyle CH_3}{\underset{\vert}{CH}}-$	*sec*-butyl[a]	$CH_3-CH_2-\underset{\displaystyle CH_3}{\underset{\vert}{C}}HOH$	*sec*-butyl alcohol
$(CH_3)_3C-$	*tert*-butyl[b]	$(CH_3)_3C-OH$	*tert*-butyl alcohol

[a]*sec* = secondary
[b]*tert* = tertiary

In the IUPAC system, the names of the saturated alcohols are derived from the names of the corresponding paraffin, or alkane, hydrocarbons by changing the ending *-e* of the name of the hydrocarbon to *-ol*, thus:

CH_4	methane	CH_3OH	methanol
C_2H_6	ethane	C_2H_5OH	ethanol

If there are more than two carbon atoms in the alcohol, and if the hydroxyl group is on a carbon at the end of the chain, a 1- is placed before the name of the alcohol; if it is on the carbon atom next to the end carbon atom, a 2- is placed before the name; etc. This number locates the position of the hydroxyl group in the alcohol molecule. In branched-chain alcohols, the longest chain of carbons containing the alcohol group is chosen, and the alcohol group is given precedence over the alkyl substituents in numbering this chain. Note the application of these rules in assigning IUPAC names to the following alcohols:

$$\overset{3}{CH_3}—\overset{2}{CH_2}—\overset{1}{CH_2OH} \qquad \text{1-propanol}$$

$$\overset{3}{CH_3}—\overset{2}{CHOH}—\overset{1}{CH_3} \qquad \text{2-propanol}$$

$$\overset{4}{CH_3}—\overset{3}{CH_2}—\overset{2}{CH_2}—\overset{1}{CH_2OH} \qquad \text{1-butanol}$$

$$\overset{4}{CH_3}—\overset{3}{CH_2}—\overset{2}{CHOH}—\overset{1}{CH_3} \qquad \text{2-butanol}$$

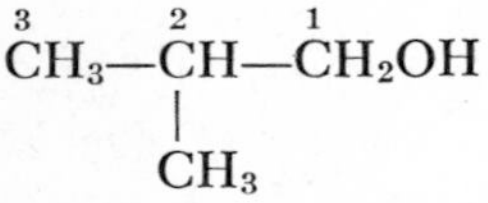

$$\overset{3}{CH_3}—\underset{\displaystyle CH_3}{\underset{|}{\overset{2}{CH}}}—\overset{1}{CH_2OH} \qquad \text{2-methyl-1-propanol}$$

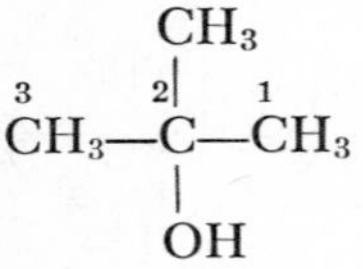

$$\overset{3}{CH_3}—\underset{\displaystyle OH}{\underset{|}{\overset{\displaystyle CH_3}{\overset{2|}{C}}}}—\overset{1}{CH_3} \qquad \text{2-methyl-2-propanol}$$

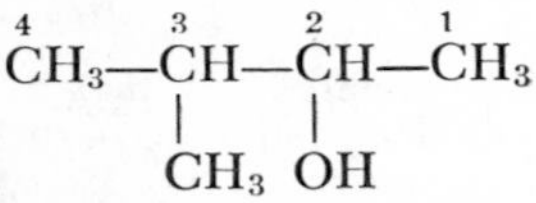

$$\overset{4}{CH_3}—\underset{\displaystyle CH_3}{\underset{|}{\overset{3}{CH}}}—\underset{\displaystyle OH}{\underset{|}{\overset{2}{CH}}}—\overset{1}{CH_3} \qquad \text{3-methyl-2-butanol}$$

REACTION OF ALCOHOLS

Alcohols Are Neutral Substances. Although the alcohols have formulas resembling the inorganic bases, they are neutral covalent compounds that do not ionize in a water solution.

Oxidation of Alcohols. All alcohols are combustible and combine with air or oxygen to form carbon dioxide and water.

With chemical oxidizing agents, primary alcohols are first oxidized to a class of organic compounds called *aldehydes* (see page 263). Ethyl alcohol is oxidized to acetaldehyde,

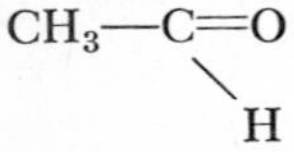

$$CH_3—\underset{\diagdown H}{C}{=}O$$

by a potassium dichromate, $K_2Cr_2O_7$, solution acidified with sulfuric acid.

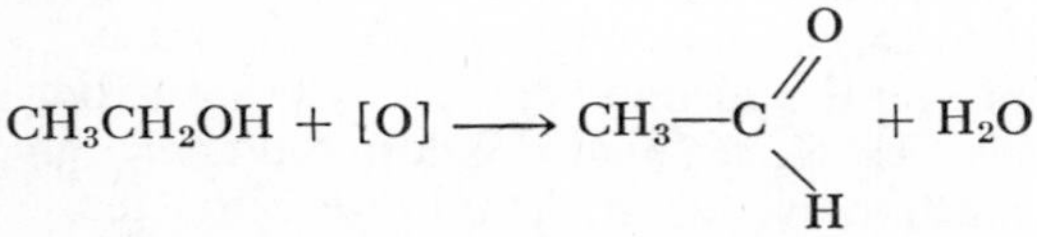

$$CH_3CH_2OH + [O] \longrightarrow CH_3—C\begin{array}{l} {/\!/}O \\ \diagdown H \end{array} + H_2O$$

The complete equation for these oxidations is complex, and to simplify it, the oxygen transferred from the oxidizing agent to the alcohol is represented as [O]. Secondary alcohols are converted to ketones (see page 263) by oxidizing agents. Thus isopropyl alcohol is oxidized to acetone, $CH_3{-}\underset{\underset{O}{\|}}{C}{-}CH_3$.

$$(CH_3)_2CHOH + [O] \longrightarrow (CH_3)_2C{=}O + H_2O$$

Tertiary alcohols are not easily oxidized.

Ester Formation (see also page 275). In the presence of a small amount of sulfuric acid, which acts as a catalyst, alcohols combine with organic acids to form esters. Thus ethyl acetate is formed by the reaction of acetic acid with ethyl alcohol:

$$CH_3CH_2OH + CH_3C({=}O){-}OH \longrightarrow \underset{\text{ethyl acetate}}{CH_3{-}C({=}O){-}O{-}CH_2{-}CH_3} + H_2O$$

METHYL ALCOHOL, OR METHANOL,

$$H{-}\underset{\underset{H}{|}}{\overset{\overset{H}{|}}{C}}{-}OH$$

This has also been called *wood alcohol.* It is a colorless liquid that is soluble in water in all proportions. Small amounts of this alcohol are still obtained by the destructive distillation of wood. Charcoal, acetone, and acetic acid are also obtained from this source. A greater portion of the commercial methyl alcohol is now prepared by a synthetic method in which carbon monoxide and hydrogen are united at a high temperature and pressure in the presence of a suitable catalyst.

$$CO + 2H_2 \longrightarrow CH_3OH$$

The resulting product of high purity is usually called *synthetic methanol.*

Methyl alcohol burns with a hot blue flame. It is an excellent solvent for shellac, gums, and resins. It is used to denature alcohol and as an antifreeze for automobile radiators. Large amounts of methanol are used in the preparation of formaldehyde.

Ingestion of small amounts of methyl alcohol or exposure to its vapors may cause blindness. If large amounts are taken internally, death may result.

Figure 17.1 A ball-and-stick model of a molecule of ethyl alcohol. (From C. W. Keenan and J. H. Wood, General College Chemistry, *4th ed., Harper & Row, 1971.)*

ETHYL ALCOHOL, OR ETHANOL, CH_3—CH_2–OH

Called *grain alcohol,* or simply *alcohol,* ethyl alcohol (Figure 17.1) is a colorless, volatile liquid that is soluble in water in all proportions. It is prepared in the laboratory by the action of yeast on glucose, or corn sugar, $C_6H_{12}O_6$. Yeast produces *zymase* (see page 370), an enzyme that converts glucose to ethyl alcohol and carbon dioxide

$$\underset{\text{glucose}}{C_6H_{12}O_6} \longrightarrow \underset{\text{ethyl alcohol}}{2C_2H_5OH} + 2CO_2$$

Much alcohol is produced industrially by the fermentation (Figure 17.2) of molasses, a by-product of sugar manufacture. Molasses contains considerable amounts of sucrose, $C_{12}H_{22}O_{11}$; an enzyme of yeast, called *invertase,* or *sucrase,* converts sucrose to glucose and fructose. The zymase of the yeast then converts the glucose and fructose to alcohol. Starch from potatoes or grain is also utilized in the manufacture of alcohol. Starch is first converted to the sugar maltose, $C_{12}H_{22}O_{11}$, by the action of malt. Malt is prepared by permitting barley grains to sprout and then killing them without destroying the enzymes present. (One of these enzymes is *diastase,* which converts starch to maltose.) Yeast is then added. An enzyme of yeast, maltase, changes the maltose to glucose, and zymase completes the fermentation.

$$\underset{\text{maltose}}{C_{12}H_{22}O_{11}} + H_2O \xrightarrow{\text{maltase}} \underset{\text{glucose}}{2C_6H_{12}O_6}$$

Ethyl alcohol is also prepared commercially by the addition of water to ethylene, using sulfuric acid as a catalyst.

$$CH_2{=}CH_2 + H_2O \xrightarrow{H_2SO_4} CH_3{-}CH_2OH$$

Alcohol is one of the most important organic substances used in the chemical industry and is an excellent solvent for a great variety of

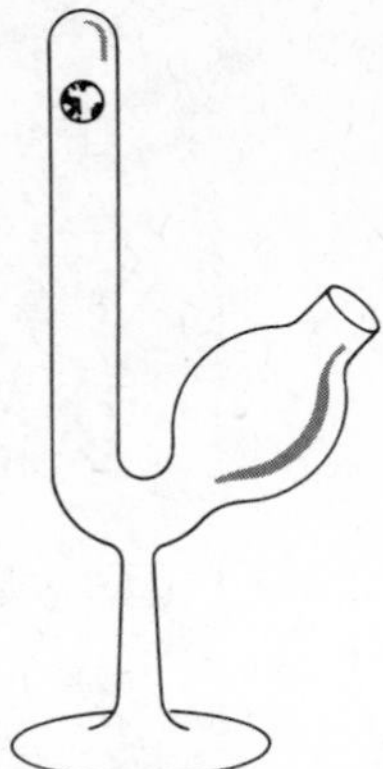

Figure 17.2 A fermentation tube. (Courtesy VWR Scientific.)

substances. Some of the uses of alcohol are the following:

1. as a solvent in the preparation of tinctures, elixirs, and spirits
2. as an antifreeze in automobile radiators
3. a 70 percent solution (by weight) as an antiseptic in the hospital
4. as a solvent for lacquers, varnishes, and shellac
5. in the preparation of toilet water, cosmetics, and perfumes
6. In the synthesis of other organic compounds, such as acetaldehyde, ether, and acetic acid, as well as synthetic rubber

Pure grain alcohol is highly taxed, because it can be used to prepare synthetic liquors. The added cost of this tax would discourage the legitimate uses of alcohol in industry. To avoid this, the government permits the addition of substances to alcohol to make it unfit for drinking. Such alcohol is called *denatured alcohol* and is tax-free. Beer, wine, and whiskey are solutions of alcohol with characteristic flavoring agents.

In fermentation, a dilute solution of alcohol is obtained which can be fractionated to 95 percent alcohol by distillation (Figure 17.3). More concentrated solutions of alcohol cannot be obtained by distillation. Most of the remaining water can be removed by chemicals such as quicklime, CaO. Anhydrous alcohol is called *absolute alcohol.*

The concentrations of alcohol solutions are often expressed in terms of *proof;* the percentage by volume is about half the proof. Thus 95 percent alcohol is approximately 190 proof.

If ingested in large quantities, alcohol causes intoxication and death. Although some consider alcohol a stimulant, the apparent stimulation is probably due to the fact that alcohol acts as a depressant of the central nervous system and destroys some of the natural inhibitions of the individual.

Figure 17.3 A commercial alcohol still. (Courtesy Publicker Industries, Inc.)

ISOPROPYL ALCOHOL, OR 2-PROPANOL, $CH_3—\underset{\displaystyle OH}{\underset{|}{CH}}—CH_3$

This alcohol, also known by the trade names Petrohol and Isopropanol, is being used as a substitute for ethyl alcohol in rubbing alcohol, shave lotions, astringents, and as an antifreeze. It is toxic when ingested internally.

ETHYLENE GLYCOL, $\begin{array}{l} CH_2OH \\ | \\ CH_2OH \end{array}$

This is a colorless, syrupy liquid and is the most important of the dihydric alcohols. It is also called *glycol* and is sold under the trade names Prestone, Zerex, and Peak as an antifreeze. It is an important industrial solvent but has no medicinal uses, because it is toxic when taken internally.

GLYCEROL, $\begin{array}{l} CH_2OH \\ | \\ CHOH \\ | \\ CH_2OH \end{array}$

This trihydric alcohol has also been called *glycerine*. It is a colorless, syrupy liquid with a slightly sweet taste, is readily soluble in water, and is nontoxic.

The most important source of glycerol is as a by-product of the manufacture of soap by the action of caustics on fats and oils. A synthetic process for its preparation is also now in use.

Glycerol is used in the preparation of hand lotions and cosmetics, is an important constituent of the glycerol suppository, and is used as a sweetening agent in prescriptions and toothpastes where a sugar would be undesirable. It is also used to moisten tobacco and has been used as an antifreeze. Because it is hygroscopic, it is added to rubber-stamp pads to prevent the pads from drying.

When treated with a mixture of concentrated nitric and sulfuric acids, glycerol is converted to a pale yellow oil, glyceryl trinitrate, which is incorrectly called nitroglycerine. This is an important explosive but is sensitive to shock. If diluted with some inert solid, such as wood pulp or sawdust, it can be handled safely and is then known as *dynamite*. Small amounts of nitroglycerine are used in medicine as a heart stimulant. When heated to high temperatures, glycerol loses water and forms acrolein, an unsaturated aldehyde.

$$CH_2{=}CH{-}C\begin{array}{l} \diagup H \\ \diagdown\!\!\diagdown O \end{array}$$

acrolein

Acrolein has an acrid penetrating odor and produces a copius flow of tears. Its formation is frequently used as a test for glycerol.

Ethers

The ethers may be considered as organic oxides, and the functional group of the ethers is —O—. The general formula for the ethers is R—O—R. The most important member of this series is diethyl ether, $CH_3—CH_2—O—CH_2—CH_3$, also called *ethyl ether,* or simply *ether.*

DIETHYL ETHER, ETHYL ETHER, OR ETHER, $CH_3CH_2—O—CH_2CH_3$

This ether is a colorless, very volatile liquid that has a slight, pleasant characteristic odor and is only slightly soluble in water. It is prepared by heating ethyl alcohol with concentrated sulfuric acid at about 140°C. The first product of this reaction is ethyl hydrogen sulfate, $CH_3—CH_2—OSO_3H$.

$$CH_3CH_2OH + \begin{matrix} HO \\ \quad\diagdown \\ \quad\quad SO_2 \\ \quad\diagup \\ HO \end{matrix} \longrightarrow \begin{matrix} CH_3—CH_2—O \\ \quad\quad\quad\diagdown \\ \quad\quad\quad\quad SO_2 \\ \quad\quad\quad\diagup \\ HO \end{matrix} + H_2O$$

This latter compound reacts with more ethyl alcohol to form ether, while sulfuric acid is regenerated.

$$\begin{matrix} CH_3—CH_2—O \\ \quad\quad\quad\diagdown \\ \quad\quad\quad\quad SO_2 \\ \quad\quad\quad\diagup \\ HO \end{matrix} + CH_3CH_2OH \longrightarrow H_2SO_4 + C_2H_5—O—C_2H_5$$

It will be noted that the overall reaction involves the loss of a molecule of water from two molecules of ethyl alcohol.

Because ether is quite volatile and combustible, great care should be taken to avoid all flames when it is in use. On standing in air, ether is partly converted to peroxides. If ether containing these peroxides is evaporated nearly to dryness, the residue containing the peroxides may explode with great violence. In general, the ethers are not very reactive chemically, and they show very few reactions besides those mentioned above.

Because ether dissolves many organic substances, such as fats and oils, it is an important solvent and is used in the extraction of oils and fats from inert materials. Until recently, ether was our most important general anesthetic, but it is now being replaced by nitrous oxide, N_2O, cyclopropane, sodium pentothal, methoxyflurane, or halothane (see page 258); general anesthesia is being partially replaced by blocking a certain area with a local anesthetic.

METHOXYFLURANE

Methoxyflurane, or 2,2-dichloro-1,1-difluoroethyl methyl ether,

$$CHCl_2—CF_2—O—CH_3$$

is a clear, colorless liquid and has been introduced under the trade name Penthrane for use in general anesthesia. It may be used for all types of surgery in patients of all age groups. Some respiratory or circulatory depression may accompany its use.

Alkyl halides

Only two groups of halogen compounds will be discussed. The first group consists of the monohalogen compounds, or alkyl halides, such as ethyl chloride, C_2H_5Cl, and methyl bromide, CH_3Br. This group of compounds may be represented by the general formula RX, where (as usual) the R group represents an alkyl group and X represents any halogen atom. The alkyl polyhalides make up the second important group of halogen compounds. Only five of these will be discussed: chloroform, carbon tetrachloride, iodoform, dichlorodifluoromethane, and halothane.

METHYL CHLORIDE, CH_3Cl, AND ETHYL CHLORIDE, C_2H_5Cl

These two compounds are colorless gases that are readily converted to the liquid state by being cooled and compressed. They are used as local anesthetics in minor skin operations. The liquid is sprayed on the spot to be anesthesized. As it evaporates rapidly, it cools the area into insensibility.

The higher alkyl halides are colorless liquids that are insoluble in water and have pleasant odors. The alkyl halides are reactive and are used as solvents and in the synthesis of many organic compounds. *n*-Butyl chloride is used as an anthelmintic, a drug used to destroy intestinal parasites.

Polyhalides

CHLOROFORM, $CHCl_3$

Chloroform is a colorless, heavy liquid, has a pleasant characteristic odor, and is insoluble in water. It is prepared commercially by reducing carbon tetrachloride with iron and dilute hydrochloric acid. Because it dissolves fats readily, it is an important solvent. It was formerly used as a general anesthetic in surgery and obstetrics. Unlike most other general anesthetics, it is not combustible. It has a depressant action on the heart.

CARBON TETRACHLORIDE, CCl_4

This is a colorless, noninflammable, heavy liquid with a pleasant characteristic odor. It is insoluble in water and is a good solvent for fats. Care should be taken not to vaporize carbon tetrachloride in a small room or confined area, because its vapors are quite poisonous. Because of its toxic properties, carbon tetrachloride has been replaced by trichloroethylene, $CCl_2{=}CHCl$, as a dry-cleaning solvent.

IODOFORM, CHI_3

This is a pale yellow solid with a characteristic persistant odor. It is prepared by the action of a solution of iodine in sodium hydroxide on ethyl alcohol. It has been used as an antiseptic for wounds but is being replaced by more efficient drugs. Its antiseptic properties may be due to the slow liberation of iodine when it is in contact with living tissue. It is still being used in the treatment of some skin diseases and skin ulcers.

DICHLORODIFLUOROMETHANE, CF_2Cl_2

This is one of the *Freons*. It is a nontoxic, noncombustible, easily liquefiable gas used in refrigeration and aerosol bombs.

HALOTHANE

Halothane, or 2-bromo-2-chloro-1,1,1-trifluoroethane,

$$CF_3{-}CHClBr$$

is a noninflammable liquid. It is available as Fluothane from Ayerst Laboratories. It is a general inhalation anesthetic that can be used for all types of surgery for patients of all ages. It is not recommended for obstetrics or for patients with known liver or biliary tract diseases.

STUDY QUESTIONS

1. What is the functional group of the alcohols? Write a general formula for the alcohols.
2. Distinguish between mono-, di-, and polyhydric alcohols. Give an example of each.
3. On what basis are alcohols classed as primary, secondary, and tertiary? Indicate whether the following alcohols are primary, secondary, or tertiary.

(a) $CH_3{-}CH_2{-}CHOH{-}CH_3$

(b) $CH_3{-}\underset{\displaystyle CH_3}{\underset{|}{CH}}{-}CH_2{-}OH$

(c) $CH_3{-}CH_2{-}OH$

(d) $CH_3{-}\overset{\displaystyle CH_3}{\overset{|}{\underset{\displaystyle CH_3}{\underset{|}{C}}}}{-}OH$

(e) $CH_3{-}\overset{\displaystyle CH_3}{\overset{|}{\underset{\displaystyle CH_3}{\underset{|}{C}}}}{-}CH_2{-}OH$

(f) $CH_3{-}\underset{\displaystyle CH_3}{\underset{|}{CHOH}}$

(g) $CH_3{-}\overset{\displaystyle CH_3}{\overset{|}{CH}}{-}CHOH{-}CH_3$

(h) $CH_3{-}CH_2{-}\overset{\displaystyle CH_3}{\overset{|}{\underset{\displaystyle OH}{\underset{|}{C}}}}{-}CH_2{-}CH_3$

4. What are the products of oxidation of a primary alcohol? Of a secondary alcohol?

5. Write equations for the reaction of (a) ethyl alcohol and (b) *n*-propyl alcohol, $CH_3CH_2CH_2OH$, with acetic acid in the presence of sulfuric acid, which catalyzes the reaction.
6. How is methanol prepared synthetically? Write an equation to illustrate this method of synthesis. What is another source of this alcohol?
7. What are enzymes? List the names of several enzymes discussed in this chapter (see also Chapter 25).
8. What three substances can serve as the starting material for the preparation of ethyl alcohol by fermentation? Describe the processes that take place in the preparation of alcohol from these substances, and indicate the enzyme required in each step.
9. What is the chief source of glycerol?
10. Discuss the action of methyl alcohol, ethyl alcohol, and glycerol when taken internally.
11. Indicate medicinal uses for the following: isopropyl alcohol, nitroglycerine, glycerol, ethyl alcohol.
12. Indicate uses other than those listed in Question 11 for ethyl alcohol, methyl alcohol, glycol, nitroglycerine, and glycerol.
13. How is glycerol identified in the laboratory?
14. What is denatured alcohol? Why was its introduction a great boon to industry?
15. What is absolute alcohol? Because it cannot be prepared by distillation of a dilute alcohol solution, how may it be prepared?
16. What concentration of ethyl alcohol is used as an antiseptic in the hospital? What is the approximate proof of this solution?
17. Write complete balanced equations for the combustion of methanol and ethanol.
18. What is the functional group of the ethers? They correspond in formula to what class of inorganic compounds? What is the general formula of the ethers?
19. Write equations for the preparation of ethyl ether from ethyl alcohol.
20. Describe the physical properties of ether.
21. What other ether besides ethyl ether is used as a general anesthetic? List three other compounds that are also used for this purpose.
22. What precautions must be taken in handling ether?
23. Disscuss the chemistry, properties, and uses of methoxyflurane and halothane. What undesired symptoms can accompany the use of methoxyflurane? When is the use of halothane as a general anesthetic "contraindicated"?
24. List the two major uses of ether.
25. Are ethers active chemically? What class of organic compounds do they most closely resemble in their chemical reactions?
26. Complete the following statements:

    alcohol + organic acid $\longrightarrow$
    primary alcohol + oxidizing agent $\longrightarrow$
    olefin + hydrogen $\longrightarrow$
    olefin + hydrogen iodide $\longrightarrow$

27. What is the product of the oxidation of the alcohol $CH_3—CH_2—CHOH—CH_3$?
28. Write a general formula for an alkyl halide.
29. Describe the physical properties of (a) methyl chloride, (b) higher alkyl halides, (c) chloroform, (d) carbon tetrachloride, (e) iodoform.

30. Which of the organic halogen compounds described in this chapter can be used as a refrigerant? As an antiseptic? As a fat solvent? As a local anesthetic?
31. What disadvantages does chloroform exhibit when used as a general anesthetic? What one advantage does it possess?
32. Why is dichlorodifluoromethane an excellent refrigerant?
33. Upon what principle does the local anesthetic action of methyl and ethyl chloride depend?
34. Write structural formulas for all of the monochloro-substitution products of normal butane.
35. Write structural formulas for all of the dichloro-substitution products of isobutane.
36. Which of the substances of column II will be required in the synthesis of the compounds listed in column I?

I	*II*
Acetylene	$CH_3—CH_2—OH$
Ethyl alcohol	H_2SO_4
Ethylene	$CH_2{=}CH_2$
Ethyl ether	$CH_3—CHOH—CH_3$
Acetaldehyde	CaC_2
	$C_6H_{12}O_6$ (glucose)
	H_2O
	$KMnO_4$
	Yeast

37. Give common and IUPAC names for the following alcohols: (a) CH_3-$CH_2CH_2CH_2OH$, (b) CH_3CH_2OH, (c) $CH_3CH_2CHOHCH_3$.
38. Assign IUPAC names to all of the alcohols of Question 3 above.

chapter 18

aliphatic aldehydes and ketones

18

The *aldehydes* and *ketones* are oxidation products of alcohols. The functional group of the aldehydes is the aldehyde group,

$$-C\begin{matrix} \diagup H \\ \diagdown\!\!\diagdown O \end{matrix}$$

and these compounds have the general formula

$$R-C\begin{matrix} \diagup H \\ \diagdown\!\!\diagdown O \end{matrix}$$

The carbonyl group,

$$\begin{matrix} \diagdown \\ \diagup \end{matrix}C=O$$

is the functional group of the ketones, and this class of compounds can be represented by the general formula

$$\begin{matrix} R \diagdown \\ R' \diagup \end{matrix}C=O$$

The simpler sugars contain either an aldehyde or a ketone group, and a study of the aldehydes and ketones will be an aid in the study of the sugars, because these two functional groups are responsible, in part, for the chemical properties of these sugars. The aldehydes and ketones in themselves have many uses, as will be enumerated below.

General method of preparing aldehydes and ketones

Aldehydes are prepared by the oxidation of primary alcohols. If the vapor of methyl alcohol reacts with cupric oxide, CuO, formaldehyde is formed:

$$CH_3OH + CuO \longrightarrow Cu + H_2O + H{-}C\begin{matrix} \diagup H \\ \diagdown\!\!\diagdown O \end{matrix}$$

formaldehyde

Oxidation of a secondary alcohol produces a ketone. Thus if isopropyl alcohol is oxidized by acidified solutions of potassium permanganate, $KMnO_4$, or potassium dichromate, $K_2Cr_2O_7$, acetone is formed:

$$\begin{matrix} CH_3 \diagdown \\ CH_3 \diagup \end{matrix} CHOH + [O] \longrightarrow \begin{matrix} CH_3 \diagdown \\ CH_3 \diagup \end{matrix} C{=}O + H_2O$$

acetone

The naming of aldehydes and ketones

COMMON NAMES

The aldehydes are commonly named from the acids they produce when oxidized. Thus the aldehyde that is oxidized to formic acid is called *formaldehyde*, while the one oxidized to acetic acid is called *acetaldehyde*.

The simplest member of the ketone series, CH_3COCH_3, is called *acetone*. The other members of this series are named by indicating the radicals attached to the carbonyl group followed by the word ketone. Thus $CH_3{-}CO{-}CH_2{-}CH_3$ is methyl ethyl ketone, while $CH_3{-}CH_2{-}CO{-}CH_2{-}CH_3$ is diethyl ketone.

IUPAC NAMES

In the IUPAC system of nomenclature, aldehydes are named by changing the *-e* ending of the parent alkane with the same number of carbon atoms to *-al*. Thus acetaldehyde is called *ethanal*. It is not necessary to use a number to locate the position of the aldehyde group, because it must always be at the end of the chain, that is, on the #1 carbon atom.

In this system of nomenclature, ketones are named by changing the *-e* ending of the corresponding hydrocarbon to *-one*. Thus acetone is known as *propanone*. In ketones, it may be necessary to indicate the position of the carbonyl group on the carbon chain. The proper IUPAC name for $CH_3{-}CO{-}CH_2{-}CH_2{-}CH_3$ is 2-pentanone.

The following are examples of the IUPAC names assigned to branched-chain aldehydes and ketones:

$$\overset{3}{C}H_3—\overset{2}{C}H(CH_3)—\overset{1}{C}HO$$ methylpropanal

$$\overset{4}{C}H_3—\overset{3}{C}H_2—\overset{2}{C}H(C_2H_5)—\overset{1}{C}HO$$ 2-ethylbutanal

$$\overset{5}{C}H_3—\overset{4}{C}H(CH_3)—\overset{3}{C}H_2—\overset{2}{C}O—\overset{1}{C}H_3$$ 4-methyl-2-pentanone

(Note that precedence is given to the carbonyl group in numbering the longest chain of a ketone.)

Oxidation of aldehydes and ketones

Aldehydes are readily oxidized even by very mild oxidizing agents, and organic acids are produced. Thus formaldehyde on oxidation yields formic acid:

$$H—C(H)=O + [O] \longrightarrow \underset{\text{formic acid}}{H—C(=O)—OH}$$

Note that the aldehydes are intermediate oxidation products in the conversion of alcohols to organic acids:

$$H—C(H)(H)—OH \xrightarrow{[O]} H—C(H)=O \xrightarrow{[O]} H—C(=O)—OH$$

Aldehydes are oxidized by such mild oxidizing agents as Benedict's solution and Tollens' reagent. Benedicts solution is a solution of copper sulfate, sodium citrate, and sodium carbonate. In an alkaline solution, cupric ions, Cu^{2+}, will precipitate as cupric hydroxide, $Cu(OH)_2$. In the presence of the citrate ions from the sodium citrate, a complex is formed between these ions and the copper ions, and cupric hydroxide does not precipitate. For all practical purposes, Benedict's solution may be considered as a basic solution of cupric oxide, CuO. In the presence of a reducing agent, such as an aldehyde, cupric ions are reduced to cuprous ions, Cu^+, and a yellow to red precipitate of cuprous oxide, Cu_2O, is formed. If sufficient formaldehyde is present, the cupric ions may be reduced to metallic copper. The latter reaction is not usually exhibited by the other aldehydes.

$$H{-}\overset{H}{\overset{|}{C}}{=}O + 2CuO \longrightarrow H{-}\overset{O}{\overset{\|}{C}}{-}OH + \underline{Cu_2O}$$

$$H{-}\overset{H}{\overset{|}{C}}{=}O + CuO \longrightarrow H{-}\overset{O}{\overset{\|}{C}}{-}OH + \underline{Cu}$$

Tollens' reagent is prepared by adding a dilute ammonium hydroxide solution to a solution of silver nitrate until the brown precipitate of silver oxide, Ag_2O, that first forms just dissolves. The solution now contains silver ammonia complex ions, $Ag(NH_3)_2{}^+$, but for practical purposes it may be considered an alkaline solution of silver oxide. In the presence of an aldehyde, Tollens' reagent is reduced to free metallic silver. If the reduction is performed in a clean test tube, the resulting silver will deposit as a silver mirror.

$$H{-}\overset{H}{\overset{|}{C}}{=}O + Ag_2O \longrightarrow H{-}\overset{O}{\overset{\|}{C}}{-}OH + \underline{2Ag}$$

Ketones are not oxidized by mild oxidizing agents, and they give negative tests with Benedict's solution and Tollens' reagent. If vigorous oxidizing agents attack ketones, the carbon-to-carbon chain of the latter will be split, and products that contain fewer carbons than the original ketone will be formed.

OTHER REACTIONS OF ALDEHYDES AND KETONES

Aldehydes and ketones are chemically reactive substances. Of the many reactions that they undergo, only two, excluding the oxidation reactions considered above, will be discussed in this text.

Both aldehydes and ketones condense with phenylhydrazine, $C_6H_5NHNH_2$, to form crystalline, easily purified derivatives called *phenylhydrazones.* The following equation represents the formation of acetone phenylhydrazone:

$$(CH_3)_2C{=}O + H_2N{-}NH{-}C_6H_5 \longrightarrow (CH_3)_2C{=}N{-}NH{-}C_6H_5 + H_2O$$

acetone phenylhydrazone

These phenylhydrazones are used in identifying aldehydes and ketones and have played a significant role in establishing the structure of carbohydrates.

In acidic media, aldehydes condense with one or two molecules of an alcohol to form unstable hemiacetals and stable acetals, respectively. Hemiacetal and acetal linkages are involved in many carbohydrate structures, and they are discussed on page 308.

Formaldehyde, or methanal, $H{-}\overset{H}{\overset{/}{C}}{=}O$

This is a colorless gas with a pungent, biting odor. It is soluble in water, and a 40 percent aqueous solution is known as *formalin.*

Formaldehyde is prepared by oxidizing methyl alcohol with cupric oxide, CuO:

$$CH_3OH + CuO \longrightarrow H{-}\overset{H}{\overset{/}{C}}{=}O + H_2O + Cu$$

A mixture of methyl alcohol vapors and air forms formaldehyde in the presence of a hot copper or silver gauze (temperature above 400°C):

$$2CH_3OH + O_2 \longrightarrow 2H{-}\overset{H}{\overset{/}{C}}{=}O + 2H_2O$$

Formaldehyde reduces both Benedict's solution and Tollens' reagent. It can be converted by slow heating of formalin to a polymer *paraformaldehyde,* $(HCHO)_x$ where x is some large, undetermined number. *Polymers* are large molecules that are formed by the union of many small molecules called *monomers.* Paraformaldehyde, pressed into candles and burned, is used as a fumigant and disinfectant. Formalin is used as a preservative of biological specimens, as a disinfectant, and also in the preparation of synthetic resins and plastics. One of the most important examples of these prepared from formaldehyde is Bakelite, a reaction product of formaldehyde with phenol (carbolic acid).

If an aqueous solution of formaldehyde and ammonia is heated to dryness, a colorless solid called *hexamethylene tetramine* is formed.

$$6HCHO + 4NH_3 \longrightarrow (CH_2)_6N_4 + 6H_2O$$

This substance is used as a urinary disinfectant and is frequently used in the treatment of infections of the kidney. If the urine is acid, this drug will be decomposed slowly into formaldehyde, which must be the active disinfectant. If the urine is not acid, it is necessary to administer sodium dihydrogen phosphate, NaH_2PO_4, to the patient. Hexamethylene tetramine is also known as *Urotropine* or *Methenamine* and is used in the preparation of plastics and explosives and in the vulcanization of rubber.

Acetaldehyde, or ethanal, $CH_3—C(H)=O$

This gas is readily condensed to a colorless liquid. It has an odor resembling that of apples and is prepared by the oxidation of ethyl alcohol. Paraldehyde, $(CH_3CHO)_3$, is a *trimer* of acetaldehyde; that is, it is formed by the union of three molecules of the latter. Paraldehyde is used as a sedative (a depressant of the central nervous system) and a hypnotic (a sleep-producing drug). It produces an unpleasant breath that lasts for hours.

Chloral, or trichloroacetaldehyde, $CCl_3—C(H)=O$

This is a colorless, oily liquid with a characteristic odor. Chloral combines with water to form a colorless, crystalline solid called *chloral hydrate.*

$$CCl_3—C(H)=O + H_2O \longrightarrow CCl_3—C(H)(OH)—OH$$

chloral hydrate

This is one of the few instances in organic chemistry in which a stable compound contains two hydroxyl groups attached to the same carbon atom. Chloral hydrate is used as a hypnotic.

Acetone, or propanone, $CH_3—CO—CH_3$

This is a sweet-smelling, colorless, mobile liquid and is soluble in water in all proportions. It is prepared by the oxidation of isopropyl alcohol. Commercially, it is also obtained by the fermentation of cornstarch by a special strain of bacteria called the *Weizmann bacteria. n*-Butyl alcohol, ethyl alcohol, carbon dioxide, and hydrogen are other products formed in this fermentation.

Acetone is an important solvent and is used to remove lacquers and nail polish. It occurs in large amounts in the urine and blood of patients suffering from diabetes mellitus, and is responsible for the characteristic acetone breath in this disease. Acetoacetic acid,

$$CH_3—C(=O)—CH_2—C(=O)—OH$$

which is both a ketone and an organic acid, and beta-hydroxybutyric acid

$$CH_3—CHOH—CH_2—\overset{\overset{O}{/\!/}}{C}—OH$$

are also abnormal products present in the blood and urine of diabetics.

STUDY QUESTIONS

1. What are the functional groups of the aldehydes and ketones? Write general formulas for these two classes of organic compounds and a structural formula for the simplest member of each of these classes.
2. Why is the study of aldehydes and ketones useful before one begins the study of the chemistry of foods?
3. Write structural formulas for the first major product of oxidation of the following alcohols: *n*-butyl alcohol, $CH_3CH_2CH_2CH_2OH$; isopropyl alcohol, $CH_3CHOHCH_3$; secondary butyl alcohol, $CH_3CH_2CHOHCH_3$; *n*-propyl alcohol, $CH_3CH_2CH_2OH$; isobutyl alcohol, $(CH_3)_2CHCH_2OH$.
4. Aldehydes are intermediate products of the oxidation of __________ to __________. Give examples.
5. Discuss the following concerning Benedict's solution and Tollens' reagent: (a) method of preparation; (b) effective chemical agent present in these reagents; (c) reaction of aldehydes and ketones with these reagents; (d) the equations for the reaction of formaldehyde with these reagents.
6. Assign common and IUPAC names to the following: (a) CH_3CH_2CHO; (b) $CH_3CH_2COCH_3$; (c) CH_3CHO; (d) CH_3COCH_3; (e) $CH_3CH_2CH_2$-CHO; (f) $CH_3CH_2COCH_2CH_3$; (g) $CH_3CH_2CH_2COCH_3$.
7. How are the following prepared commercially: acetone, formaldehyde, acetaldehyde, and hexamethylene tetramine?
8. What is a polymer? What polymeric substance is discussed in this chapter? Give an example of a trimer.
9. Indicate important uses for the following: Phenol, chloral hydrate, paraldehyde, acetone, formalin, paraformaldehyde, formaldehyde, hexamethylene tetramine.
10. What three abnormal products appear in the blood and urine of diabetics?
11. Chloral hydrate is an exception to what important rule of organic chemistry?
12. Assign IUPAC names to the following branched-chain aldehydes and ketones.

   (a) $CH_3—\underset{\underset{CH_3}{|}}{CH}—CH_2—CHO$

   (b) $CH_3—\underset{\underset{CH_3}{|}}{\overset{\overset{CH_3}{|}}{C}}—CHO$

   (c) $CH_3—CH_2—CO—CH(CH_3)_2$

   (d) $CH_3—CH_2—CO—CH_2—\underset{\underset{CH_3}{|}}{CH}—CH_3$

13. Write a structural formula for the phenylhydrazone that is formed when phenylhydrazine condenses with acetaldehyde.

chapter 19

organic acids and esters

19

Organic acids

The functional group of the organic acids is the carboxyl group,

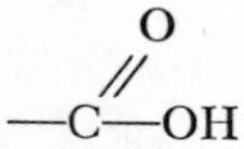

$$-\overset{\overset{\displaystyle O}{/\!/}}{C}-OH$$

which is often written as $-CO_2H$ or $-COOH$. The general formula of the acids is

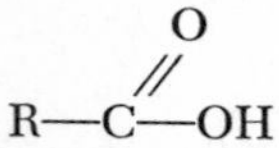

$$R-\overset{\overset{\displaystyle O}{/\!/}}{C}-OH$$

The series of acids derived from the saturated hydrocarbons by replacing a $-CH_3$ group by a carboxyl group is called the *saturated fatty acid series,* because some of the higher members of this series are obtained from the animal and vegetable fats and oils. The saturated fatty acids are the most important group of organic acids and will be discussed first, followed by a discussion of some important acids from other series.

The saturated fatty acids and their salts

NOMENCLATURE OF THE SATURATED FATTY ACIDS

Common names for these acids were usually assigned at the time of their discovery and are usually derived from some occurrence or characteristic property of these acids. The common names and formulas for some of the important members of this series follow:

formic acid, $H-CO_2H$
acetic acid, CH_3-CO_2H
propionic acid, $CH_3-CH_2-CO_2H$
butyric acid, $CH_3-(CH_2)_2-CO_2H$
valeric acid, $CH_3-(CH_2)_3-CO_2H$
palmitic acid, $CH_3-(CH_2)_{14}-CO_2H$
stearic acid, $CH_3-(CH_2)_{16}-CO_2H$

In the IUPAC system, these acids are named by changing the *-e* ending of the saturated hydrocarbon with the same number of carbon atoms to *-oic,* followed by the word *acid:*

methanoic acid, $H—CO_2H$
ethanoic acid, $CH_3—CO_2H$
butanoic acid, $CH_3(CH_2)_2—CO_2H$
hexadecanoic acid, $CH_3—(CH_2)_{14}—CO_2H$

Although most of the naturally occurring fatty acids possess a continuous chain of carbon atoms, the following examples illustrate the assignment of IUPAC names to branched-chain, as well as substituted, acids:

$$\overset{3}{C}H_3—\underset{\displaystyle CH_3}{\overset{2}{C}H}—\overset{1}{C}OOH$$ methylpropanoic acid

$$\overset{4}{C}H_3—\underset{\displaystyle CH_3}{\overset{3}{C}H}—\overset{2}{C}H_2—\overset{1}{C}OOH$$ 3-methylbutanoic acid

$$\overset{6}{C}H_3—\overset{5}{C}H_2—\overset{4}{C}H_2—\underset{\displaystyle OH}{\overset{3}{C}H}—\overset{2}{C}H_2—\overset{1}{C}OOH$$ 3-hydroxyhexanoic acid

$$\overset{3}{C}H_2Cl—\overset{2}{C}H_2—\overset{1}{C}OOH$$ 3-chloropropanoic acid

Substituted acids are also frequently assigned common names in which a substituent on a carbon atom adjacent to the carboxyl group is an α substituent. If it is on the second carbon, it is a β substituent; if on the third carbon from the carboxyl group, it is a γ substituent. The following examples illustrate this type of nomenclature:

$$CH_3—\underset{\displaystyle NH_2}{\overset{\alpha}{C}H}—COOH$$ α-aminopropionic acid (alanine)

$$CH_3—\underset{\displaystyle OH}{\overset{\beta}{C}H}—CH_2—COOH$$ β-hydroxybutyric acid

$$Cl\overset{\gamma}{C}H_2CH_2CH_2COOH$$ γ-chlorobutyric acid

GENERAL METHODS OF PREPARATION AND PROPERTIES OF ORGANIC ACIDS

Organic acids may be prepared by the action of concentrated sulfuric acid on the salts of the acids, usually followed by distillation of the free acid if it is volatile or by filtration if it is a water-insoluble solid.

$$CH_3—CO_2Na + H_2SO_4 \longrightarrow CH_3—CO_2H + NaHSO_4$$

The fatty acids may also be prepared by the oxidation of the appropriate primary alcohol or aldehyde (see pages 263–265).*

In an aqueous solution, the organic acids ionize to form an organic anion and a hydrogen ion. Because this ionization is slight, they are weak acids.

$$CH_3—CO_2H \rightleftharpoons \underset{\text{acetate ion}}{[CH_3—CO_2]^-} + H^+$$

Like the inorganic acids, the organic acids will neutralize a base to form a salt and water. Thus acetic acid reacts with sodium hydroxide to form sodium acetate and water:

$$CH_3—CO_2H + NaOH \longrightarrow CH_3—CO_2Na + H_2O$$

Another very important reaction of organic acids is *esterification,* which is the formation of an ester by the interaction of an organic acid and an alcohol. A small amount of concentrated sulfuric acid is usually added to catalyze this reaction.

$$\underset{n\text{-propyl alcohol}}{CH_3—CH_2—CH_2—O}\boxed{H + H—O}\underset{\text{formic acid}}{—\overset{\overset{\displaystyle O}{\|}}{C}—H} \longrightarrow$$

$$\underset{n\text{-propyl formate}}{CH_3—CH_2—CH_2—O—\overset{\overset{\displaystyle O}{\|}}{C}—H} + H_2O$$

FORMIC OR METHANOIC ACID, $H—CO_2H$

This acid is responsible for the painful sting of bees and some ants. Formic acid is unique in two respects. First, it consists of a hydrogen attached to a carboxyl group (all other saturated fatty acids have an alkyl group attached to the carboxyl group). Second, it contains an aldehyde group as well as an acid group, and, like the aldehydes, it is attacked by mild oxidizing agents such as Benedict's solution.

$$H—O—\boxed{C\begin{matrix}\diagup H\\ \\ \diagdown\!\!\diagdown O\end{matrix}}$$

Formic acid is a slightly stronger acid than the other fatty acids.

*Although saturated fatty acids are used as examples, their methods of preparation and properties are representative of most organic acids.

ACETIC OR ETHANOIC ACID, $CH_3—CO_2H$

This acid, of which only one of the hydrogens is acidic, is prepared by the oxidation of ethyl alcohol or acetaldehyde. Commercially, it has also been obtained as a by-product of the destructive distillation of wood. A commercial synthesis of acetic acid uses acetylene as a starting material.

Acetic acid is the acid in vinegar, and this condiment contains from 4 to 10 percent of this acid. Small amounts of other substances are responsible, in part, for the characteristic taste and odor of vinegar. Cider vinegar is prepared by converting the sugars present in cider to alcohol by means of yeast. Then bacteria (Bacterium aceti), "mother of vinegar," catalyze the oxidation of alcohol to acetic acid. In the slow vinegar process, the air is permitted to diffuse into the cider, and this takes months. In the quick vinegar process, the cider is permitted to flow over beechwood shavings that have previously been moistened with vinegar to inoculate them with the proper bacteria, and the conversion to acetic acid is usually completed with a day or so.

Anhydrous acetic acid freezes to an icelike solid when it is cooled to 16.6°C and is therefore called *glacial acetic acid.* Lead acetate, $Pb(C_2H_3O_2)_2$, also called *sugar of lead,* is used in the treatment of some skin diseases and poison ivy. It is quite toxic when taken internally.

BUTYRIC OR BUTANOIC ACID, $CH_3—(CH_2)_2—CO_2H$

This is a colorless, oily liquid with a persistent disagreeable odor. A small amount of this acid is present in butter as an ester of glycerol. Butyric acid is responsible in part for the odor of rancid butter. It also occurs in perspiration.

PALMITIC ACID $CH_3—(CH_2)_{14}—CO_2H$, AND STEARIC ACID, $CH_3—(CH_2)_{16}—CO_2H$

These are waxy, colorless solids that are nearly odorless and are quite insoluble in water. They are present in most animal and vegetable fats and oils as glycerol esters (see page 280). Most soaps contain appreciable amounts of the alkali salts of these two acids. Stearic acid is used in making candles. Zinc stearate has been used as an antiseptic and dusting powder.

Other important organic acids

OLEIC ACID, or 9-OCTADECENOIC ACID, $CH_3—(CH_2)_7—CH{=}CH—(CH_2)_7—COOH$

This is an unsaturated acid. It is an oily liquid with a slight characteristic odor. It occurs as a glycerol ester in many fats and oils and is usually especially abundant in the latter. Olive oil and cottonseed oil contain a high proportion of this ester. Stearic, palmitic, and oleic acids are such important acids that the student should memorize their formulas. Mercuric oleate is used as an antiseptic in the treatment of skin diseases.

LINOLEIC ACID, $CH_3—(CH_2)_4—CH=$ $CH—CH_2—CH=CH—(CH_2)_7—COOH$, AND LINOLENIC ACID, $CH_3—CH_2—CH=$ $CH—CH_2—CH=CH—CH_2—CH=CH—(CH_2)_7—COOH$

Both of these acids contain 18 carbon atoms. The former contains two double bonds, while the latter contains three. The glycerol esters of these two acids occur in drying oils, such as linseed oil, and are responsible for the drying properties of these oils.

ARACHIDONIC ACID, $CH_3—(CH_2)_3—(CH_2–CH=CH)_4—(CH_2)_3—COOH$

Arachidonic acid contains 20 carbon atoms and four double bonds. It occurs in animal tissues. Linoleic, linolenic, and arachidonic acids are frequently called the *essential fatty acids*. If these acids are absent from the diet of infants, loss of weight occurs, and a scaly skin or eczema may appear. The addition of any one of these acids to the diet will generally correct or improve this condition.

OXALIC ACID, $(COOH)_2$ or

$$\begin{array}{l} \text{O} \\ \;/\!/ \\ \text{C—OH} \\ |\;\;\text{O} \\ |\,/\!/ \\ \text{C—OH} \end{array}$$

This is a dibasic acid. It is a colorless, crystalline solid and is poisonous when taken internally. Oxalic acid and its salts are mild reducing agents and are used to remove rust and some ink stains from clothing. In the collection of blood, oxalate salts are added to prevent the clotting of blood.

LACTIC ACID, OR α-HYDROXYPROPIONIC ACID,

$$\begin{array}{c} \text{COOH} \\ | \\ \text{CHOH} \\ | \\ \text{CH}_3 \end{array}$$

This is a hydroxy acid and also has the properties of an alcohol. It is formed by the action of certain bacteria on lactose or milk sugar and is present in sour milk. It is present in small amounts in the blood and is formed when muscles contract. Calcium is supplemented in the body by the addition of calcium lactate to the diet.

TARTARIC ACID, $H_2C_4H_4O_6$ or

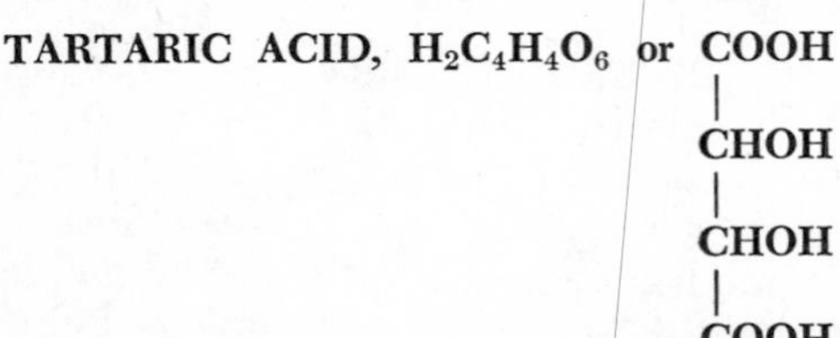

$$\begin{array}{c} \text{COOH} \\ | \\ \text{CHOH} \\ | \\ \text{CHOH} \\ | \\ \text{COOH} \end{array}$$

This acid occurs in many fruits and is particularly abundant in grapes. Potassium hydrogen tartrate, also called *cream of tartar*, is used in making baking powders. Both cream of tartar and *Rochelle salts* (potassium sodium tartrate) are used as mild cathartics. Potassium antimonyl tartrate, or *tartar emetic,* is used as an emetic and in the treatment of certain tropical diseases.

CITRIC ACID,

$$\begin{array}{l} CH_2—COOH \\ | \\ C(OH)—COOH \\ | \\ CH_2—COOH \end{array}$$

This hydroxy acid is present in the juice of citrus fruits. A solution of magnesium citrate is used as a mild laxative. Citric acid is an intermediate in the aerobic phase of carbohydrate metabolism and is the first constituent of the Krebs, or tricarboxylic acid, cycle (see pages 403–406).

PYRUVIC ACID,

$$CH_3—\overset{\overset{\displaystyle O}{\|}}{C}—COOH$$

This is an α-keto acid that occurs as an intermediate in certain metabolic processes (page 403). In the anaerobic phase of carbohydrate metabolism, it is reduced to lactic acid. An accumulation in the blood of lactic acid produces symptoms of fatigue.

Acetoacetic acid and β-hydroxybutyric acid are discussed on page 413.

Esters

The general formula for an ester is

$$R—\overset{\overset{\displaystyle O}{//}}{C}—O—R'$$

where R is the alkyl group of the acid and R′ is the alkyl group of the alcohol from which the ester has been prepared.

Esters are named as binary compounds. The first word of their name is the name of the alkyl radical of the alcohol, while the second word is formed by changing the *-ic* ending of the name of the acid from which they are formed to *-ate*. Thus

$$CH_3—CH_2—\overset{\overset{\displaystyle O}{//}}{C}—O—CH_3$$

is methyl propionate.

A knowledge of the simple esters will be of great value in the study of fats, because the latter are glycerol esters of some of the higher-molecular-weight fatty acids.

ESTERIFICATION AND PREPARATION OF ESTERS

Esters are prepared by the action of an alcohol upon an organic acid. Esterification is usually an equilibrium reaction. If a gram-molecular weight of ethyl alcohol is mixed with a gram-molecular weight of acetic acid, two-thirds of a gram-molecular weight of acetic acid and water are formed at equilibrium.

$$CH_3{-}CH_2OH + CH_3{-}CO_2H \rightleftharpoons \underset{\text{ethyl acetate}}{CH_3{-}\overset{\overset{O}{/\!/}}{C}{-}O{-}CH_2{-}CH_3} + H_2O$$

The addition of a small amount of sulfuric acid will shorten the time needed for the reaction to reach equilibrium. This equilibrium can be forced to completion by removing one of the products of the reaction, such as the low-boiling ethyl acetate, as it is formed.

PROPERTIES OF ESTERS

Most of the esters are colorless liquids with pleasant fruity odors. Esters occur in fruits and flowers and are responsible for much of their flavor and aroma. Butyl acetate and amyl acetate have the odor of bananas; ethyl butyrate, the odor of pineapples; amyl butyrate, the odor of apricots; isoamyl isovalerate, that of apples; octyl acetate, that of oranges.

Esters resemble inorganic salts in their formulas, but in properties they are entirely different. Esters are truly covalent compounds. They are formed slowly and are not soluble in water.

When esters are heated with solutions of strong bases, they are hydrolyzed to an alcohol and the salt of the organic acid. Sodium hydroxide will hydrolyze ethyl acetate to ethyl alcohol and sodium acetate:

$$CH_3{-}\overset{\overset{O}{/\!/}}{C}{-}O{-}CH_2{-}CH_3 + NaOH \longrightarrow$$

$$CH_3{-}\overset{\overset{O}{/\!/}}{C}{-}ONa + CH_3{-}CH_2OH$$

This is the essential reaction in soap making, and for this reason the basic hydrolysis of an ester is called *saponification.*

USES OF ESTERS

Esters are used in the preparation of artificial flavors and perfumes. Butyl and amyl acetate are used as solvents for lacquers and quick-drying enamels.

INORGANIC ESTERS

Esters of alcohols with inorganic acids are known; they are called *inorganic esters.* Thus ethyl alcohol and nitrous acid, HNO_2, react to form ethyl nitrite, $CH_3{-}CH_2{-}O{-}N{=}O$. An alcoholic solution of this ester is called *sweet spirits of nitre.* It is used as a diuretic and to lower fever and reduce blood presure. Amyl nitrite, $C_5H_{11}{-}O{-}$

N=O, also reduces blood pressure. It is used to reduce pain in the heart disease angina pectoris. It is volatile and is frequently furnished in small glass tubes that are readily broken in order that the vapors may be inhaled.

STUDY QUESTIONS

1. What is the functional group of the acids? What is the name of this group of atoms? Write a general formula for an organic acid and an ester.
2. What are the saturated fatty acids?
3. Give the structural formulas, the common names, and the IUPAC names for the first four members of the saturated fatty acid series. What is the name of the 16-carbon acid? The 18-carbon acid?
4. Write an equation for the preparation of propionic acid from its potassium salt. Describe another method of preparing propionic acid.
5. What chemical change occurs when an organic acid is dissolved in water? Represent this by an equation. Are organic acids highly dissociated into ions in aqueous solution? Are they strong acids? What determines the strength of an acid?
6. Write equations for the reaction of propionic acid with the following: (a) zinc metal, (b) potassium hydroxide, (c) lime, CaO, (d) sodium carbonate.
7. Discuss the occurrence of formic, acetic, butyric, palmitic, and stearic acids.
8. In what two ways does formic acid differ from the other acids of the fatty acid series? Can you give a reason why formic acid reduces Benedict's solution?
9. List the various methods that are used for the industrial production of acetic acid.
10. What is the chemical composition of vinegar, and how is it prepared? How does the quick vinegar process differ from the slow process?
11. What is glacial acetic acid, and why is it so called?
12. What is sugar of lead? For what is it used? What caution must be taken in its use?
13. List uses of stearic acid and its salts.
14. Discuss the structure, occurrence, and properties of oleic acid.
15. What two chemical substances are responsible for the drying properties of linseed oil.
16. What are the essential fatty acids? What symptoms may appear if they are lacking in the diet of infants?
17. Discuss the occurrence and uses of oxalic, tartaric, and citric acids.
18. If a gram-molecular weight of ethyl alcohol and a gram-molecular weight of glacial acetic acid are mixed and the mixture is allowed to stand until no further change occurs, how much ethyl acetate, water, ethyl alcohol, and acetic acid will be present? This is an example of what type of reaction? Why is concentrated sulfuric acid added in the preparation of ethyl acetate? Does it change the yield of ethyl acetate at equilibrium? How can the esterification reaction be made to go to completion?
19. What class of inorganic compounds do the esters resemble in name and formula? Is there any resemblance between the physical and chemical properties of these two classes of compounds? Explain fully.

20. What is the characteristic odor of most esters? Where are they found, and for what purposes are they most generally used?
21. What is the most important use of butyl and amyl acetate?
22. What are inorganic esters? What is sweet spirit of nitre, and what use is made of it in medicine? What other ester of nitrous acid is used for similar purposes?
23. Why should one study the chemistry of esters before beginning the study of the chemistry of foods?
24. Write structural formulas for the following esters: (a) ethyl acetate, (b) methyl butyrate, (c) *n*-propyl formate, (d) isopropyl acetate.
25. Write equations for the preparation of the esters listed in Question 24. What is the name of this general reaction?
26. Write equations for the saponification of the esters listed in Question 24.
27. Assign IUPAC names to the following organic acids:

(a) $CH_3—CH_2—CH_2—CH(CH_3)—COOH$ (with the CH_3 group attached below the CH)

(b) $CH_3—CH(CH_3)—CH_2—CH(CH_3)—COOH$ (with a CH_3 group attached below each CH)

(c) $CH_3—CH(OH)—COOH$ (with the OH attached below the CH)

(d) $CH_3—CH_2—CHCl—COOH$

(e) $CH_3—CCl_2—COOH$

28. Indicate the class to which the following organic compounds belong, that is, whether they are alcohols, aldehydes, etc.

$CH_3—CHO$
$CH_3—CH_2—CH_3$
$CH_3—COO—CH_2—CH_3$
$H—COOH$
$CH_3—CO—CH_3$

$CH_2{=}CH_2$
$CH_3—CH_2—O—CH_2—CH_3$
$CH_3—CH_2Br$
$CH{\equiv}CH$
$H—COO—CH_3$

29. Indicate over the arrow the reagents required to accomplish the following transformations:

alcohol $\longrightarrow$ aldehyde $\longrightarrow$ acid $\longrightarrow$ ester
alcohol $\longrightarrow$ ether
alcohol $\longrightarrow$ olefin

30. What is the physiological or pathological significance of the following organic acids: citric acid; lactic acid; pyruvic acid; acetoacetic acid; β-hydroxybutyric acid?

chapter 20

---

lipids, fats and oils, and soaps

20

Lipids are of wide occurrence and are one of the three major types of organic components of cells. *Lipids* are those components of foods or cells that are insoluble in water but soluble in fat solvents, such as ether, benzene, chloroform, alcohol, and acetone. Lipids also possess two other characteristics: They are either esters or are substances capable of forming esters. They occur in plant and animal tissue and are utilizable by the living organism.

The following is the usually accepted classification of the lipids:

1. Simple lipids produce fatty acids and alcohols, exclusively, upon hydrolysis.
   a. Animal and vegetable fats and oils, which, upon hydrolysis, produce glycerol and fatty acids.
   b. Waxes, which produce monohydric alcohols or sterols and fatty acids when hydrolyzed.
2. Compound lipids upon hydrolysis produce some other product or products besides fatty acids and alcohols.
   a. Glycerophosphatides produce glycerol, fatty acids, phosphoric acid, and a nitrogen-containing base when hydrolyzed. Lecithins and cephalins are the most important members of this class.
   b. Sphingolipids upon hydrolysis produce a fatty acid and a nitrogen-containing basic alcohol, either sphingosine or dihydrosphingosine. Phosphoric acid and choline are also products of the hydrolysis of the sphingomyelins, while galactose or glucose is a third product of the hydrolysis of the glycolipids.

The lipids classified above are saponifiable substances. Sterols, which are discussed on page 287, constitute the major portion of the unsaponifiable matter that is extracted by organic solvents from foods or cells.

The animal and vegetable fats and oils

Esters of the trihydric alcohol glycerol with the higher-molecular-weight fatty acids are called *triglycerides* (Table 20.1). The animal

TABLE 20.1 ACID COMPONENTS OF FATS AND OILS

	Component Acids (Percent)[a]					
Triglyceride	*Myristic*	*Palmitic*	*Stearic*	*Oleic*	*Linoleic*	*Linolenic*
Fats						
Butter	7–10	24–26	10–13	30–40	4–5	
Lard	1–2	28–30	12–18	40–50	6–7	
Tallow	3–6	24–32	20–25	37–43	2–3	
Edible oils						
Corn oil	1–2	8–12	2–5	19–49	34–62	
Cottonseed oil	0–2	20–25	1–2	23–35	40–50	
Olive oil		9–10	2–3	83–84	3–5	
Peanut oil		8–9	2–3	50–65	20–30	
Safflower oil		6–7	2–3	12–14	75–80	0.5–0.15
Soybean oil		6–10	2–5	20–30	50–60	5–11

[a]Totals less than 100 percent indicate the presence of lower or higher acids in small amounts.

Source: Robert J. Ouellette, *Introductory Organic Chemistry,* Harper & Row, New York, 1971.

and vegetable fats and oils are mixtures of these glycerides. The formation of glyceryl tristearate, also called *tristearin,* from glycerol and three molecules of stearic acid is represented by the following equation:

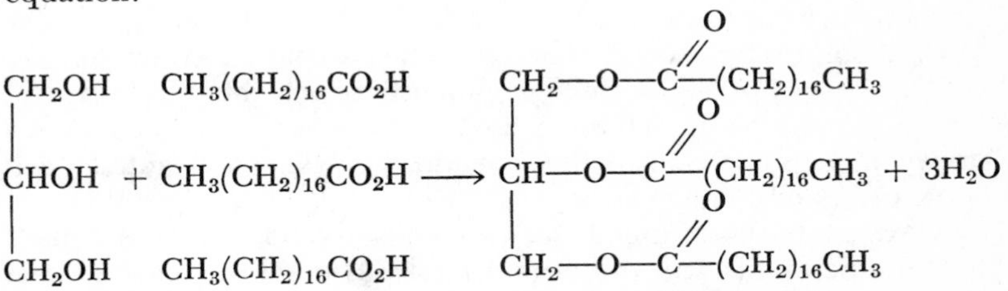

$$\begin{array}{l} CH_2OH \\ | \\ CHOH \\ | \\ CH_2OH \end{array} \begin{array}{l} CH_3(CH_2)_{16}CO_2H \\ \\ +\ CH_3(CH_2)_{16}CO_2H \\ \\ CH_3(CH_2)_{16}CO_2H \end{array} \longrightarrow \begin{array}{l} CH_2{-}O{-}C({=}O){-}(CH_2)_{16}CH_3 \\ | \\ CH{-}O{-}C({=}O){-}(CH_2)_{16}CH_3 \\ | \\ CH_2{-}O{-}C({=}O){-}(CH_2)_{16}CH_3 \end{array} + 3H_2O$$

tristearin

Other triglycerides are glyceryl tripalmitate, also called *tripalmitin,* and glyceryl trioleate, also called *triolein.* Glyceryl tributyrate occurs in small amounts in butter.

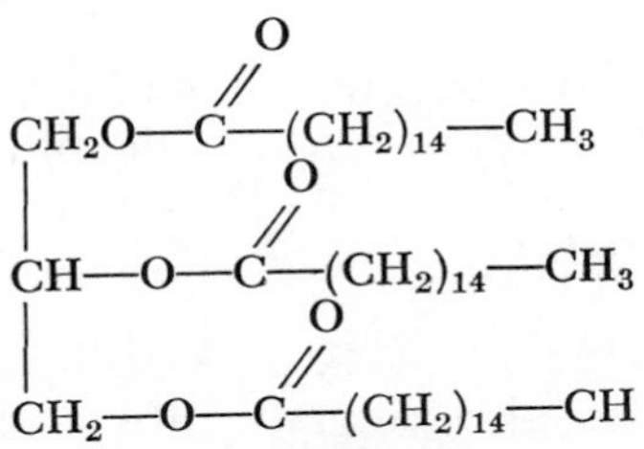

$$\begin{array}{l} CH_2O{-}C({=}O){-}(CH_2)_{14}{-}CH_3 \\ | \\ CH{-}O{-}C({=}O){-}(CH_2)_{14}{-}CH_3 \\ | \\ CH_2{-}O{-}C({=}O){-}(CH_2)_{14}{-}CH \end{array}$$

glyceryl tripalmitate (tripalmitin)

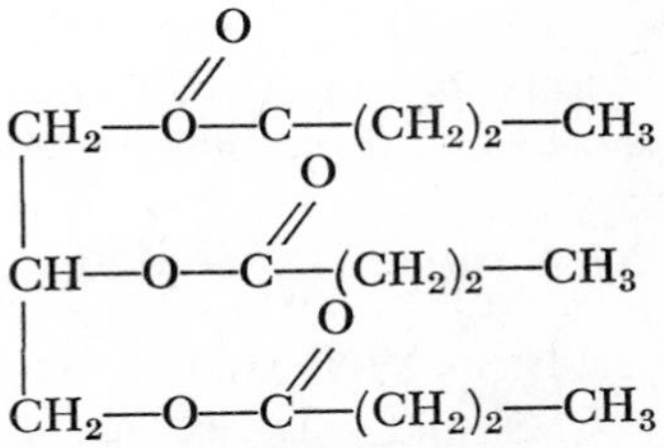

$$\begin{array}{l} CH_2{-}O{-}C({=}O){-}(CH_2)_2{-}CH_3 \\ | \\ CH{-}O{-}C({=}O){-}(CH_2)_2{-}CH_3 \\ | \\ CH_2{-}O{-}C({=}O){-}(CH_2)_2{-}CH_3 \end{array}$$

glyceryl tributyrate (tributyrin)

$$
\begin{array}{l}
CH_2{-}O{-}\overset{\overset{\displaystyle O}{/\!/}}{C}{-}(CH_2)_7{-}CH{=}CH{-}(CH_2)_7{-}CH_3 \\
| \\
CH{-}O{-}\overset{\overset{\displaystyle O}{/\!/}}{C}{-}(CH_2)_7{-}CH{=}CH{-}(CH_2)_7{-}CH_3 \\
| \\
CH_2{-}O{-}\overset{\overset{\displaystyle O}{/\!/}}{C}{-}(CH_2)_7{-}CH{=}CH{-}(CH_2)_7{-}CH_3
\end{array}
$$

triolein

The above glycerol esters are *simple glycerides;* that is, each molecule of the ester produces a single fatty acid upon hydrolysis. It was believed at one time that the fats and oils were mixtures of these simple glycerides. Recent investigations indicate that most fat molecules are *mixed glycerides;* that is, two or three different fatty acids are produced when a single molecule of the fat is hydrolyzed. The following formula is typical for a mixed glyceride:

$$
\begin{array}{l}
CH_2{-}O{-}\overset{\overset{\displaystyle O}{/\!/}}{C}{-}(CH_2)_{14}CH_3 \\
| \\
CH{-}O{-}\overset{\overset{\displaystyle O}{/\!/}}{C}{-}(CH_2)_7{-}CH{=}CH{-}(CH_2)_7CH_3 \\
| \\
CH_2{-}O{-}\overset{\overset{\displaystyle O}{/\!/}}{C}{-}(CH_2)_{16}CH_3
\end{array}
$$

Upon hydrolysis, such a molecule produces a molecule each of glycerol, palmitic acid, oleic acid, and stearic acid.

There is little distinction chemically between fats and oils. The former are solids at room temperature, whereas the latter are liquids. The fats are composed of higher proportions of the glycerides of higher-molecular-weight saturated fatty acids, whereas the oils have a relatively high proportion of the glycerides of the unsaturated acids (such as oleic and linoleic acids).

The vegetable and animal fats and oils should not be confused with the mineral oils and the essential oils. *Mineral oils* are mixtures of the saturated hydrocarbons. The *essential oils* are volatile, oily liquids that are distilled from certain aromatic plants. They are of a varied chemical nature, have characteristic odors, and are frequently used as perfumes and flavors. Some examples of this latter class of substances are oil of spearmint, oil of wintergreen, oil of cloves, oil of orange, and turpentine.

Fats and oils are insoluble in water but are soluble in the fat solvents, such as carbon tetrachloride, chloroform, ether, gasoline, and petroleum ether. The use of petroleum solvents and trichloroethylene in dry cleaning is based on these facts.

SAPONIFICATION AND HYDROLYSIS OF FATS AND OILS

Fats and oils are hydrolyzed to glycerol and fatty acids by superheated steam or by certain enzymes of the digestive tract.

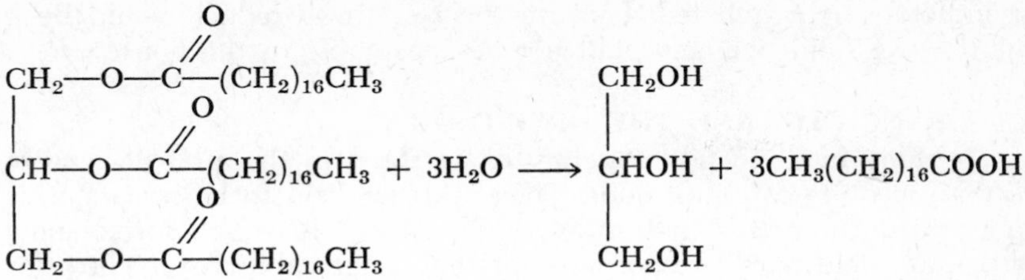

$$\begin{array}{l} CH_2{-}O{-}\overset{O}{\overset{\|}{C}}{-}(CH_2)_{16}CH_3 \\ | \\ CH{-}O{-}\overset{O}{\overset{\|}{C}}{-}(CH_2)_{16}CH_3 \\ | \\ CH_2{-}O{-}\overset{O}{\overset{\|}{C}}{-}(CH_2)_{16}CH_3 \end{array} + 3H_2O \longrightarrow \begin{array}{l} CH_2OH \\ | \\ CHOH \\ | \\ CH_2OH \end{array} + 3CH_3(CH_2)_{16}COOH$$

If this hydrolysis is accomplished by means of alkalis, it is called *saponification*, and glycerol and the alkali salts of the fatty acid are obtained.

$$C_3H_5[CH_3(CH_2)_{16}CO_2]_3 + 3NaOH \longrightarrow C_3H_5(OH)_3 + 3CH_3(CH_2)_{16}CO_2Na$$

The alkali salts of certain fatty acids are commonly known as *soaps*.

HYDROGENATION OF OILS

Stearic acid contains two more hydrogen atoms than oleic acid. Thus if two hydrogen atoms are added to oleic acid, stearic acid is formed. Introduction of six atoms of hydrogen to a molecule of the oil triolein will convert it to the solid fat tristearin, a process called *hydrogenation*, or *hardening of the oil*. Hydrogenation of the oil is usually accomplished by suspending a finely divided nickel catalyst in it and introducing hydrogen gas under pressure into the reaction mixture, which has been heated to about 180°C.

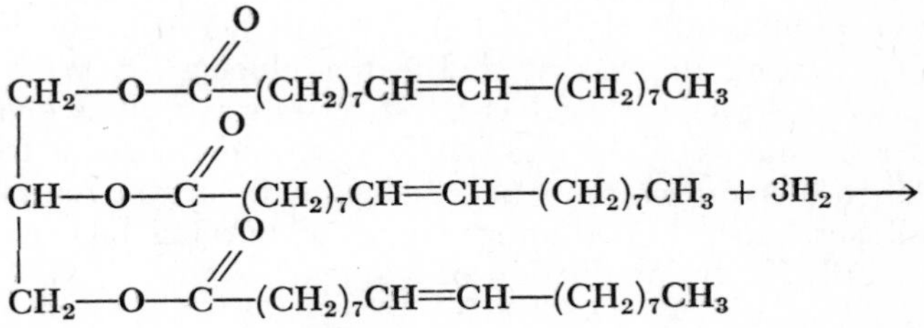

$$\underset{\text{triolein}}{\begin{array}{l} CH_2{-}O{-}\overset{O}{\overset{\|}{C}}{-}(CH_2)_7CH{=}CH{-}(CH_2)_7CH_3 \\ | \\ CH{-}O{-}\overset{O}{\overset{\|}{C}}{-}(CH_2)_7CH{=}CH{-}(CH_2)_7CH_3 \\ | \\ CH_2{-}O{-}\overset{O}{\overset{\|}{C}}{-}(CH_2)_7CH{=}CH{-}(CH_2)_7CH_3 \end{array}} + 3H_2 \longrightarrow$$

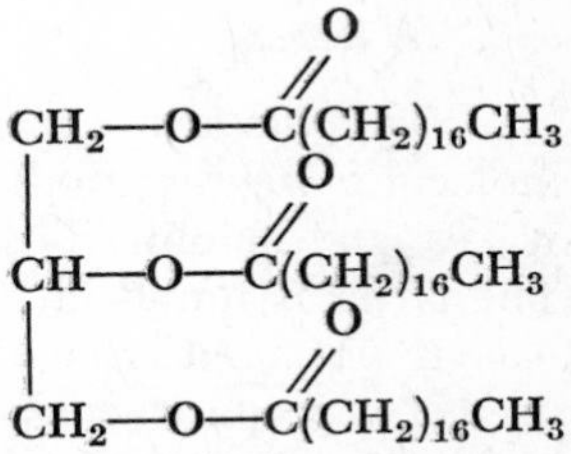

$$\underset{\text{tristearin}}{\begin{array}{l} CH_2{-}O{-}\overset{O}{\overset{\|}{C}}(CH_2)_{16}CH_3 \\ | \\ CH{-}O{-}\overset{O}{\overset{\|}{C}}(CH_2)_{16}CH_3 \\ | \\ CH_2{-}O{-}\overset{O}{\overset{\|}{C}}(CH_2)_{16}CH_3 \end{array}}$$

By this process, cottonseed oil and soybean oil, which have not been well accepted as cooking oils, are converted to solids resembling lard. Crisco and Spry are commercial products of the hydrogenation of these oils. In actual practice, the unsaturated substance is not completely hydrogenated, because the resulting product would be a hard waxy solid, which would not be acceptable to the housewife.

RANCIDITY AND THE DRYING OILS

Many fats and oils if permitted to stand in the air will acquire a disagreeable taste and odor. They are then said to be *rancid.* This may be due to some hydrolysis of the fat or oil, which forms some free fatty acid partly responsible for the disagreeable taste. The odor and taste of rancid butter are due primarily to small amounts of butyric acid liberated from the tributyrin present. In fats containing unsaturated constituents, rancidity may also be due to polymerization and oxidation of the multiple bonds. Products called *antioxidants,* such as hydroquinone, which prevent this oxidation from occurring are now added to kitchen shortenings.

The drying oils, such as linseed oil and tung oil, contain appreciable amounts of highly unsaturated glycerides. On exposure to air, these oils become oxidized, and they harden into hornlike materials. For this reason, the drying oils are important constituents of most exterior house paints. This oxidation of drying oils will occur in rags saturated with them. Sufficient heat is often generated in this oxidation to ignite the rags; many fires are caused by neglecting to destroy oily paint rags.

THE TEST FOR FATS AND OILS: THE ACROLEIN TEST

If a fat is heated to a high temperature, a small amount of glycerol is formed by hydrolysis of the fat. The glycerol will then decompose to produce the unsaturated aldehyde acrolein, the vapors of which are irritating, unpleasant, and tear producing (see page 255). This formation of acrolein is used as a test for fats as well as glycerol. Some solid potassium acid sulfate, $KHSO_4$, is added to the fat because a higher temperature can be reached with a more abundant formation of acrolein. The irritating vapors of burning fats and oils are due primarily to the production of acrolein.

Waxes

The waxes are mixtures of esters of fatty acids with the higher-molecular-weight monohydric alcohols and sterols. Some important waxes are lanolin (obtained from wool), beeswax, and spermaceti. The latter is found in the skull of certain species of whale. Carnauba wax is obtained from the leaves of certain tropical trees. Waxes are used to prepare cosmetics, furniture and automobile polish, and ointments and salves.

Compound lipids: glycerophosphatides

The glycerophosphatides are derivatives of the phosphatidic acids, the esterification product of glycerol with two molecules of fatty acid and a molecule of phosphoric acid. These can be assigned the following structure:

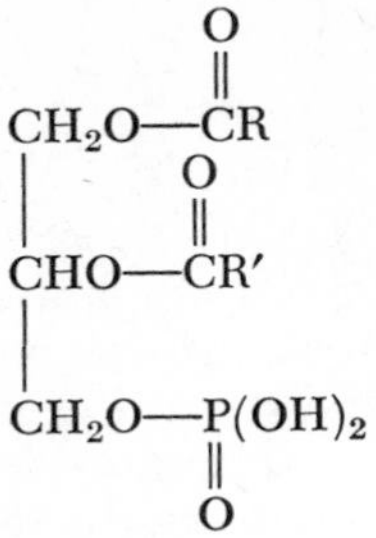

$$\begin{array}{l} \qquad\qquad\quad O \\ \qquad\qquad\quad \| \\ CH_2O{-}CR \\ |\qquad\quad\;\; O \\ |\qquad\quad\;\; \| \\ CHO{-}CR' \\ | \\ | \\ CH_2O{-}P(OH)_2 \\ \qquad\qquad\quad \| \\ \qquad\qquad\quad O \end{array}$$

phosphatidic acids

Although phosphoric acid is indicated as esterified with an end primary alcohol group of glycerol in the above formula, it may also be esterified with the middle or secondary alcohol group. There are two important types of glycerophosphatides, namely, the *lecithins*, in which the nitrogen base choline, and the *cephalins*, in which the nitrogen base ethanolamine or the amino acid serine (page 320), are esterified with the phosphoric acid portion of phosphatidic acids. The phosphatides occur in all tissues of the body but are most abundant in the nerve tissue and the brain. Egg yolk is an excellent source of lecithins; the blood platelets are rich in cephalins; the glycerophosphatides also occur in plant tissues.

LECITHINS

Upon hydrolysis of a molecule of a lecithin, two molecules of fatty acids and a molecule each of glycerol, phosphoric acid, and a nitrogen-containing base, hydroxyethyltrimethylammonium hydroxide, are formed. The base is better known as *choline* and has the formula

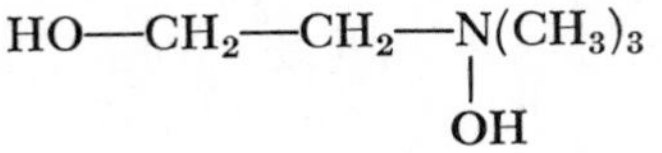

$$\begin{array}{l} HO{-}CH_2{-}CH_2{-}N(CH_3)_3 \\ \qquad\qquad\qquad\quad\;\; | \\ \qquad\qquad\qquad\quad OH \end{array}$$

There are a number of lecithins which differ primarily in the nature of the fatty acids that are present. The following general formula will represent the composition of a typical lecithin, where R and R′ represent the alkyl radicals of the fatty acids present.

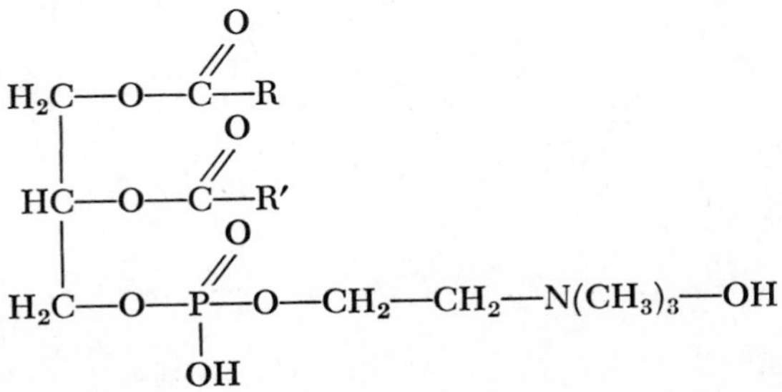

$$\begin{array}{l} \qquad\qquad\qquad O \\ \qquad\qquad\quad // \\ H_2C{-}O{-}C{-}R \\ \;\; |\qquad\qquad\; O \\ \;\; |\qquad\quad\;\; // \\ HC{-}O{-}C{-}R' \\ \;\; |\qquad\qquad\; O \\ \;\; |\qquad\quad\;\; // \\ H_2C{-}O{-}P{-}O{-}CH_2{-}CH_2{-}N(CH_3)_3{-}OH \\ \qquad\qquad\; | \\ \qquad\qquad OH \end{array}$$

The student may find it advantageous to represent the structure of lecithins by the following block formula:

glycerol	—	fatty acid		
	—	fatty acid		
	—	phosphoric acid	—	choline

The lecithins are insoluble in water but are excellent emulsifying agents and form stable emulsions with water. Neutral fats are converted to lecithins in the body, and the latter may be the essential intermediate products in the oxidation of fats to carbon dioxide and water. The lecithins may also serve as a source of phosphorus for the formation of new tissue and in the transport of fats in the body.

CEPHALINS

The cephalins resemble the lecithins in structure, except that the nitrogen-containing base ethanolamine, also called *colamine,* or the amino acid serine replaces choline.

$$HO{-}CH_2{-}CH_2{-}NH_2$$

ethanolamine

$$HO{-}CH_2{-}\underset{\displaystyle NH_2}{\underset{|}{CH}}{-}COOH$$

serine

Some of the cephalins appear to be essential for the normal coagulation of blood. They may also serve as a source of phosphoric acid in the living organism.

Some now prefer the name phosphatidyl choline to lecithin and phosphatidyl ethanolamine or phosphatidyl serine to cephalin.

Sphingolipids

Those compound lipids that are hydrolyzed to sphingosine or dihydrosphingosine rather than glycerol are known as *sphingolipids.* The most important of these are the sphingomyelins and the glycolipids, or cerebrosides.

SPHINGOMYELINS

These are hydrolyzed to a molecule each of nitrogen-containing alcohol, either *sphingosine* or *dihydrosphingosine,* fatty acid, phosphoric acid, and choline. Their structure may be illustrated by a block formula.

sphingosine or dihydrosphingosine	—	phosphoric acid	—	choline
|				
fatty acid				

Sphingosine and dihydrosphingosine are 18-carbon atom dihydric alcohols that contain an amino, NH_2, group. Sphingosine contains a double bond between two carbon atoms. The structures assigned to these two alcohols are

$$CH_3—(CH_2)_{12}CH{=}CH—CHOH—CHNH_2—CH_2OH$$

sphingosine

and

$$CH_3—(CH_2)_{14}—CHOH—CHNH_2—CH_2OH$$

dihydrosphingosine

The sphingomyelins serve as an integral part of the myelin sheath of nerve fibers. However, the role they play in living tissues has not as yet been clearly elucidated.

Niemann-Pick's disease is a familial, congenital disorder in infants in which phospholipids, especially sphingomyelins, accumulate in the brain, liver, and spleen. Its typical symptoms are anemia and leukocytosis. Fortunately, it is of rare occurrence, because it is usually fatal within the first two years of life.

GLYCOLIPIDS, OR CEREBROSIDES

These are hydrolyzed to a molecule each of a fatty acid, sphingosine or dihydrosphingosine, and a sugar. The sugar most commonly found in glycolipids is galactose, and for this reason they are frequently known as *galactolipids*. However, glucose has been isolated from a few glycolipids. The block formula

```
| sphingosine or dihydrosphingosine | — | galactose or glucose |
                 |
           | fatty acid |
```

may be used to represent the structure of glycolipids. [*Note:* No phosphoric acid is obtained upon hydrolysis of this class of compound lipids.]

Four glycolipids, kerasin, phrenosin, nervone, and oxynervone, that apparently differ only in the fatty acid present have been isolated from brain tissue. Because glycolipids are so abundant in the brain, they have also been called *cerebrosides*. The role of the glycolipids in body functions has not been established. In the rare disorder known as *Gaucher's disease,* the liver and spleen become greatly enlarged due to an excessive accumulation of cerebrosides in these organs. This disease appears in childhood but is not usually fatal.

Sterols

The sterols are solid, high-molecular-weight, tetracylic alcohols. The most abundant animal sterol is cholesterol, which occurs in all living animal tissue and is particularly abundant in nervous tissue and in the brain. The following formula has been assigned to cholesterol by Windaus.

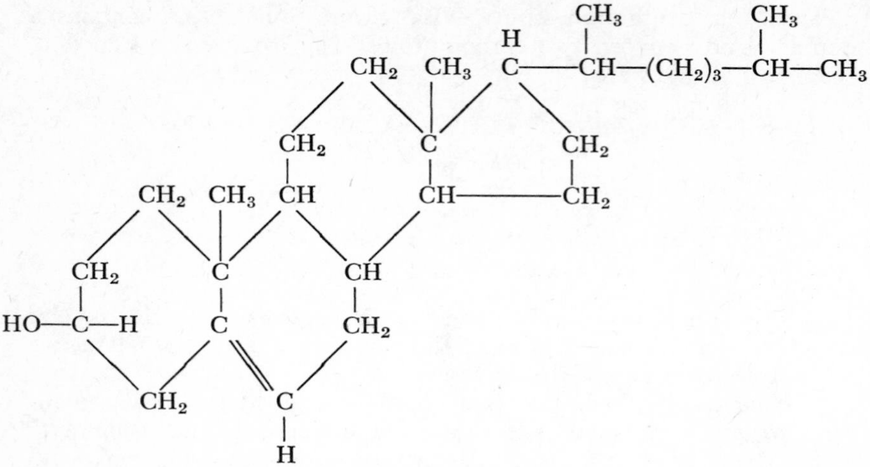

```
                                             CH3            CH3
                                       H      |              |
                        CH2   CH3      C ─── CH ─(CH2)3── CH ── CH3
                       /    \   |    /   \
                      /      \  |   /     \
                   CH2         C           CH2
                    |          |            |
       CH2  CH3   CH          CH ───────── CH2
      /    \  |  /   \       /
     /      \ | /     \     /
  CH2          C         CH
   |           |          |
HO─C─H         C         CH2
     \        / \\       /
      \      /   \\     /
       CH2           C
                     |
                     H
```

cholesterol

Ergosterol, another important sterol, was first discovered in ergot, a fungus of rye. It is also present in yeast and certain mushrooms. Upon irradiation with ultraviolet light, ergosterol is converted to a number of products, one of which is calciferol, or vitamin D_2. Irradiated ergosterol is sold under the trade name *Viosterol.*

Soaps

Soaps are mixtures of the salts of the higher-molecular-weight organic acids, such as stearic, oleic, and palmitic acids. Soaps are prepared by heating a fat with a caustic solution, a process called *saponification.* Glycerol is a by-product of soap manufacture.

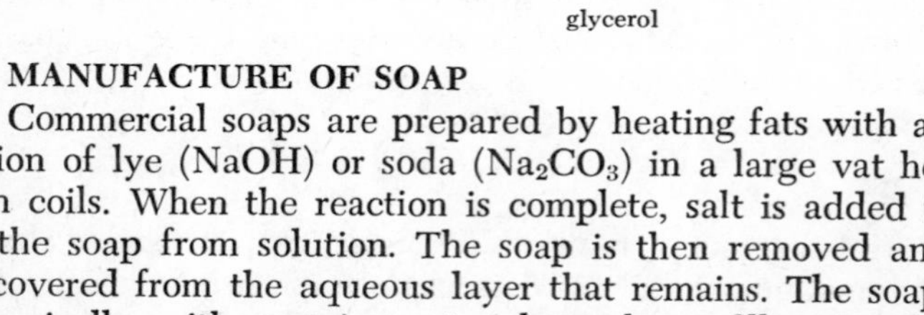

```
               O
              //
CH2──O──C──R                     CH2OH
              O                    |
             //                    |
CH──O──C──R + 3NaOH ──→ CHOH + 3R──CO2Na
              O                    |           a soap
             //                    |
CH2──O──C──R                     CH2OH
                                glycerol
```

MANUFACTURE OF SOAP

Commercial soaps are prepared by heating fats with an aqueous solution of lye (NaOH) or soda (Na_2CO_3) in a large vat heated with steam coils. When the reaction is complete, salt is added to precipitate the soap from solution. The soap is then removed and glycerol is recovered from the aqueous layer that remains. The soap is mixed mechanically with certain materials such as fillers, perfumes, and coloring matter and is then pressed into cakes.

Sodium soaps, such as sodium stearate, are hard soaps, whereas potassium soaps, such as potassium oleate, are soft soaps. An alcoholic solution of a potassium soap often used in hospitals is called *tincture of green soap*. Olive oil is used in making castile soap.

CLEANSING ACTION OF SOAPS

Soaps reduce surface tension and have the power to stabilize a suspension of oil as fine droplets in water. Such a suspension is called an *emulsion*. Most dirt or soil on our skin is covered with a thin film of oil; water will not remove it, because it cannot wet this oily layer. Addition of soap to the water wets the surface and emulsifies this oil as fine droplets which are easily washed away, along with the dirt and soil.

PROPERTIES OF SOAPS

Soaps exhibit three objectional chemical properties: Most soaps are insoluble in salt solutions and for this reason cannot be used in ocean water. In a strong acid solution, they are decomposed, and the free fatty acid is liberated as an oil or waxy solid.

$$CH_3(CH_2)_{16}CO_2Na + HCl \longrightarrow \underset{\text{stearic acid}}{\underline{CH_3(CH_2)_{16}CO_2H}} + NaCl$$

For this reason they can be used only in neutral or alkaline solutions. The calcium and magnesium salts of stearic, palmitic, and oleic acids are insoluble in water. Hard water (page 95) contains inorganic salts of calcium and magnesium in solution, and a large amount of soap has to be wasted in such water in order to soften it so that suds can be obtained.

$$CaCl_2 + 2CH_3(CH_2)_{16}CO_2Na \longrightarrow 2NaCl + \underline{[CH_3(CH_2)_{16}CO_2]_2Ca}$$

(See Chapter 6 for chemical methods that can be employed in softening hard water.)

SYNTHETIC DETERGENTS, OR SYNDETS

These are synthetic products that are used as cleansing agents in place of soap and do not show many of the disadvantages of soap. For instance, they do not usually form precipitates with hard water, because their calcium and magnesium salts are usually moderately to quite soluble in water, and they do not leave the dulling film on glassware that is characteristic of soaps. One of these detergents is Dreft. This contains the sodium salt of alkyl hydrogen sulfates, $R—OSO_3H$. These salts are prepared by treating an alcohol, such as lauryl alcohol, with sulfuric acid and neutralizing the resulting product with sodium hydroxide.

$$\underset{\text{lauryl alcohol}}{CH_3—(CH_2)_{10}—CH_2OH} + H_2SO_4 \longrightarrow$$

$$\underset{\text{lauryl hydrogen sulfate}}{CH_3—(CH_2)_{10}—CH_2—OSO_3H} + H_2O$$

$$CH_3\text{—}(CH_2)_{10}\text{—}CH_2\text{—}OSO_3H + NaOH \longrightarrow CH_3\text{—}(CH_2)_{10}\text{—}CH_2\text{—}OSO_3Na + H_2O$$

sodium lauryl sulfate

Many of the synthetic detergents, or syndets, on the market today are sodium salts of alkylbenzenesulfonic acids. At one time, these were prepared by condensing an olefin of the structure

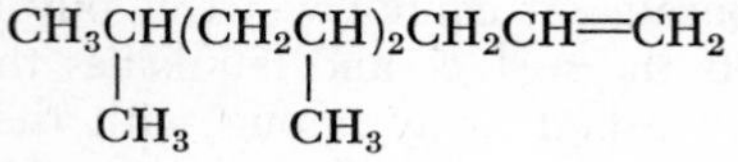

$$\underset{\substack{|\\CH_3}}{CH_3CH}(CH_2\underset{\substack{|\\CH_3}}{CH})_2CH_2CH{=}CH_2$$

(prepared by converting propylene to a tetramer) with benzene and sulfonating the resulting alkylbenzene to an alkylbenzenesulfonic acid:

$$\underset{\substack{|\\CH_3}}{CH_3CH}(CH_2\underset{\substack{|\\CH_3}}{CH})_2CH_2CH{=}CH_2 \xrightarrow[\text{anhyd. } AlCl_3]{C_6H_6}$$

$$\underset{\substack{|\\CH_3}}{CH_3CH}(CH_2\underset{\substack{|\\CH_3}}{CH})_2CH_2CH_2CH_2\text{—}C_6H_5 \xrightarrow{\text{conc. } H_2SO_4} \xrightarrow{NaOH}$$

$$\underset{\substack{|\\CH_3}}{CH_3CH}(CH_2\underset{\substack{|\\CH_3}}{CH})_2CH_2CH_2CH_2\text{—}C_6H_4\text{—}SO_3^-\ Na^+$$

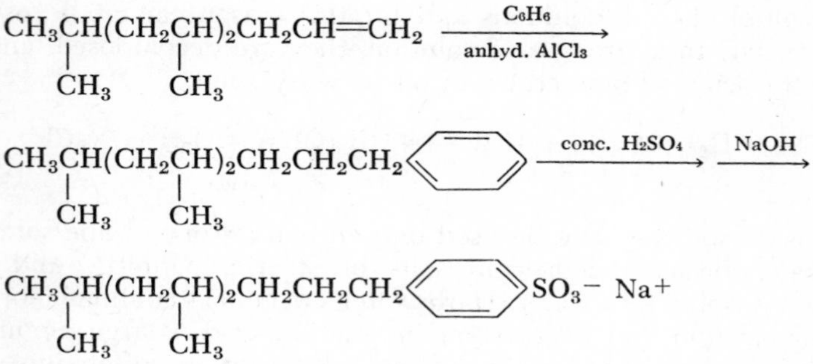

The use of these detergents that contain a highly branched alkyl side chain has led to serious pollution in our streams. Unlike soaps and the *n*-alkyl sulfate type detergents, these are not biodegradable by the bacteria in septic tanks and those present in treated sewage. This has frequently led to the production of a very great amount of sudsing in streams and even in drinking water. Incidentally, bacterial degradation of soaps and detergents likely occurs by a mechanism quite similar to the β-oxidation of fatty acids proposed by Knoop (page 408).

Because a number of municipalities have passed laws to prevent the sale of these nonbiodegradable detergents, the manufacturers of these synthetic detergents are preparing alkylbenzenesulfonate detergents with a straight-chain alkyl side group or with only one methyl group branched from this chain, because straight-chain alkylbenzenesulfonate detergents with the structure

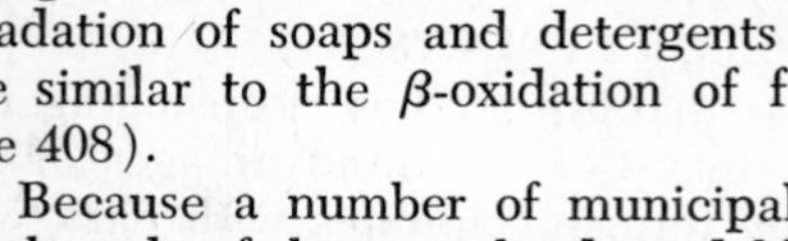

$$CH_3(CH_2)_9\underset{\substack{|\\CH_3}}{CH}\text{—}C_6H_4\text{—}SO_3^-Na^+$$

are extensively biodegradable.

STUDY QUESTIONS

1. Describe the chemical nature of the animal and vegetable fats and oils.
2. How do the mineral and essential oils differ from the animal and vegetable oils?
3. Distinguish between fats and oils.
4. Write structural formulas for tripalmitin and triolein.
5. Distinguish between simple and mixed glycerides. Write a structural formula to illustrate each. Indicate by equation the saponification of each of these by sodium hydroxide.
6. In what solvents are the animal and vegetable fats and oils soluble?
7. Which has the higher percentage of glycerides derived from saturated fatty acids, fats or oils?
8. What is saponification? Why is it so named?
9. What means can be used to hydrolyze fats and oils?
10. What is hydrogenation, and how is this process performed commercially on vegetable oils?
11. Write an equation for the chemical reaction that occurs when cottonseed oil (triolein) is hydrogenated.
12. What commercial applications are made of the products that result from the hydrogenation of the vegetable oils?
13. What disagreeable product is formed when butter becomes rancid? How is it formed?
14. What types of chemical reactions occur when fats become rancid?
15. In what way may the rancidity of fats be prevented?
16. Why is linseed oil used in the preparation of many exterior house paints?
17. Why should oily rags be destroyed after use?
18. Describe the common laboratory test for fats.
19. What tear-producing substance is formed when fats are overheated?
20. What characteristics must a substance possess to be classed as a lipid?
21. To what class of lipids do the waxes belong? What do they form when hydrolyzed?
22. List some common waxes. For what purposes are waxes used?
23. How do the compound lipids differ from the simple lipids? What products are formed when the glycerophosphatides are hydrolyzed? When sphingolipids are hydrolyzed?
24. What are the two important classes of glycerophosphatides? How do they differ in structure?
25. What purposes are served by lecithins in the body?
26. What is the most important function of cephalins in the body? What other function may they serve?
27. What types of substances are formed when the sphingolipids are hydrolyzed? Where do the sphingomyelins and the glycolipids occur most abundantly in the body? Why are the glycolipids also called cerebrosides? Galactolipids? What is characteristic of Gaucher's disease? Niemann-Pick's disease?
28. What is the chemical nature of the sterols? What is the most abundant animal sterol, and where does it occur in the body?
29. What is ergosterol? Calciferol? Viosterol? Which of these is a D vitamin?
30. What are the raw products and by-products of soap manufacture? What is the chemical composition of soaps? Describe briefly the method used to prepare soap.

31. Distinguish between the composition of hard and soft soaps. What is tincture of green soap? Castile soap?
32. What undesirable properties are usually possessed by ordinary soaps?
33. What substances in water are responsible for temporary hardness? Permanent hardness? Why is hard water softened? Discuss briefly some methods that are used to soften hard water. (See Chapter 6.)
34. Discuss the synthetic detergents from the following standpoints: (a) chemical composition of one of the best-known classes of these; (b) method used to synthesize a typical detergent; (c) advantages of these substances as cleansing agents as compared to ordinary soaps; (d) commercial names of some of these products.
35. Write equations for the saponification of (a) triolein and (b) tripalmitin.
36. Write an equation to represent the decolorization of a bromine solution by oleic acid.
37. Indicate whether the following are simple or compound lipids: waxes, cephalins, glycolipids, vegetable fats.
38. Write structural formulas for the products of hydrogenation and saponification of triolein.

chapter 21

carbohydrates

21

The carbohydrates are a group of naturally occurring substances including the sugars, starch, cellulose, dextrins, and gums. These occur chiefly in plants and serve as their supporting structure and reserve food supply.

Most of the carbohydrates may be represented by the general formula $C_n(H_2O)_m$, in which the ratio of hydrogen to oxygen is the same as in water. This led the early chemists to consider them as hydrates of carbon; hence their class name. However, some sugars, such as rhamnose ($C_6H_{12}O_5$), do not fit the above general formula. Furthermore, the carbohydrates do not have the properties associated with true hydrates.

Carbohydrates are now defined as *polyhydroxy aldehydes* or *ketones,* or *substances that are hydrolyzed to these simpler substances.* The simpler members of this class exhibit many of the typical properties of the alcohols and either aldehyde or ketone properties.

Classification of the carbohydrates

The classification of carbohydrates is based on the number of simple sugar molecules that they will form upon hydrolysis. *Monosaccharides* are the simplest carbohydrates and are not decomposed into simpler ones by hydrolysis. *Disaccharides* form two molecules of monosaccharides on hydrolysis. Each molecule of a *polysaccharide* produces three or more molecules of monosaccharides, and usually the number of these simple molecules produced is considerably larger than three.

Monosaccharides that contain an aldehyde group are called *aldoses,* while those that contain a ketone group are called *ketoses.* A further classification of monosaccharides depends on the number of carbon atoms in each molecule of the sugar. Those with four carbon atoms are called *tetroses;* those with five, *pentoses;* those with six *hexoses.* Glucose is an aldohexose; that is, it is an aldehyde sugar with six carbon atoms. Fructose is a ketohexose, while arabinose is an aldopentose.

The term *sugar* is usually applied to the mono- and disaccharides.

A general test for carbohydrates

The reagent for the Molisch test, a general test for carbohydrates, consists of an alcoholic solution of α-naphthol. This reagent is mixed with a solution or suspension of the carbohydrate. A test tube containing the solution is held in an inclined position while cold concentrated sulfuric acid is carefully poured into it. If the substance being tested is a carbohydrate, a red-violet colored ring will form at the juncture of the test solution and the lower sulfuric acid layer.

Monosaccharides

Glucose, fructose, and galactose are the most important monosaccharides and are hexoses. The pentose xylulose occurs in the urine of certain individauls in the pathological condition known as *pentosuria.* Other pentoses are obtained by the hydrolysis of certain polysaccharides.

STRUCTURE OF MONOSACCHARIDES

Glucose can be represented by the following structural formula:

```
        1
   H—C=O
     |2
   H—C—OH
     |3
  HO—C—OH
     |4
   H—C—OH
     |5
   H—C—OH
     |6
     CH2OH
```

It has been established that only a small amount of glucose in solution has the straight-chain formula and that most of it exists in a cyclic structure, which is now being represented by the following two structures:

```
     OH                                 6
     |1                                 CH2OH
   H—C——————————┐                       |5
     |2          |                      C——————————O
   H—C—OH        |             H       /|           \        H
     |3          |             |4     / H            \       |1
  HO—C—H         O     or      C                       C
     |4          |             | \      OH        H   /  |
   H—C—OH        |             OH \     |3        |2 /   OH
     |5          |                 C——————————C
   H—C——————————┘                  |          |
     |6                            H          OH
     CH2OH
```

For convenience, the carbon atoms of glucose have been numbered. The aldehyde or potential aldehyde group is numbered 1. Glucose differs from the other aldohexoses by the position of the hydroxyl groups about the carbon atoms numbered from 2 to 5. There are 16 isomeric aldohexoses including glucose. One of these isomers of glucose is galactose.

Fructose is a ketohexose and is represented by the following open-chain formula (although fructose exists predominately in a cyclic form when it is dissolved in water):

$$\begin{array}{c} CH_2OH \\ | \\ C{=}O \\ | \\ HO{-}C{-}H \\ | \\ H{-}C{-}OH \\ | \\ H{-}C{-}OH \\ | \\ CH_2OH \end{array}$$

For a more extensive discussion concerning the structure of the monosaccharides, see pages 301–311.

CHEMICAL PROPERTIES OF THE MONOSACCHARIDES

Combustion. Upon complete combustion, the monosaccharides are oxidized to carbon dioxide and water:

$$C_6H_{12}O_6 + 6O_2 \longrightarrow 6CO_2 + 6H_2O$$

About 4 kcal of heat per gram are produced in the combustion of hexoses. The oxidation of carbohydrates in the human body serves as an important source of heat and muscular energy.

Fermentation. Glucose in the presence of yeast is converted to ethyl alcohol and carbon dioxide, a reaction called *fermentation.* The yeast cells produce an enzyme (p. 370), zymase, which catalyzes this conversion.

$$C_6H_{12}O_6 \xrightarrow{\text{zymase}} 2CO_2 + 2C_2H_5OH$$

Fructose is also fermented to alcohol and carbon dioxide, while galactose and the pentoses are either fermented slowly or not at all by yeast.

Reduction Tests. The monosaccharides reduce such mild oxidizing agents as Fehling's and Benedict's solution, alkaline solutions of cupric oxide, and Tollens' reagent, an alkaline solution of silver oxide (see page 264). Under even mild conditions, no single product can be isolated when glucose has been oxidized by these reagents. Under carefully controlled conditions, however, some investigators have isolated glucosone, $CH_2OH(CHOH)_3—CO—CHO$, as apparently the first oxidation product of glucose.

A convenient modification of Benedict's test involves the use of Clinitest Reagent Tablets. Their composition and a suitable procedure for their use as a reduction test for carbohydrates are described in the laboratory manual. These tablets contain citric acid and caustic soda (NaOH), so unlike Benedict's reagent, no external heating is necessary, because the temperature required for the test is attained through the heat of neutralization as sodium citrate is formed from this acid and base.

Glucose occurs in the urine in appreciable quantities in certain pathological disorders, such as diabetes mellitus, and this reduction test, using Benedict's solution or Clinitest tablets, is used to detect its presence. If small amounts of glucose are present, the blue color of these reagents changes to green, whereas appreciable amounts produce a yellow to orange-red precipitate of cuprous oxide.

The pentoses also reduce these reagents, and some pentoses will give a positive test in the cold.

SPECIFIC MONOSACCHARIDES

Glucose, which is also called *dextrose* or *grape sugar,* is a colorless crystalline solid which is soluble in water but insoluble in most organic solvents, such as ether. It is about as sweet as sucrose. Glucose occurs in the juice of many fruits, especially in grape juice, and is also formed in the hydrolysis of many di- and polysaccharides. Normal blood contains about 0.1 percent of this sugar, but only small traces of it occur in the urine of a normal individual. However, in certain pathological conditions, glucose occurs in urine in appreciable amounts. The other hexoses when ingested are usually changed to this sugar in the liver.

Glucose reduces Fehling's and Benedict's solution and is fermented by yeast. Commercially, it is prepared by the acid hydrolysis of *cornstarch* and for this reason is sometimes called *corn sugar.* Much of the commercial glucose on the market is not the pure sugar but rather contains appreciable amounts of dextrins and maltose.

Fructose, also called *levulose* or *fruit sugar,* occurs in the juice of many fruits and is present in honey. It is one of the products of the hydrolysis of sucrose. Inulin, a polysaccharide obtained from the roots of some artichokes and dahlias, produces fructose almost exclusively on hydrolysis.

Fructose, which is sweeter than sucrose, is fermented by yeast; although it is a ketone, it is a reducing agent, reducing both Benedict's solution and Tollens' reagent. This is typical, however, of those ketones that possess an alcohol group on a carbon atom adjacent to a carbonyl group.

The sugar *galactose* does not occur free in nature but is formed as a product of the hydrolysis of lactose and certain polysaccharides called *galactans.* It is also a hydrolysis product of most of the glycolipids. It is a reducing sugar and is fermented slowly or not at all by baker's yeast.

The *pentoses* are five-carbon atom monosaccharides. The four most important of these are arabinose, formed in the hydrolysis of gum arabic and cherry gum; xylose, obtained in the hydrolysis of corn cobs, bran, wood, and straw; and ribose and deoxyribose, hydrolysis products of the nucleic acids which occur in the nuclei and cytoplasm of cells. A pentose, xylulose, is excreted in the urine of a few individuals in a pathological condition known as *pentosuria*. The individual with pentosuria experiences no ill effects, but this pentose may be easily confused with glucose (because these are both reducing sugars) and lead to a false diagnosis of diabetes mellitus. To detect the presence of pentoses in solution, the solution is acidified with hydrochloric acid, and phloroglucinol is added. If pentoses are present, a red color will appear; glucose and other sugars will produce no color with this reagent. This is called *Tollens' pentose test*. Galactose is purported to give a color with this reagent. Orcinol, when heated with a hydrochloric acid solution of a pentose, produces a green color or precipitate. This is known as *Bial's test*.

$$
\begin{array}{cccc}
\mathrm{CHO} & \mathrm{CHO} & \mathrm{CHO} & \mathrm{CH_2OH} \\
| & | & | & | \\
\mathrm{H{-}C{-}OH} & \mathrm{H{-}C{-}OH} & \mathrm{CH_2} & \mathrm{C{=}O} \\
| & | & | & | \\
\mathrm{HO{-}C{-}H} & \mathrm{H{-}C{-}OH} & \mathrm{H{-}C{-}OH} & \mathrm{H{-}C{-}OH} \\
| & | & | & | \\
\mathrm{HO{-}C{-}H} & \mathrm{H{-}C{-}OH} & \mathrm{H{-}C{-}OH} & \mathrm{HO{-}C{-}H} \\
| & | & | & | \\
\mathrm{CH_2OH} & \mathrm{CH_2OH} & \mathrm{CH_2OH} & \mathrm{CH_2OH} \\
\text{arabinose} & \text{ribose} & \text{deoxyribose} & \text{xylulose}
\end{array}
$$

Note that the deoxy sugar contains one less oxygen than the sugar from which it is derived.

Disaccharides

These sugars produce two molecules of monosaccharides on hydrolysis. The three common sugars of this class are sucrose, lactose, and maltose, which are isomers with the molecular formula $C_{12}H_{22}O_{11}$.

SUCROSE, $C_{12}H_{22}O_{11}$, OR CANE SUGAR

This sugar occurs in the juice of many fruits and vegetables and is a constituent of honey. It is a colorless, crystalline solid, soluble in water but not in organic solvents, and it is quite sweet. Commercially, it is obtained from sugar cane and sugar beets. The juice of the sugar cane contains about 20 percent sucrose and is removed from the cane by means of heavy rollers. This juice is then evaporated under reduced pressure to prevent scorching; the sugar crystallizes, leaving the molasses, which is then removed from the sugar. This sugar still has a brown color and is known as *brown sugar*. It is dissolved in water and the brown color is removed by filtering this solution through some form of decolorizing carbon, such as animal charcoal. The resulting clear, colorless solution is again evaporated under reduced pressure until crystallization of the sugar occurs. The juice of the sugar beet contains about 14 percent sucrose, which is obtained from it in the same general manner as from sugar cane. Maple sugar consists of sucrose and some other substances that give it its characteristic flavor.

Sucrose is hydrolyzed to a mixture of glucose and fructose by acids or by enzymes such as invertase (also called sucrase), which occurs in the intestinal tract and is produced by yeast. This mixture of sugars obtained by the hydrolysis of sucrose is called *invert sugar.*

$$\underset{\text{sucrose}}{C_{12}H_{22}O_{11}} + H_2O \longrightarrow \underset{\text{glucose}}{C_6H_{12}O_6} + \underset{\text{fructose}}{C_6H_{12}O_6}$$

Yeast ferments sucrose. The invertase of the yeast converts the sucrose to invert sugar, which is then converted to alcohol by the zymase of the yeast. Sucrose will not reduce Fehling's and Benedict's solution; therefore the glucose and fructose residues in sucrose must be united through the aldehyde group of glucose and the carbonyl group of fructose.

LACTOSE, $C_{12}H_{22}O_{11}$, OR MILK SUGAR

This sugar occurs to the extent of 4 to 5 percent in cow's milk and 6 to 8 percent in human milk. It is a colorless, amorphous powder that is only slightly soluble in water. Unlike the other sugars, it is nearly tasteless. For the latter reason, this sugar is employed in high-caloric special diets.

Lactose is hydrolyzed by acids and certain enzymes to a molecule each of glucose and galactose.

$$\underset{\text{lactose}}{C_{12}H_{22}O_{11}} + H_2O \longrightarrow \underset{\text{glucose}}{C_6H_{12}O_6} + \underset{\text{galactose}}{C_6H_{12}O_6}$$

Lactose is not fermented by yeast but is a reducing sugar because it possesses a free aldehyde group. Its presence in the urine rather than glucose may be established by its nonfermentability by yeast. Certain bacteria convert lactose to lactic acid, a fermentation reaction responsible for the souring of milk.

MALTOSE, $C_{12}H_{22}O_{11}$, OR MALT SUGAR

This sugar is a product of the partial hydrolysis of starch. The conversion of starch to maltose is accomplished by an enzyme, diastase, present in sprouted barley grains or *malt*. Enzymes present in the digestive juices, such as the salivary amylase of saliva, also convert starch to maltose. Maltose is a sweet sugar and is soluble in water. It is hydrolyzed to two molecules of glucose,

$$\underset{\text{maltose}}{C_{12}H_{22}O_{11}} + H_2O \longrightarrow \underset{\text{glucose}}{2C_6H_{12}O_6}$$

and is fermented by yeast. Yeast produces an enzyme, maltase, which converts maltose to glucose, and the latter is then attacked by the zymase of yeast. Maltose is a reducing sugar and must contain one aldehyde group that is not involved in the linkage of the two glucose units.

Polysaccharides

The polysaccharides, when completely hydrolyzed, produce a number of molecules of monosaccharides. Pentosans are those polysaccharides that produce pentoses upon hydrolysis. They occur abundantly in wood, bran, and certain gums. The hexosans produce hexoses when hydrolyzed and will occupy most of our attention. Some important hexosans are starch, glycogen, dextrins, and cellulose. The hexosans are represented by the formula $(C_6H_{10}O_5)_x$, where x is some large undetermined number.

GLYCOGEN, $(C_6H_{10}O_5)_x$

Glycogen is a colorless powder that forms an opalescent solution in water. It is also called *animal starch* because it occurs abundantly in the liver and serves as a reserve food supply for animals. It occurs in shellfish and is abundant in the oyster and in the muscle of the scallop.

Glycogen is hydrolyzed to glucose by acids. The liver is able to synthesize glycogen from glucose, a process called *glycogenesis*. If insufficient glucose is being ingested to meet the demands of the body, glycogen stored in the liver is converted to glucose, a process called *glycogenolysis* (see page 399).

Glycogen resembles starch in structure. The molecules of glycogen are probably larger but, being more highly branched, consist of shorter chains than starch molecules. Glycogen produces a red color with iodine.

STARCH, $(C_6H_{10}O_5)_x$

This polysaccharide is found chiefly in the seeds, tubers, and roots of plants. Starch usually occurs as granules that have different shapes depending on the source of the starch. It is therefore possible to identify the source of the starch by a microscopic examination. Starch is the most important food polysaccharide and is an important constituent of our diet.

In Europe, starch is usually prepared from potatoes, whereas in the United States, corn serves as its most important source. Cornstarch is a major source of glucose and of syrups, such as Karo syrup. Upon complete hydrolysis, starch is converted to glucose, but there are many intermediate products of hydrolysis, such as the dextrins and maltose. It is hydrolyzed either by acids or enzymes. The salivary amylase of the saliva, certain enzymes called *amylases* present in the small intestine, and the diastase of malt convert starch to maltose. Starch must therefore be made up of glucose units.

Starch is not soluble in water, but if heated in water, the granules rupture and form a colloidal suspension. Starch forms a characteristic deep blue-black color with iodine, which disappears when the starch–iodine mixture is warmed and returns when it cools.

CELLULOSE, $(C_6H_{10}O_5)_x$

Cellulose composes the greater portion of wood, paper, flax, and cotton. Acid-washed cotton is nearly pure cellulose. Cellulose serves as a structural and supporting material for plants. It cannot be digested by the digestive system of man, but serves to give bulk to the feces and thus prevents constipation. Some bacteria are capable of utilizing a portion of cellulose.

If cellulose is treated with concentrated hydrochloric acid and is then diluted and boiled, it is hydrolyzed to glucose.

OTHER POLYSACCHARIDES

The *dextrins* are produced by the partial hydrolysis of starch. They form sticky solutions with water and are used in the preparation of mucilage. Dextrins are components of infant foods. They usually produce a red color with iodine.

Inulin is a polysaccharide found in the roots of the Jerusalem artichoke and in the tuber of dahlias. It is hydrolyzed to fructose by acids.

The pentosans produce pentoses upon hydrolysis (see page 298). *Agar-agar* is a galactan and is hydrolyzed to galactose. It is not digested in the human alimentary tract and is used in the treatment of constipation, because it swells in water and gives bulk to the feces. It is also used in the preparation of bacterial cultures.

The *pectins* are complex materials that resemble the polysaccharides. Appreciable amounts of arabinose, galactose and galacturonic acid, an oxidation product of galactose, are produced during the hydrolysis of the pectins. Apples and citrus fruits are important sources of commercial pectin, which is used in the preparation of jams and jellies. When pectins are present in (or added to) fruit juices, gelling will occur if fruit acids and sugar are also present.

More on the structures of the monosaccharides

The formulas for the open chain and a cyclic structure of glucose was indicated on page 295. The following sections, in which the structures of the monosaccharides are discussed, have been included for those instructors who may desire to pursue this subject further. This discussion should make much more understandable the Fischer and Haworth types of structural formulas for the cyclic structures of these sugars and their derivatives discussed in the sections on the metabolism of carbohydrates (pages 398–405) and the nucleic acids and their hydrolysis products (pages 438–444).

TRIOSES

The simplest aldotriose (page 294) is glyceraldehyde,

$$
\begin{array}{l}
{}^{1}\mathrm{CHO} \\
\quad | \\
{}^{2}\mathrm{CHOH} \\
\quad | \\
{}^{3}\mathrm{CH_2OH}
\end{array}
$$

It is to be noted that in glyceraldehyde, carbon atom #2 contains four unlike substituents, namely H, OH, CHO, and CH_2OH. Such a carbon atom that is attached to four unlike univalent substituents is called an *asymmetric carbon atom.* Van't Hoff suggested not only that a carbon is tetrahedral when joined to four univalent atoms or groups of atoms (page 218) but that if the four substituents are unlike, that is, if the carbon atom is asymmetric, there are two different arrangements of substituents, called *configurations,* about such a carbon atom. Tetrahedra are used below to indicate the configuration of the two isomeric glyceraldehydes, called *dextro*-glyceraldehyde and *levo*-glyceraldehyde:

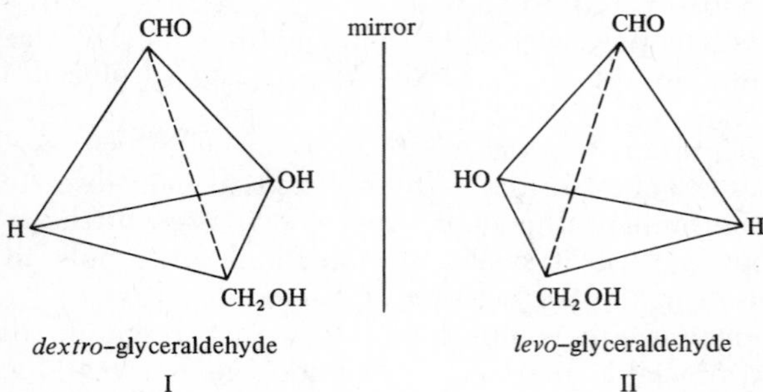

dextro-glyceraldehyde I *levo*-glyceraldehyde II

It is to be noted that II is the mirror image of I. Examination of models of these two structures will establish not only that they are mirror images of each other, but also that one cannot be superimposed upon its mirror image structure (Figure 21.1). Such isomers as I and II are called *mirror-image isomers, enantiomorphs* or *enantiomers.* The dextrorotatory isomer, which can also be designated as *dextro-*, *d-*, or +, rotates the plane of plane-polarized light to the right (clockwise), while the levorotatory isomer, also designated as *levo-*, *l,* or −, rotates the plane of plane-polarized light to the left (counterclockwise). Mirror-image isomers or enantiomorphs have the same physical properties (such as melting point, boiling point, density, and solubility) except for their effect on plane-polarized light. Chemical properties are also identical.

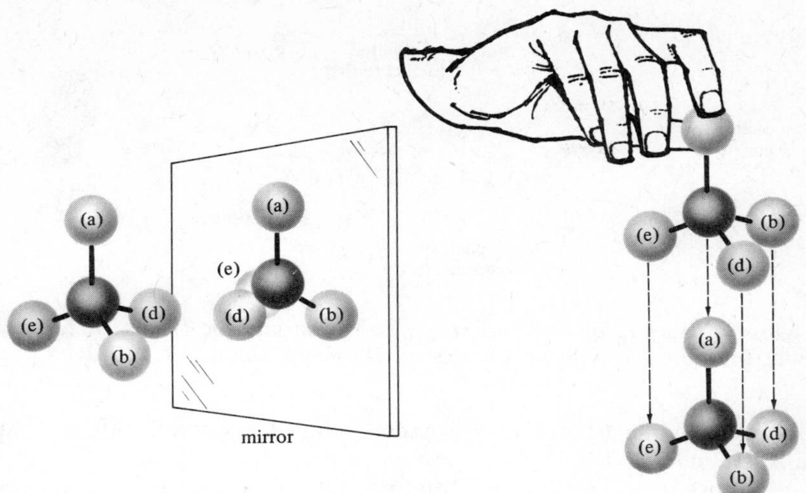

Figure 21.1 A demonstration of the nonsuperimposability of mirror-image isomers. (From Linstromberger, Organic Chemistry, *2nd ed., Heath, Lexington, Mass., 1970.)*

PLANE-POLARIZED LIGHT

Ordinary light may be considered as made up of waves vibrating in all planes perpendicular to its line of propagation, as indicated in Figure 21.2 (line of propagation approaches point *A* directly from the rear). By the use of suitable optical systems, such as tourmaline crystals, a Nicol prism, or a polaroid lens, the component of the light vibrating in only one plane is transmitted through these optical systems.

Light that vibrates in a single plane perpendicular to its line of propagation is said to be *plane polarized.* As previously noted, dextrorotatory optical isomers rotate the plane of plane-polarized light

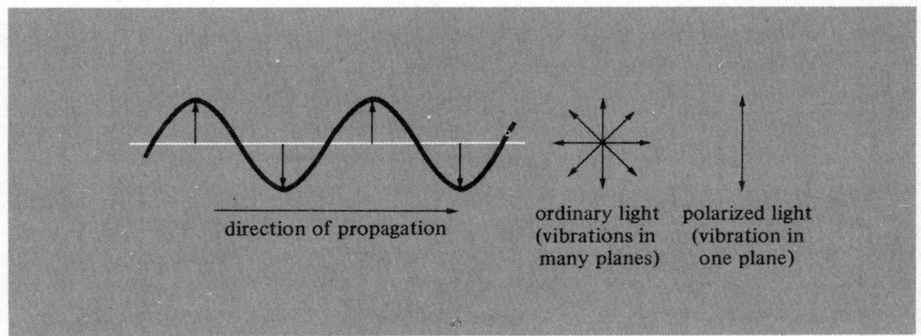

Figure 21.2 Representation of light as a wave motion. (From R. J. Ouellette, Introductory Organic Chemistry, *Harper & Row, N.Y., 1971.)*

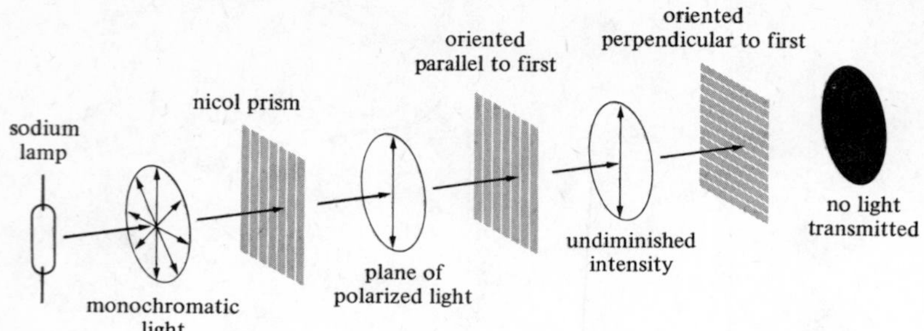

***Figure 21.3** Effect of nicol prisms on light transmission. (From R. J. Ouellette,* Introductory Organic Chemistry, *Harper & Row, N.Y., 1971.)*

clockwise, while levorotatory isomers rotate it counterclockwise (Figures 21.3 and 21.4).

Substances capable of rotating the plane of plane-polarized light are said to be *optically active.* Isomers such as *d*- and *l*-glyceraldehyde that differ in their effect on plane-polarized light are known as *optical isomers.* Optical isomers represent a type of *stereoisomerism.* Stereoisomers do not differ in the order of linkages of atoms or groups of atoms in a molecule, but differ in the three-dimensional arrangement of atoms or groups of atoms about certain carbon atoms.

If two of the four groups attached to a carbon atom are identical, as indicated in the following structure,

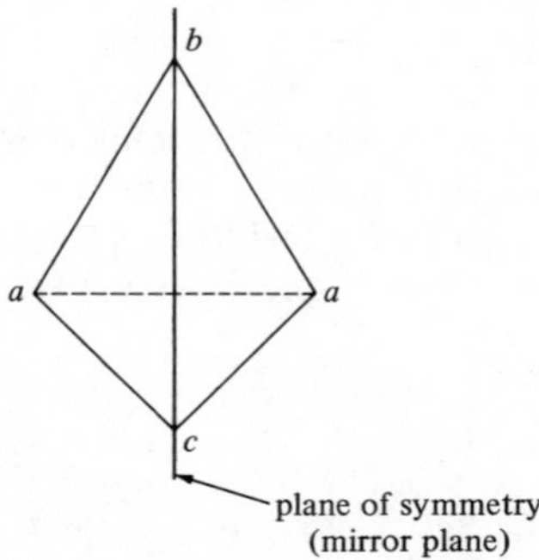

a *plane of symmetry* will divide the molecule into two equal portions, one the mirror image of the other. In such molecules, mirror-image structures are superimposable and *identical.* Such a molecule will be optically inactive.

At Emil Fischer's suggestion, it is the usual practice to employ planar projections for the three-dimensional structures as indicated above for the two isomeric glyceraldehydes. The following planar projection formulas are usually used to represent the structures of

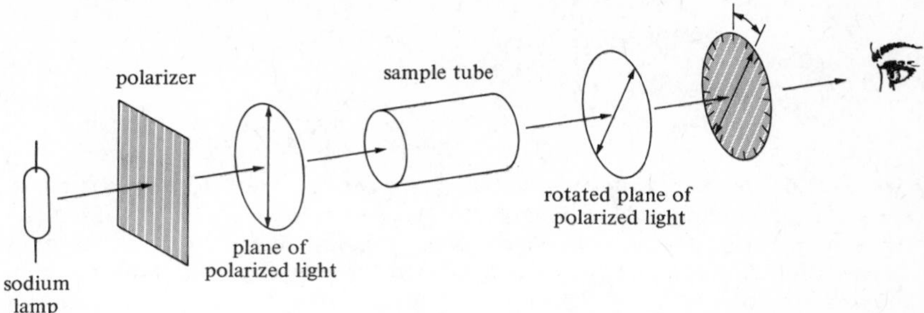

***Figure 21.4** Schematic representation of a polarimeter. (From R. J. Ouellette,* Introductory Organic Chemistry, *Harper & Row, N.Y., 1971.)*

the two isomers of glyceraldehyde:

```
    CHO                 CHO
     |                   |
H—C*—OH           HO—C*—H
     |                   |
    CH2OH               CH2OH
```

l-glyceraldehyde   *d*-glyceraldehyde

In these projection formulas, the asymmetric carbon, marked with an asterisk, is in the plane of the paper, while CHO and CH_2OH are below this plane and H and the OH are above. Solid lines are also used in place of the broken lines in these planar projection formulas. However, it will still be understood that the CHO and CH_2OH groups are below the plane of the paper (Figures 21.5, 21.6, and 21.7).

A phosphate ester of glyceraldehyde, as well as a phosphate ester of the ketotriose dihydroxyacetone, appears in the glycolytic cycle in glucose metabolism (pages 402–403). Because dihydroxyacetone,

```
CH2OH
|
C=O
|
CH2OH
```

unlike glyceraldehyde, contains no asymmetric carbon atoms, it exists as a single structure.

TETROSES

The introduction of a second asymmetric carbon atom doubles the number of optically active isomers. Thus there are four optically active aldotetroses, namely (+) and (−)-erythrose and (+) and (−)-threose, represented by the following projection formulas:

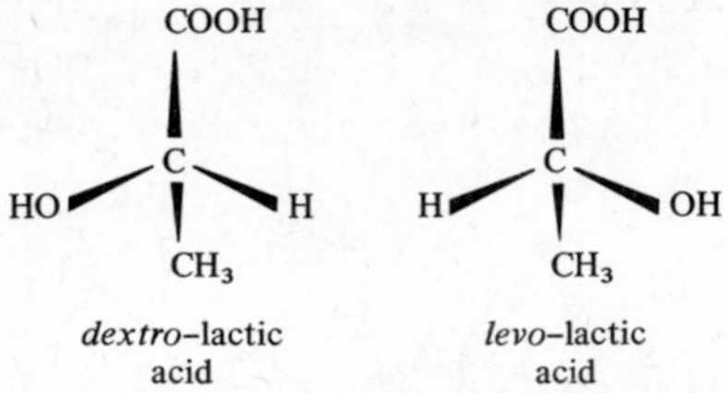

COOH
C
HO
H
CH3
COOH
C
H
OH
CH3
dextro-lactic acid
levo-lactic acid

Figure 21.5 The two structures indicate the configuration of the two enantiomers (enantiomorphs) *or mirror image isomers of lactic acid. Since the points of the wedges are directed away from the viewer, vertical wedges indicate that the valence bonds extend from the asymmetric carbon atom, which lies in the plane of the paper, to groups that lie below the plane of the paper. The horizontal wedges extend from the asymmetric carbon atom to groups that lie above the plane of the paper.*

COOH
OH
C
H
CH3
COOH
H—C—OH
CH3

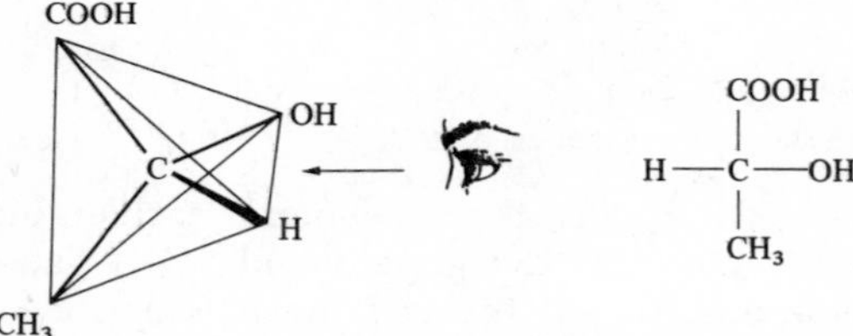

Figure 21.6 *Projection of lactic acid. (From R. J. Ouellette,* Introductory Organic Chemistry, *Harper & Row, N.Y., 1971.)*

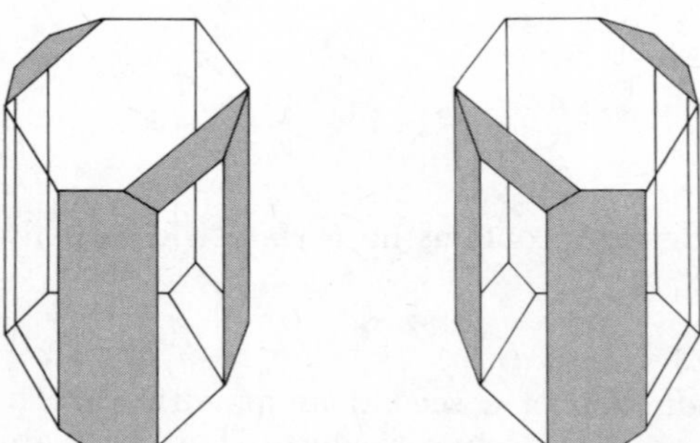

Figure 21.7 *Schematic representation of right- and left-handed crystals of sodium ammonium tartrate.*

(−)-erythrose	(+)-erythrose	(−)-threose	(+)-threose
CHO	CHO	CHO	CHO
H—C—OH	HO—C—H	HO—C—H	H—C—OH
H—C—OH	HO—C—H	H—C—OH	HO—C—H
CH_2OH	CH_2OH	CH_2OH	CH_2OH

(+) erythrose is the enantiomer of (−) erythrose, while at the same time, (+) threose and (−) threose are said to be enantiomorphic. The threoses are optical isomers of but not enantiomorphs of (−) erythrose. Optical isomers that are not enantiomorphs are called *diastereoisomer.* Whereas enantiomorphs exhibit similar physical properties except for their effect on plane-polarized light, diastereoisomers exhibit differences in such physical properties as melting and boiling points, solubility, and density. They produce unequal rotations of plane-polarized light.

PENTOSES

Because the introduction of each new asymmetric carbon atom into a molecule will double the number of possible optically active isomers, van't Hoff has suggested the formula

$$2^n = \text{number of active isomers}$$

to determine the number of possible active isomers that contain *n different* asymmetric carbon atoms.

Based upon van't Hoff's rule, 2^3, or eight, stereoisomeric aldopentoses (which can exist as four racemic modifications) should exist. All eight (whose formulas and names follow) are known.

(−)-ribose	(+)-ribose	(−)-arabinose	(+)-arabinose
CHO	CHO	CHO	CHO
H—C—OH	HO—C—H	HO—C—H	H—C—OH
H—C—OH	HO—C—H	H—C—OH	HO—C—H
H—C—OH	HO—C—H	H—C—OH	HO—C—H
CH_2OH	CH_2OH	CH_2OH	CH_2OH

(+)-xylose	(−)-xylose	(+)-lyxose	(−)-lyxose
CHO	CHO	CHO	CHO
H—C—OH	HO—C—H	H—C—OH	HO—C—H
HO—C—H	H—C—OH	H—C—OH	HO—C—H
H—C—OH	HO—C—H	HO—C—H	H—C—OH
CH_2OH	CH_2OH	CH_2OH	CH_2OH

HEXOSES

Only three aldohexoses, (+)-glucose, (+)-mannose, and (+)-galactose, are obtained from natural sources.

```
     CHO                  CHO                  CHO
     |                    |                    |
  H—C—OH             HO—C—H               H—C—OH
     |                    |                    |
 HO—C—H              HO—C—H              HO—C—H
     |                    |                    |
  H—C—OH              H—C—OH             HO—C—H
     |                    |                    |
  H—C—OH              H—C—OH              H—C—OH
     |                    |                    |
    CH2OH                CH2OH                CH2OH
```

(+)-glucose (+)-mannose (+)-galactose

The other thirteen stereoisomers have been prepared synthetically.

On pages 295–296, it was indicated that (+)-glucose, as well as the ketohexose (−)-fructose, exists in solution primarily in cyclic structures. The cyclic structures are in equilibrium with a small concentration of the open-chain structures. All hexoses, as well the pentoses, can exist in cyclic structures.

To fully understand the nature of these cyclic structures of the sugars, one must consider the nature of hemiacetal and acetal formation by aldehydes. If acetaldehyde is treated with ethyl alcohol in the presence of a mineral acid, such as HCl, a hemiacetal of slight stability will be formed, thus:

$$CH_3C(=O)H + C_2H_5OH \xrightarrow{HCl} CH_3\text{—}C(H)(OH)(OCH_2CH_3)$$

a hemiacetal

A mole of this hemiacetal may react with another mole of alcohol (in the presence of HCl) to form the very stable acetal:

$$CH_3\text{—}C(H)(OH)(OCH_2CH_3) + C_2H_5OH \xrightarrow{HCl} CH_3\text{—}C(H)(OCH_2CH_3)(OCH_2CH_3)$$

an acetal

Emil Fischer observed that when glucose reacted with methyl alcohol and HCl, only one molecule of methyl alcohol reacted with a molecule of glucose to form a stable product called *methyl glucoside,* which must have an acetal structure, which Fischer represented as

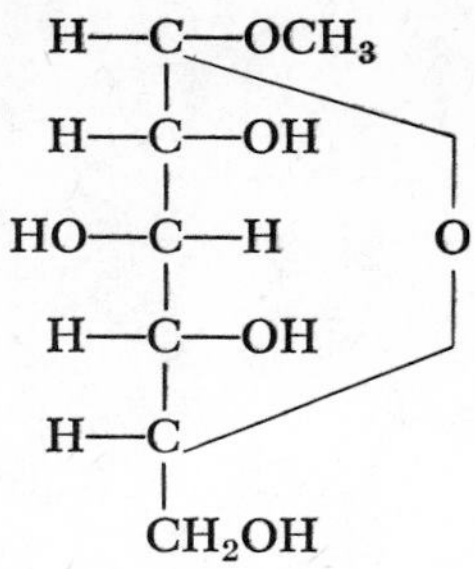

$$
\begin{array}{l}
\mathrm{H{-}C{-}OCH_3} \\
\quad | \\
\mathrm{H{-}C{-}OH} \\
\quad | \\
\mathrm{HO{-}C{-}H} \qquad \mathrm{O} \\
\quad | \\
\mathrm{H{-}C{-}OH} \\
\quad | \\
\mathrm{H{-}C} \\
\quad | \\
\mathrm{CH_2OH}
\end{array}
$$

methyl glucoside

Glucose must then be able to exist as a cyclic hemiacetal structure, which Fischer represented as

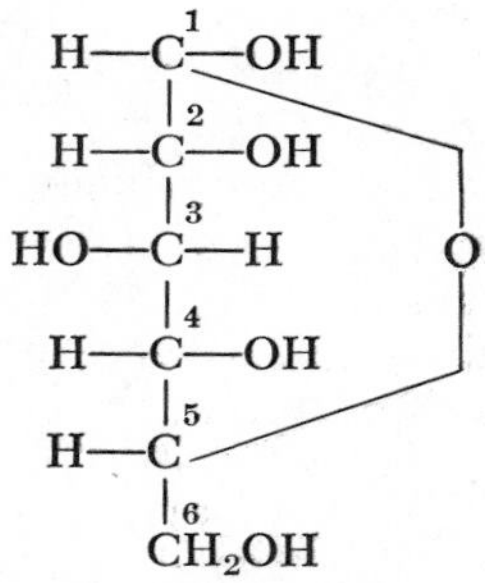

$$
\begin{array}{l}
\mathrm{H{-}\overset{1}{C}{-}OH} \\
\quad |_{2} \\
\mathrm{H{-}C{-}OH} \\
\quad |_{3} \\
\mathrm{HO{-}C{-}H} \qquad \mathrm{O} \\
\quad |_{4} \\
\mathrm{H{-}C{-}OH} \\
\quad |_{5} \\
\mathrm{H{-}C} \\
\quad |_{6} \\
\mathrm{CH_2OH}
\end{array}
$$

Note that this hemiacetal structure has been formed by the reaction of the hydroxyl group on carbon #5 with the aldehyde group on carbon #1.

In the hemiacetal structure of glucose and the acetal structure of methyl glucoside, a new asymmetric carbon atom on carbon #1 has been generated. Crystalline α-glucose, in which the OH group is to the right of carbon atom #1, is obtained when glucose is crystallized from a hot alcohol solution and can be assigned the formula

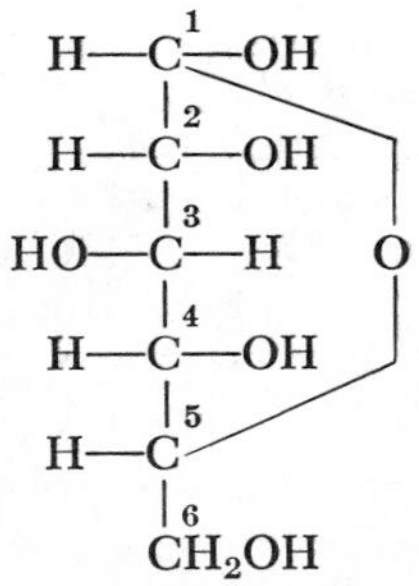

$$
\begin{array}{l}
\mathrm{H{-}\overset{1}{C}{-}OH} \\
\quad |_{2} \\
\mathrm{H{-}C{-}OH} \\
\quad |_{3} \\
\mathrm{HO{-}C{-}H} \qquad \mathrm{O} \\
\quad |_{4} \\
\mathrm{H{-}C{-}OH} \\
\quad |_{5} \\
\mathrm{H{-}C} \\
\quad |_{6} \\
\mathrm{CH_2OH}
\end{array}
$$

α-glucose

The diastereoisomeric β-glucose is assigned the formula

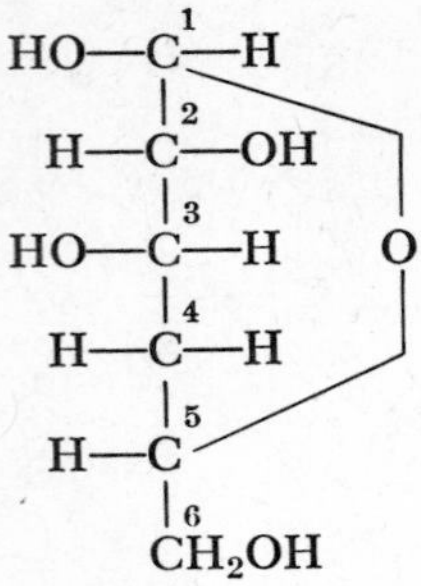

```
       1
HO—C—H ─────┐
   |2        |
 H—C—OH      |
   |3        |
HO—C—H       O
   |4        |
 H—C—H       |
   |5        |
 H—C ────────┘
   |6
  CH2OH
```

β-glucose

The formulas for the open-chain and cyclic hemiacetal structures of glucose and the other sugars, as employed so far in this discussion, were suggested by Emil Fischer. Haworth has objected to these types of cyclic formulas because the C—O bond length should be about the same as the C—C bond length and the ring of five carbon atoms and the oxygen atom should have approximately the shape of a regular hexagon. In the formulas suggested by Haworth, the primary alcohol group, CH_2OH (on carbon #6), is placed above the upper left carbon atom of the ring, while the oxygen atom is placed in the upper right position in the ring. Substituents that appear to the right in the Fischer cyclic formula appear below the plane of the ring in the Haworth formulas, while those that appear to the left in the Fischer formula appear above the plane of the ring. Thus the Haworth formulas for α- and β-glucose are represented as follows:

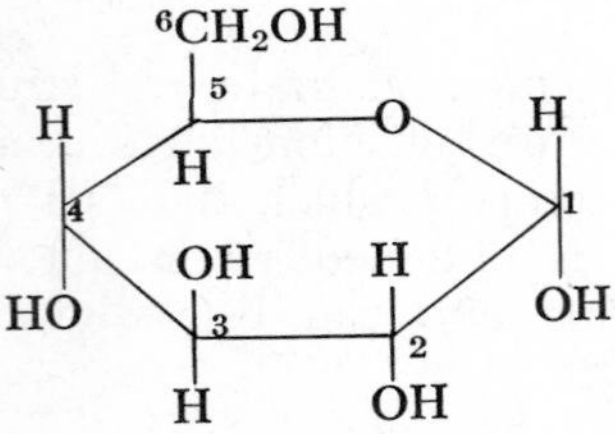

```
       6CH2OH                            6CH2OH
        |5                                |5
  H     |________O        H         H     |________O        OH
   |   / H        \       |          |   / H        \       |
   |4/             \1     |          |4/             \1     |
   |  \  OH   H    /      |          |  \  OH   H    /      |
  HO   \ |3   |   /      OH         HO   \ |3   |   /       H
        \|____|2 /                        \|____|2 /
         H    OH                           H    OH
```

α-glucose 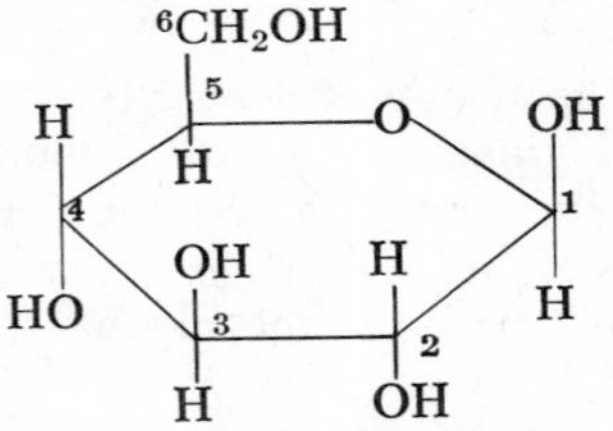 β-glucose

Note that the carbon atoms in the ring are not indicated in the above formulas and that the only difference in these two formulas is the position of the OH groups on the #1, or hemiacetal, carbon atom, which is called the *anomeric carbon.* In α-glucose, this OH group is below the plane of the ring, while in β-glucose, it is above the plane. Incidentally, in the cyclic structures for the ketoses, the #2 carbon atom will be the anomeric carbon.

If α-glucose (that is, the glucose that crystallizes from a hot alcohol solution) is dissolved in water, the initial rotation of plane-polarized light produced by this freshly prepared solution will undergo change for some period of time, whereupon the rotation will become constant. This phenomenon exhibited by some optically active compounds, in particular by many of the sugars, is known as *mutarotation.* For glucose, this is readily explained as a conversion of α-glucose to the open-chain structure, which then changes to an equilibrium mixture of α- and β-glucose and a small amount of the open-chain structure. This equilibrium, when mutarotation is complete, can be represented as

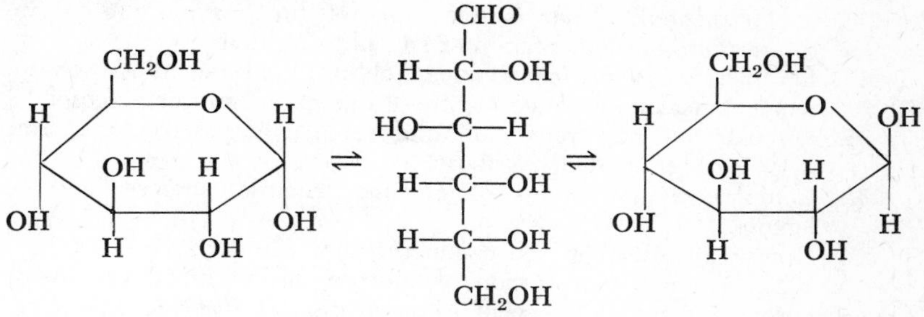

```
        CHO
         |
     H—C—OH
         |
    HO—C—H
         |
     H—C—OH
         |
     H—C—OH
         |
       CH2OH
```

α-glucose ⇌ open-chain glucose ⇌ β-glucose

In the ketohexose fructose, the alcohol group on carbon #5 can form two hemiacetal (actually hemiketal) structures with the carbonyl group on the anomeric #2 carbon. These two cyclic structures are represented by the following Haworth formulas:

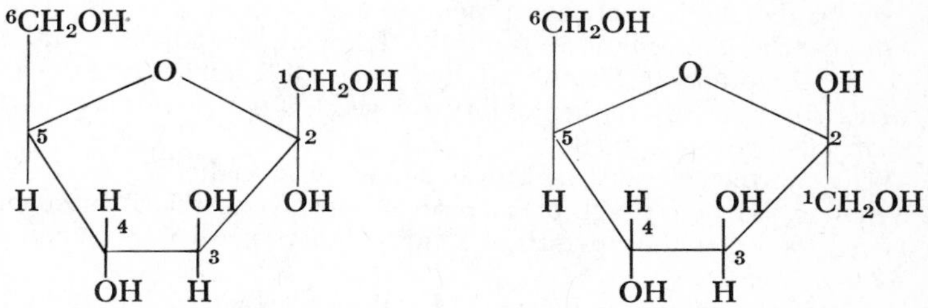

α-fructose

β-fructose

The structures of many disaccharides and polysaccharides have been established and are discussed in textbooks of organic chemistry and biochemistry.

STUDY QUESTIONS

1. What is the modern definition of carbohydrates?
2. Give several reasons why the word "carbohydrate" is a misnomer.
3. Classify the carbohydrates according to the number of simple sugar molecules formed on hydrolysis.
4. Classify monosaccharides according to (a) the type of functional groups that they possess and (b) the number of carbon atoms in the molecule.
5. Give an example of an aldohexose, an aldopentose, a ketopentose, and a ketohexose.
6. What are the three most important hexoses?
7. How many aldohexoses exist that are isomeric with glucose? How do they differ in structure from glucose?

8. Do glucose and fructose exist in solution primarily as open-chain or cyclic compounds?
9. What products are formed upon complete combustion of the monosaccharides? Write an equation for the complete combustion of fructose. How much heat is generated when 1 g of a carbohydrate burns? Of what significance is this fact in nutrition?
10. Which of the monosaccharides are fermented by yeast? Which are not fermented? What is the name of the enzyme present in yeast responsible for fermentation of sugars to alcohol?
11. Do all monosaccharides reduce Fehling's and Benedict's solutions? Of what clinical importance are these reduction tests of glucose?
12. Discuss the occurrence of glucose, galactose, fructose, arabinose, xylose, ribose, and deoxyribose.
13. Compare the sweetness of glucose, fructose, sucrose, maltose, and lactose.
14. How is glucose prepared commercially?
15. What is the concentration of glucose in the blood and urine of a normal individual? In what common disease does the concentration of glucose in the blood and urine increase?
16. To what sugar are the hexoses changed in the liver?
17. By what other names are glucose and fructose known?
18. In what chemical reaction does fructose differ from the simple ketones? What portion of the fructose molecule is responsible for this difference?
19. What is pentosuria? Do any ill effects accompany this condition? Are pentoses fermented by yeasts? Describe three methods that may be used to distinguish between a pentose and glucose.
20. Discuss the disaccharides from the following standpoints: (a) occurrence; (b) products formed on hydrolysis; (c) ability to reduce Benedict's solution; (d) fermentability by yeast; (e) the various substances that will hydrolyze them.
21. What enzymes of yeast hydrolyze sucrose and maltose?
22. What is the percentage of sucrose in sugar cane juice? In sugar beet juice? Describe the operations required to prepare pure sucrose from these sources.
23. What explanation can be given for the fact that maltose and lactose are reducing disaccharides whereas sucrose is a nonreducing sugar?
24. What chemical change is responsible for the souring of milk?
25. What is malt? What important enzyme does it contain?
26. What are the products of the hydrolysis of glycogen, starch, inulin, dextrins, cellulose, pentosans, agar-agar, and pectins?
27. What color does iodine produce with glycogen, dextrins, and starch?
28. By what other name is glycogen known? How does it differ in structure from starch? Discuss its occurrence. What purpose does it serve in the animal organism?
29. What is glycogenesis? Glycogenolysis?
30. What purpose do starch and cellulose serve in the plant?
31. How may the plant source of starch be determined?
32. What are some of the intermediate hydrolysis products of starch?
33. What enzymes catalyze the hydrolysis of starch to maltose?
34. What purpose do cellulosic materials serve in the diet of the human?
35. Name some commercial substances that are primarily cellulose.
36. Mention some uses of agar-agar, the dextrins, and the pectins.
37. What are some of the important sources of the pectins, and what unusual property do they possess?

38. Explain clearly the following terms, which have been introduced in the study of carbohydrates: aldose, disaccharide, hexose, reducing sugar, glycogenolysis.
39. How do stereoisomers differ from the structural isomers that have been studied previously?
40. How does plane-polarized light differ from ordinary light?
41. How is plane-polarized light produced?
42. How does a dextrorotatory isomer differ from its levorotatory isomer in physical properties? Which of the physical properties are identical?
43. What is meant by the following terms that relate to optical isomerism: asymmetric carbon atoms, enantiomorph, plane of symmetry, levorotatory isomer, diastereoisomer, configuration?
44. Give examples of enantiomorphs, an asymmetric carbon atom, diastereoisomers, and a compound with a plane of symmetry.
45. In the compound $CHICl—SO_3H$, which possesses a single asymmetric carbon atom, indicate the two possible configurations (arrangement of substituents) about the asymmetric carbon atom.
46. Based on the number of asymmetric carbon atoms that are present in a molecule, indicate the formula that may be used to determine the maximum number of optically active isomers possible for this structure. What is the maximum number of optically active isomers that possess (a) three asymmetric carbon atoms and (b) five asymmetric carbon atoms?
47. By the use of Fischer planar-projection formulas, indicate the configuration of

    (a) (−)-ribose
    (b) its enantiomorph
    (c) a diastereoisomer of (−)-ribose
    (d) (+)-glucose

48. What evidence can be cited that glucose exists in solution primarily as a cyclic "hemiacetal" structure? Which is the anomeric carbon in α-glucose? In β-fructose?
49. Most sugars exhibit the phenomenon called mutarotation. Describe this phenomenon. Can an explanation be given for it?
50. (+) Mannose differs from (+) glucose only in the configuration of its #2 carbon atom. Write configurational formulas for the open-chain structures of these two sugars.
51. Indicate the number of asymmetric carbon atoms in the following formulas by labeling them with an asterisk; then predict the total number of possible configurations for each.

$$\begin{array}{l} \mathrm{COOH} \\ | \\ \mathrm{CHOH} \\ | \\ \mathrm{CHBr} \\ | \\ \mathrm{CH_3} \\ \text{(a)} \end{array} \qquad \begin{array}{l} \mathrm{CH_2Br} \\ | \\ \mathrm{CHOH} \\ | \\ \mathrm{CHOH} \\ | \\ \mathrm{CHOH} \\ | \\ \mathrm{CHOH} \\ | \\ \mathrm{CH_3} \\ \text{(b)} \end{array} \qquad \begin{array}{l} \mathrm{CH_2COOH} \\ | \\ \mathrm{CHOH} \\ | \\ \mathrm{CH_2COOH} \\ \text{(c)} \end{array} \qquad \begin{array}{l} \mathrm{COOH} \\ | \\ \mathrm{CHOH} \\ | \\ \mathrm{CH_2} \\ | \\ \mathrm{COOH} \\ \text{(d)} \end{array} \qquad \begin{array}{l} \mathrm{CHO} \\ | \\ \mathrm{CHOH} \\ | \\ \mathrm{CHOH} \\ | \\ \mathrm{CHOH} \\ | \\ \mathrm{CH_3} \\ \text{(e)} \end{array}$$

$CH_3—CHBr—CH_2—CO—CH_3$
(f)

$CH_3—CHCl—CH_2—CHOH—COOH$
(g)

52. Encircle the following configurations that represent optically inactive structures. In each of these, indicate the plane of symmetry present in the molecule.

```
  CH2OH           COOH          COOH          CHO
   |               |             |             |
H—C—OH          H—C—OH          CH2        H—C—OH
   |               |             |             |
  CH2OH          HO—C—H         C=O        H—C—OH
  (a)              |             |             |
                  COOH           |            COOH
                  (b)           CH2           (d)
                                 |
                                COOH
                                (c)

  COOH            COOH            COOH
   |               |               |
H—C—OH          H—C—OH          H—C—OH
   |               |               |
H—C—OH          H—C—OH          H—C—OH
   |               |               |
  COOH          H—C—OH         HO—C—H
  (e)              |               |
                  COOH            COOH
                  (f)             (g)
```

Note! Even though a molecule contains asymmetric carbon atoms, if it contains a plane of symmetry it will be optically inactive.

chapter 22

some simple nitrogen-containing organic compounds

22

Amines, amides, amino acids, and urea will be discussed in this chapter. A knowledge of the first three of these will be a great aid in the study of the proteins (Chapter 24). Urea is an end product of the metabolism of the proteins in the body and the most abundant organic constituent of the urine.

Amines

These may be considered as derivatives of ammonia, NH_3, in which one or more hydrogen atoms of ammonia are replaced by alkyl groups. If only one hydrogen of ammonia is replaced by an alkyl group, the amine is classed as a *primary amine;* if two hydrogens are replaced, it is a *secondary amine;* and if all three are replaced, it is a *tertiary amine.* The general formula for each class of amines with an example of each follows:

primary amine	RNH_2	$C_2H_5NH_2$	ethylamine
secondary amine	R_2NH	$(CH_3)_2NH$	dimethylamine
tertiary amine	R_3N	$(C_2H_5)_3N$	triethylamine

The NH_2 group of a primary amine is called the *amino group.*

The lower-molecular-weight amines are colorless liquids or gases that are soluble in water and have either a fishlike or ammoniacal odor. Unlike ammonia, they will burn in air.

Like ammonia gas, the aliphatic amines dissolve in water to form weak basic solution that turn red litmus blue:

$$NH_3 + H_2O \rightleftharpoons NH_4^+ + OH^-$$
$$CH_3NH_2 + H_2O \rightleftharpoons CH_3NH_3^+ + OH^-$$

Ammonia and amines react with acids to form salts that are generally colorless crystalline solids.

$$NH_3 + HCl \longrightarrow NH_4^+Cl^-$$
$$CH_3NH_2 + HCl \longrightarrow CH_3NH_3^+Cl^-$$

The organic chemist usually writes the formula of the latter compound $CH_3NH_2 \cdot HCl$ and calls it *methylamine hydrochloride.*

Some amines are formed by the putrefaction of proteins. These amines were previously thought to produce ptomaine poisoning and are frequently called *ptomaines.* However, it is now recognized that ptomaine poisoning is produced by bacterial toxins.

Amides

The amides may be considered as derivatives of organic acids in which the OH of the carboxyl group is replaced by the NH_2 group. The general formula for the amides is

$$R{-}\overset{\overset{\large O}{/\!/}}{C}{-}NH_2$$

Amides are named by replacing the *-ic* ending of the corresponding acid by the suffix *-amide.*

Acetamide,

$$CH_3{-}\overset{\overset{\large O}{/\!/}}{C}{-}NH_2$$

is a typical amide. It is a colorless crystalline solid that is quite soluble in water and, like most simple amides, is a neutral substance.

Amides are hydrolyzed by both acids and bases. When heated with hydrochloric acid, acetamide forms ammonium chloride and acetic acid:

$$CH_3CONH_2 + H_2O + HCl \longrightarrow CH_3CO_2H + NH_4Cl$$

Sodium hydroxide converts it to ammonia gas and sodium acetate:

$$CH_3CONH_2 + NaOH \longrightarrow CH_3CO_2Na + NH_3$$

An amide-like linkage, called the *peptide linkage,* is present in the peptides and the proteins (page 353).

Urea, or carbamide, $NH_2{-}\overset{\overset{\large O}{/\!/}}{C}{-}NH_2$

This is the diamide of carbonic acid,

$$HO{-}\overset{\overset{\large O}{\|}}{C}{-}OH$$

Urea is a colorless, crystalline solid that is soluble in water but nearly insoluble in ether. It has a slightly bitter taste and is obtained commercially by heating a mixture of carbon dioxide and ammonia gas under high pressure.

$$2NH_3 + CO_2 \longrightarrow NH_2\text{—}CO\text{—}NH_2 + H_2O$$

Its preparation from ammonium cyanate is of historical interest (see Chapter 15).

Urease, the enzyme obtained from the jack bean and from soybeans, catalyzes the hydrolysis of urea to carbon dioxide and ammonia:

$$NH_2\text{—}CO\text{—}NH_2 + H_2O \xrightarrow{\text{urease}} 2NH_3 + CO_2$$

Acids convert urea to an ammonium salt and carbon dioxide, while bases convert it to a carbonate and ammonia:

$$NH_2\text{—}CO\text{—}NH_2 + 2HCl + H_2O \longrightarrow 2NH_4Cl + CO_2$$
$$NH_2\text{—}CO\text{—}NH_2 + 2NaOH \longrightarrow Na_2CO_3 + 2NH_3$$

Urea, unlike the amides, is slightly basic (usually no reaction with litmus) and reacts with concentrated nitric and oxalic acids to form colorless, crystalline precipitates of the composition $NH_2\text{—}CO\text{—}NH_2 \cdot HNO_3$ and $(NH_2\text{—}CO\text{—}NH_2)_2 \cdot H_2C_2O_4$, respectively.

If solid urea is carefully heated above its melting point, in part, a molecule of ammonia is lost from two molecules of urea, and biuret, $NH_2\text{—}CO\text{—}NH\text{—}CO\text{—}NH_2$, is formed:

$$NH_2\text{—}CO\text{—}\boxed{NH_2 + NH}\boxed{_2}\text{—}CO\text{—}NH_2 \longrightarrow NH_2\text{—}CO\text{—}NH\text{—}CO\text{—}NH_2 + NH_3$$

biuret

Biuret, but not urea, produces a violet color when treated with a dilute solution of copper sulfate, which is then made slightly alkaline. This is known as the *biuret test* and is used as a test for the peptide linkages in proteins.

Nitrous acid, HNO_2, converts urea to free nitrogen and carbon dioxide:

$$O{=}C\begin{cases}\boxed{N}=\boxed{H_2 \quad O}=\boxed{N}\text{—}O\text{—}H \\ \qquad\qquad + \\ \boxed{N}=\boxed{H_2 \quad O}=\boxed{N}\text{—}O\text{—}H\end{cases} \longrightarrow 2N_2 + 3H_2O + CO_2$$

This reaction has been used to estimate the amount of urea in a solution by measuring the amount of nitrogen gas formed. Because this does not give accurate results, urea is usually determined quantitatively by a special color reaction or by hydrolyzing it with urease and measuring the amount of ammonia formed in its hydrolysis.

Urea is used as a source of nitrogen in commercial fertilizers and in the production of many useful plastics.

Amino acids

These are amphoteric substances (see page 138) that is, they possess the properties of both a weak acid and a weak base. Their amphoteric properties derive from the fact that they contain at least one amino and one carboxyl group in each molecule. Thus they are able to react with both acids and bases to form salts. The reactions of glycine, aminoacetic acid (NH_2CH_2COOH), with hydrochloric acid and with sodium hydroxide are represented by the following equations:

$$H_2N—CH_2—CO_2H + NaOH \longrightarrow H_2N—CH_2—CO_2Na + H_2O$$
$$H_2N—CH_2—CO_2H + HCl \longrightarrow (NH_3—CH_2—CO_2H)^+ + Cl^-$$

Glycine is the simplest amino acid. Alanine, another important amino acid, may also be called α-aminopropionic acid.

$$\overset{\beta}{CH_3}—\overset{\alpha}{CH}(NH_2)—CO_2H$$

alanine

β-Alanine or β-aminopropionic acid is

$$\overset{\beta}{CH_2}(NH_2)—\overset{\alpha}{CH_2}—CO_2H$$

Note that the letters of the Greek alphabet are commonly used to designate the position of a substituent in an organic acid. If the substituent is on the carbon atom adjacent to the carboxyl group, it is said to be an α-substituent, if on the next carbon it is called a β-substituent, etc.

The α-amino acids are, with few exceptions, the only amino acids that occur in nature. Proteins (see page 350) may be considered as natural polymers of α-amino acids to which they are hydrolyzed by acids, bases, and enzymes.

The following is a listing of the α-amino acids that have been well established as the final hydrolysis products of the naturally occurring proteins. The monoamino-monocarboxylic acids are amphoteric, and because each molecule contains a single basic and a single acid group, they are nearly neutral substances. The monoamino-dicarboxylic acids are slightly acidic, whereas the diamino-monocarboxylic acids are slightly basic substances.

MONOAMINO-MONOCARBOXYLIC ACIDS
Aliphatic Acids

$H_2N—CH_2—COOH$ glycine, or aminoacetic acid

$$CH_3—\underset{\displaystyle NH_2}{\underset{|}{CH}}—COOH$$
alanine, or α-aminopropionic acid

$$HO—CH_2—\underset{\displaystyle NH_2}{\underset{|}{CH}}—COOH$$
serine, or β-hydroxy-α-aminopropionic acid

$$CH_3—\underset{\displaystyle OH}{\underset{|}{CH}}—\underset{\displaystyle NH_2}{\underset{|}{CH}}—COOH$$
threonine, or β-hydroxy-α-aminobutyric acid

$$CH_3—\underset{\displaystyle CH_3}{\underset{|}{CH}}—\underset{\displaystyle NH_2}{\underset{|}{CH}}—COOH$$
valine, or α-aminoisovaleric acid

$$CH_3—\underset{\displaystyle CH_3}{\underset{|}{CH}}—CH_2—\underset{\displaystyle NH_2}{\underset{|}{CH}}—COOH$$
leucine, or α-aminoisocaproic acid

$$CH_3—CH_2—\underset{\displaystyle CH_3}{\underset{|}{CH}}—\underset{\displaystyle NH_2}{\underset{|}{CH}}—COOH$$
isoleucine, or α-amino-β-methylvaleric acid

Aromatic Acids (See Chapter 23)

$$\begin{array}{ccccc} & CH{=}CH & & & \\ / & & \backslash & & \\ CH & & & C—CH_2—\underset{\displaystyle NH_2}{\underset{|}{CH}}—COOH \\ \backslash\!\backslash & & // & & \\ & CH—CH & & & \end{array}$$
phenylalanine, or α-amino-β-phenylpropionic acid

$$\begin{array}{ccccc} & CH{=}CH & & & \\ / & & \backslash & & \\ HOC & & & C—CH_2—\underset{\displaystyle NH_2}{\underset{|}{CH}}—COOH \\ \backslash\!\backslash & & // & & \\ & CH—CH & & & \end{array}$$
tyrosine, or α-amino-β-(p-hydroxy)-phenylpropionic acid

$$\begin{array}{ccccccccc} & CI{=}CH & & & & CI{=}CH & & & \\ / & {\scriptstyle 5'}\quad{\scriptstyle 6'} & \backslash & & / & {\scriptstyle 5}\quad{\scriptstyle 6} & \backslash & & \\ HOC_{4'} & & & {}_1C—O—C_4 & & & & {}_1C—CH_2—\underset{\displaystyle NH_2}{\underset{|}{CH}}—COOH \\ \backslash\!\backslash & {\scriptstyle 3'}\quad{\scriptstyle 2'} & // & & \backslash\!\backslash & {\scriptstyle 3}\quad{\scriptstyle 2} & // & & \\ & CI—CH & & & & CI—CH & & & \end{array}$$

thyroxine, or 3,5-diodo-4-(3′,5′-diodo-4′-hydroxy-phenoxy)-phenylalanine

$$\begin{array}{ccccc} & CI{=}CH & & & \\ / & {\scriptstyle 5}\quad{\scriptstyle 6} & \backslash & & \\ HOC_4 & & & {}_1C—CH_2—\underset{\displaystyle NH_2}{\underset{|}{CH}}—COOH \\ \backslash\!\backslash & {\scriptstyle 3}\quad{\scriptstyle 2} & // & & \\ & CI—CH & & & \end{array}$$
iodogorgoic acid, or 3,5-diodotyrosine

HETEROCYCLIC ACIDS

```
    CH
  //  \
CH     C———C—CH2—CH—COOH
|      ||  ||     |
CH     C   CH     NH2
  \\  / \  /
    CH   N
         |
         H
```

tryptophan, or α-amino-β-3-indolepropionic acid

```
H2C———CH2
 |     |
H2C    CH—COOH
  \   /
   NH
```

proline, or pyrollidine-α-carboxylic acid

```
HO—CH———CH2
    |    |
  H2C    CH—COOH
     \  /
      NH
```

hydroxyproline, or γ-hydroxypyrollidine-α-carboxylic acid

Note: Neither proline nor hydroxyproline contains an amino group, but each contains a basic imino group, =NH, as a part of the ring. This group may be considered as being on an α-carbon atom to the carboxyl group.

```
     CH
   / 2 \\
HN1     3N
 |5     4|
HC=====C—CH2—CH—COOH
              |
              NH2
```

histidine, or β-4-imidazoyl-α-aminopropionic acid

MONOAMINO-DICARBOXYLIC ACIDS

```
COOH
|
CH2
|
CH—NH2
|
COOH
```

aspartic acid, or α-aminosuccinic acid

```
COOH
|
CH2
|
CH2
|
CH—NH2
|
COOH
```

glutamic acid, or α-aminoglutaric acid

DIAMINO-MONOCARBOXYLIC ACIDS

$$H_2N—(CH_2)_4—\underset{\displaystyle NH_2}{\underset{|}{CH}}—COOH$$

lysine, or α,ξ-diaminocaproic acid

$$H_2N—\underset{\displaystyle NH}{\underset{\|}{C}}—NH—(CH_2)_3—\underset{\displaystyle NH_2}{\underset{|}{CH}}—COOH$$

arginine, or δ-guanidyl-α-aminovaleric acid

$$H_2N—CH_2—\underset{\displaystyle OH}{\underset{|}{CH}}—(CH_2)_2—\underset{\displaystyle NH_2}{\underset{|}{CH}}—COOH$$

hydroxylysine, or α-ξ-diamino-δ-hydroxycaproic acid

$$H_2N—\underset{\displaystyle O}{\underset{\|}{C}}—NH—CH_2—CH_2—\underset{\displaystyle NH_2}{\underset{|}{CH}}—COOH$$

citrulline, or δ-carbamido-α-aminovaleric acid

Citrulline does not occur as a hydrolysis product of proteins but is an important intermediate in the formation of urea in the liver.

Sulfur-Containing Amino Acids

$$\begin{array}{l} S—CH_2—\underset{\displaystyle NH_2}{\underset{|}{CH}}—COOH \\ | \\ S—CH_2—\underset{\displaystyle NH_2}{\underset{|}{CH}}—COOH \end{array}$$

cystine, or di-(β-thio-α-aminopropionic acid)

$$HS—CH_2—\underset{\displaystyle NH_2}{\underset{|}{CH}}—COOH$$

cysteine, or β-thiol-α-aminopropionic acid

$$CH_3—S—CH_2—CH_2—\underset{\displaystyle NH_2}{\underset{|}{CH}}—COOH$$

methionine, or γ-methylthio-α-aminobutyric acid

Methionine and cysteine are monoamino-monocarboxylic acids, while cystine is a diamino-dicarboxylic acid.

A list of the amino acids whose presence in the diet is essential for satisfactory growth is given on page 420.

Abbreviations (symbols) for the amino acids are listed later in Table 24.1.

STUDY QUESTIONS

1. To what inorganic substance are the amines related?
2. Upon what basis are the amines divided into the three classes—primary, secondary, and tertiary? Write a general formula for each of the classes of amines, and give an example of each.
3. Describe the physical properties of the lower-molecular-weight amines.
4. Describe the properties of a solution of diethylamine in water. Write

an equation to represent the reaction that occurs when this amine is dissolved in water.

5. Write an equation for the reaction of hydrobromic acid with *n*-propylamine and for the reaction of hydrochloric acid with trimethylamine.
6. Are the amines that are formed in the putrefactive decay of proteins responsible for ptomaine poisoning?
7. Write a general formula for the amides. The amides are derived from what class of organic compounds?
8. Write a formula for and name the amide derived from propionic acid.
9. Write equations for the hydrolysis of propionamide by sodium hydroxide and sulfuric acid.
10. What functional groups do the amino acids contain in common? Write a structural formula for the simplest of the amino acids. What is its common name?
11. The amino acids are amphoteric. Explain, using suitable equations as illustrations.
12. The α-amino acids are obtained upon hydrolysis of what important class of organic substances?
13. How do the β-amino acids differ structurally from the α-amino acids?
14. How is urea related to carbonic acid?
15. Write an equation for (a) Wöhler's synthesis of urea and (b) the commercial synthesis of urea.
16. What substances can be used to catalyze the hydrolysis of urea? What products are formed from urea in each case?
17. What is the color test for biuret? How is biuret formed from urea? What other substances will give a positive biuret test?
18. How may the amount of urea in a solution be determined quantitatively?
19. What is the nature and chemical composition of the precipitates of urea with concentrated nitric and oxalic acids?
20. For what purposes is urea used commercially?
21. Name and give formulas for the amino acids that have been isolated from the hydrolysis products of the proteins. Distinguish between monoamino-monocarboxylic acids, diamino-monocarboxylic acids, and monoamino-dicarboxylic acids with regard to structure and properties.

chapter 23

---

cyclic organic compounds

23

Because present-day knowledge concerning cyclic organic compounds is quite extensive, only the most important of these compounds will be discussed in this chapter. Their formulas and properties, as well as some of their uses, especially those of importance in medical practice, will be briefly indicated. Much of the information furnished in this chapter should serve as a background to the study of pharmaceutical chemistry and pharmacology.

Of the carbocyclic compounds (see page 221), the aromatic group is the most important, and most of this discussion will be focused on these compounds. Only a few alicyclic compounds merit discussion. Many of the heterocyclic compounds have unusual physiological properties, and some are important medicinals. Some, like the alkaloids, are usually quite poisonous, while others play important roles in the proper functioning of the body.

Aromatic compounds

Benzene may be considered the parent hydrocarbon from which all aromatic compounds are derived. All of these compounds possess at least one grouping of atoms characteristic of benzene, namely a ring of six carbon atoms joined by alternate single and double bonds (but see page 328). As has been previously mentioned, some of the first aromatic compounds studied have a pleasant, perfumelike odor; hence the name aromatic. However, this is not characteristic of all of these compounds. In a number of reactions, the aromatic compounds react quite differently to the corresponding aliphatic compounds. Many useful substances besides drugs are aromatic, such as dyes, explosives, flavoring agents, perfumes, and plastics. A few aromatic compounds occur normally in the body. Thus tyrosine and phenylalanine (page 320), which are amino acids, are constituents of many proteins.

BENZENE, C_6H_6

Benzene consists of a ring of six carbon atoms, with a hydrogen atom attached to each. Furthermore, these six hydrogens are equivalent positionwise, and there is but a single monosubstitution product of benzene (see page 328).

To satisfy the tetravalency of carbon, Kekulé assigned the

structure

```
          H
          |
          C
        //  \
   H—C        C—H
      |       ||
   H—C        C—H
        \\  /
          C
          |
          H
```

to benzene, in which three double bonds alternate with three single ones. According to this formula, two 1,2- or *ortho*-dichlorobenzenes should be isolatable; in one the two chlorine atoms are attached to two carbon atoms joined by a single bond and in the other to two carbon atoms joined by a double bond, as indicated in the following two structural formulas:

```
          Cl                          Cl
          |                           |
          C                           C
        //  \                       /  \\
   H—C        C—Cl             H—C        C—Cl
      |       ||                  ||       |
   H—C        C—H              H—C        C—H
        \\  /                       \  //
          C                           C
          |                           |
          H                           H
```

Because there is only one *ortho*-dichlorobenzene, Kekulé suggested that benzene alternates between the two structures

```
          H                      H
          |                      |
          C                      C
        //  \                  /  \\
   H—C        C—H         H—C        C—H
      |       ||     ⇌       ||       |
   H—C        C—H         H—C        C—H
        \\  /                  \  //
          C                      C
          |                      |
          H                      H
```

Such a structure for benzene appears to be faulty, because benzene does not exhibit the typical properties of an unsaturated hydrocarbon, such as ethylene, but rather those of a saturated hydrocarbon. It does not decolorize a permanganate solution or a bromine solution, and it generally undergoes substitution rather than addition reactions (see page 330).

In such a substance as benzene for which two nearly equivalent structures can be written, such as the two Kekulé structures above, which differ only in the position of pairs of electrons, neither of these structures but rather a single intermediate structure, called a "resonance hybrid," represents the true structure. It is difficult to represent this in a conventional structural formula. The two Kekulé structures above are usually designated as "contributing forms" of the resonance hybrid. If benzene is a resonance hybrid, all six carbon-to-carbon bonds should be of equal length. It has been established that all six of these bonds have a length of 1.39 Å, a distance intermediate between the carbon–carbon single bond (1.54 Å) and the carbon–carbon double bond (1.34 Å).

As with all substances that exist as resonance hybrids, benzene exhibits a much greater stability, that is, it reacts more like a saturated hydrocarbon, than one would predict for either of the two contributing Kekulé structures.

In order to emphasize the resonating hybrid structure of benzene, some today prefer to represent benzene as

rather than as a hexagon in which three double bonds alternate with three single bonds, as represented in the following structure:

In using either of these formulas to represent benzene, it is to be remembered that a carbon atom is present at each corner of the hexagon, and in benzene and its derivatives, if not otherwise indicated, a hydrogen atom is attached to each carbon atom.

Benzene is a colorless, oily liquid with a slight, somewhat pleasant, odor. One of the most important sources of benzene has been coal tar. If soft coal is destructively distilled, coke (Figure 23.1), a combustible gas called *coal gas,* ammonia, and a tarry material called *coal tar* are obtained. The coal tar is fractionally distilled to furnish benzene, toluene, xylenes, naphthalene, anthracene, phenol, the cresols, and small amounts of other substances. Until recent years, coal tar was the only important source of benzene. Today benzene is also prepared as a by-product of the cracking of heavy petroleum oils in the production of gasoline.

Benzene is a symmetrical molecule, and only one compound can be formed when a single hydrogen atom is replaced by another atom, irrespective of which of the six hydrogen atoms are thus replaced. *There is one and only one monosubstitution product of benzene, because the six hydrogen atoms of benzene are equivalent.*

Figure 23.1 Discharging coke from a coke oven. (Courtesy Koppers Company, Inc.)

Chlorobenzene is an example of such a monosubstitution product of benzene.

chlorobenzene

If two hydrogens of benzene are replaced by other atoms or groups, then three isomeric disubstitution products are possible. If the two groups are on adjacent carbon atoms, it is then called the *ortho* isomer. If substituents are on the first and third carbon atoms, it is called the *meta* isomer. The isomer with substituents on the first and fourth carbon atoms is called the *para* isomer. The three dichloro derivatives of benzene are called *ortho-* or *o*-dichlorobenzene, *meta-* or *m*-dichlorobenzene, and *para-* or *p*-dichlorobenzene; their formulas are

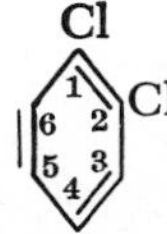

o-dichlorobenzene

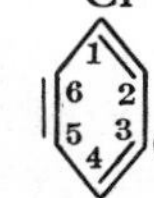

m-dichlorobenzene

p-dichlorobenzene

It is also possible to number the carbon atoms of benzene from 1 to 6 and to name the three compounds above as 1,2-dichlorobenzene, 1,3-dichlorobenzene, and 1,4-dichlorobenzene, respectively. *p*-Dichlorobenzene is used to destroy or repel moths and other insects and is known commercially as Parachlor or Paradow.

Because benzene acts as if it is saturated, it will not decolorize dilute potassium permanganate solutions, nor will it react with bromine in the cold or in the absence of catalysts or sunlight. In the presence of certain catalysts, such as iron, it will combine with bromine at room temperature to produce bromobenzene and hydrogen bromide:

$$C_6H_6 + Br_2 \xrightarrow{[Fe]} C_6H_5Br + HBr$$

bromobenzene

The production of hydrogen bromide indicates that this is a substitution and not an addition reaction. At high temperatures and in the presence of sunlight and the absence of a catalyst, bromine will slowly combine with benzene to form hexabromocyclohexane.

If benzene is heated with a mixture of nitric and sulfuric acids, a hydrogen atom is replaced by the NO_2 group, called the *nitro group,* and nitrobenzene is formed. This is called a *nitration reaction* and is typical of aromatic compounds.

$$C_6H_6 + HONO_2 \xrightarrow{H_2SO_4} C_6H_5NO_2 + H_2O$$

nitrobenzene

If benzene is heated with concentrated sulfuric acid, a hydrogen atom is replaced by the SO_3H group, called the *sulfonic acid group,* and benzenesulfonic acid is formed. This is called a *sulfonation reaction.*

$$C_6H_6 + HO\text{—}SO_3H \longrightarrow C_6H_5SO_3H + H_2O$$

sulfuric acid

benzenesulfonic acid

Benzene is an important solvent and is used in the synthesis of other aromatic compounds.

THE HIGHER HOMOLOGS OF BENZENE

The monomethyl substitution product of benzene is called *toluene,* while the three isomeric dimethyl derivatives are known as *o-*, *m-*, and *p-xylene.*

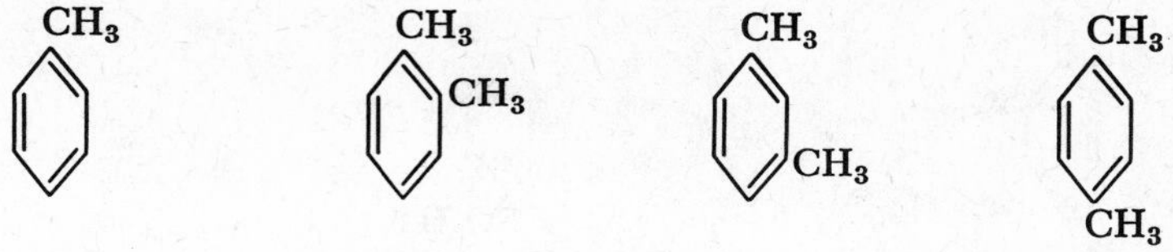

CH_3 — toluene; CH_3, CH_3 — o-*xylene*; CH_3, CH_3 — *m*-xylene; CH_3, CH_3 — *p*-xylene

These four hydrocarbons are colorless, oily liquids. Toluene is obtained from coal tar and as a by-product of the cracking of heavy petroleum oils. The three xylenes are obtained from coal tar as a mixture called *xylol.* Xylol is a good solvent for oils and is used in cleaning slides and optical lenses of microscopes.

In toluene, the methyl group is called the *side chain,* while the benzene ring is called the *nucleus.*

Toluene, unlike benzene, is oxidized by concentrated solutions of strong oxidizing agents, such as potassium permanganate. The aromatic nucleus remains intact, but the methyl group is oxidized to a carboxyl group, and benzoic acid is formed.

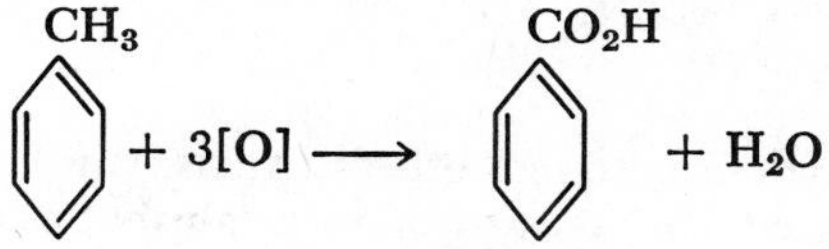

$$C_6H_5CH_3 + 3[O] \longrightarrow C_6H_5CO_2H + H_2O$$

benzoic acid

Nitration of toluene produces a mixture predominately of *o*- and *p*-nitrotoluene.

CH_3, NO_2 — *o*-nitrotoluene; CH_3, NO_2 — *p*-nitrotoluene

Further nitration of toluene produces trinitrotoluene, or TNT.

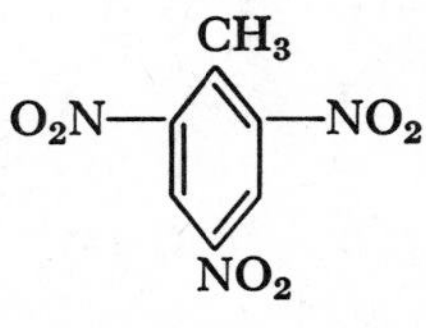

CH_3, O_2N—, —NO_2, NO_2

TNT

This is an important explosive. It is not sensitive to shock or jarring and must be exploded by a detonator, such as mercuric fulminate.

Upon sulfonation, toluene forms a mixture of *o*- and *p*-toluenesulfonic acids.

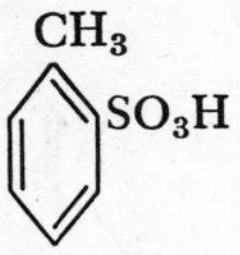

CH_3 / SO_3H (ortho-substituted benzene ring)

o-toluenesulfonic acid

CH_3 / SO_3H (para-substituted benzene ring)

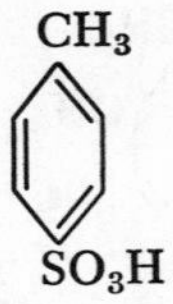

p-toluenesulfonic acid

Saccharin, which has the formula

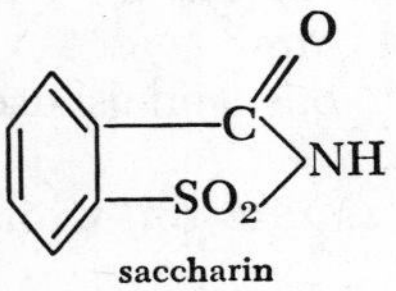

(benzene ring fused to —C(=O)— and —SO_2— bridged by >NH)

saccharin

is derived from *o*-toluenesulfonic acid. It is sweeter than sugar and is used as a sweetening agent, especially by diabetic patients; it has no food value.

Toluene is used as a solvent and in the preparation of benzoic acid, dyes, drugs, explosives, etc.

AROMATIC ALCOHOLS

Aromatic alcohols, aldehydes, ketones, ethers, acids, and esters are well known. They resemble the corresponding aliphatic compounds in their formulas and many of their reactions.

Aromatic alcohols contain a hydroxyl group attached to a side chain; the best known of these is benzyl alcohol.

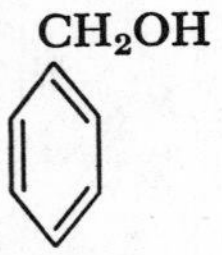

CH_2OH (on benzene ring)

benzyl alcohol

This is a colorless, oily liquid with a faint, pleasant odor and shows typical alcohol reactions. As an example, it reacts with organic acids to form esters. The following equation illustrates the synthesis of benzyl acetate.

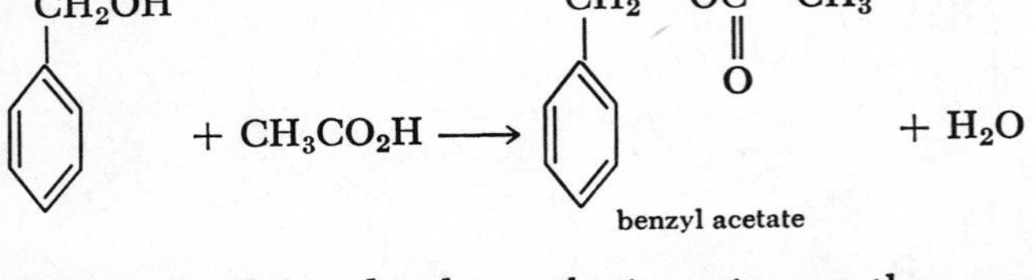

$$C_6H_5CH_2OH + CH_3CO_2H \longrightarrow C_6H_5CH_2{-}O\overset{\displaystyle O}{\overset{\|}{C}}{-}CH_3 + H_2O$$

benzyl acetate

Benzyl alcohol exhibits local anesthetic action on the mucous surface (topical anesthesia) and on injection, but it is too irritating to be used for this purpose. Benzyl alcohol and its esters are used in the manufacture of perfumes.

PHENOLS

Those aromatic compounds in which a hydroxyl group is attached directly to a carbon atom of an aromatic nucleus exhibit properties that for the most part are different from those of typical alcohols. Such compounds fall into a separate class called *phenols*, which are weakly acidic compounds, being less acidic than carbonic acid. They react with bases to form salts called *phenoxides*. As an example, sodium hydroxide converts the simplest member of this series, called *phenol*, or *carbolic acid*, to sodium phenoxide.

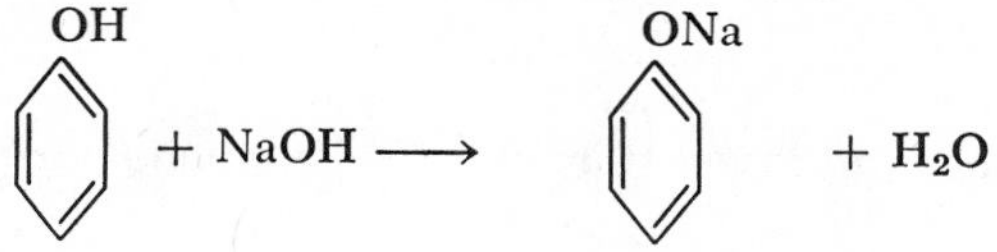

OH (phenol) $+ NaOH \longrightarrow$ ONa (sodium phenoxide) $+ H_2O$

phenol sodium phenoxide

Most of the phenols are good antiseptics, but their use for this purpose is somewhat limited because of their toxicity.

Phenol is a colorless, crystalline solid that possesses a characteristic odor. On standing, it turns pink to red due to oxidation. It is an excellent antiseptic and germicide. Pure phenol is caustic to the skin and is quite toxic when taken internally. Skin areas that come in contact with pure phenol or concentrated solutions of it should be washed with alcohol immediately. As an antiseptic, phenol is used as a very dilute solution. Most homologs of phenol are superior to it because they are less toxic and are good antiseptics. Phenol is a starting material for the synthesis of dyes, drugs, explosives, and disinfectants. Plastics, such as Bakelite, are prepared by the condensation of phenol and formaldehyde.

The three isomeric methylphenols are called *o*-, *m*-, and *p*-cresol.

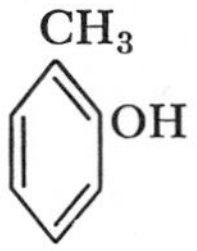

CH_3 OH — *o*-cresol

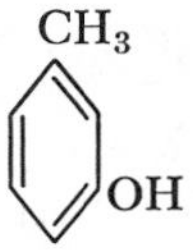

CH_3 OH — *m*-cresol

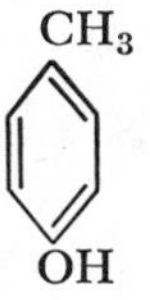

CH_3 OH — *p*-cresol

A mixture of these three cresols is usually known as *cresol*. Lysol is an emulsion of the cresols in water, in which soap is employed to stabilize the emulsion. The cresols are stronger antiseptics than phenol and are less toxic.

Thymol, a phenolic compound obtained from oil of thyme, is used as an antiseptic in mouthwashes and in dentistry.

Resorcinol, or *m*-dihydroxybenzene,

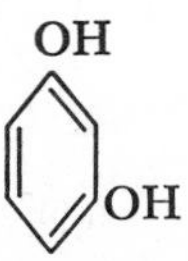

OH OH

resorcinol

is an antiseptic but is inferior to phenol. A derivative of resorcinol hexylresorcinol, or S.T. 37, is used as a general antiseptic, a germicide, and a urinary antiseptic.

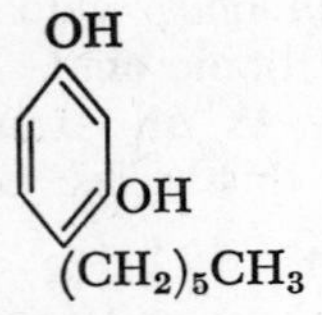

OH

OH

$(CH_2)_5CH_3$

hexylresorcinol

The final product of nitration of phenol is trinitrophenol, or picric acid.

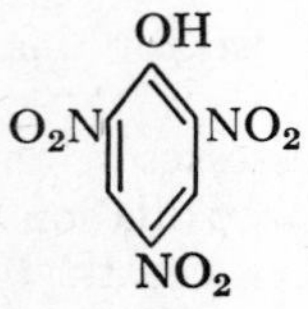

OH

O_2N NO_2

NO_2

picric acid

It is a yellow solid and is used as an explosive because it is not usually sensitive to shocks and jars. It is also used as an antiseptic, particularly in the treatment of burns, because it coagulates the protein on the surface of the burn and prevents the loss of blood serum.

AROMATIC AMINES AND ANILINE

Although three classes of aromatic amines are known, only one, a primary amine called *aniline,* is important enough to be discussed. It has the formula

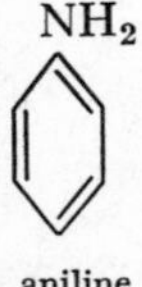

aniline

and may be considered as derived from ammonia by replacing a hydrogen atom by the group C_6H_5, called the *phenyl group.* Aniline is a colorless liquid which on standing soon turns red or dark brown. It has a distinctive, unpleasant odor and is insoluble in water but dissolves in dilute acid solutions. Both liquid aniline and its vapors are toxic.

Aniline is prepared by the reduction of nitrobenzene. In the laboratory, tin and hydrochloric acid are used as the reducing agent. Industrially, iron, steam, and a small quantity of hydrochloric acid are employed.

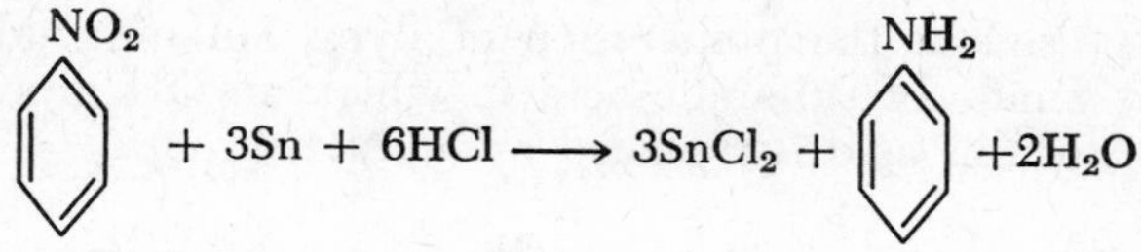

$$C_6H_5NO_2 + 3Sn + 6HCl \longrightarrow 3SnCl_2 + C_6H_5NH_2 + 2H_2O$$

Aniline, like the aliphatic amines, is basic and reacts with acids to form salts. It is, however, a much weaker base than the aliphatic amines. The following equation represents the formation of aniline hydrochloride.

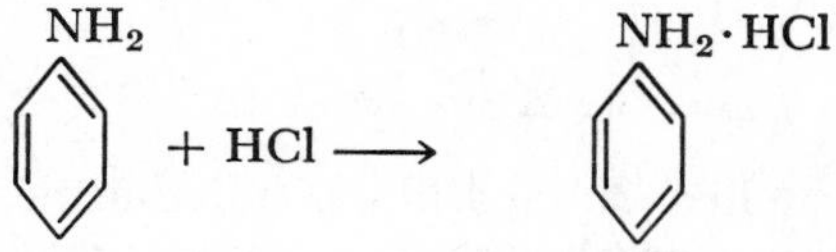

$$C_6H_5NH_2 + HCl \longrightarrow C_6H_5NH_2 \cdot HCl$$

aniline hydrochloride

If aniline is heated with concentrated acetic acid, a substituted amide called *acetanilide* is formed.

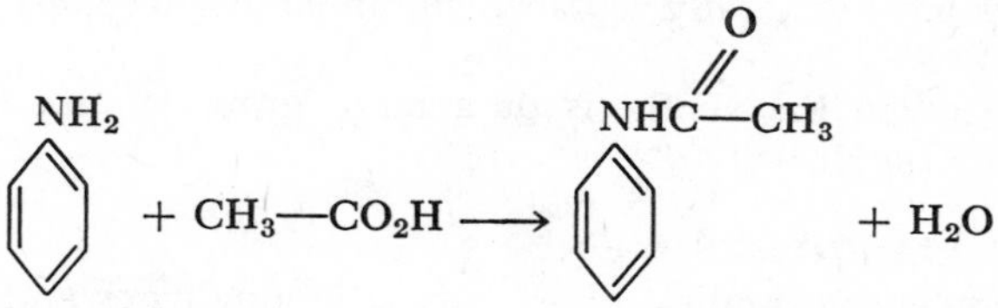

$$C_6H_5NH_2 + CH_3{-}CO_2H \longrightarrow C_6H_5NHC(=O){-}CH_3 + H_2O$$

acetanilide

This is a colorless, odorless, crystalline solid. It is an important drug and is used as an analgesic (reduces pain) and an antipyretic (reduces fever) in the treatment of headaches, neuralgia, and fevers. Phenacetin has a structure similar to acetanilide and has similar uses. It is less toxic than acetanilide.

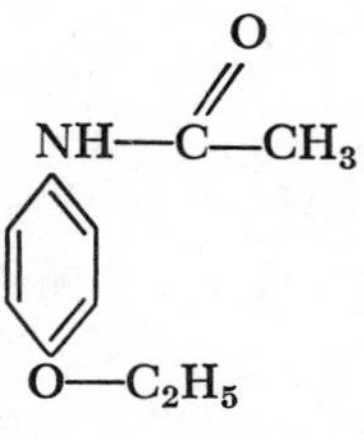

$$C_2H_5{-}O{-}C_6H_4{-}NH{-}C(=O){-}CH_3$$

phenacetin

If heated with concentrated sulfuric acid, aniline is converted to sulfanilic acid (*p*-aminobenzenesulfonic acid).

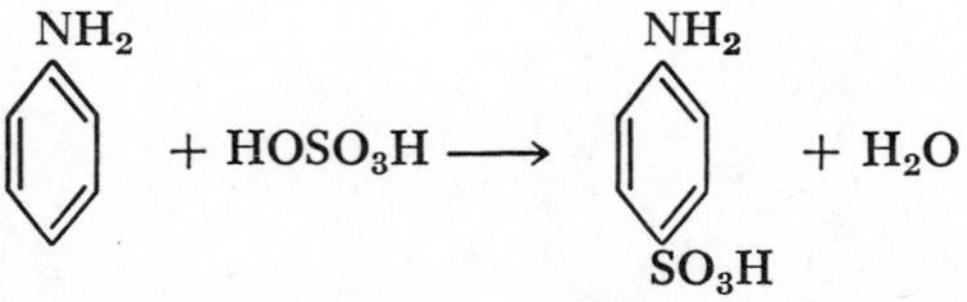

$$C_6H_5NH_2 + HOSO_3H \longrightarrow p\text{-}NH_2C_6H_4SO_3H + H_2O$$

sulfanilic acid

Sulfanilic acid is used in the preparation of dyes. Sulfanilamide, a sulfa drug, is the amide of sulfanilic acid in which an OH group of the SO_3H portion of this acid is replaced by an NH_2 group.

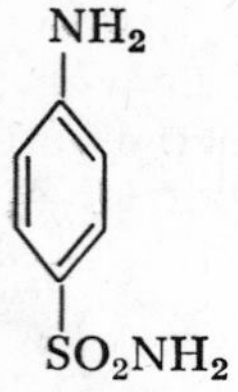

NH_2

SO_2NH_2

sulfanilamide

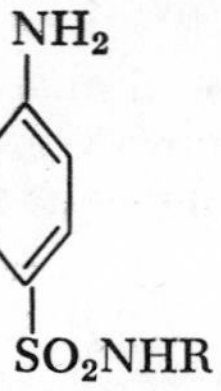

NH_2

SO_2NHR

general formula of a sulfa drug

Other sulfa drugs are sulfaguanidine, sulfapyridine, sulfadiazine, and sulfathiazole, in which R of the general formula above is guanidine and the heterocyclic molecules pyridine, diazine, and thiazole, respectively. They are used in the treatment of streptococcus infections and in combating "strep" throat, pneumonia, gonorrhea, dysentery, etc. These latter drugs have generally replaced sulfanilamide, because they are less toxic.

Aniline is used in the synthesis of a large group of important dyes, called the *aniline* or *azo dyes.*

AROMATIC ALDEHYDES

Benzaldehyde is the simplest of the aromatic aldehydes and consists of an aldehyde group attached to a benzene ring. It is a colorless, oily liquid with an almond-like odor and is used in the preparation of flavoring agents, perfumes, and cosmetics. Vanillin, another aromatic aldehyde, has a pleasant odor and is used in the preparation of flavoring agents and perfumes.

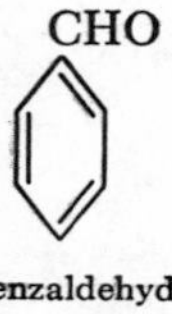

benzaldehyde 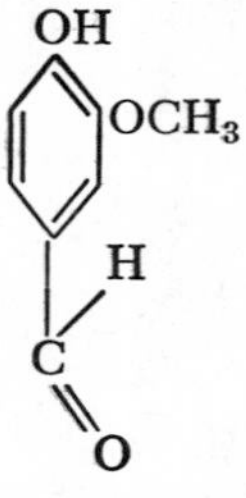 vanillin

AROMATIC ACIDS

The simplest aromatic acid is benzoic acid.

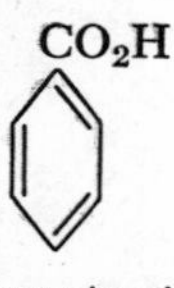

benzoic acid

It is a colorless, crystalline solid, is insoluble in cold water, and, because it is an acid, is neutralized by bases. Sodium hydroxide converts it to sodium benzoate, which is used as a preservative of certain foods (such as catsup).

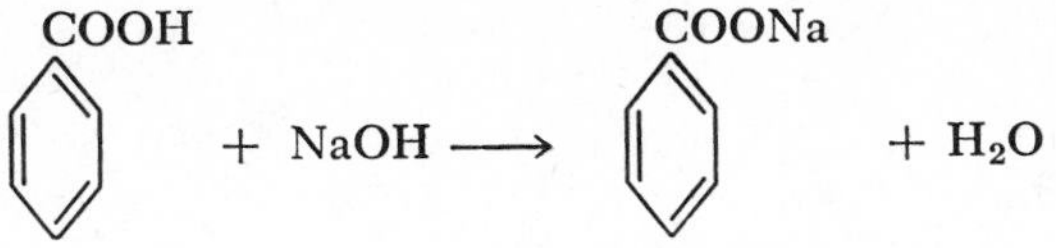

$$C_6H_5COOH + NaOH \longrightarrow C_6H_5COONa + H_2O$$

sodium benzoate

Salicylic acid is both an acid and a phenol, because it has a carboxyl and a hydroxyl group attached to the benzene ring in an ortho position.

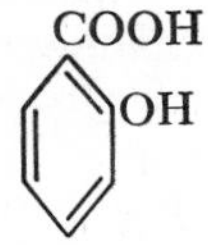

COOH, OH (ortho) on benzene ring

salicylic acid

It is a colorless, crystalline solid, and many of its derivatives have important applications in medicine. Sodium hydroxide converts it to sodium salicylate, which is used as a food preservative and in the treatment of rheumatism and arthritis. However, it irritates the stomach and sometimes causes nausea and vomiting.

The acetyl derivative of salicylic acid is known as *acetylsalicylic acid*, or *aspirin*. (CH_3CO— is called the *acetyl group*.)

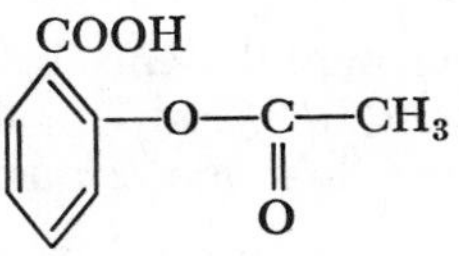

COOH, —O—C(=O)—CH_3 on benzene ring

aspirin

Aspirin is an analgesic and an antipyretic, and large quantities of it are used in the treatment of colds, neuralgia, headaches, and minor pains.

The methyl ester of salicylic acid, known as *artificial oil of wintergreen*, is used in liniments and in perfumes and flavorings.

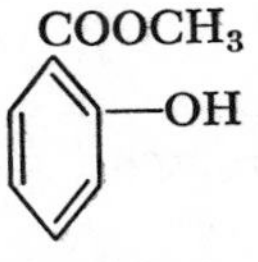

$COOCH_3$, —OH on benzene ring

methyl salicylate

The phenol ester of salicylic acid, also called *salol*, is used as an intestinal antiseptic. It passes through the stomach unchanged and is slowly hydrolyzed to phenol and salicylic acid in the small intestine. These latter substances are both good antiseptics. Salol is also used to coat pills that are to pass through the stomach unchanged but that are decomposed by the bases in the intestines to liberate their contents. Such pills are known as *enteric pills.*

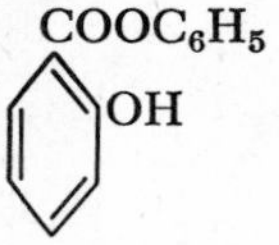

$COOC_6H_5$

OH

phenyl salicylate, or salol

POLYNUCLEAR AROMATIC HYDROCARBONS

These are hydrocarbons in which two or more benzene nûclei are fused together. The three most important of these, naphthalene, anthracene, and phenanthrene, are obtained from coal tar. Some of the more complex hydrocarbons of this series are capable of producing cancer in humans and are called *carcinogenic hydrocarbons.*

Naphthalene, $C_{10}H_8$, is a white, crystalline solid with a characteristic odor. It is readily purified by sublimation. It is used as a moth repellent and in the synthesis of many important dyes. The following formula indicates the structure and method of numbering the carbon atoms in naphthalene:

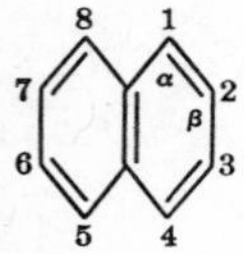

8 1 7 α 2 β 6 3 5 4

There are only two monosubstituted isomers of naphthalene, because the 1, 4, 5, and 8 positions are identical and the 2, 3, 6, and 7 positions are identical. The former positions are usually called the α positions while the latter are called the β positions.

Two important derivatives of naphthalene are α- and β-naphthol. These have the structures

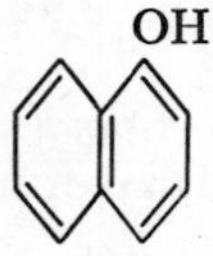

OH

OH

alpha-naphthol 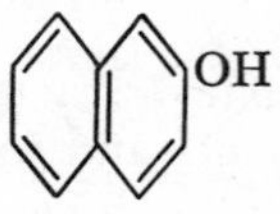 beta-naphthol

and are true phenolic compounds. Both these compounds are used as external and internal antiseptics and in the synthesis of important dyes.

Anthracene, $C_{14}H_{10}$,

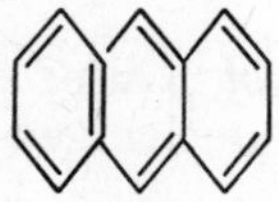

is a colorless, crystalline solid with a blue fluorescence and is used primarily in the synthesis of dyes. Alizarin is an important dye obtained from anthracene.

Phenanthrene, $C_{14}H_{10}$,

is an isomer of anthracene. It has no important medicinal uses but is of interest because a portion of the molecules of cholesterol, ergosterol, vitamin D, the sex hormones, and some alkaloids have a phenanthrene-type structure.

OTHER AROMATIC COMPOUNDS USED AS MEDICINALS

Procaine or novocaine has the following formula:

$$H_2N-C_6H_4-\underset{\underset{O}{\|}}{C}-OCH_2CH_2N(C_2H_5)_2$$

It is a synthetic product and is used as a substitute for cocaine because it is a powerful local anesthetic when injected subcutaneously. Procaine is less toxic than cocaine and, unlike the latter, is not habit forming. Many compounds which are similar in structure to procaine and which exhibit local anesthetic action have been synthesized in the laboratory.

Adrenaline, or epinephrine, is obtained from the adrenal glands (page 517) and can be prepared synthetically. It has the structure

$$(HO)_2C_6H_3-\underset{\underset{OH}{|}}{CH}-CH_2-\underset{\underset{CH_3}{|}}{NH}$$

This was the first hormone to be isolated. It constricts blood vessels and raises the blood pressure. It is used medicinally to control hemorrhage and to revive a heart which has ceased beating. Ephedrine and benzedrine are similar in structure and action to adrenaline. They dilate the bronchi of the lungs and are used to give relief to those suffering from asthma, hay fever, and similar respiratory ailments. Benzedrine can be used by inhalation. Ephedrine also dilates the pupils of the eyes.

Some years ago, Ehrlich was seeking a drug that would be effective in the treatment of spirochetal infections. The 606th compound synthesized by him had the properties he was seeking. It has been called arsphenamine, salvarsan, or 606 and has been quite effective in the treatment of syphilis. This drug is a derivative of *o*-aminophenol and contains arsenic.

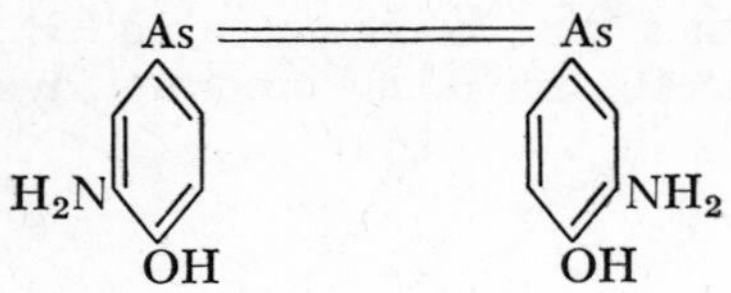

As ═══════ As

H_2N ... NH_2

OH OH

salvarsan

Neoarsphenamine, a modification of arsphenamine, has replaced it because this new drug, although less effective, is also less toxic to the patient. Arsphenamine was the first synthetic chemotherapeutic agent, and Ehrlich has been called the father of *chemotherapy.* Chemotherapy is the inhibition or destruction of harmful organisms by chemical agents without harmful effects to the host.

Pesticides

Although of a variety of structures, many pesticides, such as DDT, which is a chlorinated aromatic hydrocarbon, and 2,4D and 2,4,5T, which are derivatives of phenols, can be classed as aromatic compounds. It is for this reason that a study of pesticides concludes the discussion of aromatic compounds.

Pesticides are used to destroy obnoxious insects (insecticides), fungi (fungicides), rodents (rodenticides), and weeds (herbicides) in order to protect humans and domestic animals, as well as fields and forests.

HERBICIDES

Herbicides are those chemical substances that can be utilized for the destruction of plants. The synthetic compounds 2,4-dichlorophenoxyacetic acid, or 2,4D,

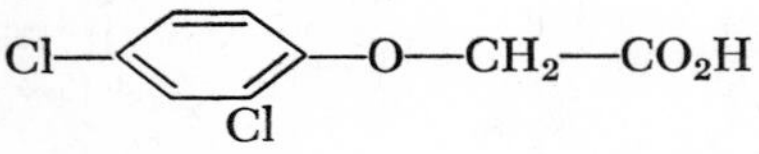

Cl—C_6H_3(Cl)—O—CH_2—CO_2H

2,4D

used for postemergence weed control, and 2,4,5-trichlorophenoxyacetic acid, or 2,4,5T,

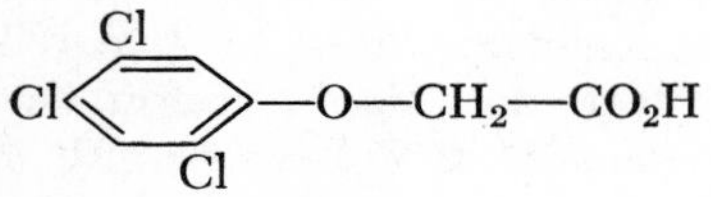

$Cl_3C_6H_2$—O—CH_2—CO_2H

2,4,5T

used in brush control (particularily as a defoliant in Vietnam), are usually employed as esters or salts. They are effective herbicides for many broadleaf plants. The initial rapid growth that these substances induce results in the death of the plant due to the rapid depletion of needed nutrients.

INSECTICIDES

The insecticidal properties of DDT, or 1,1,1-trichloro-2,2-bis(*p*-chlorophenyl)ethane were discovered by Paul Müller of Geigy & Company in Switzerland in 1939. It was first used as a larvacide to control malaria in India and prevented a typhus epidemic in Naples during World War II.

DDT is a polychlorinated aromatic hydrocarbon produced by the condensation of chlorobenzene with chloral, thus:

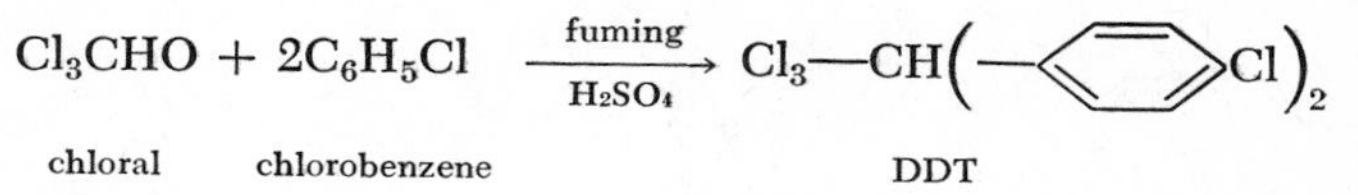

$$\underset{\text{chloral}}{Cl_3CHO} + \underset{\text{chlorobenzene}}{2C_6H_5Cl} \xrightarrow[H_2SO_4]{\text{fuming}} \underset{\text{DDT}}{Cl_3\text{—}CH(\text{—}C_6H_4Cl)_2}$$

It is used as both an argicultural and a household spray.

Although DDT has proved a very effective insecticide, certain problems have now been associated with its use, so a universal ban on its use is possible. It is not usually metabolized in living organisms and appears to be concentrated in the liver of animals at the end of a food chain. DDT appears to effect the metabolism of calcium in birds, resulting not only in retarded production of smaller eggs, but also in a decrease in the thickness of the shell, resulting in many more broken eggs.

The chlorination of benzene, which is catalyzed by light, yields four isomers, known as α-, β-, γ-, and δ-hexachlorocyclohexane. This mixture of isomers is sold under such trade names as Gammexane, Gammane, and 666. The insecticidal properties of this mixture depend on its content of the γ-isomer. If the γ-isomer is concentrated to 99 percent purity, an effective broad spectrum insecticide is obtained which is sold under the trade name Lindane.

The structure of the γ-isomer of hexachlorocyclohexane is

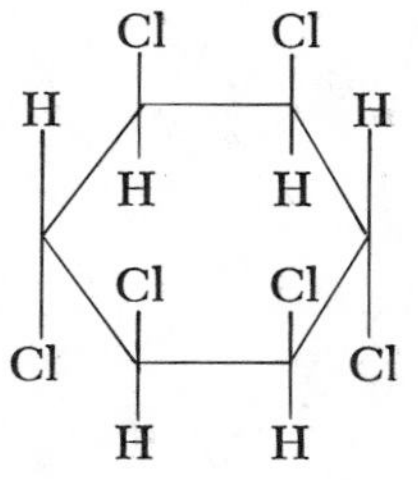

Cl Cl
H H
H H
Cl Cl
Cl Cl
H H

Other polychlorinated hydrocarbons used as broad spectrum insecticides are Aldrin, Dieldrin, Endrin, and Chlordan.

Certain derivatives of phosphoric acid can be used as nerve gases and as insecticides because they interfere with the transmission of nerve impulses by acetyl choline and with breathing mechanisms. Important phosphate insecticides are Malathion and Parathion. Malathion has the structure

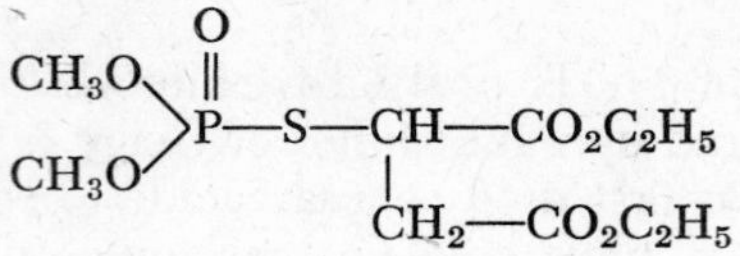

$$(CH_3O)_2P(=O)—S—CH(CO_2C_2H_5)—CH_2—CO_2C_2H_5$$

It is of low toxicity to man and higher animals but is quite lethal to mites and other insects. It is frequently used as an insecticide in dairying and in home gardens.

Parathion,

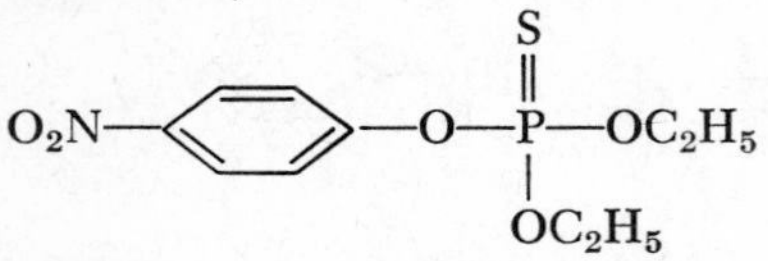

$$O_2N—C_6H_4—O—P(=S)(OC_2H_5)—OC_2H_5$$

unlike Malathion, is highly toxic to man and animals, not only when ingested but also if it contacts the skin.

Some very effective insecticides, usually of low toxicity to humans and domestic animals, are generated by and isolated from plants. One of the most potent of these natural insecticides is pyrethrum. This is obtained from the flower of the pyrethrum plant, which is grown extensively in Kenya. Pyrethrum consists of a mixture of esters of chrysanthemum monocarboxylic and dicarboxylic acids,

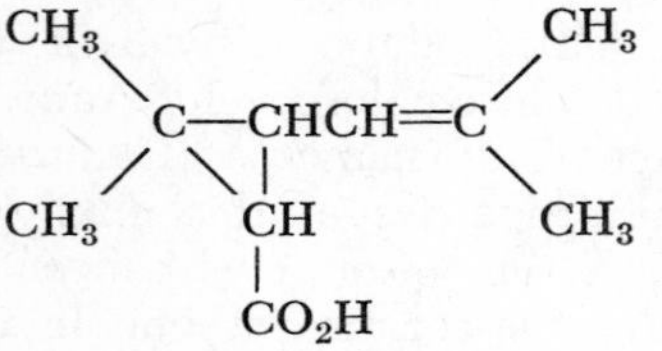

$$(CH_3)_2C—CHCH{=}C(CH_3)_2 \text{ (cyclopropane ring closed by CH—}CO_2H\text{)}$$

chrysanthemum monocarboxylic acid

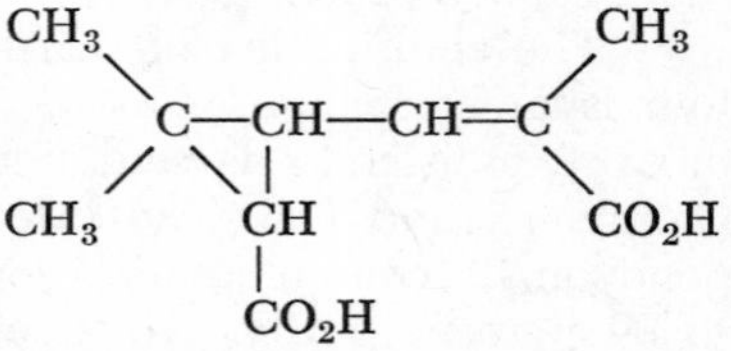

$$(CH_3)_2C—CH—CH{=}C(CH_3)CO_2H \text{ (cyclopropane ring closed by CH—}CO_2H\text{)}$$

chrysanthemum dicarboxylic acid

with two ketoalcohols pyrethrolone and cinnerolone.

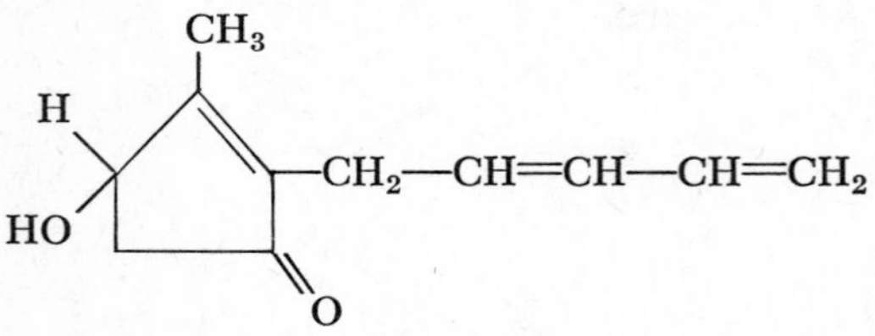

$$\text{cyclopentenone ring: } CH_3,\ H,\ HO,\ =O,\ —CH_2—CH{=}CH—CH{=}CH_2$$

pyrethrolone

$$\text{cyclopentenone ring: } CH_3,\ H,\ HO,\ =O,\ —CH_2—CH{=}CH—CH_3$$

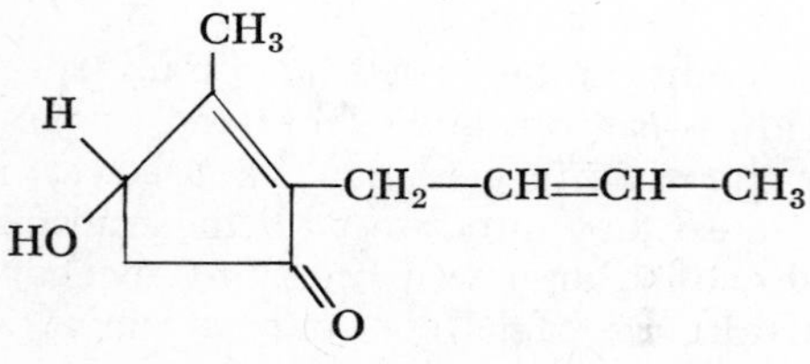

cinnerolone

In hydrocarbon solvents, pyrethrum serves as an effective spray to destroy such flying insects as house flies and fruit flies.

Extracts of the tuba plant (*Derris elliptica*) have been used by the Chinese as a fish poison and an insecticide. The active ingredient in these extracts, rotenone, is now available commercially and is effective in controlling garden, household, and pet insect pests.

As the number of undesirable side effects resulting from the use of synthetic insecticides are now becoming known, resulting in many restrictions in their use, interest in and investigations of chemosterilants and sex attractants have been greatly increased. Chemosterilants are substances that induce sterility in one sex of an insect pest, preventing its reproduction. Sex attractants are chemical substances that exert a powerful attraction for one sex of a noxious pest. These, when collected at the site of the attractant, are then destroyed by suitable insecticides that need not come in contact with the plant.

FUNGICIDES

Not only are fungicides sprayed on fruit, vegetable, and ornamental plants, but seeds or soil may be inoculated with the fungicide to prevent growth of fungi or other pathogens.

At present, the most important fungicides are metal salts of derivatives of dithiocarbamic acid,

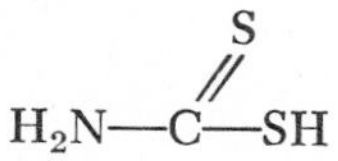

$$H_2N-\overset{\overset{S}{/\!/}}{C}-SH$$

Three of these are Maneb, Nabam, and Zineb, which are the manganese, sodium, and zinc salts of ethylene-*bis*-dithiocarbamic acid, respectively.

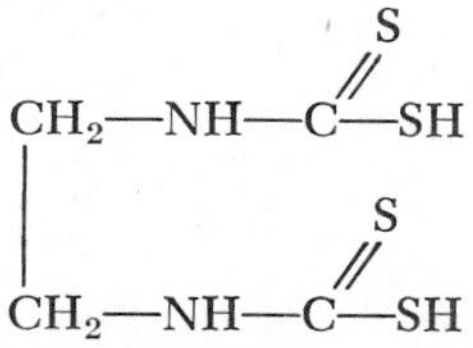

$$\begin{array}{l} CH_2-NH-\overset{\overset{S}{/\!/}}{C}-SH \\ | \\ CH_2-NH-\overset{\overset{S}{/\!/}}{C}-SH \end{array}$$

ethylene-*bis*-dithiocarbamic acid

These are broad spectrum spray fungicides for a large number of fruit, vegetable, and ornamental plants; they are also effective as soil fungicides.

RODENTICIDES

Three popular rodenticides, ANTU, or α-naphthylthiourea,

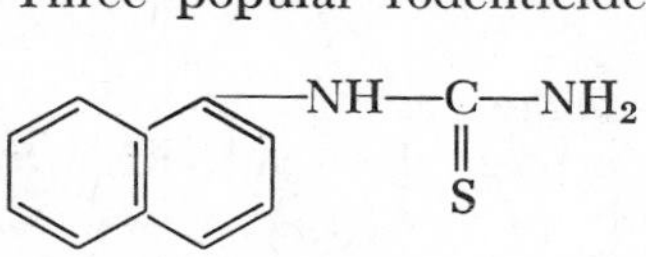

$$C_{10}H_7-NH-\underset{\underset{S}{\|}}{C}-NH_2$$

ANTU

sodium fluoroacetate (or 1080), CH_2FCO_2Na, and Warfarin,

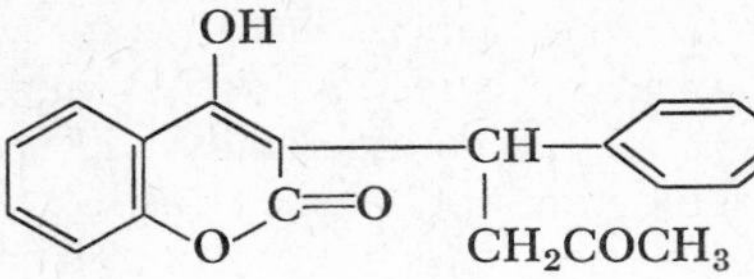

OH

C=O CH

O CH_2COCH_3

warfarin

are potent rat killers. They are also toxic to man and domestic animals. ANTU attacks the lungs and drives the rat into the open. Warfarin is a potent blood anticoagulant.

Alicyclic compounds

These are homocyclic compounds that resemble the corresponding aliphatic compounds in most of their properties. In cyclopropane, C_3H_6, three carbon atoms are placed at the corners of a triangle.

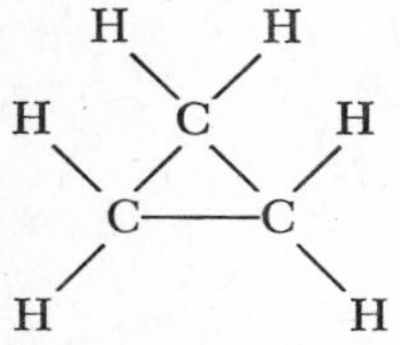

H H

H C H

C—C

H H

cyclopropane

It is an inflammable gas that is used as a general anesthetic. Vitamin A and the terpenes are alicyclic compounds.

Camphor is a bicyclic (two rings) ketone with the formula

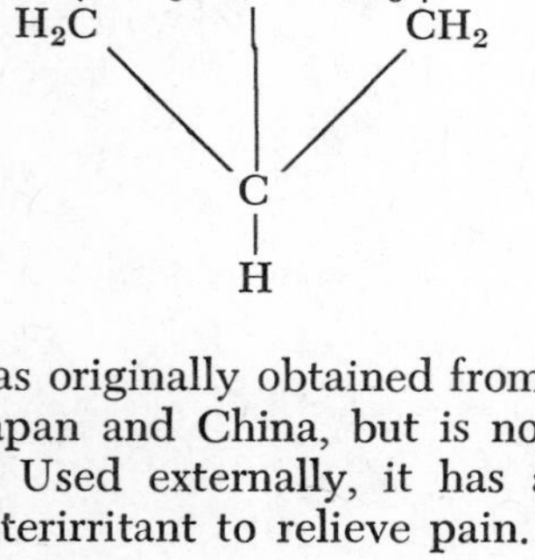

CH_3

C

H_2C C=O

CH_3—C—CH_3

H_2C CH_2

C

H

It was originally obtained from *Cinnamomum camphora,* which grows in Japan and China, but is now produced synthetically from turpentine. Used externally, it has antiseptic properties and is used as a counterirritant to relieve pain. Internally, it acts as a heart stimulant.

Heterocyclic compounds

Heterocyclic compounds (see also page 222) are cyclic compounds that contain one or more atoms other than carbon as members of the ring. Many of the compounds of this class occur in plants and animals, and many are important medicinals.

Four simple cyclic compounds that contain one nitrogen atom in the ring are pyrrole, pyridine, quinoline, and indole.

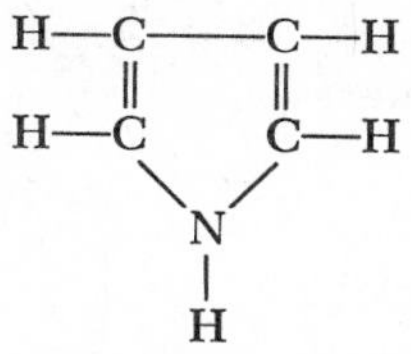

```
H—C———C—H
  ||      ||
H—C      C—H
   \    /
     N
     |
     H
```

pyrrole

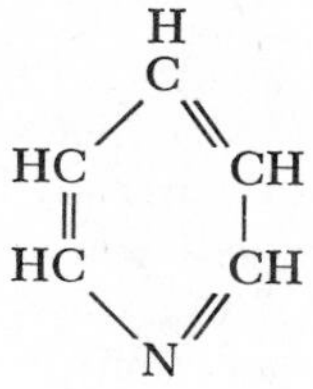

```
      H
      C
    /   \\
 HC       CH
 ||       |
 HC       CH
    \   //
      N
```

pyridine

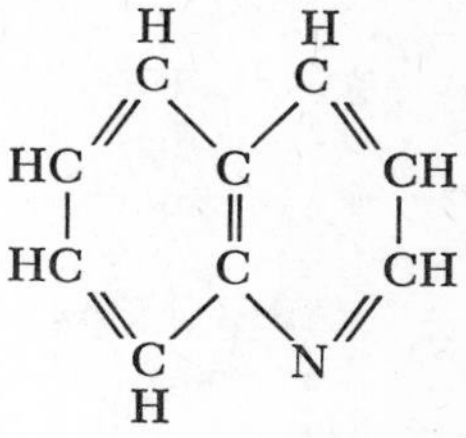

```
      H       H
      C       C
   //   \   /   \\
 HC       C       CH
 |        ||      |
 HC       C       CH
   \\   /   \   //
      C       N
      H
```

quinoline

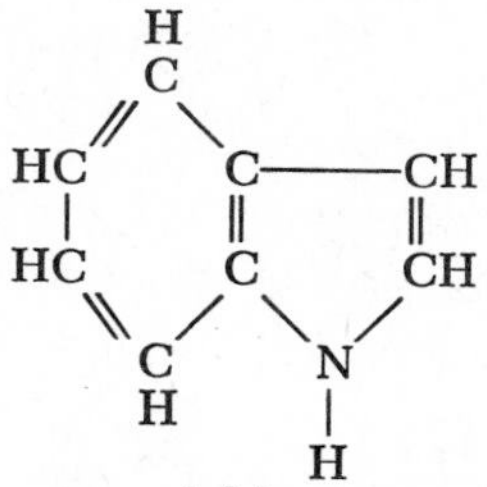

```
      H
      C
   //   \
 HC       C———CH
 |        ||    ||
 HC       C     CH
   \\   /   \   /
      C       N
      H       |
              H
```

indole

In heme, a decomposition product of hemoglobin, four pyrrole rings are joined by four methine (—CH=) groups to form a larger, 16-membered heterocyclic ring. The following structure has been assigned to heme. Note that an iron atom (in the ferrous state) lies in the center of the heme molecule attached by two normal covalent bonds and two coordinating covalent bonds to the pyrrole nitrogen atoms.

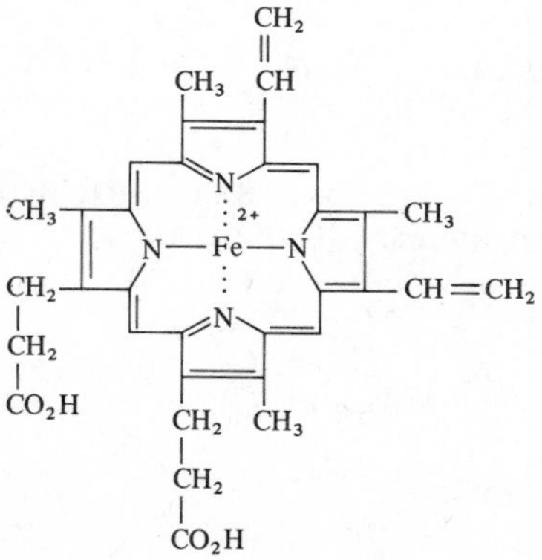

```
               CH2
               ||
         CH3   CH
             N
             : 2+
CH3      N—Fe—N      CH3
CH2          :       CH=CH2
|            N
CH2
|
CO2H
         CH2   CH3
         |
         CH2
         |
         CO2H
```

Heme

The pyridine group appears in nicotinic acid (one of the B vitamins that prevents pellagra), in sulfapyridine (page 336), and in certain alkaloids, such as nicotine and coniine. The following alkaloids contain a quinoline or a reduced quinoline nucleus: strychnine, morphine, codeine, heroin, and quinine. Tryptophan (page 321), an amino acid, skatole, a product of intestinal putrefaction, and the dye indigo contain an indole nucleus. Indole itself occurs as a product of intestinal putrefaction.

PYRIMIDINES AND PURINES

For a discussion of these two important types of heterocyclic nitrogen compounds, see page 439–440.

BARBITURATES OR BARBITALS

These are heterocyclic compounds formed by replacing two hydrogen atoms of barbituric acid by two alkyl groups.

```
    HN—C=O              HN—C=O
    |   |               |   |
  O=C   CH2           O=C   CRR'
    |   |               |   |
    HN—C=O              HN—C=O
```

barbituric acid — general formula for the barbitals

Some of the important barbitals and the nature of the two alkyl groups present in them are:

	R	R'
Veronal, or barbital	C_2H_5—	C_2H_5—
Phenobarbital, or luminal	C_2H_5—	C_6H_5—
Amytal	C_2H_5—	$(CH_3)_2CH—CH_2—CH_2$—
Pentothal, or nembutal	C_2H_5—	$CH_3—CH_2—CH_2—CH(CH_3)$—

The *barbitals* are synthetic drugs used as hypnotics and sedatives and frequently as a premedication before the use of general anesthesia in surgery. In very large doses, they are poisonous and dangerous. They also appear to be cumulative poisons; that is, on continued use, their concentration in the body will increase to the toxic level. These drugs should be used only upon the advice of a physician.

ALKALOIDS

The *alkaloids* in general have the following characteristics:

1. They are obtained from plants.
2. They contain nitrogen atoms as constituents of a cyclic ring.
3. They are either amines or basic nitrogen compounds that form colorless crystalline salts with acids. These salts are the form in which most of the alkaloids are employed in medicine.

4. With the exception of nicotine and coniine, which are liquids, they are all solids.
5. Each alkaloid possesses some characteristic physiological property, and many are quite toxic.

Some important alkaloids are listed below.

1. Nicotine, which is found in tobacco, is an important insecticide.
2. Coniine, which is quite toxic, is obtained from hemlock.
3. Strychnine is employed for its stimulating action on the nervous system. An overdose will produce convulsions and death.
4. Quinine, which is obtained from cinchona bark, is used in the treatment of malaria.
5. Morphine, codeine, and heroin are obtained from opium, the dried juice of certain types of poppies. These alkaloids are used to relieve pain but are habit forming.
6. Atropine is obtained from the poisonous belladonna plant and is used in optometry because it has the power to dilate the pupils of the eyes. It has also been used as an antispasmodic.
7. Cocaine is obtained from the leaves of the cocoa tree. It has been used as a local anesthetic but has been replaced by synthetic products because it is habit forming.
8. Scopolamine paralyzes the nervous system. In small doses, it is used as a sedative and hypnotic. When administered with morphine, it produces a loss of memory (*twilight sleep*). It is a mydriatic and has been used in place of atropine in ophthalmology.

STUDY QUESTIONS

1. Discuss the structure of benzene. How does the chemist usually abbreviate its formula?
2. Does benzene show the typical reactions of an unsaturated hydrocarbon?
3. What are two characteristic substitution reactions exhibited by benzene and related aromatic compounds? What are the names of the products formed from benzene in each of these reactions?
4. Toluene and the three isomeric xylenes are homologs of benzene. Write structural formulas for these four compounds.
5. How do the phenols and the true aromatic alcohols differ in structure? Give an example of each of these classes.
6. What property do most phenols possess that make them so valuable in the fields of nursing and medicine?
7. How is aniline prepared in the laboratory and industrially?
8. How does aniline compare in strength as a base with the aliphatic amines?
9. Describe and write equations to illustrate the reaction of phenols with bases.
10. List the names, formulas, and uses of some medicinal agents derived from aniline.

11. Write structural formulas for benzoic acid and salicylic acid.
12. Write equations to demonstrate the fact that benzoic and salicylic acids are typical organic acids.
13. What uses are made of the sodium salts of benzoic and salicylic acids?
14. Write structural formulas for and indicate uses of three other important derivatives of salicylic acid.
15. What are enteric pills?
16. In toluene, the methyl group is called the ______________, while the rest of the molecule is called the ______________.
17. Benzene is not usually attacked by strong oxidizing agents, whereas toluene is usually oxidized by them. Indicate the latter oxidation by equation.
18. How many monochloro-substitution products of benzene exist? How many dichloro-substitution products? Name the latter.
19. Write structural formulas for naphthalene, *p*-cresol, aspirin, aniline, benzoic acid, benzyl alcohol, phenol, anthracene, toluene, nitrobenzene, salicylic acid, *m*-xylene, picric acid, sodium salicylate, TNT, benzenesulfonic acid, acetanilide, *p*-dichlorobenzene, salol, hexylresorcinol, *o*-nitrotoluene, sulfanilamide, benzaldehyde, aniline hydrochloride, methyl salicylate, and phenanthrene.
20. Give medical uses for six compounds listed in Question 19.
21. Give other uses for six compounds listed in Question 19.
22. Complete and balance the following equations:

salicylic acid + NaOH $\longrightarrow$
aniline + HCl $\longrightarrow$
benzene + H_2SO_4 (concentrated) $\longrightarrow$
phenol + KOH $\longrightarrow$
toluene + HNO_3 + H_2SO_4 $\longrightarrow$
benezene + Cl_2 (presence of an iron catalyst) $\longrightarrow$
oxidation of toluene $\longrightarrow$
p-nitrotoluene + HCl + Sn $\longrightarrow$
aniline + $H \cdot C_2H_3O_2$ $\longrightarrow$

23. Distinguish between the following terms: pesticide, fungicide, rodenticide, herbicide, insecticide.
24. What are 2,4D and 2,4,5T? Describe their action on plants.
25. How is DDT prepared commercially? As we continue to use DDT as an insecticide, we are finding many unwanted problems resulting from its use. Describe some of these.
26. Why is there an ever-increasing interest in chemosterilants and sex attractants?
27. When chlorinated using light, benzene is converted to four isomeric hexachlorocyclohexanes. Which of these isomers is a potent insecticide? Under what trade name has it been sold as an insecticide?
28. Although many of the insecticides in use are synthesized, some important insecticides are isolated from plant sources. Describe these. What advantage do they usually have over the synthetic insecticides?
29. What is the most important fungicide now in use?
30. What three important rodenticides are now in use?
31. Besides the spraying of plants in the field, what other uses are made of fungicides?
32. Describe the characteristics and importance of the alkaloids and the barbitals.

chapter 24

---

proteins

24

The *proteins* (G. *proteios,* or first) are complex substances that occur in all living cells and are essential for life. They are also one of the three important classes of foodstuffs. The other two, the fats and carbohydrates, have already been discussed (Chapters 20 and 21).

All proteins contain carbon, hydrogen, oxygen, and nitrogen. Many also contain sulfur, while a few possess phosphorus and iodine. The proteins serve several purposes in the body. Their most important function is their utilization in the building of new tissue and the replacement of old, worn-out tissue. Proteins are also used in the synthesis of such important substances as the enzymes, pigments, and hormones. There is little if any storage of reserve proteins in the body. Any excess of protein over that needed for the purposes mentioned above is oxidized to furnish heat and mechanical energy to the body.

Hydrolysis of proteins

Proteins are hydrolyzed by acids, bases, and certain enzymes. Except for the conjugated class (page 352) of proteins, hydrolysis of a protein produces a mixture of α-amino acids (page 319), exclusively. Conjugated proteins are usually first cleaved to two fragments, one of which will retain protein properties and produce α-amino acids on hydrolysis. The other fragment will be nonprotein in nature and is called the *prosthetic group.*

The molecular weight of proteins

Because each molecule of a protein is hydrolyzed to a large number of molecules of amino acids, the proteins must be considered as polymers (page 266). The approximate molecular weights of proteins have been determined and, as can be expected, are very large numbers. The molecular weight of egg albumin has been estimated to be 40,000; that of hemoglobin, 68,000; while that of casein is between 75,000 and 375,000. Some proteins obtained from viruses have molecular weights of 10,000,000 or more (Figure 13.1).

Because of their very large size, proteins do not form true solutions but are either insoluble or form colloidal dispersions in water. The protein are very large molecules that can be separated from inorganic salts by the process called *dialysis*, because they are unable to pass through the fine pores of certain membranes (see Chapter 13).

Classification of proteins

The classification of proteins is based on several characteristics of these substances, such as the nature of the products formed upon hydrolysis, solubility of the protein (rather, the ability to disperse as a colloid in the liquid media rather than be a true solution), and heat coagulability.

Some biochemists recognize three classes of proteins: (1) the *simple proteins* that are naturally occurring and produce α-amino acids exclusively as their final hydrolysis products, (2) *conjugated proteins* that produce a moiety that is hydrolyzed exclusively to α-amino acids and a nonprotein moiety called the prosthetic group, and (3) the *derived proteins* that are the hydrolysis products and the modification products of the other classes of proteins. Other biochemists do not recognize the derived proteins as a legitimate class of proteins.

SIMPLE PROTEINS

The class of simple proteins is further divided into subgroups, based on the solubility of the protein in different solvents and on whether it is coagulated by heat.

The *albumins* constitute the most important group of simple proteins. These are soluble in water and dilute salt solutions but are precipitated by a saturated solution of ammonium sulfate. The albumins are present in egg white, milk, and blood. They are coagulated by heat.

The *globulins* are insoluble in water but soluble in dilute salt solutions. They are insoluble in half-saturated ammonium sulfate solution and are coagulated by heat. Examples of this class are the globulin of the blood serum and egg white and the fibrinogen of the blood.

The *histones* are strongly basic proteins that are insoluble in ammonium hydroxide but soluble in water. They are hydrolysis products of the nucleoproteins.

The *globins* are sometimes classed as histones but appear to be a separate group, because they are not basic nor are they precipitated by ammonium hydroxide. An example of this class is the globin formed in the hydrolysis of hemoglobin.

The *albuminoids* (also called *scleroproteins*) are insoluble in water and all neutral solvents and are not coagulated by heat. Examples of this class are the keratin of hair, skin, hoofs, and nails; the collagen of the bone and cartilage; and the elastin of the yellow elastic fibers of the connective tissue.

CONJUGATED PROTEINS

These are also divided into subgroups, but this division depends upon the chemical nature of the prosthetic group formed upon their hydrolysis.

Phosphoproteins are converted to phosphoric acid and amino acids upon complete hydrolysis. The casein of milk is one of the important members of this class.

Glycoproteins produce carbohydrates or derivatives of the carbohydrates and amino acids on complete hydrolysis. The mucin of saliva is a glycoprotein.

Chromoproteins produce a pigment, as well as amino acids, upon complete hydrolysis. The hemoglobin of the blood belongs to this class because upon hydrolysis the iron-containing pigment heme is produced.

Nucleoproteins occur in both the nuclei and in the cytoplasm of the cell. Careful hydrolysis of nucleoproteins produces a simple protein molecule, usually a histone, and a nucleic acid. Deoxyribonucleic acid, DNA, which is hydrolyzed to the pentose sugar deoxyribose, phosphoric acid, the two pyrimidines, cytosine and thymine, and the two purines, adenine and guanine, is a hydrolysis product of nucleoproteins that occurs primarily in cell nuclei. Ribonucleic acid, RNA, the hydrolysis product of nucleoproteins that occur predominantly in the cell cytoplasm but to a smaller extent in the cell nuclei, is hydrolyzed to the pentose sugar ribose, phosphoric acid, the two pyrimidenes, cytosine and uracil, and the two purines, adenine and guanine. There is a detailed discussion of these nucleic acids and their hydrolysis products in Chapter 28.

DERIVED PROTEINS

The coagulated proteins that are formed by heating or treatment of a simple protein with alcohol, acids, or bases belong to this class. Gelatin, a conversion product of the albuminoid collagen, is a derived protein.

Still other members of this class are the hydrolysis products of the native proteins. Some of these, in order of decreasing complexity, are proteoses, peptones, polypeptides, dipeptides, and amino acids. Proteoses are soluble in water but are precipitated in a saturated solution of ammonium sulfate. They are not coagulated by heat. The peptones are similar in properties to the proteoses but are not precipitated by a saturated solution of ammonium sulfate. The polypeptides are small fragments of the proteins composed of three or more molecules of amino acids, while dipeptides contain only two of these. The amino acids are the final hydrolysis products of the proteins.

Although it has been recently established that the proteoses and peptones are actually mixtures of polypeptides, dipeptides, and amino acids, these classifications have been retained for convenience.

The structure of the peptides and the proteins

In the dipeptide, the two amino acids must be joined through the amino group of one and the carboxyl group of the other amino acid with the loss of a molecule of water:

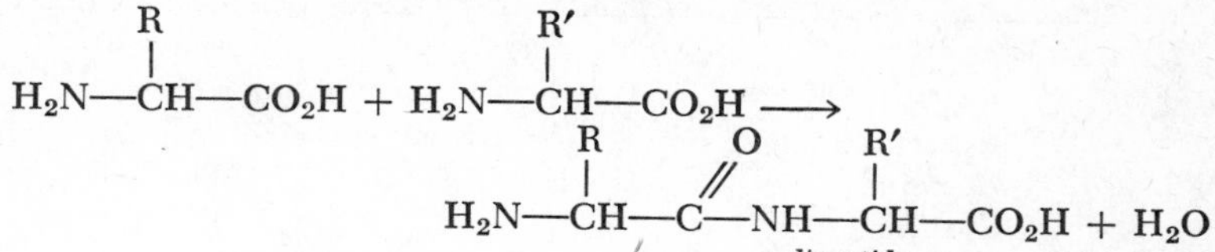

$$H_2N\text{—}\underset{\displaystyle |}{\overset{\displaystyle R}{C}}H\text{—}CO_2H + H_2N\text{—}\overset{\displaystyle R'}{\underset{}{C}}H\text{—}CO_2H \longrightarrow H_2N\text{—}\overset{\displaystyle R}{C}H\text{—}\overset{\displaystyle O}{\overset{\displaystyle /\!/}{C}}\text{—}NH\text{—}\overset{\displaystyle R'}{C}H\text{—}CO_2H + H_2O$$

a dipeptide

The group responsible for the linkage of the two amino acids in the dipeptide is

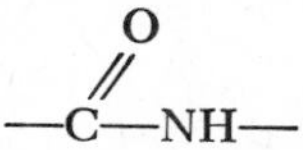

$$\text{—}\overset{\displaystyle O}{\overset{\displaystyle /\!/}{C}}\text{—}NH\text{—}$$

It resembles the functional group of the amides and is called the *peptide linkage.* The dipeptide still contains a free amino and a carboxyl group. It can react with both acids and bases and, like the simple amino acids, is an amphoteric substance (see page 138).

If the amino group of another molecule of an amino acid reacts with the carboxyl group of this dipeptide, or if the carboxyl group of an amino acid reacts with the free amine group of this dipeptide, a *tripeptide,* which could be represented by the following formula, would be formed.

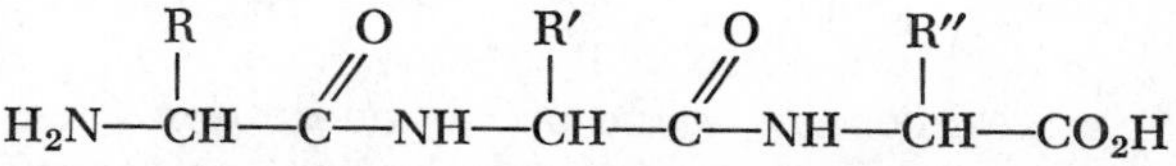

$$H_2N\text{—}\overset{\displaystyle R}{C}H\text{—}\overset{\displaystyle O}{\overset{\displaystyle /\!/}{C}}\text{—}NH\text{—}\overset{\displaystyle R'}{C}H\text{—}\overset{\displaystyle O}{\overset{\displaystyle /\!/}{C}}\text{—}NH\text{—}\overset{\displaystyle R''}{C}H\text{—}CO_2H$$

It contains two peptide linkages, and because it contains a free amino and a free carboxyl group, it too will be amphoteric.

Glycylalanyglycine, a typical tripeptide, has the following structure:

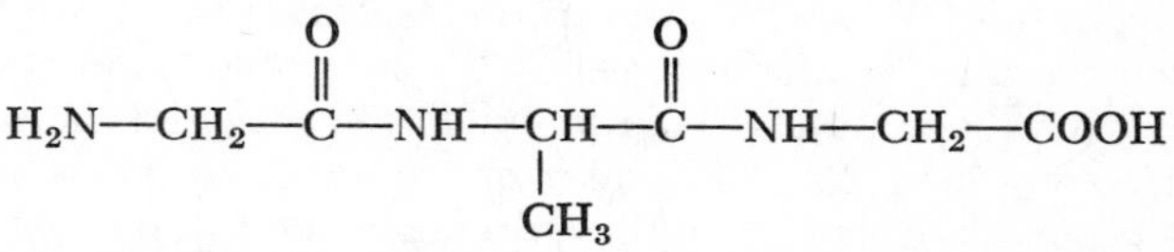

$$H_2N\text{—}CH_2\text{—}\overset{\displaystyle O}{\overset{\|}{C}}\text{—}NH\text{—}\underset{\displaystyle CH_3}{\underset{|}{C}}H\text{—}\overset{\displaystyle O}{\overset{\|}{C}}\text{—}NH\text{—}CH_2\text{—}COOH$$

This tripeptide can be readily represented by the use of residue symbols as

Gly-Ala-Gly

The accepted abbreviations or symbols for the amino acids are listed in Table 24.1.

If molecules of amino acids (Table 24.2) now reacted with the ever-growing polypeptide molecule and this process were repeated over and over again hundreds of times, then a very large molecule resembling the proteins would be formed. This type of structure is now visualized for the proteins. Such a large molecule will be amphoteric, because there will be an amino group at one end of the

TABLE 24.1 ALPHABETICAL LISTING OF THE AMINO ACIDS AND THEIR ACCEPTED ABBREVIATIONS (SYMBOLS)

Amino Acid	*Symbol*
Alanine	Ala
Arginine	Arg
Aspartic acid	Asp
Cysteine	Cy-SH
Cystine	Cy-S-S-Cy
Glutamic acid	Glu
Glycine	Gly
Histidine	His
Hydroxyproline	Hyp
Isoleucine	Ile
Leucine	Leu
Lysine	Lys
Methionine	Met
Phenylalanine	Phe
Proline	Pro
Serine	Ser
Threonine	Thr
Tryptophan	Trp
Tyrosine	Tyr
Valine	Val

molecule and a carboxyl group at the other end. Because the side groups of a polypeptide, represented as R, R′, and R″ in the tripeptide formula above, may also contain acidic and basic groups, these side groups may also account for an appreciable amount of the amphoteric properties of the proteins. In fact, native proteins are amphoteric; this explains their action as buffers in the blood, because they will quickly react with any excess of acid or base.

The polypeptide nature of the proteins has now been well established, particularly by the synthesis of a number of native proteins, including a polypeptide prepared by Dr. Klaus Hofmann and associates, who combined 23 amino acid molecules. This polypeptide possesses the typical properties of the adrenocorticotrophic hormone ACTH (page 521). The enzyme ribonuclease, a polypeptide formed by combining 124 amino acid molecules, was synthesized by Merrifield and colleagues in 1969.

One of the great wonders of nature is that a living cell can take the right number and kind of amino acids from a mixture and build a certain type of protein characteristic of the cell or needed by the cell and can reject those amino acids not needed for this synthesis. A plausible mechanism by which the cell arranges the amino acids in the proper order is discussed in Chapter 28.

The peptide bonds are considered the "primary" structure in proteins (Figure 24.1). Other types of bonding also occur in proteins, such as disulfide bonds, hydrogen bonds, and ionic bonds.

TABLE 24.2 NUMBER OF COMPONENT AMINO ACIDS IN PROTEINS

Amino Acid	*Human Insulin*	*Horse Myoglobin*	*Pepsin*	*Egg Albumin*
Glycine	4	13	29	19
Alanine	1	15		12
Serine	3	6	40	36
Threonine	3	7	28	16
Valine	4	6	21	28
Leucine	6	22	27	32
Isoleucine	2		28	25
Proline	1	5	16	14
Cysteine			2	5
Cystine	3		2	1
Methionine		2	4	16
Phenylalanine	3	5	13	21
Tyrosine	4	2	16	9
Tryptophan		2	4	3
Aspartic acid	3	10	41	32
Glutamic acid	7	19	28	52
Arginine	1	2	2	15
Lysine	1	18	2	20
Histidine	2	7	9	2

Source: Robert J. Ouellette, *Introductory Organic Chemistry,* Harper & Row, New York, 1971.

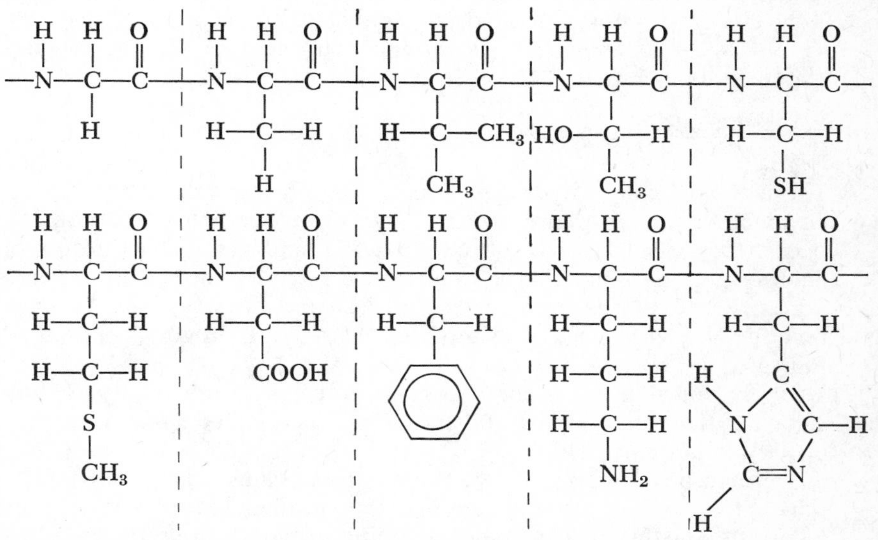

***Figure 24.1** A portion of a protein molecule, showing amino acid residues joined by peptide bonds.*

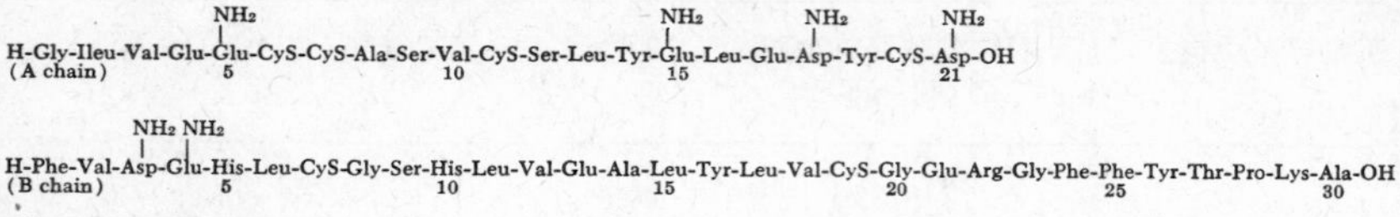

```
                    NH2                                   NH2        NH2        NH2
                    |                                     |          |          |
H-Gly-Ileu-Val-Glu-Glu-CyS-CyS-Ala-Ser-Val-CyS-Ser-Leu-Tyr-Glu-Leu-Glu-Asp-Tyr-CyS-Asp-OH
(A chain)           5                  10                 15                    21

             NH2 NH2
              |   |
H-Phe-Val-Asp-Glu-His-Leu-CyS-Gly-Ser-His-Leu-Val-Glu-Ala-Leu-Tyr-Leu-Val-CyS-Gly-Glu-Arg-Gly-Phe-Phe-Tyr-Thr-Pro-Lys-Ala-OH
(B chain)         5                   10                  15                  20                  25                  30
```

***Figure 24.2** Amino acid sequences in the A and B chains of beef insulin.*

One of the amino acids, cysteine (page 322), possesses a thiol (SH) group and can be readily oxidized to cystine (page 322) thus:

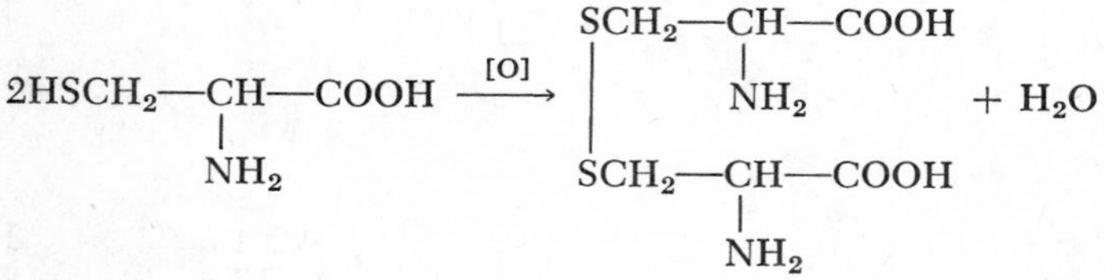

$$\underset{\text{cysteine}}{2HSCH_2{-}\underset{\displaystyle NH_2}{\underset{|}{CH}}{-}COOH} \xrightarrow{[O]} \underset{\text{cystine}}{\begin{array}{l} SCH_2{-}\underset{\displaystyle NH_2}{\underset{|}{CH}}{-}COOH \\ | \\ SCH_2{-}\underset{\displaystyle NH_2}{\underset{|}{CH}}{-}COOH \end{array}} + H_2O$$

If two thiol-bearing side chains of cysteine appear in a polypeptide or protein and are oxidized, a closed loop will be formed. Two polypeptides may be linked by the oxidation of a thiol group in each to a disulfide linkage. In insulin, the structure of which has been established by F. Sanger, two polypeptide chains, chain *A* consisting of 21 amino acids and chain *B* consisting of 30 amino acids, are joined by two disulfide linkages, while in chain *A*, a closed loop containing a disulfide bond is also present. These linkages of chains and the amino acid sequence of beef insulin are illustrated in Figure 24.2.

Hydrogen atoms bonded to highly electronegative atoms such as N, O, or F may form a bridging to another similar electronegative atom thus:

$$—N—H---O—$$

These bonds, called hydrogen bonds, are much weaker than typical covalent bonds. In many proteins, hydrogen bonding very probably occurs between the NH of one peptide and the C=O group of another peptide bond. These to a great extent determine the "secondary" structure of the proteins. Figure 24.3 indicates typical hydrogen bonding between two polypeptide chains. (Hydrogen bonding is indicated by dashed lines.)

Hydrogen bonding appears to be the linkage primarily responsible for the formation of the α-helical and pleated-sheet structures of fibrous proteins, which are discussed on pages 357–358.

Ionic bonds between two polypeptide chains due to salt formation between basic groups, such as the ϵ-amino group that appears in the side chain of lysine, and an acidic group, such as the carboxyl group that appears in the side chain of a glutamic acid residue, represents another type of bonding. The following shows such a type

$$
\begin{array}{ccccccccc}
 & O & & R & & H & & O & \\
 & \| & & | & & | & & \| & \\
\diagdown & C & \diagdown & CH & \diagdown & N & \diagdown & C & \diagdown \\
CH & & N & & C & & CH & & N \\
| & & | & & \| & & | & & | \\
R & & H & & O & & R & & H \\
 & & \vdots & & \vdots & & & & \vdots \\
R & & O & & H & & R & & O \\
| & & \| & & | & & | & & \| \\
CH & & C & & N & & CH & & C \\
\diagup & N & \diagup & CH & \diagup & C & \diagup & N & \diagup \\
 & | & & | & & \| & & | & \\
 & H & & R & & O & & H &
\end{array}
$$

Figure 24.3 *Interchain bonding in proteins due to hydrogen bonding between the peptide groups of parallel polypeptide chains. Dashed lines indicate the hydrogen bonds that link peptide groups.*

of bonding between two polypeptide chains:

$$
\begin{array}{l}
\quad | \\
\quad C{=}O \\
\quad | \\
\quad NH \\
\quad | \\
H{-}C{-}(CH_2)_4NH_2 \\
\quad | \\
\quad C{=}O \\
\quad | \\
\quad NH \\
\quad |
\end{array}
+
\begin{array}{l}
\quad | \\
\quad C{=}O \\
\quad | \\
\quad NH \\
\quad | \\
H{-}C{-}(CH_2)_2CO_2H \\
\quad | \\
\quad C{=}O \\
\quad | \\
\quad NH \\
\quad |
\end{array}
\longrightarrow
$$

$$
\begin{array}{l}
\quad | \\
\quad C{=}O \\
\quad | \\
\quad NH \\
\quad | \\
H{-}C{-}(CH_2)_4\overset{+}{N}H_3 \\
\quad | \\
\quad C{=}O \\
\quad | \\
\quad NH \\
\quad |
\end{array}
+
\begin{array}{l}
\quad | \\
\quad C{=}O \\
\quad | \\
\quad NH \\
\quad | \\
H{-}C{-}(CH_2)_2CO_2^- \\
\quad | \\
\quad C{=}O \\
\quad | \\
\quad NH \\
\quad |
\end{array}
$$

Note that proton transfer from the glutamic acid residue to the lysine residue of another polypeptide chain forms oppositely charged ions in adjacent chains.

In 1951, Pauling and Corey suggested that the polypeptide chain of α-keratin, the natural unstretched protein of wool, exists in a helical structure in which there are 3.6 amino acid residues in each turn of the helix. Such a helical structure appears to be primarily

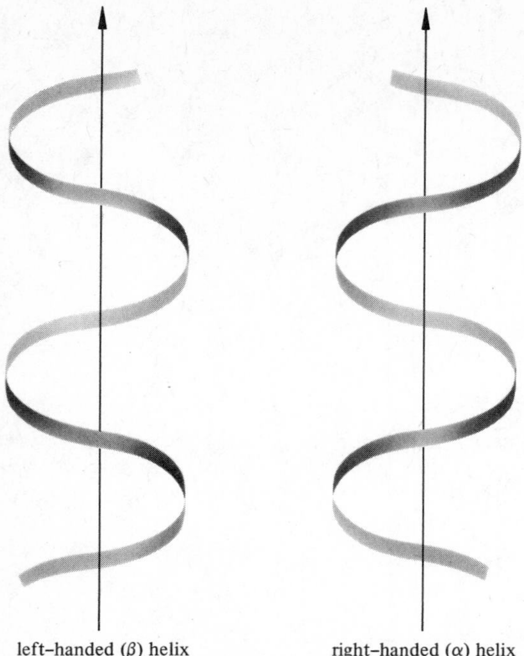

Figure 24.4 α and β helices.

stabilized by hydrogen bonding of a carbonyl (C═O) group of one peptide bond to an imide (N—H) group of a peptide bond three residues further along in the peptide chain. The α-helical structure of α-keratin is illustrated in Figures 24.4 and 24.5.

If the α-keratin of wool is steamed and dried under tension, it maintains its stretched length. In these stretched fibers of wool, called β-keratin, peptide chains run in the same direction held together in what has been called a parallel-chain, pleated-sheet structure (Figure 24.6), which is stabilized by hydrogen bonds between the peptide bonds of adjacent parallel peptide chains.

The fibrous protein of silk, called *fibroin,* also appears to be a pleated-sheet structure of parallel polypeptide chains. However, adjacent polypeptide chains, which are also primarily stabilized by hydrogen bonding between the chains, run in opposite directions. Thus this is called an *antiparallel-chain, pleated-sheet structure* (Figure 24.6)

Such globular proteins as hemoglobin, albumin, and the myoglobins of muscular tissue appear to consist of polypeptide chains, sections of which are present as small helical segments. The looping and folding of the portions of the peptide chain between these helical portions of the chain, considered the *tertiary structure* of the protein, are probably responsible for the compact structure typical of globular proteins.

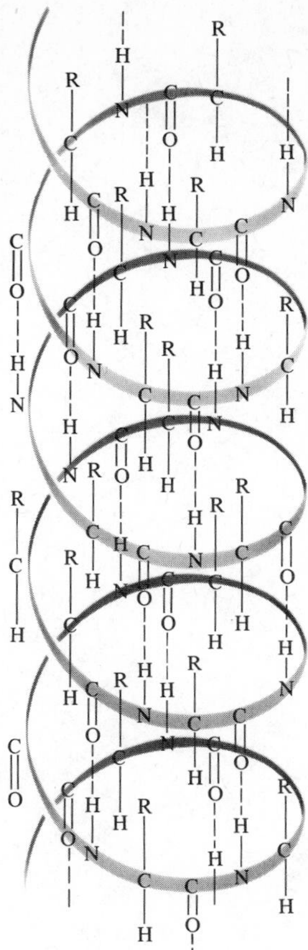

Figure 24.5 Hydrogen bonding in an α helix. (From R. J. Ouellette, Introductory Organic Chemistry, *Harper & Row, N.Y. 1971.)*

It has now been established that native active hemoglobin consists of four separate peptide chains bound together. Surface active forces are likely responsible for the union of the four chains, and no covalent-type bonding exists between them. Two of these chains are identical and have been designated as α-chains consisting of 141 residues, while the other two are designated as β-chains and consist of 146 residues. A number of other proteins appear to consist of several discrete chains; the manner in which these multiple peptide chains are fitted together in the native protein is now designated as the *quaternary structure* of such proteins.

THE ISOELECTRIC POINT

It is now believed that in an aqueous solution at some certain pH, which is different for each amino acid, the amino acid molecule possesses both a positive and a negative charge and exists as a *dipolar ion* of the general structure.

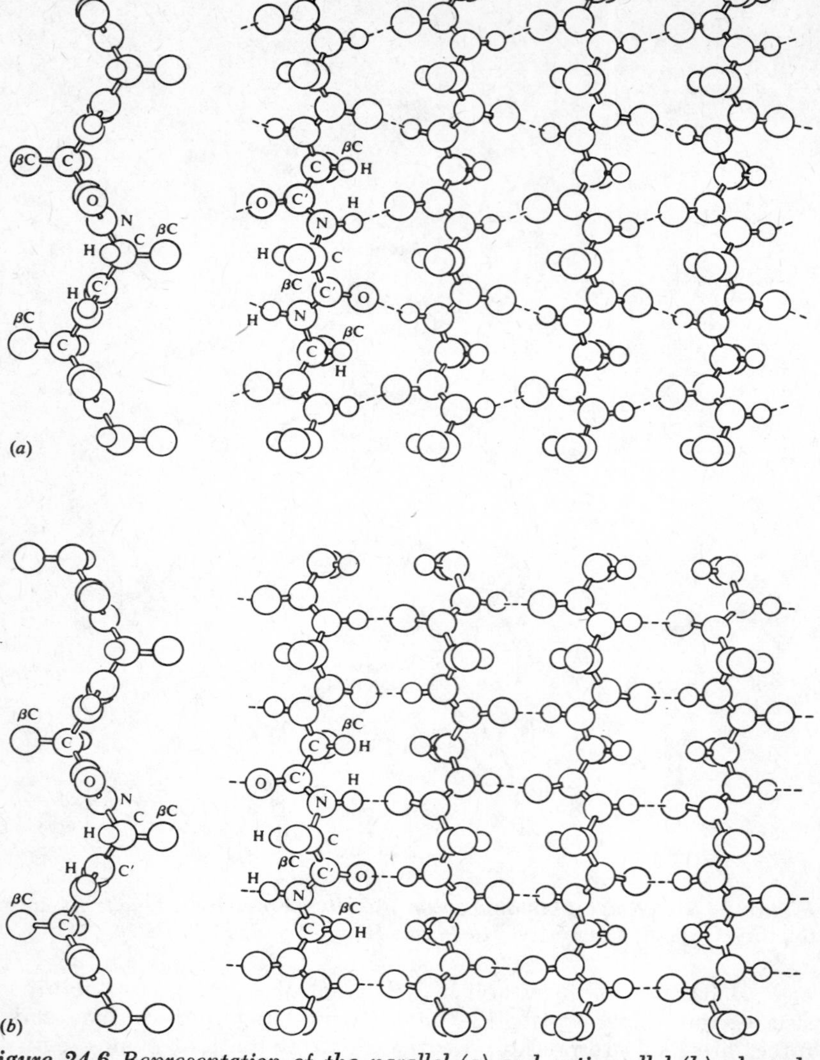

***Figure 24.6** Representation of the parallel (a) and antiparallel (b) pleated-sheet structures of polypeptides. (L. Pauling and R. B. Corey,* Proc. Nat. Acad. Sci., U.S., *37: 729 (1951). From Mahler and Cordes,* Basic Biological Chemistry, *Harper & Row, N.Y., 1968.)*

$$H_3\overset{+}{N}-\underset{}{\overset{R}{\overset{|}{C}}}H-C\begin{matrix}\nearrow O \\ \searrow O^-\end{matrix}$$

This pH is called the *isoelectric point* of the amino acid, and the amino acid, when at this pH, will not migrate in an electric field. In a solution that is more acid than the isoelectric point, the amino acid will have the structure

$$H_3\overset{+}{N}-\underset{}{\overset{R}{\overset{|}{C}}}H-CO_2H$$

and it will migrate to the cathode, whereas in a solution more basic than the isoelectric point, it will have the structure

$$H_2N-\overset{R}{\overset{|}{C}}H-C\begin{matrix}\nearrow O \\ \searrow O^-\end{matrix}$$

and it will migrate to the anode in an electric field.

At a certain pH, the number of positive and negative charges on a protein molecule will equal each other. The protein will not migrate in an electric field, will exhibit its least stability, and will be most easily precipitated at this point. This characteristic pH is called the isoelectric point of the protein. The isoelectric point of casein is at a pH of 4.6, while that of the hemoglobin of horse blood is at pH 6.8. Note that the isoelectric point of both amino acids and proteins are not usually at the neutral point (pH 7).

Precipitation and denaturation of proteins

Usually, if proteins are precipitated from solution or if they are coagulated and are then redispersed in the solvent, the redispersed protein will not have the same properties as the original protein. In such cases, the protein resulting from the precipitation is called a *denatured protein.* Careful precipitation of certain proteins by ammonium sulfate, sodium sulfate, magnesium sulfate, or sodium chloride will not denature them; that is, they can be redissolved in water without any change of properties from the original protein.

Heat and alcohol will coagulate many proteins. The sterilization of dressings, etc., in an autoclave (Figure 24.7) depends on the coagulation of the protein in any bacteria that may be present. In sterilization by alcohol, a 70 percent solution by weight is preferred to the 95 percent solution. The latter only coagulates the proteins at the surface of the bacterium without killing it. The former concentration of alcohol diffuses throughout the entire bacterium and thus effectively destroys it. It is common practice to disinfect the surface of the skin with a 70 percent alcohol solution before obtaining blood for blood transfusions or before the insertion of needles into blood vessels for the injection of drugs.

Heavy metallic salts, such as silver, lead, and mercury, precipitate the proteins by the formation of insoluble protein salts. Thus silver nitrate combines with proteins and precipitates silver proteinate. The antiseptic action of silver nitrate (lunar caustic) and

Figure 24.7 An autoclave used for sterilization. (Courtesy Chas. Pfizer & Co., Inc.)

mercuric chloride (corrosive sublimate) depends on the precipitation of the proteins present in the bacteria. Most of the heavy metal salts are toxic when taken internally because they precipitate the proteins of the cells that they contact. Egg white and milk are used as antidotes for heavy metal poisoning, because the proteins present in them combine with the metal ions to form an insoluble solid that is not readily absorbed into the bloodstream through the walls of the stomach and intestines. Sodium bicarbonate is frequently administered with egg white; this gives a negative charge to the protein, which as an anion can combine more effectively with a heavy metal cation. It is necessary to remove the resulting insoluble material from the stomach by the use of an emetic or stomach pump; if it is not removed, the digestive juices will destroy the protein and liberate the metallic salts, which will be absorbed by the digestive system, and toxicity will only have been delayed.

Certain acids, such as trichloroacetic acid, phosphotungstic acid, tannic acid, and picric acid, have been used to precipitate the alkaloids (page 346). For this reason, they have been called *alkaloidal reagents*. They also precipitate proteins and are used in the preparation of protein-free blood filtrates prior to routine analysis. Tannic acid, present in appreciable amounts in the bark of certain trees, coagulates the proteins present in the hides of cattle and converts these to tough leather. Both picric and tannic acids are used in the treatment of burns. They combine with the protein in the exposed areas, forming a leathery coating that affords some protection and stops the loss of body fluids. The loss of this latter fluid is the most important cause of shock and the fatalities resulting from burns.

Certain types of radiation will also denature or coagulate proteins. X rays and the emanations from radium attack the proteins of cancerous tissue more rapidly than that of normal tissue and are used in cancer therapy.

The color reactions of the proteins

Certain color tests can be used for the qualitative detection of proteins. The production of a color with the reagent depends on the presence of characteristic groups or of certain amino acid residues in the protein. It is necessary to perform several of these tests before deciding whether a substance is a protein.

BIURET TEST

This test consists in making a protein solution alkaline and then adding a very dilute solution of copper sulfate. Proteins will produce a rose-pink to violet color with this reagent. Biuret also gives a positive test with this reagent; hence the name (page 318). This test appears to depend on the presence of a minimum of two peptide linkages in each molecule. Urea and amino acids do not give this test.

XANTHOPROTEIC TEST

This literally means *yellow protein test*. Concentrated nitric acid produces a yellow color with proteins. The resulting yellow color is changed to orange when treated with alkalis. The student working in the laboratory has probably noticed the yellow color produced when nitric acid is dropped on the skin. This test depends on the presence in the protein of amino acids that possess a benzene ring, such as phenylalanine, tyrosine, and tryptophan. (See pages 320 and 321 for the formulas of these amino acids.)

MILLON TEST

Millon's reagent is a solution of mercurous nitrate and mercuric nitrate in nitric acid. If Millon's reagent is added to a protein, a white precipitate forms which upon boiling changes through yellow to a brick-red color. This test depends on the presence of tyrosine in the protein.

HOPKINS-COLE TEST

The reagent for this test consists of an aqueous solution of either glyoxylic acid or its magnesium salt.

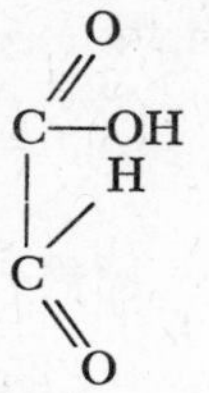

```
     O
    //
   C—OH
   |  H
   | /
   C
    \\
     O
```

glyoxylic acid

If the protein solution containing this reagent is layered over concentrated sulfuric acid, a violet-colored ring will form at the junction of the two liquids. This test depends on the presence of tryptophan in the protein.

NINHYDRIN TEST

When boiled with ninhydrin (triketohydrindene hydrate), proteins and most amino acids produce a blue to purple color. This test depends on the presence of free amino groups in the protein or amino acid.

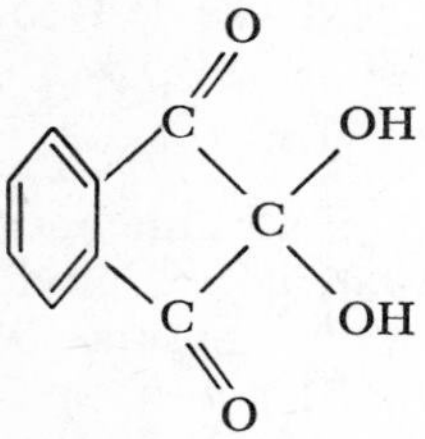

```
            O
           //
          C       OH
   /\    / \     /
  |  |--    C
   \/    \ /     \
          C       OH
           \\
            O
```

ninhydrin

STUDY QUESTIONS

1. What chemical elements are most frequently found in the proteins?
2. What are the three most important classes of foodstuffs?
3. Are the proteins of low or high molecular weight? Does a protein molecule produce only one kind of amino acid upon hydrolysis?
4. How may protein be separated from low-molecular-weight impurities?
5. Does albumin form a true solution when it dissolves in water?
6. How are the proteins utilized in the body?
7. List the three classes of proteins. How may the three classes be differentiated?
8. On what properties does the division of the simple proteins into subgroups depend?
9. Describe the properties and occurrence of the albumins, globulins, histones, globins, and albuminoids. To what class of proteins do these belong?
10. Upon what basis are the conjugated proteins divided into subgroups?
11. Name four subgroups of the conjugated proteins. What nonprotein substances are formed upon the hydrolysis of each of these?
12. Indicate an important occurrence for each subclass of the conjugated proteins.

13. To what class of proteins do the coagulated proteins belong? What is the source of gelatin?
14. What are dipeptides? Polypeptides?
15. What three types of substances may be used to catalyze the hydrolysis of proteins?
16. How are the proteoses and peptones related to the simple proteins? How are proteoses separated from peptones?
17. Write a general formula for a dipeptide, a polypeptide. What linkage joins the amino acid residues in the di- and polypeptides?
18. How are the proteins related in structure to the polypeptides? Why are the polypeptides and proteins, as well as the simple amino acids, amphoteric?
19. List some proteins that have been synthesized. As far as the structure of proteins goes, what is the significance of these syntheses?
20. What unusual power do the living cells possess in their synthesis of proteins from amino acids?
21. What is considered the primary linkage in proteins?
22. What are hydrogen bonds? How do they stabilize the structure of polypeptides and proteins?
23. Ionic bonds probably occur between certain peptide chains. Describe this type of linkage. Do these appear to be important in the structure of proteins?
24. The oxidation of cysteine residues in polypeptide chains probably leads to interpolypeptide-chain linkages, as well as loops in some polypeptide chains. Name this important secondary linkage in proteins. Illustrate how it is formed. Relate the presence of this type of linkage to the recently established structure of insulin.
25. Describe the α-helical structure of wool α-keratin and the parallel-chain, pleated-sheet structure of wool β-keratin.
26. What is the typical structure of fibroin, the protein of silk?
27. What is the probable structure of the globular proteins, such as albumin and hemoglobin?
28. Write the formula for the ion that is formed when glycine, H_2NCH_2-COOH, is treated with an acid; with a base.
29. What is the isoelectric point of an amino acid? Of a protein? What is the structure of glycine at its isoelectric point? What is a dipolar ion?
30. How do the properties of a protein at its isoelectric point differ from its properties when in solution of other pH values?
31. Is the isoelectric point of a protein identical with the neutral point?
32. What is a denatured protein?
33. What types of substances will usually precipitate proteins without denaturing them?
34. Why does heat act as a good sterilizing agent?
35. Why is a 70 percent by weight alcohol solution more effective as a disinfectant than a 95 percent solution?
36. What heavy metal salts are frequently used to precipitate proteins? How do these salts react with proteins?
37. Egg white and milk serve as valuable antidotes for heavy metal poisoning. Explain their action. Why must an emetic be used to make these antidotes effective?
38. What are the alkaloidal reagents, and why are they so called?
39. What role does tannic acid play in the preparation of leather?

40. What is one of the most important causes of fatalities resulting from burns? How do tannic and picric acids aid in the treatment of burns?
41. Name five color tests that are used in identifying proteins. How is each test performed? What portion of protein molecules is responsible for each of these tests?
42. Why must more than one of the protein color tests be used to decide whether a substance is a protein?
43. Write the structural formulas for phenylalanine, tyrosine, and tryptophan. Which of the color tests depend on the presence of these amino acids in the protein?

chapter 25

---

enzymes

25

Enzymes are organic catalysts elaborated by living organisms. As previously discussed, catalysts are substances that change the rate of a chemical reaction without undergoing any apparent change themselves. Catalysts increase the speed of a chemical reaction (see Chapter 8).

It is to be noted that catalysts are not able to initiate chemical reactions. Catalysts can only speed up a reaction that is capable of occurring in their absence. In many cases, the reaction in the absence of an enzyme or other catalyst proceeds so slowly that it might be surmised that no reaction is occurring. Enzymes and, for that matter, catalysts in general do not change the proportions of reactants and products at the point of a chemical equilibrium. Therefore it is sometimes possible, under the proper set of conditions, to reverse the usual reactions catalyzed by enzymes. Thus proteinlike substances have been isolated after a protein-splitting enzyme has been in contact with the amino acids prepared from a previous protein hydrolysis. However, usually at the equilibrium point, such a small amount of reactant remains from an enzymic reaction that most such reactions may be considered as going to completion.

Life processes are dependent on the presence of enzymes, because enzymes play a prominent part in the chemical reactions that occur in the cell. Life would not be possible in their absence.

Until 1897, there was a great deal of controversy concerning whether or not the action of enzymes could occur only in the presence of the organism elaborating them. In that year, the Buchner brothers demonstrated that the action of the enzyme was independent of the cells forming it. They separated zymase from the yeast cell and showed that it would still ferment glucose in the absence of these cells. In 1926, Sumner prepared crystalline urease, and over 100 enzymes have now been obtained in crystalline form (Figure 25.1).

Enzymes are protein in nature, and, like any other protein, they are coagulated or denatured when they are heated. All enzymes are inactivated under such conditions.

Substances undergoing change due to the presence of an enzyme are called *substrates,* while the products of an enzymatic reaction are called *end products.* In the conversion of starch to maltose by the

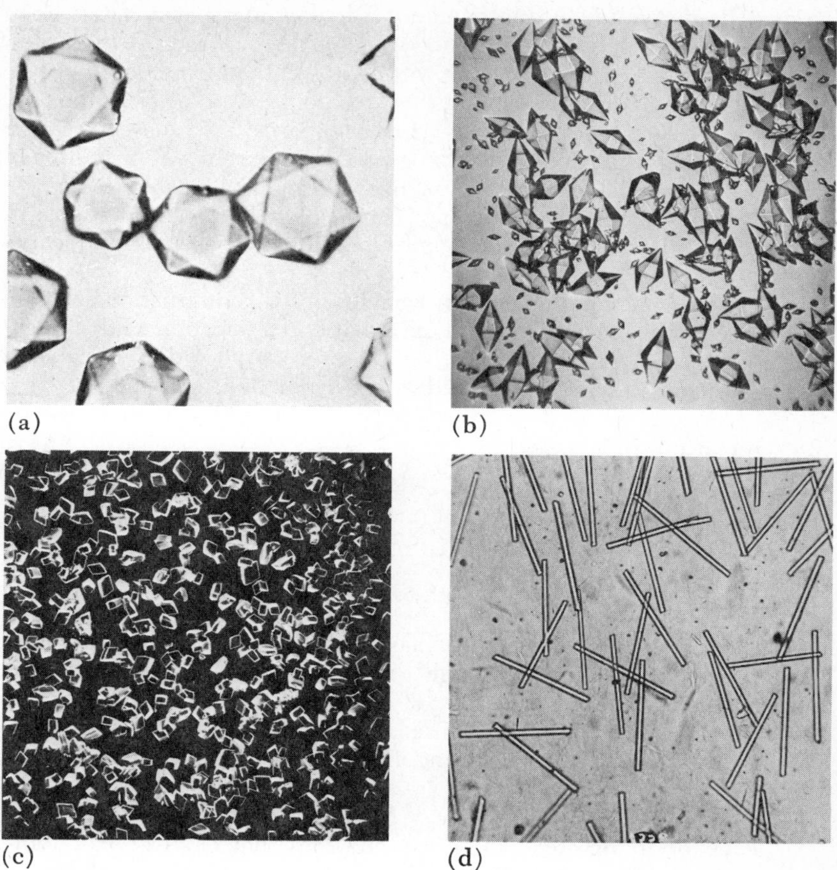

Figure 25.1 (a) Crystalline urease. (From Sumner and Somers, Chemistry and Methods of Enzymes, *3rd ed., Academic, N.Y., 1953.) (b) Crystalline pepsin. (From Northrop, Kunitz, and Herriott,* Crystalline Enzymes, *rev. ed., Columbia University Press, N.Y., 1948.) (c) Crystalline chymotrypsin. (Courtesy Dr. Moses Kunitz. From Kleiner and Orten,* Biochemistry, *7th ed., Mosby, St. Louis, 1966.) (d) Crystalline phosphorylase, from rabbit muscle. (Courtesy Dr. Carl F. Cori. From Green and Cori,* Journal of Biological Chemistry, *151, 1943.)*

catalytic action of the diastase of malt, the starch is the substrate, and maltose is the end product.

Enzyme nomenclature

Many of the early recognized enzymes, such as trypsin, rennin, and pepsin, were given names ending in *-in,* and these names are still retained today. It is now the usual practice to add the ending *-ase* to a prefix derived from the name of the substrate that is acted upon or the type of reaction that is catalyzed by the enzyme. As examples, maltase catalyzes the hydrolysis of maltose to glucose, and urease catalyzes the hydrolysis of urea to ammonia and carbon dioxide. Lipids are hydrolyzed in the presence of lipases. Hydrolases catalyze hydrolytic reactions, dehydrogenases catalyze the removal of hydrogen from a substrate, and oxidases oxidize hydrogen atoms to water.

Some so-called enzymes now appear to be combinations of enzymes. Thus zymase is a complex mixture of enzymes and coenzymes (page 406).

In 1961, the Commission on Enzymes of the International Union of Biochemistry recommended a systematic classification and nomenclature for all known enzymes. In this recommendation, they suggested that all enzymes be placed in six major divisions as follows:

I. oxido-reductases
II. transferases
III. hydrolases
IV. lyases
V. isomerases
VI. ligases

Each enzyme is further subclassified and subsubclassified under its appropriate division. Although the Commission has recommended a new systematic nomenclature, it appears to be more practical in this text to use the previously used trivial names of the enzymes.

More information concerning these six official enzyme divisions follows:

I. Oxido-reductases. These are enzymes that catalyze oxidation-reduction reactions. These are briefly described on pages 374–375.

II. Transferases. These are enzymes that catalyze the transfer of a characteristic group from one substrate to another. Transaminases transfer an amino group from an amino acid to a keto acid (page 416). Kinases transfer phosphate groups. Kinase is a term also used to designate organic activators or proenzymes (page 373). Hexokinase transfers a phosphate group from adenosine triphosphate (ATP) to glucose to form glucose-6-phosphate (page 399). Phosphorylases can catalyze the transfer of phosphate groups to polysaccharides, such as starch or glycogen, to form hexose phosphates (page 400).

III. Hydrolases. These enzymes catalyze the hydrolysis of numerous substances by water and are described on pages 372–373.

IV. Lyases. Lyases are enzymes that catalyze the removal of groups of atoms from substrates nonhydrolytically. The decarboxylases are an important group of lyases. They liberate carbon dioxide by attacking a carboxyl group. Amino acid decarboxylases convert amino acids to amines. Carbonic anhydrase (page 473) reversibly catalyzes the decomposition of carbonic acid to carbon dioxide and water.

Hydrases catalyze the reversible addition of water or its removal from a substrate. Thus fumarase can catalyze the addition of water to fumaric acid to form malic acid (page 408) or the removal of water from malic acid to form fumaric acid.

$$\underset{\text{fumaric acid (trans)}}{HOOC—CH{=}CH—COOH} + H_2O \rightleftharpoons \underset{\text{malic acid}}{HOOC—CH_2—CHOH—COOH}$$

V. Isomerases. These enzymes catalyze various types of isomerizations. Thus phosphohexoisomerase converts glucose-6-phosphate to fructose-6-phosphate (page 404), whereas phosphoglucomutase isomerizes glucose-6-phosphate to glucose-1-phosphate reversibly (page 399).

Phosphotriose isomerase catalyzes the following interconversion (page 404):

$$\underset{\text{dihydroxyacetonephosphate}}{\begin{matrix} CH_2OPO_3H_2 \\ | \\ C{=}O \\ | \\ CH_2OH \end{matrix}} \rightleftharpoons \underset{\text{3-phosphoglyceraldehyde}}{\begin{matrix} CH_2OPO_3H_2 \\ | \\ CHOH \\ | \\ CHO \end{matrix}}$$

VI. Ligases. These catalyze the linking together of two molecules that will be coupled with the cleavage of some pyrophosphate bond such as occurs in adenosine triphosphate (ATP).

Acetylthiokinase catalyzes the transfer of acetyl groups to coenzyme A (CoASH), a reaction in which ATP is converted to AMP (adenosine monophosphate) and pyrophosphoric acid (P—P):

$$\text{acetate} + \text{CoASH} + \text{ATP} \longrightarrow \text{Acetyl—CoA} + \text{AMP} + \text{P—P}$$

If the student desires a more detailed discussion of the official classification and nomenclature of enzymes, he will find the following references useful:

1. E. E. Conn and P. K. Stumpf, *Outlines of Biochemistry*, 2nd ed., Wiley, New York (1966), pages 150–152.
2. H. R. Mahler and E. H. Cordes, *Basic Biological Chemistry*, Harper & Row, New York (1968), pages 175–179.
3. H. R. Mahler and E. H. Cordes, *Biological Chemistry*, 2nd ed., Harper & Row, New York (1971), pages 326–334.
4. M. Florkin and E. H. Stotz (eds.), *Comprehensive Biochemistry*, vol. 13, Elsevier, New York (1964).

Enzymes that catalyze hydrolytic reactions

Two important types of reactions catalyzed by enzymes are hydrolysis and oxidation-reduction reactions. Hydrolytic reactions involve the addition of water to a substrate, with a simultaneous decomposition of the substrate into two or more products. These hydrolytic reactions can be represented by the following general equation:

$$A\text{—}B + \mathrm{HOH} \longrightarrow \mathrm{H}\text{—}A + B\text{—}\mathrm{OH}$$

Such reactions mostly occur during digestion of food in the gastrointestinal tract, because the complex foods must be converted to smaller molecules before these substances can be absorbed through the walls of the intestines and into the bloodstream.

Enzymes that catalyze hydrolytic reactions are called *hydrolases.* Some classes of these enzymes are given below.

1. *Carbohydrases.* These enzymes catalyze the hydrolysis of polysaccharides and disaccharides into simpler sugars.
   (a) Amylases hydrolyze starch to dextrins and maltose. Some amylases are the salivary amylase of saliva, the diastase of malt, and the amylopsin of the pancreatic juice.
   (b) Sucrase or invertase catalyzes the hydrolysis of sucrose to a mixture of glucose and fructose, a mixture called invert sugar.
   (c) Maltase catalyzes the hydrolysis of maltose to glucose.
   (d) Lactase catalyzes the hydrolysis of lactose to glucose and galactose.
2. *Esterases.* These catalyze the hydrolysis of esters to alcohols and acids.
   (a) Lipases catalyze the hydrolysis of fats to glycerol and fatty acids. Examples of lipases are the gastric lipase of the stomach, the vegetable lipase of the castor bean, and the steapsin of the pancreatic juice.
   (b) Phosphatases catalyze the hydrolysis of phosphoric acid esters to phosphoric acid and alcohols. Thus these enzymes hydrolyze glycerophosphate to glycerol and phosphoric acid.
3. *Proteases.* These catalyze the hydrolysis of the natural and derived proteins to simpler derived proteins and amino acids.
   (a) Peptidases catalyze the conversion of polypeptides and dipeptides to amino acids. Because they attack peptide bonds adjacent to free amino or carboxyl groups, they are now being designated as *exopeptidases.* Examples of exopeptidases are the aminopeptidase and carboxypeptidase of the small intestine.
   (b) Proteinases catalyze the hydrolysis of proteins to derived proteins. Because this group of enzymes preferentially attack an interior peptide bond, they are now known as

endopeptidases. The pepsin of the stomach converts proteins primarily to proteoses and peptones. Trypsin and chymotrypsin of the pancreatic juice convert proteins, proteoses, and peptones to polypeptides. The rennin of the stomach converts casein to paracasein. Papain of the paw-paw and the bromelin of pineapples are vegetable proteinases.

4. *Amidases* catalyze the hydrolytic cleavage of nonprotein or nonpeptide C–N linkages. Urease of the jack bean and soybeans catalyzes hydrolysis of urea to carbon dioxide and ammonia.
5. *Nucleases* catalyze the hydrolysis of the nucleic acids. This hydrolysis occurs through several stages until pyrimidines, purines, phosphoric acid, and pentoses are obtained (pages 438–441). A different nuclease is required for each stage of the hydrolysis.

Proenzymes or zymogens

Some enzymes are first produced in an inactive form called a *proenzyme* or a *zymogen* and must be activated by some other substance before they can exhibit their catalytic activity.

The following are some examples of these proenzymes or zymogens:

1. Pepsinogen, a product of the gastric mucosa, is converted to active pepsin by the hydrochloric acid present in the gastric juice.
2. Trypsinogen, a product of the pancreas, is changed into active trypsin by enterokinase in the duodenum.
3. Chymotrypsinogen is also produced by the pancreas. When this proenzyme comes in contact with trypsin in the small intestine, it is converted into active chymotrypsin.

Organic activators, such as enterokinase, are frequently called *kinases.*

Coenzymes

After a proenzyme has been converted to the active enzyme, the presence of the activating substance is no longer required. Some enzymes require the presence of some other substance to exhibit their activity. Such enzymes are now being called *apoenzymes,* and if organic, the activating substance is called a *coenzyme.* If this substance is removed, the apoenzyme is inactive. Coenzymes do not have a protein nature, and therefore, unlike the enzymes, they are not inactivated by heat.

Phosphopyridine nucleotides, flavin nucleotides, the cytochromes, and cytochrome oxidase are coenzymes that activate the enzymes that catalyze oxidation-reduction reactions in the living tissues (see pages 374–375). Cocarboxylase contains thiamin or vitamin B_1 (page 495) and is a coenzyme required in the oxidation of carbohydrates. Coenzyme A is a derivative of pantothenic acid (page 499) and is required in the transfer of acetic acid and acetyl, CH_3CO—, groups from one molecule to another.

Enzymes that catalyze oxidation-reduction reactions

The many enzymes that catalyze oxidation-reduction reactions are frequently designated as *oxido-reductases*. Oxidation (a more general definition has been given in Chapter 12 for oxidation and reduction) is the gain of oxygen or loss of hydrogen from a substance, whereas reduction involves the addition of hydrogen or the removal of oxygen from a substance. Oxidation of one substance is always accompanied by a reduction of some other substance.

Those enzymes that catalyze oxidation-reduction reactions are quite important, because they are responsible for the production of heat and muscular energy in the body.

Much progress has been made concerning our knowledge of this type of enzyme, although it is far from complete. Systems involving these enzymes are usually much more complex than are those of the hydrolases, and coenzymes are frequently required.

Four important types of oxido-reductases are dehydrogenases, oxidases, peroxidases, and catalases.

Dehydrogenases catalyze the removal of hydrogen of a substrate to an easily reducible substance, which must also be present for the reaction to proceed. Phosphopyridine nucleotides, flavin nucleotides, and the cytochromes are important coenzymes for the dehydrogenases. The phosphopyridine nucleotide, nicotinamide adenine dinucleotide, NAD, is a combination of the purine adenine, the vitamin nicotinic acid amide, the pentose ribose, and phosphoric acid and may be represented by the following simplified formula:

```
[adenine]—[ribose]—[phosphoric acid]—[phosphoric acid]
                                            |
                 [nicotinic acid amide]—[ribose]
```

Nicotinamide adenine dinucleotide phosphate, NADP, contains a third molecule of phosphoric acid. These two coenzymes were formerly known as diphosphopyridine nucleotide, DPN, and triphosphopyridine nucleotide, TPN, respectively.

In the presence of a suitable dehydrogenase, the phosphopyridine nucleotides accept hydrogen from a substrate and may transfer it to a flavin nucleotide. The latter are coenzymes composed of the vitamin riboflavin and for this reason are usually yellow in color. Two important flavin coenzymes are flavin mononucleotide (FMN) and flavin adenine dinucleotide (FAD). These coenzymes appear to accept hydrogen atoms from the phosphopyridine nucleotides and transfer them to the cytochromes.

The cytochromes resemble hemoglobin closely because they contain heme, the prosthetic group of hemoglobin, or a hemelike molecule, combined with characteristic proteins. The main function of the cytochromes appears to be the transfer of hydrogen atoms from the flavin nucleotides to cytochrome oxidase.

Oxidases activate molecular oxygen so that it will combine easily with a substrate. Thus cytochrome oxidase, which resembles the cytochromes in composition, activates oxygen so that it may combine with the hydrogen atoms transferred from the cytochromes, with the production of water. Catalase catalyzes the decomposition of hydrogen peroxide to water and oxygen.

$$2H_2O_2 \xrightarrow{\text{catalase}} 2H_2O + O_2$$

Peroxidases also catalyze the decomposition of hydrogen peroxide, as well as organic peroxides. However, in the case of this latter type of enzyme, some easily oxidizable substance must be present to take up the oxygen formed if the reaction is to proceed. Catalase and the peroxidases appear to be conjugated proteins and generally contain a hemelike molecule closely associated with a simple protein molecule.

The aromatic diamine benzidine

$$H_2N\text{—}C_6H_4\text{—}C_6H_4\text{—}NH_2$$

is converted to an intense blue-green colored dye upon oxidation. Hydrogen peroxide is not a strong enough oxidizing agent to convert benzidine to the colored compound, but in the presence of a peroxidase closely associated with the hemoglobin of the blood, the oxygen of the hydrogen peroxide is easily transferred to and oxidizes the benzidine. This is a very sensitive qualitative test, because only a very small quantity of blood is required to give positive results.

Specificity of enzymes

Enzymes are specific in their action; that is, they act on a single substance or on a group of closely related substances. Urea is the only substance hydrolyzed by urease (absolute specificity). Lipases hydrolyze lipids but not proteins. The proteinase pepsin preferentially attacks a protein molecule at the peptide bond that joins two aromatic amino acid units, such as phenylalanine or tyrosine (page 320). Another proteinase, trypsin, attacks the peptide bond of the protein adjacent to the free amino group of lysine or the guanido group of arginine (page 322). Chymotrypsin preferentially attacks a peptide bond adjacent to an aromatic amino acid in the protein. Aminopeptidases attack and remove the amino acid with the free amino group, while carboxypeptidases correspondingly attack and remove the amino acid with a free carboxyl group of the polypeptide attacked.

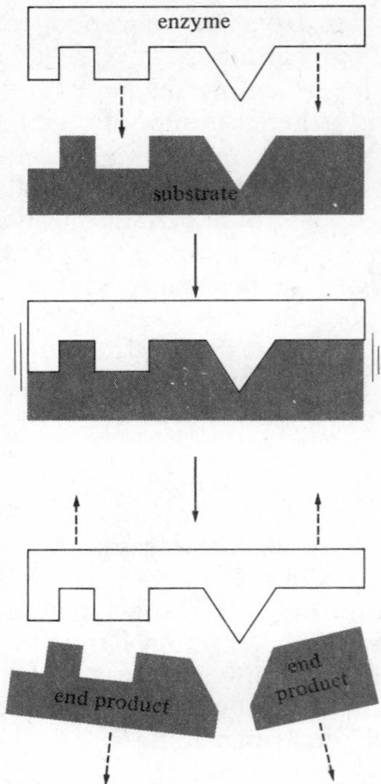

Figure 25.2 Lock-and-key model of an enzyme-catalyzed reaction.

The specificity of enzymes must depend on the following two factors:

1. An enzyme E combines with a substrate S to form a transitory enzyme–substrate complex ES. The latter complex rapidly decomposes to end product P with the regeneration of the enzyme:

   $$E + S \rightleftharpoons ES \rightarrow E + P$$

2. Enzymes contain active sites. The substrate must possess a structure in which certain characteristic groups must be able sterically to bond to the active sites on the enzyme to form the intermediate enzyme–substrate complex. This corresponds to the original lock and key theory of enzyme action that was proposed by E. Fischer (Figure 25.2). Apparently, when the active sites of an enzyme are bonded to certain groups of atoms in the substrate, the bonding of these groups in the substrate are weakened or modified so that the substrate can be readily converted to its characteristic end products.

Enzymes will act only between certain temperatures. Below these temperatures, the rate of the reaction will be so slow that it may hardly be noted. At higher temperatures, the enzymes are inactivated

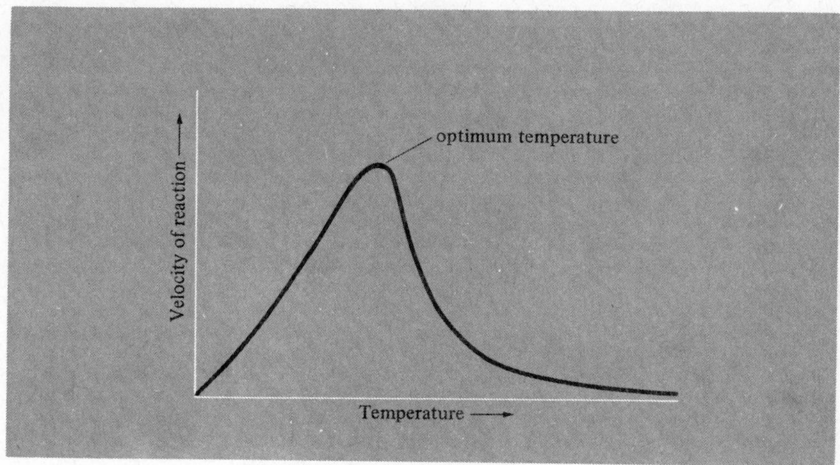

Figure 25.3 Effect of an increase in temperature on the rate of an enzyme-catalyzed reaction.

because the protein portion of the enzyme is coagulated or denatured. There is one temperature at which the enzyme will show its greatest activity, and this temperature will be different for different enzymes. This temperature is called the *optimum temperature* for the enzyme. The optimum temperature for most enzymes of the body will be between 37° and 45°C (see Figure 25.3).

There is also a pH at which an enzyme will exhibit its greatest reactivity. This is called the *optimum* pH for the enzyme. The action of the enzyme is usually limited to a certain pH range. At acidities or basicities outside this range, the enzyme shows no activity. Pepsin, an enzyme of the stomach, shows no activity at a pH above 4 or below 0.1; its optimum pH is about 2. Trypsin, an enzyme of the pancreatic juice, has an optimum pH of about 8 (see Figure 25.4).

Within reasonable limits, the rate of an enzyme-catalyzed reac-

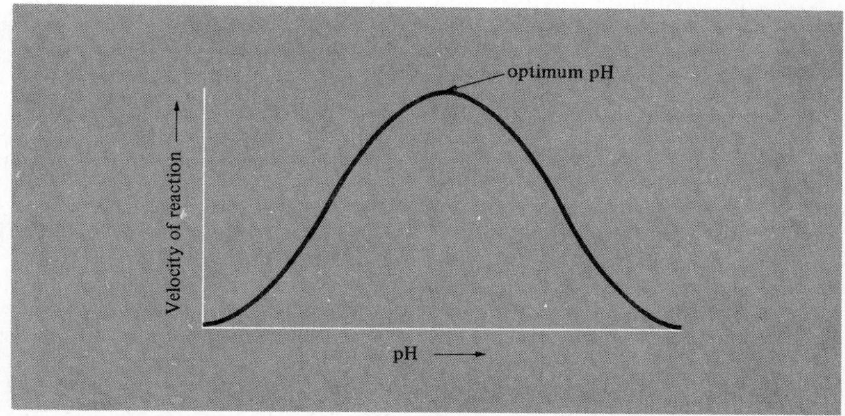

Figure 25.4 Effect of pH on the rate of an enzyme-catalyzed reaction.

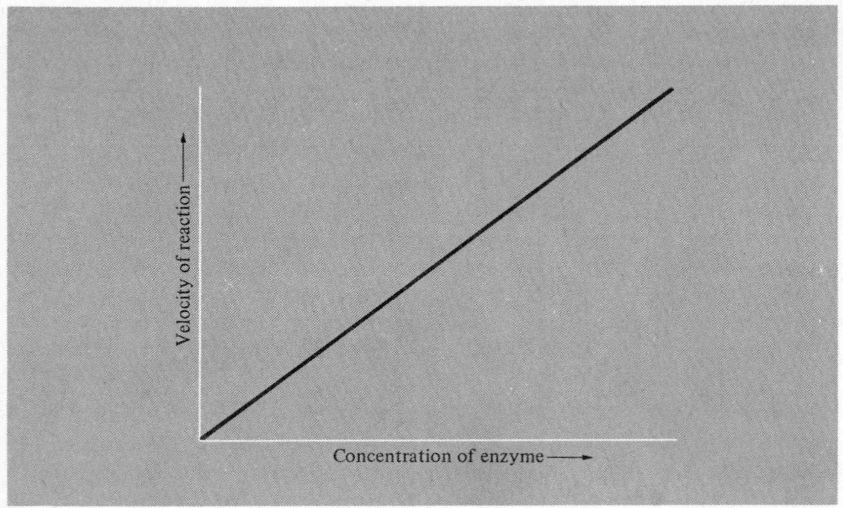

Figure 25.5 Rate of an enzyme-catalyzed reaction vs. the concentration of the enzyme.

tion will be proportional to the concentration of the enzyme, as illustrated in Figure 25.5. At low concentrations of substrate, the rate of an enzyme-catalyzed reaction is also proportional to the concentration of the substrate. As the concentration of the substrate is increased, the increase in rate diminishes and a maximum rate of reaction, indicated at *A* in Figure 25.6, is attained. The initial attainment of this maximum rate must correspond to a concentration of substrate in which the substrate has combined with all of the active sites on the enzyme molecules present.

Enzyme inhibitors

Certain chemicals will inhibit the action of an enzyme. The salts of the heavy metals, such as silver, mercury, and gold, inhibit enzymes by precipitating the protein of the enzyme. Hydrocyanic acid, HCN, and its salts and carbon monoxide, CO, will inhibit many enzymes. Some enzymes are inhibited by sodium fluoride, NaF. Some antiseptics, but not all, will also inhibit enzyme action. Toluene is frequently used to preserve enzyme systems because it acts as an antiseptic and prevents the growth of bacteria but does not usually damage enzymes.

Competitive or metabolic inhibitors appear to combine with an enzyme and prevent some cell reactions by preventing (competitively) the formation of some essential substrate–enzyme complex. These former may also be called *antimetabolites.* Thus many bacteria require the presence of *p*-aminobenzoic acid (PABA) for growth, and the therapeutic action of sulfonamides (compounds with a somewhat similar structure to PABA) appears to be due to a very favorable competition for an enzyme essential to a growth-promoting reaction.

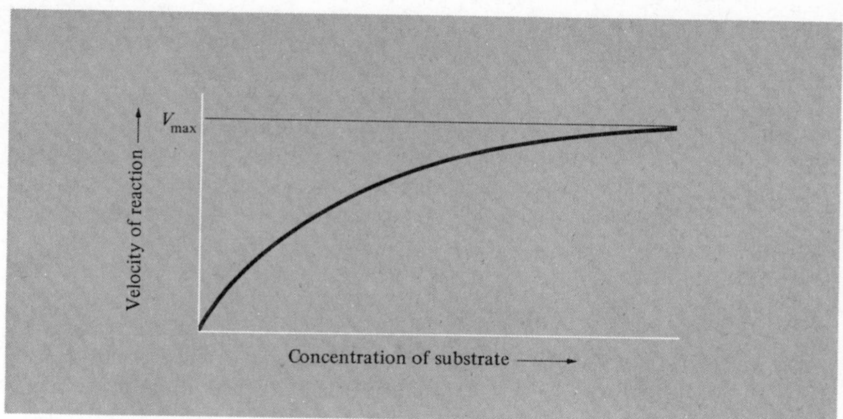

Figure 25.6 Rate of an enzyme-catalyzed reaction vs. concentration of substrate. Note that the reaction reaches a maximum rate at V_{max}.

Antibiotics

Antibiotics are chemical substances produced by molds, bacteria, yeast, or actinomycetes from certain substances present in the media in which they are growing. The antibiotics have the power of inhibiting the growth of other microorganisms. Whereas many of these substances are bacteriostatic (prevent the growth and multiplication of bacteria), some are bactericidal; that is, they destroy bacteria.

It is conjectured that the antibiotics act as antimetabolites by combining with and inactivating some substance or substances that are essential for the proper functioning of certain enzyme systems in the organism, thus preventing the normal activity, growth, and reproduction of the affected organism.

PENICILLIN

This was one of the first antibiotics to be described and is still an important one in use today. However, there are a number of pathogenic bacteria that are unaffected by penicillin, and some bacteria appear to develop an immunity or "fastness" to it.

In 1929 at the University of London, Fleming was preparing some agar cultures of *Staphylococcus aureus*, an organism that frequently appears in human boils. These cultures became contaminated with other bacteria and molds. Whenever one of the less common molds, *Penicillium notatum*, appeared in a culture, the staphylococcus organism did not grow (Figure 25.7). Fleming grew pure cultures of this mold and observed that some substance present in the culture media would inhibit the growth of certain microorganisms but was nontoxic to animals. Because it was produced by a *Penicillium* mold, he called it penicillin. Little note was made of Fleming's discovery for another ten years. In 1940, Florey and a group of his associates at the University of Oxford reinvestigated and isolated penicillin in a crude form. A number of groups in Britain and the United States

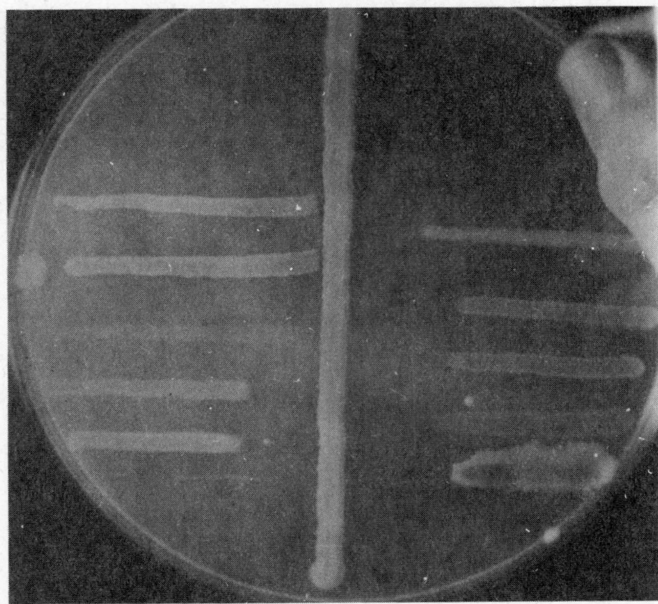

Figure 25.7 *Antagonistic effect of actinomycetes on bacteria. An actinomycete was grown on an agar plate containing nutrient material. After growth had been established, different bacteria were streaked at a right angle to, and right up to, the actinomycete streak. The picture, taken after 24 hours additional incubation to allow the bacteria to grow, shows antibiotic production by the actinomycete. Some bacteria are resistant to the antibiotic: they grew all the way up to the actinomycete; others show different degrees of inhibition. (Courtesy Bristol Laboratories, Syracuse, N.Y.)*

took up this investigation, and soon the commercial production of this antibiotic in a highly active form was a reality (Figure 25.8).

There are, in fact, several penicillins. The penicillin that the *Penicillium* synthesizes depends on the presence of certain organic compounds in the culture media. The general formula now assigned to the penicillins is

```
(CH3)2—CH———CH—CO2H
       |      |
       S      N
        \   /   \
          C       C=O     O
        /   \   /        //
      H       C———NH—C—R
            /
          H
```

general formula for the penicillins

In pencillin G, R is $—CH_2C_6H_5$, whereas in penicillin F it is $—CH_2CH{=}CH—CH_2—CH_3$.

The penicillins are effective against Gram-positive organisms

Figure 25.8 Large fermentation tanks used in the industrial production of antibiotics. (Courtesy Bristol Laboratories, Syracuse, N.Y.)

(organisms that are stained by Gram's dye) and a few Gram-negative organisms. They have proved effective in the treatment of gonorrhea, pneumonia, peritonitis, syphilis, meningitis, diphtheria, gas gangrene, many pus infections, and certain types of boils. The penicillins are not effective in the treatment of malaria, tuberculosis, cholera, dysentery, plague, infantile paralysis, influenza, and measles.

STREPTOMYCIN

This antibiotic has been isolated by Waksman at Rutgers from *Streptomyces griseus.* A provisional structure suggested by Folkers is

```
                    H                    H
     NH             C─────────O──────────C                     ┌──C─H
     ‖            /   \                / |            O        |  H |
NH2─C─NH─CH        CHOH               /  |                     | CH3NCH
           |        |               / H─C──────────/          |    |
         HOHC      CHOH            |  H  |                    |  HCOH
            \      /               |   \ |                    |    |
              CH                   |    C─C─OH                |  HOCH
              |                    |  //  |                   O    |
              NH─C─NH2             |  O   |                   └───CH
                 ‖                 O──────CH                       |
                 NH                       |                      CH2OH
                                          CH3
```

proposed structure for streptomycin

Streptomycin is effective in the treatment of dysentery, undulant fever, some urinary tract infections, and typhoid fever. It is effective against the microrganisms of human tuberculosis *in vitro* and has given remarkable protection to a group of test guinea pigs inoculated with an active culture of the tuberculosis pathogens. Some optimistic reports have been given concerning the treatment of human tuberculosis by streptomycin, but it is of little value in the treatment of enteritis and influenza. Because it is resistant to the action of gastric juice, it can be used in the treatment of some gastric infections.

Dihydrostreptomycin, a hydrogenation product of streptomycin, is replacing it as an effective therapeutic agent because it has a lower toxicity.

CHLOROAMPHENICOL, OR CHLOROMYCETIN

This antibiotic, which is isolated from the cultures in which *Streptomyces venezulae* are grown, is quite effective against many rickettsial and viral infections.

Rebstock and associates at Parke, Davis and Co. have synthesized this antibiotic in the laboratory and have assigned it the following structure:

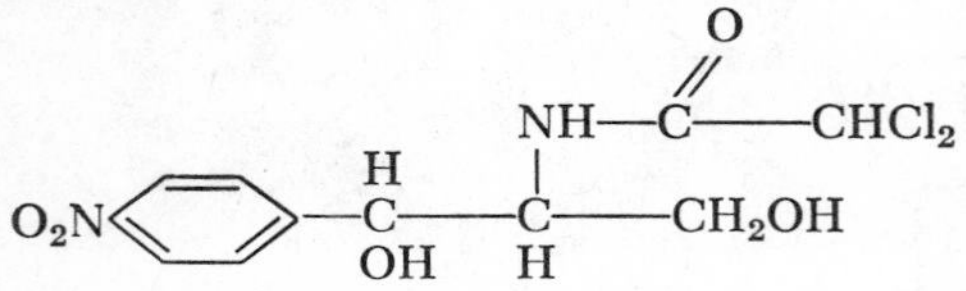

```
                                O
                               //
                      NH—C——CHCl2
              H        |
O2N<benzene>—C————C————CH2OH
             OH       H
```

chloroamphenicol, or chloromycetin

It exhibits a very low toxicity to experimental animals and is effective when given by mouth. It is rapidly absorbed into the body and is excreted by the kidneys, a small portion unchanged but the major portion as an inactive nitro compound. It appears to be effective against influenza, whooping cough, typhoid fever, and typhus.

THE TETRACYCLENES

These are a group of generally useful antibiotics whose structures have been established; they characteristically contain four fused six-membered rings. The best known of these are tetracycline; chlorotetracycline, or aureomycin, obtained from *Streptomyces aureofaciens*; and oxytetracycline, or terramycin, obtained from S. *rimosus*. The tetracyclenes are usually active against both Gram-positive and Gram-negative organisms. They are useful in the treatment of human tularemia, psittacosis, viral pneumonia, typhus, urinary infections, malaria, syphilis and gonorrhea, skin lesions, and peritonitis.

Other antibiotics that have been used as therapeutic agents are bacitracin, tyrothricin, and erythromycin.

STUDY QUESTIONS

1. Define and give examples for each of the following: catalyst, enzyme, substrate, end product, coenzyme, activator, kinase, proenzyme, enzyme inhibitor, apoenzyme.

2. Are the following true or false?
   (a) All catalysts speed up chemical reactions.
   (b) The action of enzymes is not reversible.
   (c) Enzymes do not change the proportion of reactants and products at the equilibrium point.
   (d) Enzymes are capable of initiating chemical reactions.
   (e) Catalysts are usually recovered unchanged at the end of a chemical reaction.
   (f) Life is possible in the absence of catalysts.
   (g) Pasteur discovered the fact that all enzymes can act in the absence of living cells and tissues that produce them.
   (h) Proteins appear to be an essential part of enzymes, and enzymes become inactive if the protein material closely associated with them is denatured.
   (i) Enzymes are not deactivated at high temperatures.
   (j) Many enzymes have been obtained in a cystalline condition.
3. What principle is now followed in naming enzymes?
4. What are the two most important general types of reactions occurring in the body which are catalyzed by enzymes?
5. Define hydrolysis, oxidation, and reduction. What general names are given to enzymes that catalyze hydrolytic reactions? Oxidation-reduction reactions? Transfer reactions? Decarboxylations? Hydrations? Isomerizations?
6. Make a chart and indicate (a) whether the enzyme is a carbohydrase, esterase, protease, or amidase; (b) source or occurrence of the enzyme; (c) substrate; (d) end products of the action of the enzyme for the following enzymes: pepsin, steapsin, salivary amylase, phosphatase, urease, amylopsin, maltase, sucrase, diastase, lactase, trypsin, rennin, aminopeptidase, carboxypeptidase.
7. Are enzymes specific in their action? Give examples.
8. What types of enzymes are most commonly found in our digestive system? In muscular tissue?
9. Name the four important classes of oxido-reductases. What reactions do these classes of enzymes catalyze? How do the resulting enzyme systems differ in complexity from the hydrolytic enzymes?
10. What enzyme catalyzes the decomposition of hydrogen peroxide to oxygen? How does this enzyme differ from the peroxidases?
11. On the presence of what type of enzyme does the benzidine test for the detection of blood depend? Describe the test.
12. Why do enzymes become inactive above and below certain ranges of temperature?
13. What types of reactions are catalyzed by the following classes of enzymes: lyases; isomerases; ligases; transferases. Name and describe the action of one enzyme from each of these four classes.
14. What is meant by the optimum temperature of enzyme action? Optimum pH of enzyme action? What is the optimum pH for the action of pepsin? The optimum temperature for the action of most of the enzymes of the human body falls in what temperature range?
15. From the following numbers, pick the optimum pH for the enzymes listed:

    2 4 6.8 8.5 12 14

    pepsin ______ salivary amylase ______ trypsin ______

16. By a suitable word or phrase, identify the following:

    (a) an enzyme that converts polypeptides to amino acids
    (b) the substance in the small intestine that activates trypsinogen
    (c) an enzyme that catalyzes the transfer of hydrogen from a substrate to an easily reducible substance

17. What are antibiotics? What types of organisms produce them?
18. What roles did Florey and Fleming play in the development of penicillin?
19. How do the various penicillins vary in their composition?
20. What diseases have been successfully combated by penicillin?
21. Streptomycin has shown promise in the treatment of what important disease that is resistant to the action of penicillin? What other diseases have responded favorably to its action?
22. What organism produces chloromycetin? Discuss the chemical structure of this antibiotic. What atoms or group of atoms present in this antibiotic are unusual for natural products? What diseases have responded to treatment by chloromycetin?
23. Discuss the clinical uses of chloro- and oxytetracycline.

chapter 26

the chemistry of digestion

26

The fats, carbohydrates, and proteins are too complex to be absorbed through the intestinal wall and must be converted to simpler substances in the digestive tract. During digestion, fats are converted to glycerol and fatty acids, carbohydrates to monosaccharides, and proteins to simple amino acids. Digestion consists of all those reactions in which the more complex foods are hydrolyzed into simpler substances so that they may be readily absorbed into the bloodstream.

Digestion occurs in the mouth, stomach, and small intestine (Figure 26.1). The enzymes that are essential for digestion are excreted from glands or cells in the walls of the digestive tract or from the pancreas (Table 26.1).

Digestion in the mouth

Saliva is produced by certain glands located in the inner surface of the mouth and under the tongue. It is a colorless, viscous, and somewhat slippery fluid, about 99.4 percent water, in which the following are also present: the enzyme salivary amylase (ptyalin), mucin (a glycoprotein), small amounts of other proteins, urea and uric acid, small traces of such enzymes as maltase, lipase, and catalase, and some protein-cleaving enzymes.

Salivary amylase (ptyalin) is a starch-splitting enzyme and is the most important digestive enzyme in the mouth. It acts by splitting molecules of maltose from the starch molecule. The starch is first converted to soluble starch, then to a dextrin called *erythrodextrin,* which gives a red color with iodine, then to *achroodextrin,* which gives no color with iodine, and finally to maltose. The extent of the digestion of starch by salivary amylase can be followed by the addition of iodine. At first the iodine produces a blue-black color. As digestion continues, this color changes to red and as the reaction continues, the test solution becomes colorless. The other enzymes present in saliva are of little or no importance. They may assist in removing food particles from the teeth by digesting them.

The *mucin* of saliva is a glycoprotein. It is a ropy liquid and gives saliva its viscous properties. It serves to lubricate food and thus aids in swallowing.

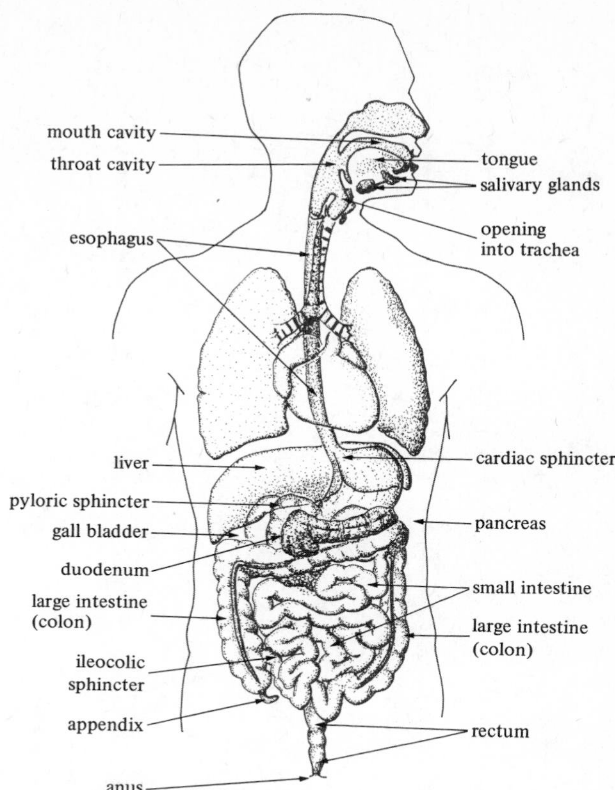

Figure 26.1 *Diagram of the human digestive system. The glands that secrete digestive juices into the alimentary tract are to be especially noted.*

The pH of saliva is between 6.4 and 6.9. The optimum pH for the action of salivary amylase is about 6.8, although it will be active in a pH range from 4 to 9. The saliva serves to moisten and lubricate food and also to start the digestion of starch. The action of salivary amylase may continue in the stomach for some time, but stops when the pH of the stomach contents drop below 4 due to the admixture of the food with the hydrochloric acid of the gastric juice.

Digestion in the stomach

The fluid secreted by the gastric wall is called *gastric juice.* Although gastric juice is secreted into the stomach at all times, its flow is stimulated by the sight or odor of food or by the presence of food in the mouth. The important constituents of gastric juice are the enzymes rennin and gastric lipase, the proenzyme pepsinogen, hydrochloric acid, and intrinsic factor.

The proenzyme *pepsinogen* is converted into active pepsin when

TABLE 26.1 IMPORTANT ENZYMES OF THE DIGESTIVE SYSTEM

Enzyme	*Class*	*Source*	*Substrate*	*End Products*
Salivary amylase	Amylase	Saliva	Starch	Dextrins and maltose
Pepsin	Proteinase	Gastric juice	Native protein	Proteoses, peptones
Rennin	Proteinase	Gastric juice	Casein	Paracasein
Trypsin	Proteinase	Pancreatic juice	Proteins, proteoses, peptones	Polypeptides, amino acids
Chymotrypsin	Proteinase	Pancreatic juice	Proteins, proteoses, peptones	Polypeptides, amino acids
Steapsin (pancreatic lipase)	Lipase	Pancreatic juice	Fats	Glycerol, fatty acids, mono- and diglycerides
Amylopsin (pancreatic amylase)	Amylase	Pancreatic juice	Starch	Maltose

Lactase	Carbohydrase	Pancreatic juice	Lactose	Glucose, galactose
Maltase	Carbohydrase	Pancreatic juice	Maltose	Glucose
Sucrase	Carbohydrase	Pancreatic juice	Sucrose	Glucose, fructose
Carboxypeptidase	Peptidase	Pancreatic juice	Polypeptides	Dipeptides and amino acids
Aminopeptidase	Peptidase	Intestinal juice	Polypeptides	Dipeptides and amino acids
Dipeptidase	Peptidase	Intestinal juice	Dipeptides	Amino acids
Lactase	Carbohydrase	Intestinal juice	Lactose	Glucose, galactose
Maltase	Carbohydrase	Intestinal juice	Maltose	Glucose
Sucrase	Carbohydrase	Intestinal juice	Sucrose	Glucose, fructose
Intestinal lipase	Lipase	Intestinal juice	Fats	Glycerol, fatty acids, mono- and diglycerides

it comes in contact with the hydrochloric acid of the stomach. Pepsin is a proteolytic enzyme (proteinase) and converts proteins to proteoses and peptones. Its optimum pH is about 2, and it becomes inactive at a pH of 4 or greater. Pepsin can precipitate casein, but it is not nearly as efficient as rennin.

Rennin is particularly abundant in the stomach of the young mammal. Commercial rennin is obtained from the stomach of the calf. There is some evidence that rennin is not present in the stomach of the human adult. Rennin converts the casein of milk to paracasein, which is precipitated by the calcium ions in milk. If the calcium ions of the milk are removed before the addition of rennin, no precipitation of casein will occur. The coagulation of the casein of milk serves two purposes. First, the curdled milk will remain in the stomach longer because it is a solid; thus protein digestion will be more complete. Second, the coagulated proteins are easier to digest than native proteins.

There is still considerable doubt concerning the origin of *gastric lipase*. Some consider that it is produced in the walls of the stomach, whereas other investigators claim that it is regurgitated from the small intestine. Because the contents of the stomach have a pH of 1 to 2 and the fats present are not emulsified, little or no fat digestion occurs in the stomach. Significant fat hydrolysis occurs in a basic media and when the fat is emulsified.

Gastric juice contains appreciable amounts of hydrochloric acid and has a pH of 1.6 to 1.8. However, when it becomes mixed with foods in the stomach, the pH is raised to about 4 due to the buffer action of the food. Hydrochloric acid serves several useful functions in the stomach: It is an antiseptic and prevents bacterial fermentation and the growth of bacteria; it converts inactive pepsinogen into active pepsin; and it denatures and coagulates proteins so that they may be digested readily.

A substance of unknown structure called the *intrinsic factor* is secreted by certain cells of the stomach wall. Vitamin B_{12} present in the food must react with this factor before it can be absorbed by the intestinal mucosa. Vitamin B_{12} is required in the formation of red blood cells; pernicious anemia (page 472) results from its nonabsorption.

In *hypochlorhydria*, the amount of hydrochloric acid in the stomach is abnormally low; in *achlorhydria*, there is a complete absence of it. Hypochlorhydria and achlorhydria are common in patients suffering from pernicious anemia and stomach cancer. In the latter condition, large amounts of lactic acid are present in the stomach due to bacterial fermentation, and the stomach often empties slowly due to the obstruction of the passage of the stomach contents into the small intestine.

Hyperchlorhydria is a term used to designate an abnormally large concentration of hydrochloric acid in the stomach. This occurs in individuals suffering from hypertension, gastric or duodenal ulcers, and gastritis, or inflammation of the stomach walls.

Digestion in the small intestine

After food leaves the stomach, it enters the small intestine, which is divided into three sections: the *duodenum,* the most important from the standpoint of digestion, the *jejenum,* and the *ileum,* The food material entering the small intestine is called *chyme* and is acid in reaction, but it soon becomes alkaline, because the different juices present in the intestine with which it is mixed are alkaline.

There are three different juices that enter the duodenum, all of which are alkaline in reaction: pancreatic juice, which is produced by the pancreas, an organ that lies close to the duodenum; intestinal juice, which is produced in the intestinal mucosa; and bile, which enters from the gall bladder. The first two contain sodium bicarbonate, $NaHCO_3$, which is responsible for their alkalinity, and many important enzymes. Bile contains no enzymes but plays an important role in digestion. Incidentally, the pancreas produces another secretion, called *insulin,* which is directly absorbed into the bloodstream and is necessary for the proper utilization of carbohydrates in the body. This secretion belongs to a group of substances called *hormones* (see page 513).

Pancreatic juice contains two proenzymes, *trypsinogen* and *chymotrypsinogen.* A kinase (page 373) called enterokinase, present in the intestinal juice, converts typsinogen into active *trypsin.* Chymotrypsinogen is activated by trypsin to form *chymotrypsin.* These two enzymes, trypsin and chymotrypsin, convert proteins, proteoses, and peptones to polypeptides and some amino acids. A pancreatic lipase called *steapsin* is the most important fat-splitting enzyme in the body. It converts emulsified fats into fatty acids and glycerol; however, the triglycerides may be absorbed, in part, as the partially hydrolyzed mono- and diglycerides. The emulsification of fats is aided by the presence of bile and by the alkalinity of the small intestine. A pancreatic amylase called *amylopsin,* the most important amylase of the body, converts starch to maltose. Small amounts of maltase, sucrase, and lactase are present in the pancreatic juice, and these convert maltose, sucrose, and lactose to monosaccharides. The *carboxypeptidase* of the pancreatic juice degrades polypeptides to dipeptides and amino acids by splitting off the amino acid residue containing a free carboxyl group at the end of the polypeptide chain.

The intestinal juice contains a maltase, a sucrase, a lactase, and possibly a lipase. An *aminopeptidase* and a *dipeptidase* present in this juice complete the hydrolysis of the polypeptides and dipeptides to amino acids. Enzymes that catalyze the hydrolysis of the nucleoproteins are also present in the intestinal juice. Recall that enterokinase, a kinase that activates trypsinogen, is also present.

Bile

Bile is continuously manufactured in the liver and is stored in the gall bladder. In the gall bladder, some substances that are of value to the body are reabsorbed and there is absorption of water, which results in a concentration of the bile. When foods, particularly fats, enter the small intestine, the wall of the gall bladder is induced to contract and the bile is expelled into the duodenum by way of the bile duct.

Bile is a yellowish-brown to green viscous liquid, alkaline in reaction, and quite bitter in taste. It may be considered as both a secretion and an excretion. The three most important constituents of bile are the bile salts, the bile pigments, and cholesterol; it contains no digestive enzymes.

The most important bile salts are the sodium salts of glycocholic and taurocholic acid. Glycocholic acid is formed by the union of glycine, while taurocholic acid is the product of the union of taurine (β-amino-ethanesulfonic acid) with cholic acid and related bile acids.

$H_2N—CH_2—COOH$

glycine

$CH_2—SO_3H$
$|$
$CH_2—NH_2$

taurine

CH_3, OH, $CH—CH_2—CH_2—COOH$, CH_3, CH_3, —OH, HO, H

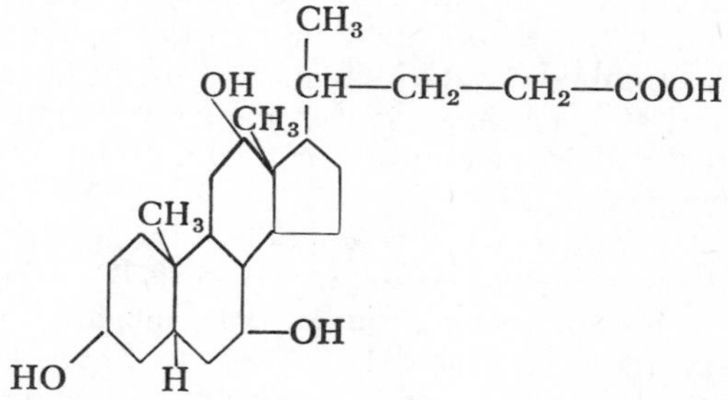

cholic acid

Cholic acids resemble the sterols (page 287), such as cholesterol, in structure. The bile salts have the power of lowering surface tension; they assist in the digestion of fats by readily emulsifying them so that each resulting small fat particle is readily attacked by the steapsin of the pancreatic juice. In obstructive jaundice, in which bile does not enter the small intestine, not only fats but also fat-coated carbohydrates and proteins are not digested. The bile salts also appear to assist in the absorption of the fatty acids through the walls of the intestines. They may form a complex with the fatty acids during absorption. After absorption, the bile salts are separated from the fatty acids and are carried by the blood to the liver, which reexcretes them in the bile.

Bilirubin (a red pigment) is the most important of the pigments of human bile and is a reduction product of biliverdin (green bile pigment). In the breakdown of dead red blood cells in the body, the hemoglobin present is separated into the heme and the globin. Because the body is very economical with its supply of iron, the iron is removed from heme, which is then converted to biliverdin. The iron may be reconverted to hemoglobin or may be stored as ferritin, an iron–protein complex. The liver, spleen, and bone marrow serve as storage areas for ferritin. The liver collects the resulting pigments and excretes them in the bile. In the intestines, bilirubin is reduced to *stercobilinogen* (urobilinogen) and finally to *stercobilin* (urobilin), which gives the feces its typical yellow to brown color. Some of the bile pigments, especially urobilin, are absorbed from the intestinal tract and are excreted in the urine as such or as the yellow pigment called *urochrome.*

The sterol cholesterol is excreted from the body by way of the bile and sometimes precipitates in the gall bladder. Cholesterol stones are the most common type of gall stones, although a few are composed mostly of bilirubin, calcium, or other insoluble inorganic salts.

Feces

The content of the large intestine is called the *feces.* As the content of the small intestine is discharged into the large intestine, it is in a semifluid condition and consists of indigestible material (cellulose), undigested food particles, remnants of the digestive juices, and epithelial tissue from the walls of the digestive tract.

In the large intestine, much of the water and some other substances are absorbed from the feces, while little if any digestion occurs. Salts of calcium, iron, and other metals, as well as some fatlike substances, are discharged into the large intestine through its wall. Conditions are ideal for the growth of bacteria, and usually one-fourth to one-half of the stool will consist of bacteria.

In *obstructive jaundice,* the feces may be nearly colorless due to the absence of the reduced bile pigments. The resulting stools are called *clay-colored,* or *acholic.* Blood may appear in the feces in several forms. Fresh blood may indicate rectal cancer, piles, or inflammation of the colon (colitis). If the blood enters the alimentary tract in the mouth, stomach, or small intestine, it may give the feces a tarry black appearance due to the oxidation of the hemoglobin to methemoglobin. Such blood is called *occult blood,* and its appearance may be due to a gastric or duodenal ulcer or to stomach cancer.

Absorption of fats, carbohydrates, and proteins

In the digestive process, all the carbohydrates are converted to glucose, galactose, and fructose. These monosaccharides are the only sugars that are readily absorbed through the intestinal mucosa. After entering the blood, these sugars are carried to the liver, where they are converted into glycogen and stored as such, or are passed to the blood as glucose. Some glycogen is also stored in the muscles. Before utilization in the body, glycogen is converted to glucose by the liver.

In digestion, the proteins are hydrolyzed to amino acids which are absorbed through the intestinal mucosa and into the bloodstream. The amino acids are carried by the blood to the tissues, which select those required for their building or repair. Because the body does not store proteins, any amino acids not utilized are oxidized at once, thus producing heat and other forms of energy for the body. There is still some uncertainty concerning whether small amounts of polypeptides may also be absorbed into the bloodstream.

During digestion, fats are converted into mono- and diglycerides, glycerol, and fatty acids. After passing through the intestinal wall, the glycerol passes into the bloodstream and follows the same route of metabolism as the carbohydrates. The mono- and diglycerides and the fatty acids, as well as possibly some unhydrolyzed triglycerides, enter into the intestinal *lymphs.* The bile salts appear to be essential for the absorption of fatty acids, possibly forming a complex with them, which is decomposed after absorption. The partially hydrolyzed glycerides and the fatty acids reappear immediately as resynthesized fat in the lymph, which is forced into the larger *lymphatics* and finally enters the bloodstream by way of the thoracic duct. The fat is then carried by the blood to the cells, which utilize a portion for the production of heat and other forms of energy. Any fat not immediately needed for this purpose is stored in the adipose tissue of the body.

STUDY QUESTIONS

1. Why must food be digested before it can be utilized by the body?
2. What is the percentage of water in saliva? What solid solutes are present in saliva?
3. What is the most important enzyme present in saliva? Describe the steps in the breakdown of starch by this enzyme. How can these changes be followed experimentally?
4. What type of substance is mucin? What role does it play in digestion?
5. What is the optimum pH for the action of salivary amylase? How long does this enzyme continue to act in the stomach?
6. What important enzymes are present in gastric juice? What are the substrates and the end products of the action of these enzymes? What important inorganic constituent is present as a solute in gastric juice?
7. What type of substance is pepsinogen? How is it activated?
8. The absence of what ion will prevent rennin from precipitating the casein of milk?
9. What purposes does hydrochloric acid serve in gastric juice?
10. Why is fat digestion of little importance in the stomach?
11. What are hypochlorohydria, achlorohydria, and hyperchlorohydria? What pathological or abnormal conditions usually accompany these conditions?
12. What is the normal pH of the gastric fluid?
13. Name the three portions of the small intestine. From the standpoint of digestion, what is the most important part of the small intestine? What is chyme?
14. What two important secretions are produced by the pancreas? What is the name of the digestive juice that is excreted by the walls of the small intestine?
15. What enzymes and proenzymes are present in pancreatic juice? What substances activate the proenzymes of the pancreatic juice?
16. What enzymes are present in intestinal juice?
17. Summarize the digestion of proteins, fats, and carbohydrates in the small intestine.
18. Does digestion in the small intestine occur best in an acid, neutral, or alkaline medium?

19. What three general types of substances are present in the bile? Where is bile prepared and stored?
20. Why can one call bile both a secretion and an excretion?
21. What is the chemical nature of the bile salts? What unusual properties do they possess, and what role do they play in digestion?
22. What is the origin of bilirubin? What are some of the transformation products of bilirubin? Which of these are present in the intestines and urine?
23. What chemical substances are frequently found in gallstones?
24. Discuss the absorption of the products of digestion of the proteins, carbohydrates, and fats in the intestines.
25. What is the composition of the feces?
26. What condition produces clay-colored stools?
27. Of what significance is the presence of blood in the stool?

chapter 27

---

metabolism of carbohydrates, fats, and proteins and energy metabolism

27

Under the term *metabolism* we will consider all those chemical reactions that occur from the time the digested food products are absorbed from the intestinal tract until their oxidation products are excreted.

In digestion, complex molecules are hydrolyzed to small simple molecules, which after absorption are usually converted to new complex molecules to form new tissue and replace worn-out tissue or to be used as reserve food supplies and in the synthesis of enzymes, pigments, hormones, etc. Such processes are called *anabolic processes* or *anabolism.* In *catabolic processes,* or *catabolism,* the absorbed food products and worn-out tissues are decomposed into simpler substances, such as CO_2, H_2O, and urea, with the simultaneous production of energy, and the resulting products are then excreted.

It should be remembered that metabolism includes energy transformations, as well as the transformations of chemical substances.

Normal carbohydrate metabolism

During digestion, the carbohydrates are converted to glucose, fructose, and galactose. Phosphoric acids or phosphates appear to play an important role in the absorption of these monosaccharides into the bloodstream, as well as in their conversion to glycogen in the liver. These simple sugars are carried to the liver by way of the portal vein. In the liver, they are converted by a process called *glycogenesis* into glycogen, a polysaccharide that serves as a reserve carbohydrate in the body and is stored in the liver and muscles. Upon hydrolysis, glycogen appears to be exclusively hydrolyzed to glucose, a process called *glycogenolysis.* This process occurs exclusively in the liver.

In glycogenesis, glucose is converted to glycogen in the following steps:

1. Glucose is first transformed to glucose-6-phosphate by adenosine triphosphate (ATP) and the enzyme *hexokinase.* By the transfer of a phosphate group to glucose, a process called *phosphorylation,* ATP is converted into adenosine diphosphate (ADP) (see page 401).

```
  H  OH                                       H  OH
  | /                                         | /
  C------------|                              C------------|
  |            |                              |            |
H—C—OH         |                            H—C—OH         |
  |            |          hexokinase          |            |
HO—C—H         |  + ATP ------------->     HO—C—H          |   + ADP
  |            |            Mg2+              |            |
H—C—OH         O                            H—C—OH         O
  |            |                              |            |
H—C------------|                            H—C------------|
  |                                           |
 CH2OH                                       CH2OPO3H2
glucose                                   glucose-6-phosphate
```

2. The enzyme *phosphoglucomutase* isomerizes glucose-6-phosphate to glucose-1-phosphate.

```
  H  OH                                       H  OPO3H2
  | /                                         | /
  C------------|                              C------------|
  |            |                              |            |
H—C—OH         |                            H—C—OH         |
  |            |                              |            |
HO—C—H         |                           HO—C—H          |
  |            O    ------------------>       |            O
H—C—OH         |    <------------------     H—C—OH         |
  |            |    phosphoglucomutase        |            |
H—C------------|                            H—C------------|
  |                                           |
 CH2OPO3H2                                   CH2OH
glucose-6-phosphate                       glucose-1-phosphate
```

3. The synthesis of glycogen from glucose-1-phosphate involves the conversion of the latter to uridine diphosphoglucose, UDPG, by a nucleoside triphosphate, uridine triphosphate, UTP (see page 442):

$$\text{glucose-1-phosphate} + \text{UTP} \longrightarrow \text{UDPG} + \underset{\text{pyrophosphoric acid}}{H_4P_2O_7}$$

The presence of a small starchlike chain, which we will designate as $(\text{glucose})_n$, appears to be required for the synthesis of glycogen. UDPG molecules transfer their glucose residue to the nonreducing end of this chain, thus:

$$\text{UDPG} + (\text{glucose})_n \longrightarrow \text{UDP} + \text{glycogen}$$

In glycogenolysis, glycogen is converted to glucose in the following steps:

1. Glycogen is converted to glucose-1-phosphate by the action of the enzyme glycogen phosphorylase:

$$\underset{\text{glycogen}}{(C_6H_{10}O_5)_x} + x\ H_3PO_4 \xrightarrow[\text{phosphorylase}]{\text{glycogen}} x \text{ glucose-1-phosphate}$$

2. Glucose-1-phosphate is isomerized to glucose-6-phosphate by the action of phosphoglucomutase.
3. The enzyme glucose-6-phosphatase, which occurs in the liver but not in striated muscles, irreversibly hydrolyzes glucose-6-phosphate to glucose.

In a normal individual, the fasting blood-sugar level is nearly constant at some value usually between 70 and 100 mg of glucose for each 100 ml of blood. After a carbohydrate-rich meal, the monosaccharides are absorbed into the bloodstream faster than they can be converted to glycogen by the liver. At this time, the blood-sugar level will rise appreciably above the normal level. If the rise becomes very large, at some point glucose will begin to appear in the urine; the amount of glucose in the blood at this point is called the *renal threshold for glucose.* The renal threshold occurs at about 150 to 170 mg of glucose per 100 ml of blood. In most individuals, there will then be an increased conversion of sugar to glycogen in the liver, or glucose will be oxidized to carbon dioxide and water so that the concentration of glucose in the blood will decrease to near normal values in 1 to 2 hours. In diabetics, a higher blood-sugar level will be evident and it will drop slowly, requiring a number of hours to reach its original value (Figure 27.1). This is the basis of the *sugar tolerance test,* which is a more dependable test for diabetes mellitus than the determination of the presence of sugar in the urine. In this test, a large quantity of glucose is given by mouth, and, at regular intervals thereafter, samples of blood are taken and analyzed for glucose. Although the glucose content of the blood rises rapidly in normal and diabetic patients, the blood-sugar level of the former returns to its original value much more rapidly than that of the latter.

The muscles do not serve as important storage depots for carbohydrates, but small amounts of glycogen are present in them. The muscles have the power to convert glucose of the blood to glycogen, which serves as a source of energy for muscular contraction. During fasting, the blood-sugar concentration will tend to decrease due to this loss of glucose to the muscles. During this time of fasting, the liver will be converting glycogen to glucose in order to maintain a nearly constant blood-sugar level.

ENERGY-RICH PHOSPHATE BONDS

The formation of new bonds in the conversion of compound A to compound B may release chemical energy as heat, light, etc.; or a certain amount of energy may be absorbed and stored as chemical energy. The formation of alkyl phosphate esters by the interaction of alcohols with phosphoric acid or phosphates requires the absorption

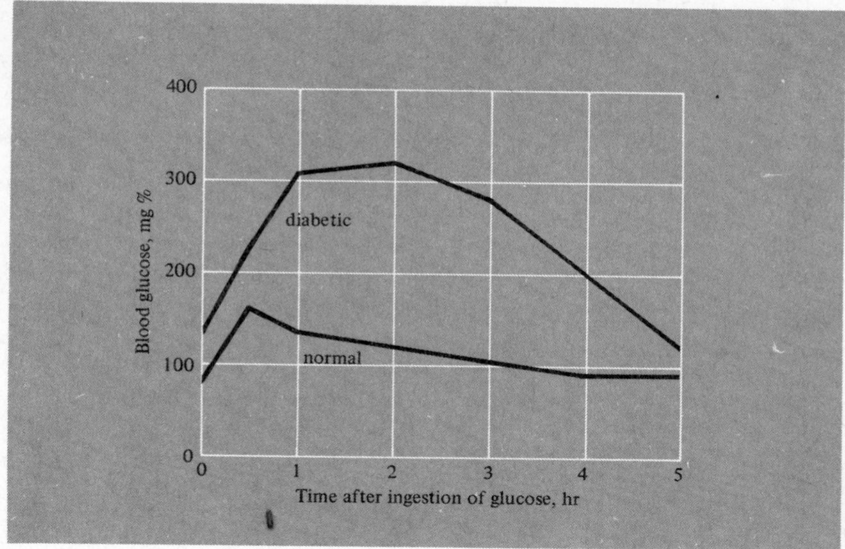

***Figure 27.1** Sugar tolerance curves of normal and diabetic human patients. The data refer to the arterial blood-sugar levels after the ingestion of 100 g of glucose. Milligram percent of glucose is the number of milligrams of glucose in each 100 ml of blood. (From Fruton and Simmonds,* General Biochemistry, *2nd ed., Wiley, N.Y., 1958.)*

of a small amount of energy to form the new phosphate bond; such a bond is called an *energy-poor bond.*

In the conversion of adenosine monophosphate or adenylic acid (AA) to adenosine diphosphate (ADP) and of adenosine disphosphate to adenosine triphosphate (ATP), about five times the energy needed to form a simple phosphate bond is required to form the new phosphate bonds; such bonds are frequently represented by a wavy line (~) and are called *energy-rich phosphate bonds.*

R—O—P(=O)(—OH)—OH

an alkyl phosphate

adenine-ribose—P(=O)(—OH)—OH

adenylic acid (AA)

adenine-ribose—P(=O)(—OH)—O ~ P(=O)(—OH)—OH

adenosine diphosphate (ADP)

adenine-ribose—P(=O)(—OH)—O ~ P(=O)(—OH)—O ~ P(=O)(—OH)—OH

adenosine triphosphate (ATP)

Thus adenosine triphosphate has two energy-rich phosphate bonds. The structural formula for the purine adenine is given on page 440.

Compounds with energy-rich phosphate bonds play an important role in the transfer of energy in the body. If glucose were oxidized in one step to carbon dioxide, little if any of the large amount of energy produced could be utilized effectively, and most would be lost as heat. The actual oxidation of glycogen or glucose occurs in a number of steps, as will be indicated below, and a number of ADP molecules absorb the energy produced at certain of these steps to form ATP. At some later time and some other location, the latter can be hydrolyzed to ADP, and the energy thus released will be able to produce useful muscular work.

The conversion of glucose to glycogen (glycogenesis) requires a certain input of energy, for glycogen contains more chemical energy than glucose. The hydrolysis of ATP is capable of releasing the energy needed for this transformation.

ANAEROBIC DEGRADATION OF GLUCOSE

The process by which glucose is converted to carbon dioxide and water (with the simultaneous production of muscular energy) is quite complex and involves many steps. For each step there is a different enzyme, which in many cases also requires specific coenzymes. The oxidation of glucose in the muscle may be divided into an anaerobic phase, usually called *glycolysis,* during which time the muscle is contracting, and an aerobic phase, during which the muscle is recovering. Although the reactions typical of the anaerobic phase will require no oxygen for their completion, oxygen may be present.

The main steps in the anaerobic decomposition of glucose are as follows:

1. Glucose, through the action of two phosphorylases, hexokinase and phosphofructokinase, is converted to fructose-1,6-diphosphate.

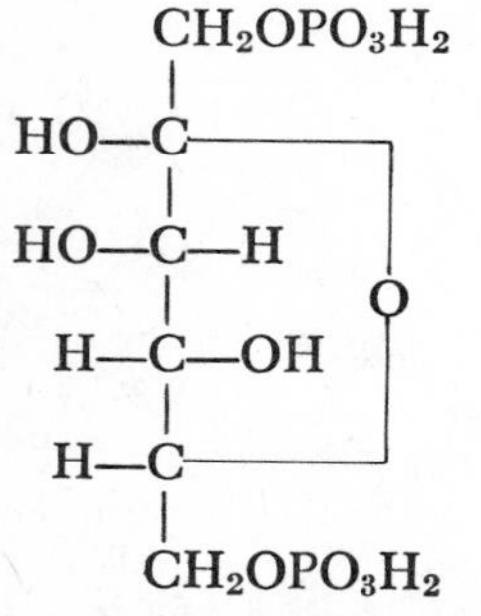

```
      CH2OPO3H2
       |
  HO—C————————┐
       |       |
  HO—C—H       |
       |       O
   H—C—OH      |
       |       |
   H—C————————┘
       |
      CH2OPO3H2
```

fructose-1,6-diphosphate

Two molecules of adenosine triphosphate furnish the two phosphates required and are converted to two molecules of adenosine diphosphate. These changes involve the loss of a certain amount of energy, as two energy-rich phosphate bonds are lost and two energy-poor hexose phosphate ester bonds are formed.

2. The fructose-1,6-diphosphate is then cleaved into two molecules of a triose phosphate.

$$\begin{array}{c} \overset{H}{\diagup} \\ C\!\!\overset{\diagdown\!\diagdown}{}O \\ | \\ H-C-OH \\ | \\ CH_2OPO_3H_2 \end{array}$$

3-phosphoglyceraldehyde, a triose phosphate

3. Through a series of reactions, the triose phosphate loses phosphoric acid and is converted by dehydrogenation into pyruvic acid, CH_3—CO—COOH.
4. The last step of the anaerobic phase is the reduction of pyruvic acid to lactic acid, CH_3—CHOH—COOH.

Generally, a number of reactions constitute each of the four steps in the anaerobic degradation of glycogen. These are summarized in Table 27.1. Enzymes and coenzymes needed for the various reactions, plus energy transfers through the intermediation of ADP and ATP, are also indicated in the table. Note that no overall oxidation-reduction in the anaerobic breakdown of glycogen to lactic acid occurs as the hydrogen atoms produced in the dehydrogenation of 3-phosphoglyceraldehyde are reutilized in the reduction of pyruvic acid to lactic acid in the last step.

The overall reaction involved in this glycolytic sequence, in which glucose is converted to lactic acid, can be represented by the equation

$$C_6H_{12}O_6 + 2ADP + 2H_3PO_4 \longrightarrow 2CH_3CHOHCOOH + 2H_2O + 2ATP$$

As indicated by this equation, the conversion of a molecule of glucose to lactic acid results in the net production of two high-energy phosphate bonds (2ADP $\longrightarrow$ 2ATP).

The citric acid cycle and aerobic degradation of pyruvic acid

Part of the lactic acid formed in the anaerobic phase described above escapes into the bloodstream and increases the concentration of this acid in the blood. Some of this lactic acid is carried to the liver, where it is converted to glycogen. Part of the pyruvic acid

TABLE 27.1 ANAEROBIC DEGRADATION OR GLYCOLYSIS OF GLUCOSE TO LACTIC ACID

Step 1

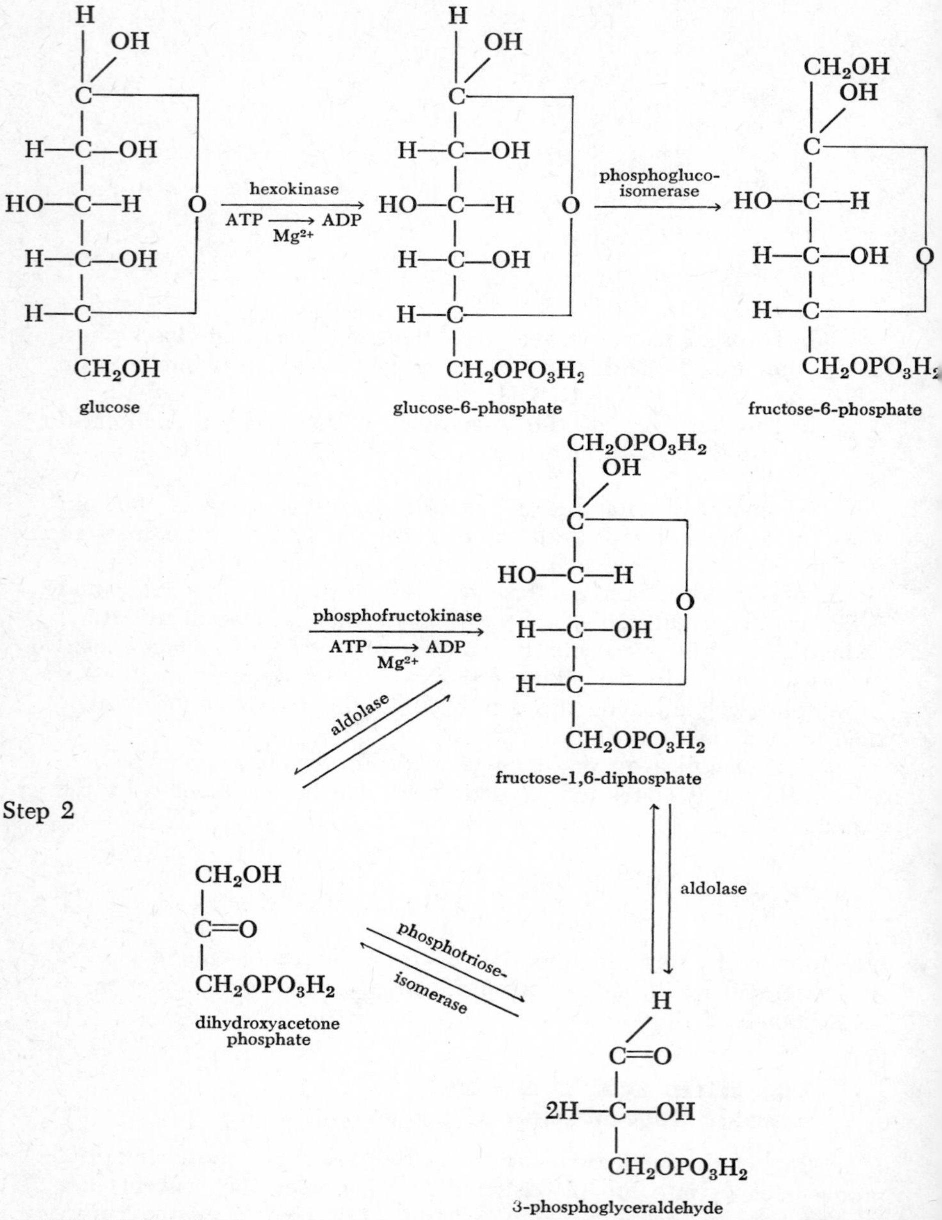

glucose —(hexokinase; ATP → ADP, Mg^{2+})→ glucose-6-phosphate —(phosphogluco-isomerase)→ fructose-6-phosphate

glucose: H, OH; C; H—C—OH; HO—C—H; H—C—OH; H—C; CH_2OH (ring O)

glucose-6-phosphate: H, OH; C; H—C—OH; HO—C—H; H—C—OH; H—C; $CH_2OPO_3H_2$ (ring O)

fructose-6-phosphate: CH_2OH, OH; C; HO—C—H; H—C—OH; H—C; $CH_2OPO_3H_2$ (ring O)

—(phosphofructokinase; ATP → ADP, Mg^{2+})→ fructose-1,6-diphosphate

fructose-1,6-diphosphate: $CH_2OPO_3H_2$, OH; C; HO—C—H; H—C—OH; H—C; $CH_2OPO_3H_2$ (ring O)

Step 2

fructose-1,6-diphosphate ⇌ (aldolase) dihydroxyacetone phosphate

fructose-1,6-diphosphate ⇌ (aldolase) 3-phosphoglyceraldehyde

dihydroxyacetone phosphate ⇌ (phosphotriose-isomerase) 3-phosphoglyceraldehyde

dihydroxyacetone phosphate: CH_2OH; C=O; $CH_2OPO_3H_2$

3-phosphoglyceraldehyde: H; C=O; 2H—C—OH; $CH_2OPO_3H_2$

TABLE 27.1 (*Continued*)

Step 3

$$\underset{\text{3-phosphoglyceraldehyde}}{\begin{array}{c}\text{H}\\ /\\ \text{C=O}\\ |\\ \text{2H—C—OH}\\ |\\ \text{CH}_2\text{OPO}_3\text{H}_2\end{array}} \;\underset{\substack{\text{NAD}^+(\rightarrow\text{NADH})\\ -2\text{H}}}{\overset{\substack{\text{H}_3\text{PO}_4\\ \text{phosphoglyceraldehyde}\\ \text{dehydrogenase}}}{\rightleftharpoons}}\; \underset{\text{1,3-diphosphoglyceric acid}}{\begin{array}{c}\text{O}\\ //\\ \text{C—O—PO}_3\text{H}_2\\ |\\ \text{2H—C—OH}\\ |\\ \text{CH}_2\text{OPO}_3\text{H}_2\end{array}}$$

$$\underset{\substack{\text{ADP} \longrightarrow \text{ATP}\\ \text{Mg}^{2+}}}{\overset{\substack{\text{3-phosphoglyceric}\\ \text{kinase}}}{\rightleftharpoons}}\; \underset{\text{3-phosphoglyceric acid}}{\begin{array}{c}\text{O}\\ //\\ \text{C—OH}\\ |\\ \text{2H—C—OH}\\ |\\ \text{CH}_2\text{OPO}_3\text{H}_2\end{array}} \;\overset{\text{phosphoglyceromutase}}{\rightleftharpoons}\; \underset{\text{2-phosphoglyceric acid}}{\begin{array}{c}\text{O}\\ //\\ \text{C—OH}\\ |\\ \text{2H—C—OPO}_3\text{H}_2\\ |\\ \text{CH}_2\text{OH}\end{array}}$$

$$\underset{\substack{-\text{H}_2\text{O}\\ \text{Mg}^{2+}}}{\overset{\text{enolase}}{\rightleftharpoons}}\; \underset{\text{phosphoenolpyruvic acid}}{\begin{array}{c}\text{O}\\ //\\ \text{C—OH}\\ |\\ \text{2C—OPO}_3\text{H}_2\\ \|\\ \text{CH}_2\end{array}} \;\underset{\substack{\text{ADP} \longrightarrow \text{ATP}\\ \text{Mg}^{2+}\\ \text{K}^+}}{\overset{\substack{\text{pyruvic}\\ \text{kinase}}}{\rightleftharpoons}}\; \underset{\text{pyruvic acid}}{\begin{array}{c}\text{O}\\ //\\ \text{C—OH}\\ |\\ \text{2C=O}\\ |\\ \text{CH}_3\end{array}}$$

Step 4

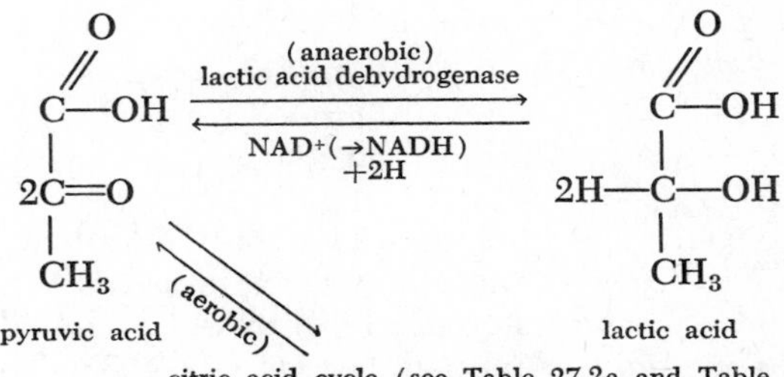

citric acid cycle (see Table 27.2a and Table 27.2b.)

ADP = adenosine diphosphate
ATP = adenosine triphosphate
NAD = oxidized nicotine adenine dinucleotide
NADH = reduced nicotine adenine dinucleotide

formed in step 3 above may not undergo reduction to lactic acid, but may be directly oxidized to carbon dioxide and water:

$$2CH_3—CO—COOH + 5O_2 \longrightarrow 6CO_2 + 4H_2O$$

However, this oxidation does not occur as a single reaction; it is believed that a complex series of reactions occur in which pyruvic acid is oxidized in steps and loses three successive molecules of carbon dioxide. Pyruvic acid is first dehydrogenated and decarboxylated to form acetyl coenzyme A, which combines with oxaloacetic acid present in the muscle to form citric acid. This citric acid is then oxidized through a number of steps, in which oxaloacetic acid is the end product. This acid can combine with acetyl coenzyme A from the breakdown of more pyruvic acid to start the cycle again. This cycle of changes was first known as the *Krebs cycle*, from the name of the first investigator to suggest it, but it is more commonly known as the *citric acid*, or the *tricarboxylic acid, cycle*. It is more thoroughly detailed in Table 27.2a and 27.2b.

Note from these tables that through the citric acid cycle the three carbon atoms of pyruvic acid are converted into three molecules of carbon dioxide. Phosphorylations accompany the citric acid cycle; that is, energy produced in the cycle is used to convert ADP to ATP. Fifteen high-energy phosphate bonds are produced in each cycle. These are noted in Table 27.2 by $\sim P$ preceded by the number of these phosphate bonds produced in each step. Hydrogens liberated in this cycle are combined with oxygen to form water. NAD, flavoproteins, the cytochromes, and cytochrome oxidase are essential for this oxidation of hydrogen (page 375).

Alcoholic fermentation

Under anaerobic conditions, yeast and some related microorganisms convert glucose to alcohol and carbon dioxide, a process called *alcoholic fermentation*. The overall reaction is

$$C_6H_{12}O_6 \longrightarrow 2C_2H_5OH + 2CO_2$$

After conversion of glucose to glucose-6-phosphate, the anaerobic degradation of glucose by these organisms proceeds through the same sequence as glycolysis of starch and glucose by animals to the stage at which pyruvic acid is formed (pages 402–403). These organisms contain in their cell fluid an enzyme, α-carboxylase, which, in the presence of the coenzyme cocarboxylase or thiamine pyrophosphate and Mg^{2+}, converts pyruvic acid to acetaldehyde and carbon dioxide, thus:

$$\underset{\text{pyruvic acid}}{CH_3COCOOH} \xrightarrow{Mg^{2+}} \underset{\text{acetaldehyde}}{CH_3CHO} + CO_2$$

The resulting acetaldehyde is then reduced to ethanol by NADH (reduced nicotine adenine dinucleotide or DPNH) in the presence of the enzyme *alcohol dehydrogenase*:

$$CH_3CHO + NADH + H^+ \longrightarrow CH_3CH_2OH + NAD^+$$

This NADH has been previously produced in the anaerobic cycle when NAD^+ was reduced in the oxidation of 3-phosphoglyceraldehyde to 3-phosphoglyceric acid (page 405):

$$\text{3-phosphoglyceraldehyde} + NAD^+ + ADP + \text{inorganic phosphate} \longrightarrow \text{3-phosphoglyceric acid} + NADH + ATP + H^+$$

Normal fat metabolism, storage of fat, and depot fat

Fats are hydrolyzed to fatty acids and glycerol in the small intestine and are absorbed through the intestinal wall into the lymph. The bile salts must play an important role in the absorption of these fatty acid molecules. In the lymph, the glycerol and fatty acid appear as resynthesized fat. The fats then enter the bloodstream through the thoracic duct and are transported to the liver, where portions are changed to phospholipids. Both the resulting phospholipids and free fats appear in the blood leaving the liver. Their concentration in the blood increases shortly after a meal rich in fats.

The fats are presented to the cells, where part is oxidized to provide heat and other forms of energy while some are used to produce certain essential cell constituents, such as the phospholipids. The excess fat not needed for the above purposes is stored under the skin and around organs and is called *depot fat*, or *adipose tissue*. It serves not only as a reserve food supply, but also to cushion the vital organs from shock. Because fat is a poor conductor of heat, it protects the interior of the body from sudden changes in outside temperature.

During fasting, the fat deposits of the body are depleted. Death from starvation will occur before all fat in the body has been removed. The remaining fat is therefore evidently essential to life and must make up an important part of the protoplasm. Such fat has been called the *constant element*, whereas fat stored as a reserve food supply is designated as the *variable element*.

Animals usually deposit a fat which is characteristic of their species and which has a melting point near their body temperature. If, however, very large quantities of a foreign fat are fed to an animal, the nature of the fat deposited by the animal may be greatly modified. If swine are fattened on corn, a solid fat characteristic of swine will be deposited, because the animal is synthesizing fat from carbohydrates. If peanuts are used to fatten them, a soft fat, which is more unsaturated and has a lower melting point than the usual type, will be deposited.

TABLE 27.2a AEROBIC CONVERSION OF PYRUVIC ACID TO CARBON DIOXIDE VIA THE CITRIC ACID CYCLE

Substrate	*Enzyme*	*Coenzymes and Other Cofactors*	*Miscellaneous Information*
Pyruvic acid ↓	Pyruvic oxidase	TPP[a], Lipoic acid, Coenzyme A, NAD[a] (→NADH), Mg^{2+}	−2H, $-CO_2$, 3 ~ P
Acetyl coenzyme A + →Oxaloacetic acid ↓	Citrate Synthase (citrogenase)		-Coenzyme A
Citric acid ↓	Aconitase	Fe^{2+}	
Isocitric acid ↓	Isocitric dehydrogenase	NAD[a] (→NADH)	−2H, $-CO_2$, 3 ~ P
α-Ketoglutaric acid ↓	α-Ketoglutaric acid oxidase	TPP[a], Lipoic acid, Coenzyme A, NAD[a] (→NADH), Mg^{2+}	−2H, $-CO_2$, 3 ~ P
Succinyl coenzyme A ↓	Succinic thiokinase		-Coenzyme A, 1 ~ P
Succinic acid ↓	Succinic dehydrogenase	FAD[a] (→$FADH_2$)	−2H, 2 ~ P
Fumaric acid ↓	Fumarase		$+H_2O$
Malic acid ↓	Malic dehydrogenase	NAD[a] (→NADH)	−2H, 3 ~ P
Oxaloacetic acid (returns to Oxaloacetic acid above)			

[a]TPP = thiamine pyrophosphate, NAD = nicotine adenine dinucleotide, FAD = flavin adenine dinucleotide.

OXIDATION OF FATS IN THE BODY CELLS

The fats are probably converted to phospholipids before they are oxidized. The resulting phospholipids are then hydrolyzed with the liberation of glycerol and fatty acids. The former is likely dehydrogenated and phosphorylated to 3-phosphoglyceraldehyde and en-

TABLE 27.2b AEROBIC DEGRADATION OF PYRUVIC ACID VIA THE CITRIC ACID CYCLE

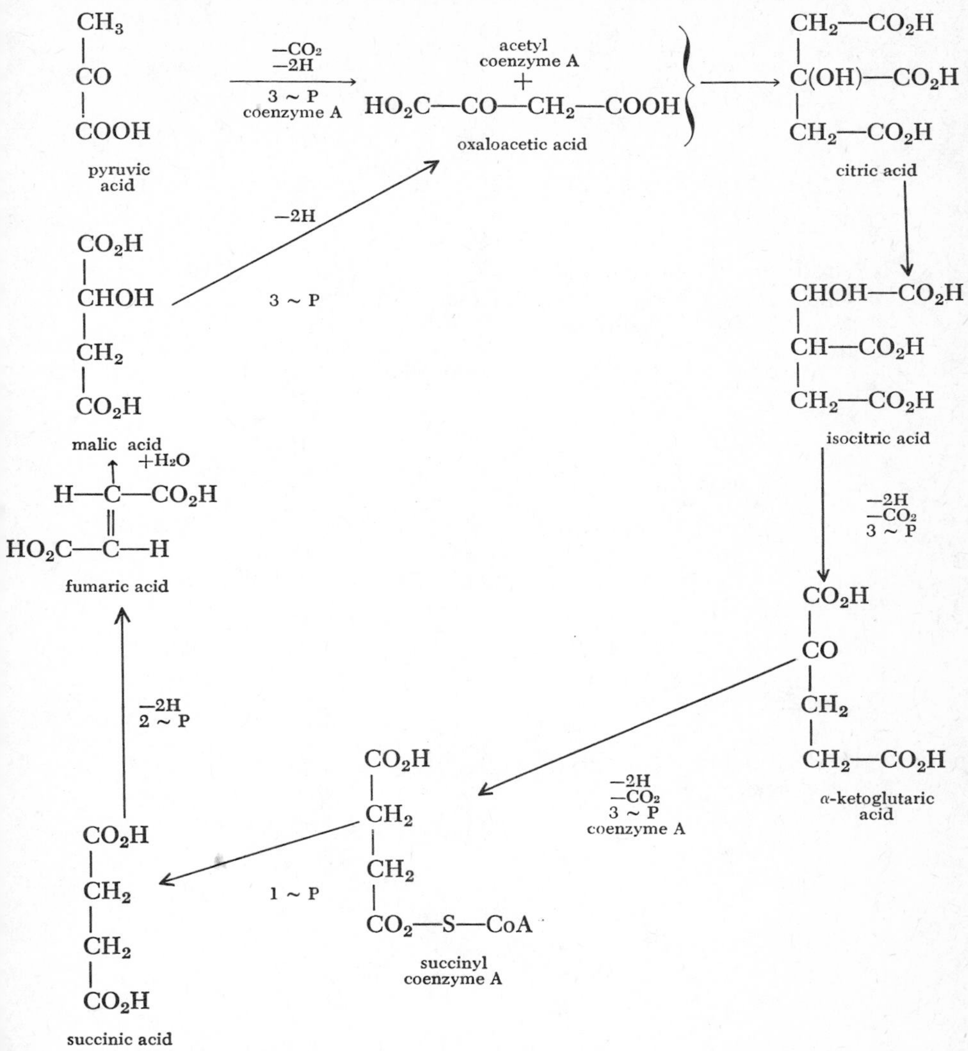

CH_3—CO—COOH (pyruvic acid) → ($-CO_2$, $-2H$, $3 \sim P$, coenzyme A) → acetyl coenzyme A + HO_2C—CO—CH_2—COOH (oxaloacetic acid) → CH_2—CO_2H / C(OH)—CO_2H / CH_2—CO_2H (citric acid)

citric acid → CHOH—CO_2H / CH—CO_2H / CH_2—CO_2H (isocitric acid)

isocitric acid → ($-2H$, $-CO_2$, $3 \sim P$) → CO_2H / CO / CH_2 / CH_2—CO_2H (α-ketoglutaric acid)

α-ketoglutaric acid → ($-2H$, $-CO_2$, $3 \sim P$, coenzyme A) → CO_2H / CH_2 / CH_2 / CO_2—S—CoA (succinyl coenzyme A)

succinyl coenzyme A → ($1 \sim P$) → CO_2H / CH_2 / CH_2 / CO_2H (succinic acid)

succinic acid → ($-2H$, $2 \sim P$) → H—C—CO_2H ‖ HO_2C—C—H (fumaric acid)

fumaric acid → ($+H_2O$) → CO_2H / CHOH / CH_2 / CO_2H (malic acid)

malic acid → ($-2H$, $3 \sim P$) → oxaloacetic acid

ters the pathway of carbohydrate metabolism, in which it is converted either to glycogen or pyruvic acid. The latter is then oxidized to carbon dioxide and water in the citric acid cycle (page 403).

In 1904, Knoop, a German biochemist, proposed a theory for the oxidation of the fatty acid molecule and presented some evidence in its support. Since that time, several other theories have been suggested, but Knoop's theory appears now to be well established. This theory is often called the *β-oxidation theory,* because the metabolism of the fatty acid involves the oxidation of the β carbon (the second carbon atom from the carboxyl group) to a carbonyl or ketone group:

$$R{-}\overset{\gamma}{C}H_2{-}\overset{\beta}{C}H_2{-}\overset{\alpha}{C}H_2{-}COOH + 2[O] \longrightarrow$$

$$R{-}CH_2{-}\overset{\overset{\displaystyle O}{\|}}{C}{-}CH_2{-}COOH + H_2O$$

In the first step in this oxidation, the fatty acid is activated by the formation of a sulfur or thio ester with coenzyme A, CoASH (page 374). This union is catalyzed by an acyl thiokinase and Mg^{2+}. To produce the high-energy bond present in the resulting acyl coenzyme A ester, ATP is simultaneously hydrolyzed to adenylic acid (AA) and pyrophosphoric acid:

$$RCH_2CH_2CO_2H + CoASH + ATP \xrightarrow[Mg^{2+}]{\text{acyl thiokinase}}$$

$$RCH_2CH_2\overset{\overset{\displaystyle O}{/\!/}}{C}{-}S{-}CoA + AA + H_4P_2O_7$$

Through three successive enzymic reactions (see Table 27.3) involving dehydrogenation, hydration and a second dehydrogenation, the acyl coenzyme A is converted to a β-ketoacyl coenzyme A of the general structure

$$R{-}\overset{\overset{\displaystyle O}{/\!/}}{C}{-}CH_2{-}\overset{\overset{\displaystyle O}{/\!/}}{C}{-}SCoA$$

Coenzyme A, in the presence of a β-ketoacyl thiolase, converts the latter to acetyl coenzyme A and an acyl coenzyme A that is a derivative of a fatty acid with two less carbon atoms than the original acid, as noted in the equation

$$R{-}\overset{\overset{\displaystyle O}{/\!/}}{C}{-}CH_2{-}\overset{\overset{\displaystyle O}{/\!/}}{C}{-}SCoA + CoASH \xrightarrow[\text{thiolase}]{\beta\text{-ketoacyl}}$$

$$R\overset{\overset{\displaystyle O}{/\!/}}{C}{-}SCoA + CH_3\overset{\overset{\displaystyle O}{/\!/}}{C}{-}SCoA$$

Now the β-oxidation of this new fatty acid coenzyme A ester will occur in the same fashion as the oxidation of the original activated fatty acid with the production of a new fatty acyl coenzyme A with four less carbon atoms than the original fatty acid. Starting with a fatty acid with an even number of carbon atoms, this process can continue until only acetyl coenzyme A molecules remain.

The resulting acetyl coenzyme A may be utilized in several fashions, including the synthesis of fatty acids (lipogenesis) and steroids, but much of it will enter the citric acid cycle (see carbohy-

TABLE 27.3 STEPS IN THE β-OXIDATION OF A FATTY ACID RESULTING IN ITS CONVERSION TO AN ACYL COENZYME A ESTER OF THE FATTY ACID WITH TWO FEWER CARBON ATOMS

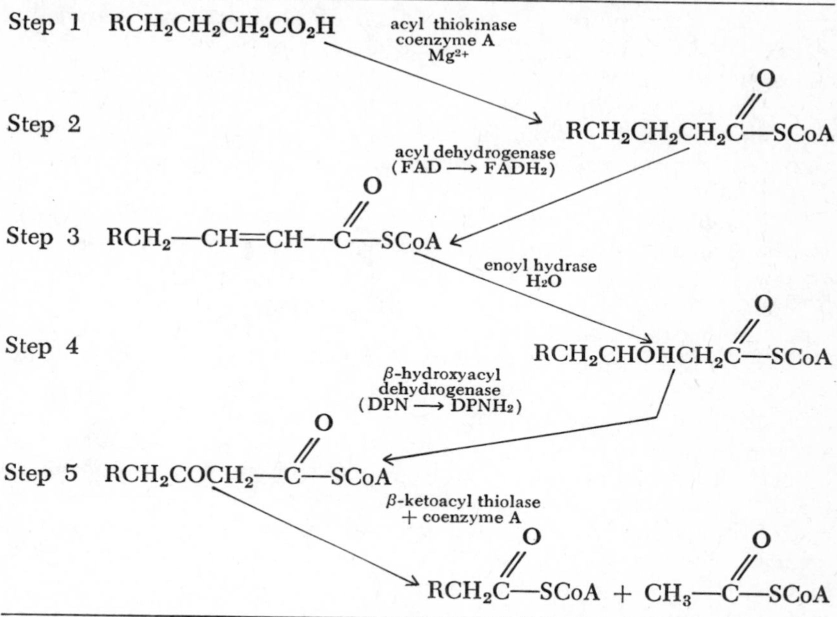

drate metabolism, pages 408–409), where it is converted to carbon dioxide and water with the production of high-energy phosphate bonds (generally as ATP).

Thus, through β-oxidation, palmitic acid will be converted to eight molecules of acetyl coenzyme A through seven cycles of the oxidation process detailed in Table 27.3.

$$\underset{\text{palmitic acid}}{C_{16}H_{33}O_2} + 8CoASH + 7O_2 \longrightarrow 8CH_3\overset{O}{\overset{\|}{C}}—SCoA + 8H_2O$$

The following equation represents the complete conversion of palmitic acid to carbon dioxide and water:

$$C_{16}H_{33}O_2 + 23O_2 \longrightarrow 16CO_2 + 16H_2O$$

Abnormal fat and carbohydrate metabolism

The pancreas produces insulin, a hormone that has a great influence on carbohydrate metabolism and the power to convert glucose present in the body to glycogen, which is stored in the liver and muscles. It also has the property of increasing the speed of the oxidation of glucose to carbon dioxide and water (see page 513).

Epinephrine (adrenaline), another hormone, is produced by the adrenal glands, which are located just above the kidneys (page 517). Epinephrine is an antagonist for insulin, because it increases the rate of glycogenolysis in the liver, and this is accompanied by an increased concentration of glucose in the blood. During emotional states such as fear or anger, more epinephrine than usual is introduced into the blood and may cause such an increase in the glucose content of the blood that glucose may appear in the urine (page 461).

The pancreas produces a hormone called *glucagon,* which is also antagonistic to insulin. It produces an increased breakdown of glycogen in the liver and an accompanying rise in the concentration of glucose in the blood (page 514).

In *pancreatic diabetes,* which is better known as *diabetes mellitus,* insulin is not produced by the pancreas, or at least there is a great reduction in its formation. Glucose is not properly metabolized, its concentration in the blood increases, and it then appears in the urine, a condition called *glucosuria.*

The most serious result of noncombustion of glucose in the body is an incomplete oxidation of fats. Normally, the fatty acids obtained from the fats are oxidized to acetyl coenzyme A, which combines with oxaloacetic acid. The citric acid that results from this combination is converted to carbon dioxide, water, and energy-rich phosphate bonds in the citric acid cycle with the regeneration of oxaloacetic acid. In the absence of carbohydrate metabolism, there is an increased metabolism of fat with an increased production of acetyl coenzyme A in the liver. The peripheral tissue to which the blood carries the acetyl coenzyme A is unable to oxidize the increased amount of it produced, and as an alternative it is recombined to form acetoacetyl coenzyme A, as represented (in a simplified fashion) by the following steps:

Step 1:

$$\underset{\text{acetyl coenzyme A}}{CH_3C(=O)\text{—SCoA}} + CO_2 \longrightarrow \underset{\text{malonyl coenzyme A}}{HO_2C\text{—}CH_2\text{—}C(=O)\text{—SCoA}}$$

Step 2:

$$\text{malonyl coenzyme A} + \text{acetyl coenzyme A} \longrightarrow \underset{\text{acetoacetyl coenzyme A}}{CH_3\text{—}C(=O)\text{—}CH_2\text{—}C(=O)\text{—SCoA}} + CO_2$$

Through three alternative mechanisms that will not be detailed here, the resulting acetoacetyl coenzyme A is "deacylated" to form free acetoacetic acid. Although a major portion of the acetoacetic acid will come from the condensation of acetyl coenzyme A, a certain portion may be expected to result directly from the oxidation of butyryl coenzyme A, an intermediate product formed in the β-oxidation of fatty acids (page 409). Part of the accumulating acetoacetic acid will be reduced to β-hydroxybutyric acid, while part will decompose into acetone and carbon dioxide. Acetoacetic acid, β-hydroxybutyric acid and acetone are called the *acetone* or *ketone bodies;* an accumulation of these substances in certain of the body fluids is called *ketosis.*

$$CH_3—CO—CH_2–COOH \xrightarrow{2[H]} \underset{\beta\text{-hydroxybutyric acid}}{CH_3—CHOH—CH_2—COOH}$$

$$CH_3—CO—CH_2–COOH \longrightarrow CH_3—CO—CH_3 + CO_2$$

Any condition that leads to a decreased oxidation of glucose in the body will produce ketosis. Starvation, liver damage, and a diet high in fats and low in carbohydrates are other causes of ketosis. In severe liver damage, glycogen is not stored in adequate amounts, and little or no glucose will be available for oxidation.

Those substances in the blood that are capable of neutralizing acids are called the *alkaline reserve of the blood.* Because two of the ketone bodies are acids, they will neutralize the alkalis of the body and reduce the alkaline-reserve of the blood. The *p*H of the blood will decrease (that is, the blood will become less basic), a condition called *acidosis.* The symptoms of acidosis are nausea, depression of the central nervous system, dehydration, coma, and finally death. Incidentally, the ketone bodies appear in the urine during ketosis.

Substances that produce ketosis, such as fats, are called *ketogenic substances.* Substances that tend to prevent ketosis, such as carbohydrates, are called *antiketogenic substances.* Some of the amino acids are ketogenic, while others will prevent ketosis and are, therefore, antiketogenic. A diet high in fats and low in carbohydrates is called a *ketogenic diet,* while one that is high in carbohydrates and low in fats is called an *antiketogenic diet.* Ketogenic diets have been employed in the treatment of epilepsy.

There is no cure for diabetes mellitus, but a person afflicted with this disease may lead a healthful and normal life. In mild cases, careful control of the diet will give relief. In many cases, insulin, which has been recovered from the pancreas of cattle, is injected under the skin in order to compensate for its deficiency in the diabetic. Insulin cannot be given by mouth because it is a polypeptide and is destroyed by the digestive juices. In some diabetics, certain sulfa drugs, such as Carbutamide (which can be given orally), lower blood glucose and can replace insulin. These either may act to inactivate enzymes that destroy insulin produced by the pancreas or can cause an increased release of insulin by the pancreas.

ESSENTIAL FATTY ACIDS

Burr and Burr* have demonstrated that rats fed on very low fat diets lose weight, produce a scaly skin and tail, and eventually die. They have shown also that linoleic, linolenic, and arachidonic acids, which are unsaturated acids, are effective in curing or preventing these conditions. These acids are now designated as the *essential fatty acids.* The skin lesions of some human subjects with eczema have been cured by use of suitable fats in the diet (page 274).

LIPOGENESIS AND CONVERSION OF CARBOHYDRATES TO FATS

It is common knowledge that if large quantities of carbohydrates are consumed, much body fat will be deposited. It has been demonstrated that animals fed low-fat diets will deposit an amount of fat in their bodies that is in excess of the amount consumed.

The process by which fatty acids are biosynthesized is called *lipogenesis.* Because this involves successive condensations of two carbon-atom fragments utilizing acetyl coenzyme A, fatty acids with an even number of carbon atoms predominate in body fat.

In the conversion of carbohydrates to fatty acids, the acetyl coenzyme A, instead of entering the citric acid cycle, is shunted into the lipogenesis sequence. The lipogenesis sequence involves first the union of acetyl coenzyme A with carbon dioxide to form malonyl coenzyme A (see section above concerning the formation of acetoacetic acid in ketosis), a reaction catalyzed by the enzyme acetyl coenzyme A carboxylase. In the next step, a molecule of acetyl coenzyme A condenses with a molecule of malonyl coenzyme A to form acetoacetyl coenzyme A.

Through steps involving successively hydrogenation, dehydration, and a second hydrogenation, acetoacetyl coenzyme A is converted to *n*-butyryl coenzyme A; the three steps appear to be the reverse of steps involved in the β-oxidation of the fatty acids (see Table 27.3) but require a set of enzymes distinctly different from the latter.

The *n*-butyryl coenzyme A can be converted to the ester of an acid that contains an additional two carbon atoms by condensation with a second molecule of malonyl coenzyme A. Conversion of the β-keto group of the product of this condensation to a methylene group (CH_2) by a repetition of the three steps indicated above produces *n*-caproyl coenzyme A (hexanoyl coenzyme A). Five more repetitions of this lipogenesis sequence, which will require an additional five molecules of malonyl coenzyme A, will produce palmitoyl coenzyme A. "Deacylation" of palmitoyl coenzyme A forms free palmitic acid. The latter, through the intermediate formation of a phosphatidic

*Burr and Burr, *Journal of Biological Chemistry, 82,* 345 (1929); *86,* 587 (1930).

acid (page 284), can be combined biosynthetically with glycerol to form a typical triglyceride.*

OBESITY

This is a condition in which an excessive amount of fat is deposited in the body as adipose tissue and the individual becomes overweight. In some cases, this may be due to a derangement of some of the endocrine organs, such as the thyroid gland, the pituitary gland, or the sex organs (pages 514, 520, and 523). In most individuals, obesity is due to eating more food than is required for the normal functions of the body or to lack of exercise. Obesity leads to a shortened life expectancy.

It is claimed that diets high in saturated fatty acids induce an increased synthesis of cholesterol (page 287), which may be accompanied by an increased deposition of cholesterol in the arteries resulting in a type of hardening of the arteries referred to as *atherosclerosis.*

The metabolism of proteins

During digestion, the proteins are hydrolyzed to the free amino acids and are absorbed as such into the blood. These are then transported to the cells and are used to build new tissue, replace worn-out tissue, and synthesize enzymes and some hormones and pigments. The cells select those amino acids needed to synthesize their characteristic protein. Those amino acids not needed, as well as the ones formed by the breakdown of worn-out tissue, are converted to ammonia, carbon dioxide, and water, with the simultaneous formation of heat and mechanical energy.

The first step in the catabolism of amino acids occurs primarily in the liver, although it may also occur to some extent in the kidneys and muscle tissue. This step involves the conversion of the amino acid to a keto acid. Two types of processes are involved in this conversion, the most important of which is *transamination.* The other, *oxidative deamination,* appears to be of lesser importance.

The process of oxidative deamination probably first involves the dehydrogenation of an amino acid to an imino acid (the group =NH is called the *imino group*).

$$\underset{\displaystyle NH_2}{R\text{—}\overset{|}{CH}\text{—}CO_2H} \xrightarrow{-2H} \underset{\displaystyle NH}{R\text{—}\overset{\|}{C}\text{—}CO_2H}$$

*In bacteria and plants, biosynthesis of fatty acids appears not to utilize coenzyme A esters but esters involving a thiol (—SH) group of a heat-stable protein that is now designated as acyl-carrier protein, or ACP. The fatty acid synthesis is initiated by a condensation of acetyl-S-ACP and malonyl-S-ACP, the latter two substances first being formed from their corresponding coenzyme A esters by a "transferase" reaction with ACP. The last step in the synthesis of a fatty acid would involve the deacylation of its ACP ester (rather than its coenzyme A ester).

Addition of water to the imino acid followed by loss of ammonia produces the keto acid.

$$R\text{—}\underset{\underset{NH}{\|}}{C}\text{—}CO_2H + H_2O \longrightarrow R\text{—}\overset{\overset{OH}{|}}{\underset{\underset{NH_2}{|}}{C}}\text{—}CO_2H$$

$$R\text{—}\overset{\overset{OH}{|}}{\underset{\underset{NH_2}{|}}{C}}\text{—}CO_2H \longrightarrow R\text{—}\underset{\underset{O}{\|}}{C}\text{—}CO_2H + NH_3$$

Although amino acid oxidases that require either FMN or FAD as coenzymes (page 374) and indiscriminately convert amino acids to keto acids are known, the only significant oxidative deamination involves the conversion of glutamic acid to α-ketoglutaric acid. This conversion is catalyzed by the enzyme glutamic acid dehydrogenase; either NAD^+ or $NADP^+$ serves as a coenzyme for this reaction.

$$\begin{array}{c} CO_2H \\ | \\ CHNH_2 \\ | \\ CH_2 \\ | \\ CH_2 \\ | \\ CO_2H \\ \text{glutamic acid} \end{array} + H_2O + NAD^+ (NADP^+) \longrightarrow$$

$$\begin{array}{c} CO_2H \\ | \\ C{=}O \\ | \\ CH_2 \\ | \\ CH_2 \\ | \\ CO_2H \\ \alpha\text{-ketoglutaric acid} \end{array} + NH_3 + NADH(NADPH) + H^+$$

Many of the amino acids are catabolized by initially transferring an amino group to a keto acid, a process called *transamination.* The keto acid resulting from this transfer then undergoes oxidative degradation to form carbon dioxide and water. Because the resulting glutamic acid can be readily oxidized to α-ketoglutaric acid by glutamic acid dehydrogenase, both alanine and aspartic acid will to a great extent transfer their amino group to this keto acid to form pyruvic acid and oxaloacetic acid, thus:

$$\begin{array}{ccccccc}
CO_2H & & CO_2H & & CO_2H & & CO_2H \\
| & & | & & | & & | \\
CHNH_2 & + & C{=}O & \longrightarrow & CHNH_2 & + & C{=}O \\
| & & | & & | & & | \\
CH_3 & & CH_2 & & CH_2 & & CH_3 \\
\text{alanine} & & | & & | & & \text{pyruvic acid} \\
 & & CH_2 & & CH_2 & & \\
 & & | & & | & & \\
 & & CO_2H & & CO_2H & &
\end{array}$$

$$\begin{array}{ccccccc}
CO_2H & & CO_2H & & CO_2H & & CO_2H \\
| & & | & & | & & | \\
CHNH_2 & + & C{=}O & \longrightarrow & C{=}O & + & CHNH_2 \\
| & & | & & | & & | \\
CH_2 & & CH_2 & & CH_2 & & CH_2 \\
| & & | & & | & & | \\
CO_2H & & CH_2 & & CO_2H & & CH_2 \\
\text{aspartic acid} & & | & & \text{oxaloacetic acid} & & | \\
 & & CO_2H & & & & CO_2H
\end{array}$$

The ammonia formed in oxidative deaminations is excreted as such by aquatic vertebrates. Birds and reptiles convert this ammonia to uric acid. Because they excrete protein nitrogen primarily as uric acid, they are said to be *uricotelic.* Other terrestrial vertebrates are ureotelic because they excrete urea that is formed by the liver from the ammonia resulting from the catabolism of the amino acids. The keto acids resulting from either transamination or oxidative deamination of amino acids are further catabolized to carbon dioxide and water. Part of these keto acids follow the same route of oxidation as the intermediates of glucose metabolism. As an example, alanine (CH_3CHNH_2COOH) is catabolized to pyruvic acid ($CH_3COCOOH$), and this substance is formed as an intermediate in carbohydrate metabolism. The resulting pyruvic acid can either be degraded to carbon dioxide and water, as in carbohydrate metabolism, or converted to glucose.

Glutamic acid is converted to α-ketoglutaric acid, whereas aspartic acid is converted to oxaloacetic acid, substances that appear in the citric acid cycle. Other antiketogenic amino acids are glycine, arginine, proline, and valine.

Leucine is deaminated to a keto acid, which is converted to both acetyl coenzyme A and acetoacetyl coenzyme A as intermediates in its oxidation to carbon dioxide and water. Leucine is designated as a ketogenic amino acid because in a diabetic it will serve, as well as fats, as a source of the ketone bodies. The amino acids tyrosine, phenylalanine, isoleucine, and lysine are both ketogenic and antiketogenic, because it has now been established that some of their intermediate products of oxidation produce ketosis whereas others allay ketosis.

The sulfur of the sulfur-containing amino acids cystine, cysteine, and methionine, is oxidized and excreted in the urine as sulfate ions and sulfate esters of certain alcohols.

Purines (page 439), including those resulting from the catabolism of the nucleoproteins, are usually oxidized to uric acid. While man, some primates, and Dalmatian coach dogs excrete uric acid, most other mammals convert uric acid to the more stable allantoin,

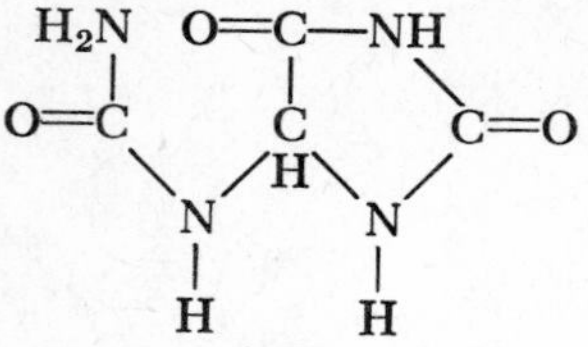

```
H2N  O=C—NH
 |     |    \
O=C    C     C=O
  \  / H \  /
   N       N
   |       |
   H       H
```

allantoin

Fish and the amphibia further convert allantoin to allantoic acid,

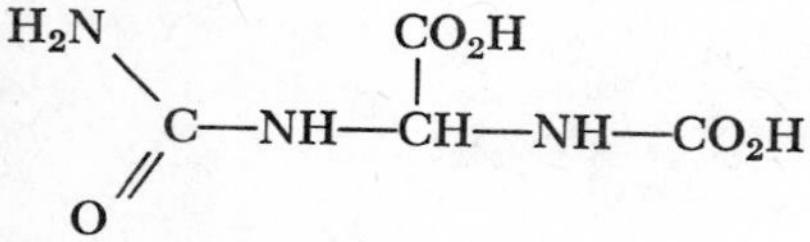

```
H2N            CO2H
   \            |
    C—NH—CH—NH—CO2H
   //
  O
```

allantoic acid

before excretion.

The student is referred to a text in biochemistry for a more extensive discussion of the metabolism of the individual amino acids, as well as creatinine and the nucleic acids.

The formation of urea in the liver

In the human, urea is produced exclusively in the liver, and its formation probably occurs through the following three steps:

Step 1: Ammonia, most probably produced by the deamination of amino acids, and carbon dioxide combine with ornithine to form citrulline.

$$H_2N\text{—}(CH_2)_3\text{—}\underset{\underset{NH_2}{|}}{CH}\text{—}COOH + NH_3 + CO_2 \longrightarrow$$

ornithine

$$H_2N\text{—}\underset{\underset{O}{\|}}{C}\text{—}NH\text{—}(CH_2)_3\text{—}\underset{\underset{NH_2}{|}}{CH}\text{—}COOH + H_2O$$

citrulline

Our present knowledge indicates that NH_3 and CO_2 do not react directly with ornithine but that the active agent for the above conversion is carbamyl phosphate, which is formed by the reaction of NH_3 and CO_2 with adenosine triphosphate:

$$CO_2 + NH_3 + 2ATP \longrightarrow H_2N\text{—}\underset{\underset{O}{\|}}{C}\text{—}OPO_3H_2 + ADP + H_3PO_4$$

carbamyl phosphate

Step 2: Citrulline combines with ammonia to form arginine. This conversion may be represented as follows:

$$\underset{\text{citrulline}}{H_2N{-}\underset{\underset{O}{\|}}{C}{-}NH{-}(CH_2)_3{-}\underset{\underset{NH_2}{|}}{CH}{-}COOH} + NH_3 \longrightarrow$$

$$\underset{\text{arginine}}{H_2N{-}\underset{\underset{NH}{\|}}{C}{-}NH{-}(CH_2)_3{-}\underset{\underset{NH_2}{|}}{CH}{-}COOH} + H_2O$$

The mechanism of this conversion is complex and not completely understood. The evidence so far indicates that the ammonia for this conversion comes from aspartic acid exclusively. Recent experiments indicate that through the intermediation of a suitable enzyme, citrulline combines with aspartic acid and that the product of this union is degraded by a second enzyme to arginine and fumaric acid. The resulting fumaric acid enters the citric acid cycle and is thereby degraded to carbon dioxide.

Step 3: In this step, arginine is acted upon by the enzyme *arginase*, which occurs in the liver, and urea and ornithine are generated:

$$\underset{\text{arginine}}{H_2N{-}\underset{\underset{NH}{\|}}{C}{-}NH{-}(CH_2)_3{-}\underset{\underset{NH_2}{|}}{CH}{-}COOH} + H_2O \longrightarrow$$

$$H_2N{-}CO{-}NH_2 + \underset{\text{ornithine}}{H_2N{-}(CH_2)_3{-}\underset{\underset{NH_2}{|}}{CH}{-}COOH}$$

The ornithine generated in this step is now available for the synthesis of more urea by the repetition of the process discussed, which therefore is often designated as the *ornithine cycle.*

After its formation in the liver, urea is carried by the blood to the kidneys, where most of it is excreted. A small amount of urea is excreted in the sweat and saliva.

Formation of urinary ammonia from amino acids

Ammonium salts formed in most of the body tissues or ingested in food are converted to urea or amino acids by the liver. The ammonia in urine is most likely formed in the kidneys, and recent studies support the view that it is formed by the deamination of amino acids.

The ammonia formed in the kidneys combines with the acid radicals of the blood. This is an important function of ammonia; otherwise these acid radicals would be excreted as sodium, potassium, and calcium salts, with the loss of much of the alkaline reserve of the body. For this reason, ammonia aids in conserving the alkaline reserves of the blood, and this mechanism delays the onset of the typical symptoms of acidosis.

In alkalosis, a condition in which the alkalinity of the blood increases abnormally, the excretion of ammonia as ammonium salts in the urine is drastically decreased in order to save the acid radicals of the blood; the decrease in amount of ammonium salts in the urine will be compensated by an increased excretion of urea.

Nitrogen equilibrium or nitrogen balance

Unlike the fats and carbohydrate, little if any protein is stored in the body. The amount of nitrogen not utilized for building or repairing tissue is rapidly excreted. In a healthy adult who is neither losing nor gaining weight, the amount of nitrogen ingested in the diet will be nearly equal to the amount of nitrogen excreted in the urine, feces, sweat, etc., and such a person is said to be in *nitrogen balance* or *nitrogen equilibrium.* Children who are growing rapidly are retaining much nitrogen for the formation of tissue, and less nitrogen will be excreted than is present in the food intake. They are said to have a *positive nitrogen balance.* In many wasting diseases, and during starvation and fevers, more nitrogen is excreted by an individual than is present in the ingested food. Such people are said to have a *negative nitrogen balance.*

The nutritionists are unable to agree upon the minimum amount of protein necessary to maintain nitrogen equilibrium in an adult. This must be due to the great variation in amino acid content of most proteins.

Essential amino acids

Feeding experiments that employed a diet consisting of all the essential nutrients but included a single protein deficient in certain amino acids indicate that certain amino acids are essential for proper growth of the animals studied. Such amino acids are now called *essential amino acids.* Experiments on human subjects on diets in which proteins were replaced with amino acids, and in which an amino acid was omitted, indicate that the following are essential amino acids for humans: valine, leucine, isoleucine, lysine, phenylalanine, tryptophan, methionine, and threonine. Arginine and histidine may also be indispensable in the diet of children. If other amino acids are absent from the diet, growth will still be normal. It is therefore concluded that the organism must be able to synthesize these from some other substances in the diet. Such amino acids are called *nonessential amino acids* and include glycine, alanine, citrul-

TABLE 27.4 SUMMARY OF THE METABOLISM OF THE DIETARY LIPIDS, CARBOHYDRATES, AND PROTEINS. THIS TABLE, IN PARTICULAR, ILLUSTRATES THE INTERRELATIONSHIP AND INTERCONVERSIONS BETWEEN THESE THREE TYPES OF FOODSTUFFS

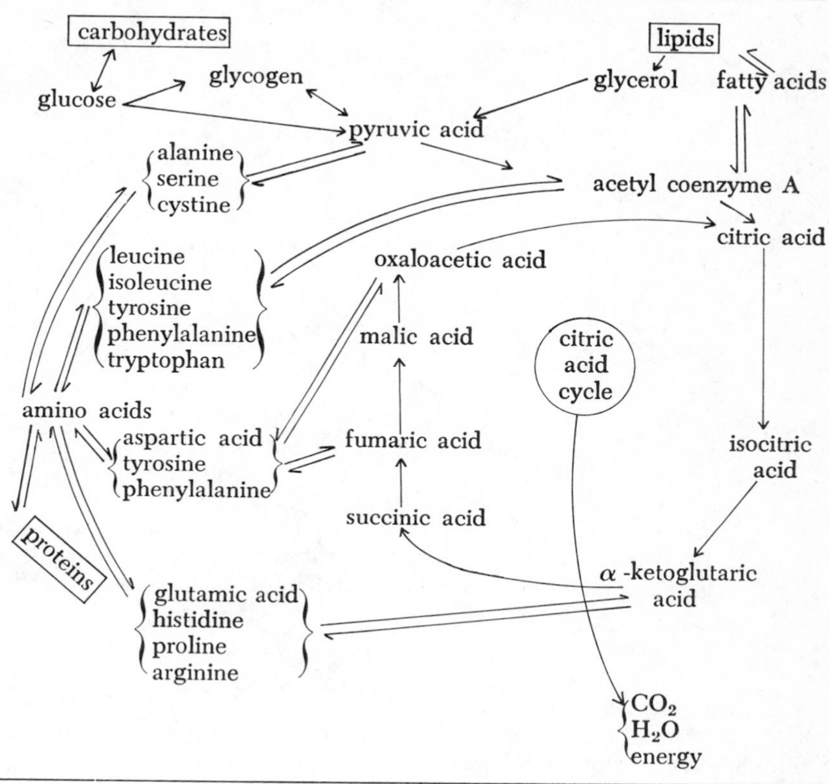

line, tyrosine, and glutamic acid. A protein that contains all the essential amino acids is called an *adequate protein,* whereas one in which these amino acids are absent is called an *inadequate protein.* The lactalbumin of milk is an adequate protein, whereas gelatin is an inadequate protein because it lacks tryptophan.

THE RELATIONSHIP BETWEEN CARBOHYDRATE, LIPID, AND PROTEIN METABOLISM

Table 27.4 summarizes the metabolism of and the interrelationship betwen the three major types of foodstuffs: carbohydrates, lipids, and proteins.

Based on previous discussions that appear in this chapter and the information presented in Table 27.4, the following observations can be made:

1. Carbohydrates serve as a source of fats by way of their conversion to acetyl coenzyme A via the glycolitic cycle (pages 402–406). In the lipogenesis sequence (page 414), the resulting acetyl coenzyme A is converted to the fatty acids required for fat synthesis.
2. Because the conversion of pyruvic acid to acetyl coenzyme A is *irreversible*, the fatty acids present in fats cannot be converted to carbohydrates via acetyl coenzyme A and a reversal of the glycolitic cycle. Thus animals are unable to convert fats to carbohydrates. Plants, some molds, and bacteria can accomplish this conversion by a metabolic cycle called the *glyoxylic cycle*, which is not detailed in this text but is described in most biochemistry texts. Incidentally, all of the steps in the citric acid cycle except one are reversible. The conversion of α-ketoglutaric acid to succinic acid is irreversible.
3. Some antiketogenic amino acids are converted to pyruvic acid, which may be either converted to glycogen or oxidized to carbon dioxide and water by way of the citric acid cycle. The ketogenic amino acids as well as the fatty acids serve as a source of acetyl coenzyme A. Thus not only may these amino acids serve as a source of newly synthesized fats in the body, but the reverse process, the conversion of fats to certain amino acids also occurs. Oxaloacetic acid and α-ketoglutaric acid, which occur in the citric acid cycle and can be withdrawn from this cycle, are transaminated to aspartic and glutamic acid, respectively. The pyruvic acid that results from the glycolytic cycle may be transaminated to the amino acid alanine.

It should now be evident that a product such as pyruvic acid, acetyl coenzyme A, or α-ketoglutaric acid that results from the catabolism of the carbohydrates will not be required to continue in the catabolic pathway in which these substances are converted to carbon dioxide, water, and high-energy ATP, but may be caught up in anabolic or biosynthetic processes in which they are converted to fatty acids and amino acids. The keto acids that result from the oxidation or transamination of the amino acids may be catabolized completely to carbon dioxide and water; however, part of these may serve as raw material for the synthesis of glycogen and the fats.

Energy metabolism

The energy produced during metabolism is measured as heat and is expressed in terms of the kilocalorie (kcal), which is the amount of heat required to change the temperature of 1 kg of water from 14.5° to 15.5°C.

The caloric value of a food can be determined in a *bomb calorimeter* (Figure 2.3). In this instrument, the sample is placed in a metal container with pure oxygen under pressure. The sample is then ignited electrically. The metal container is surrounded by water, and the heat generated by the combustion raises the temperature of the water. The number of calories of heat generated in the combustion can be determined by measuring the quantity of water in the calorimeter and the rise in its temperature.

The energy of fats, carbohydrates, and proteins in metabolism

A gram of pure fat, carbohydrate, and protein, when burned in a bomb calorimeter, produces 9.4, 4.2, and 5.6 kcal of heat, respectively. Fats and carbohydrates produce nearly the same amount of energy when they are oxidized in the body as in a calorimeter, but a protein produces only about 4 kcal. This discrepancy arises because the nitrogen in proteins is not oxidized during its metabolism in the body but is excreted as urea in the urine. An additional amount of energy is formed in the calorimeter as nitrogen atoms of the protein are oxidized to molecular nitrogen.

In calculating the calorific value of food in animal or human calorimetry, the approximate values of 9, 4, and 4 kcal are employed for a gram of fat, carbohydrate, and protein, respectively.

THE RESPIRATORY QUOTIENT

The respiratory quotient, abbreviated R.Q., is the ratio of the volume of carbon dioxide produced to the volume of oxygen used by a living organism or tissue in metabolic processes. It may be expressed by the following formula:

$$\text{R.Q.} = \frac{\text{volumes of } CO_2}{\text{volumes of } O_2}$$

The average R.Q. for the metabolism of fats, carbohydrates, and proteins has been established as 1.0, 0.7, and 0.8, respectively. Following a balanced meal, the R.Q. will usually be about 0.85, whereas at periods when digestion is not occurring it will drop to about 0.82.

BASAL METABOLIC RATE

The *basal metabolic rate,* abbreviated BMR, is defined as the amount of heat produced by the body when it is in a state of mental and physical rest and when no digestion is taking place. The BMR is usually expressed in kilocalories produced per square meter of body surface per hour.

Because a direct determination of the amount of heat produced by an individual is rather difficult, an indirect method is usually employed in which the amount of oxygen consumed in a given time is measured (Figure 27.2). Because 1 liter of oxygen is equivalent to 4825 kcal of heat (at R.Q. of 0.82), the amount of heat produced in a unit of time, such as an hour, is computed. The BMR is com-

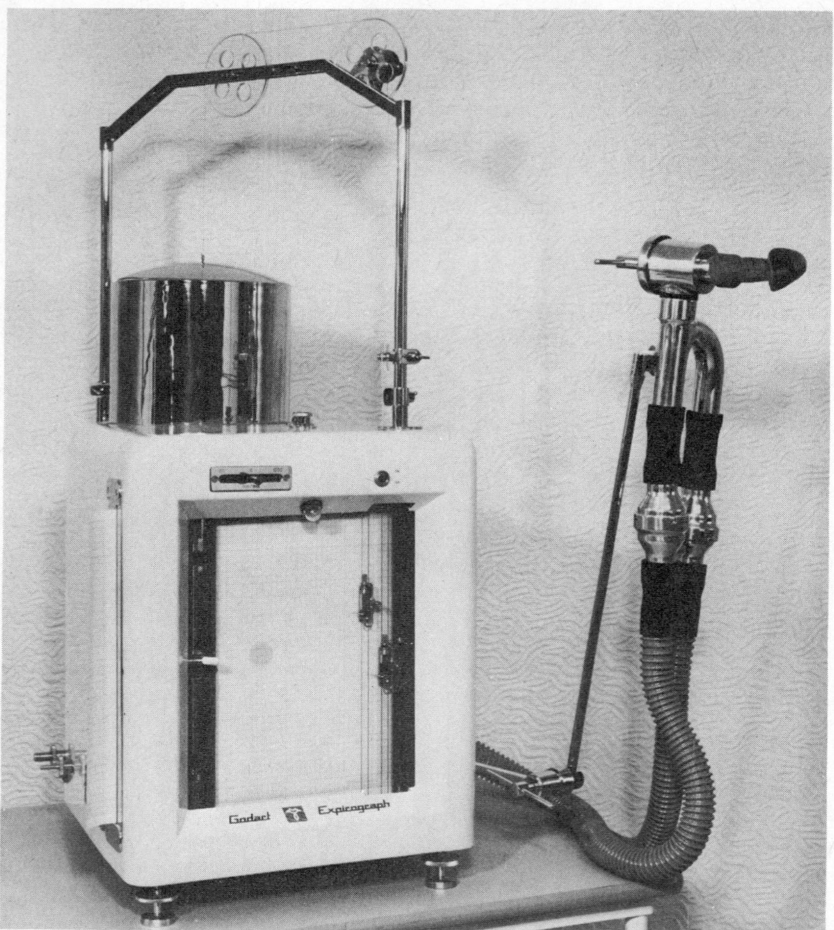

Figure 27.2 *An apparatus for determining the basal metabolic rate. (Godart Expirograph, courtesy Instrumentation Associates, Inc., N.Y.)*

pared to normal values and is expressed in percentages above or below normal. Thus if the BMR of an individual is 20 percent below the normal value, it will be reported as −20 (preferably "minus 20"). The table of normal values is based on actual tests on normal individuals of both sexes and of various weights, heights, and ages. The BMR should be determined on a patient when he awakes in the morning and before he partakes of any food. He should have had a good night's sleep and a light meal the evening before the test. For infants, the value for the BMR is frequently determined on sleeping, fed babies, but these values are still felt to be comparable to values for fasting, awake, older children and adults.

The amount of heat produced by an individual is roughly proportional to the surface area and depends little upon weight. The production of heat by a small individual pound for pound of weight will be greater than in a "heavy" person. The BMR also depends on the sex. A normal male adult has a BMR of approximately 40 kcal per square meter of surface per hour, whereas a female adult has one of approximately 37 kcal. The BMR is largest in the infant 3 to 6 months of age and decreases slowly until old age is reached.

The determination of the BMR is of great clinical value in determining abnormal activity of the thyroid gland. In hyperthyroidism (page 516), the thyroid will be overactive and the BMR will be above average. It will also be elevated during fevers and is frequently high during pregnancy or if there is an overactivity of the pituitary glands. It will be abnormally low in hypothyroidism (page 515), that is, underactivity of the thyroid glands, and also in wasting diseases, anemias, and underactivity of the pituitary glands.

STUDY QUESTIONS

1. Define metabolism, anabolic processes, and catabolic processes.
2. Summarize briefly the absorption of the products of digestion of fats, carbohydrates, and proteins through the intestinal walls.
3. In the metabolism of sugars in the body, what is meant by glycogenesis and glycogenolysis?
4. Discuss the steps involved in the conversion of (a) glucose to glycogen and (b) glycogen to glucose in the liver.
5. What is the usual range of the concentration of glucose in the blood of a normal individual? What happens to the blood-sugar level following a meal rich in carbohydrates?
6. What is meant by the renal threshold of glucose?
7. How does the concentration of glucose in the blood of a diabetic differ from that in a normal individual following a meal rich in carbohydrates? How is the sugar tolerance test carried out, and why is it an important test?
8. What is the main source of energy for the contraction of muscles? Although muscles are removing glucose from the blood and utilizing it for energy, why does the blood-sugar level remain nearly constant during fasting?
9. What is meant by energy-rich phosphate bonds? Compare the energy released when glucose-6-phosphate, the phosphate ester of glycerol, ADP, and ATP are hydrolyzed. How may the energy released in the hydrolysis of ADP and ATP be utilized?
10. Is the oxidation of glucose in the muscle tissue a simple process? Into what two phases can the degradation of muscle glycogen be divided? In which of these phases is the muscle contracting and in which one is it recovering? May oxygen be present in the anaerobic phase?
11. What is phosphorylation? Distinguish between a phosphatase and a phosphorylase.
12. Discuss briefly our present knowledge concerning the anaerobic phase of sugar metabolism in muscle tissues; the aerobic phase.
13. Describe the four main steps in the breakdown of glycogen to lactic acid in the anaerobic phase of glycogen metabolism.
14. For each molecule of glucose that is converted to two molecules of

lactic acid in the glycolytic cycle, how many molecules of ADP are converted to ATP?

15. For each molecule of pyruvic acid that is oxidized to carbon dioxide and water, how many molecules of ADP are converted to ATP?
16. How does the alcoholic fermentation of glucose differ from the glycolysis of this sugar? What organism usually induces alcoholic fermentation? What enzyme is present in this organism that is absent in the muscles of the vertebrates?
17. What uses are made of fats in the body cells?
18. What is depot fat, or adipose tissue? What other purposes besides a reserve food supply does the adipose tissue of the body serve?
19. In fat storage in the body, distinguish between the constant element and the variable element.
20. How is the glycerol that is formed by the hydrolysis of fats in the body metabolized to carbon dioxide and water?
21. How did Knoop explain the oxidation of fatty acids to carbon dioxide and water in the body cells? What is the name of the theory of fat metabolism proposed by Knoop? Why is it so called?
22. In what organs of the body are the hormones insulin, glucagon, and epinephrine produced? What is their action in the body?
23. Why does glucose sometimes appear in the urine during emotional stress?
24. What changes in the pancreas appear to be responsible for the pathological condition diabetes mellitus?
25. What other pathological conditions usually accompany incomplete carbohydrate metabolism? Explain the formation of the three ketone bodies in ketosis.
26. Name some other disorders besides diabetes mellitus that are accompanied by ketosis.
27. Why is ketosis usually accompanied by acidosis? What are the symptoms of severe acidosis?
28. Distinguish between ketogenic and antiketogenic substances.
29. Describe the mechanism by which a fatty acid is converted by oxidation to a fatty acyl coenzyme A ester with two less carbon atoms than the original fatty acid. How many times would this series of reactions have to be repeated to completely convert stearic acid to acetyl coenzyme A?
30. What symptoms will appear in rats and in humans who subsist on a diet deficient in the essential fatty acids? Name some of these essential fatty acids. What is characteristic about their structure?
31. What are the two general causes of obesity?
32. What one experiment can be cited as evidence to show that carbohydrates can be transformed into fats?
33. What is meant by lipogenesis? Why do fatty acids with an even number of carbon atoms predominate in our bodies?
34. Discuss the steps involved in the conversion of carbohydrates to fatty acids. What roles do acetyl coenzyme A and malonyl coenzyme A play in lipogenesis?
35. What is the fate of the amino acids that are not needed for the synthesis of new tissue or the replacement of worn-out tissue?
36. What changes occur in the structure of an amino acid during oxidative deamination? What happens to the ammonia that is liberated? What is transamination?
37. Why will the amino acid alanine, $CH_3—CHNH_2—COOH$, follow the

route of carbohydrate metabolism after oxidative deamination? List some other antiketogenic amino acids.
38. Why are some amino acids called ketogenic amino acids? List a few amino acids that belong to this group.
39. In a diabetic, the metabolism of an antiketogenic amino acid will be responsible for the appearance of what abnormal substance in the urine? What will be the end products of the metabolism of a ketogenic amino acid by such a patient?
40. Is urea formed in the liver by the simple union of ammonia and carbon dioxide? Describe the cycle by which it is supposedly formed in the liver of all animals that excrete urea.
41. What role does ammonia play in preventing alkalosis and acidosis?
42. When is a person in nitrogen equilibrium? Under what conditions will a negative nitrogen balance exist? A positive nitrogen balance?
43. What experiment must be performed to determine if an amino acid is essential? List some essential amino acids. What is meant by an adequate protein?
44. According to Knoop's theory of fat oxidation, through what steps will the oxidation of caproic acid, CH_3—$(CH_2)_4$—COOH, pass when it is oxidized to acetoacetic acid, CH_3—CO—CH_2—COOH, in the cells of the body?
45. In the absence of proper carbodydrate metabolism, __________ are not properly metabolized. In such cases, the end product of the metabolism of these latter substances are three substances named __________, __________, and __________ and these are called __________. Their presence in the body fluids is called __________. Two of these substances are acid, and they use the alkaline reserve of the body, producing a condition known as __________.
46. In carbohydrate metabolism, the two phases of muscular contraction are called the __________ phase and the __________ phase. __________ acid is the final product of the first phase.
47. A diet poor in carbohydrates but rich in fats is called a __________ diet.
48. What is meant by the respiratory quotient? By the BMR?
49. What are the caloric heat values of fats, carbohydrates, and proteins? What are the heat values of these food constituents when oxidized in the human body? Account for any variation noted in these values as determined in the bomb calorimeter and in human subjects.
50. Discuss the following

    (a) the interrelationship between carbohydrate and fatty acid metabolism
    (b) the interrelationship between carbohydrate and amino acid metabolism
    (c) the interrelationship between fatty acid and amino acid metabolism

51. Why are plants and some microorganisms able to convert fatty acids to glucose whereas animals are unable to do so?

chapter 28

photosynthesis and the biosynthesis of the carbohydrates

28

Introduction

Plants are capable of producing carbohydrates from the carbon dioxide of the air and water taken from the soil. Radiant energy of the sun and the chlorophyll (page 431) of the plant are necessary for this reaction, which is called *photosynthesis* (Figure 28.1). Oxygen is also produced in this reaction. The overall reaction may be represented by the equation

$$6CO_2 + 6H_2O \longrightarrow C_6H_{12}O_6 + 6O_2$$

However, it is a complex reaction, and many steps are involved before the synthesis is complete (see below).

Because animals cannot produce carbohydrates from such simple starting materials as carbon dioxide, they are dependent on plants for this important food item. Plants are apparently able to produce fats from the carbohydrates, and the latter must furnish the carbon necessary for the formation of proteins.

Photosynthesis is important, because in this reaction the radiant energy of the sun is stored as chemical energy in the carbohydrates, which are then utilized as needed. Although plants require some oxygen for their functioning, they produce a greater quantity by photosynthesis. Part of this oxygen is available for animal use.

Chlorophyll

Chlorophyll a and b are the most important of the photosynthetic pigments. Chlorophyll a occurs more abundantly and more commonly in plants than does chlorophyll b. The carotenoids may also play a role in photosynthesis.

Chlorophyll a is a tetrapyrrole (page 345) to which Willstäter has assigned the formula shown in Figure 28.2.

In pyrrole ring IV of chlorophyll a, the propionic acid side chain is esterified with the alcohol phytol, which has the formula

$$(CH_3)_2CHCH_2CH_2CH_2\underset{\displaystyle CH_3}{\underset{|}{C}}HCH_2CH_2CH_2\underset{\displaystyle CH_3}{\underset{|}{C}}HCH_2CH_2CH_2\underset{\displaystyle CH_3}{\underset{|}{C}}{=}CHCH_2OH$$

Oxygen
Water
saturated
with CO2
Water
plant

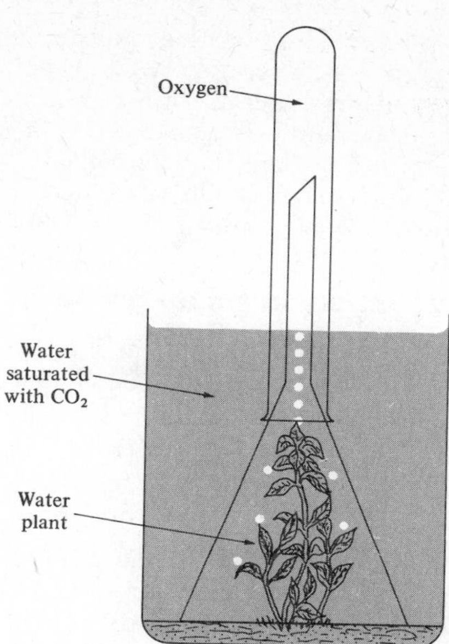

Figure 28.1 Green leaves of water plants placed in water saturated with carbon dioxide and exposed to sunlight liberate oxygen. (From Elder, Scott, and Kanda, Textbook of Chemistry, *rev. ed., Harper & Row, N.Y., 1948.)*

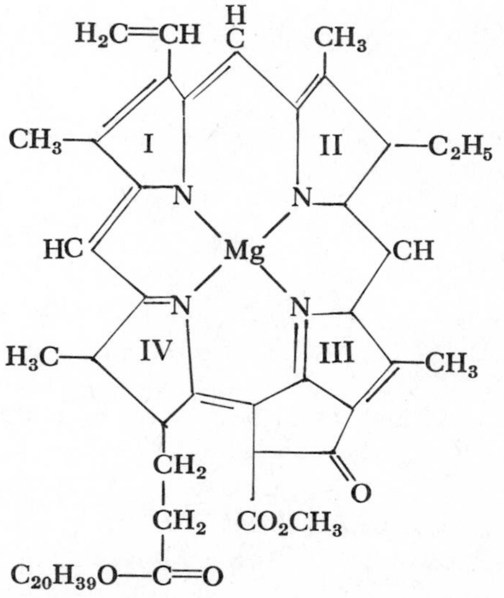

H2C=CH
H
C
CH3
CH3
I
II
C2H5
N
N
HC
Mg
CH
N
N
H3C
IV
III
CH3
CH2
O
CH2
CO2CH3
C20H39O—C=O

Figure 28.2 Structure of chlorophyll a.

In chlorophyll b, the methyl group that appears in pyrrole ring II of chlorophyll a is replaced by a formyl (—CHO) group.

Chlorophyll occurs in the grana of the chloroplast (page 449) of plant leaves. Ten to one hundred grana occur in each plant chloroplast. The grana consist of lamellae, in which layers that contain chlorophyll and lipoidal material alternate with protein layers that contain the necessary enzymes utilized in the photosynthetic process.

The light reaction in photosynthesis

At one time it was believed that in photosynthesis, carbon dioxide was reduced directly to the sugars and that it was the source of the oxygen that is evolved during the photosynthetic process, thus:

$$CO_2 + H_2O \longrightarrow \underset{\text{carbohydrate}}{(CH_2O)_x} + O_2$$

Further, formaldehyde was considered the important intermediate in the synthesis of sugars in this process, for Emil Fischer demonstrated that formaldehyde, when treated with dilute barium hydroxide, is converted, in part, to glucose.

$$\underset{\text{formaldehyde}}{6CH_2O} \longrightarrow \underset{\text{glucose}}{C_6H_{12}O_6}$$

If water labeled with ^{18}O is utilized in the photosynthetic process, all of the labeled oxygen atoms will appear in the evolved oxygen, indicating that it is water that is the source of the oxygen and not carbon dioxide.

$$C^{16}O_2 + 2H_2{}^{18}O \longrightarrow \underset{\text{carbohydrate}}{(CH_2{}^{16}O)_x} + {}^{18}O_2 + H_2{}^{16}O$$

Actually, the photosynthetic reaction can be separated into a light and a dark reaction. In the light reaction, water is photooxidized to oxygen, hydrogen ions, and electrons.

$$2H_2O \longrightarrow O_2 + 4H^+ + 4e$$

The resulting protons and electrons convert oxidized nicotine adenine dinucleotide phosphate, $NADP^+$, to the reduced state of the dinucleotide, namely NADPH. In the dark reaction, which is discussed in the next section, the resulting NADPH, through a complex set of reactions, in effect reduces carbon dioxide to hexoses.

The path of carbon in photosynthesis: the dark reaction

Melvin Calvin has established that if green algae are illuminated with a suitable source of light for a brief period and are then permitted to "fix" $^{14}CO_2$ in the dark for five seconds, the predominant labeled compound formed is phosphoglyceric acid:

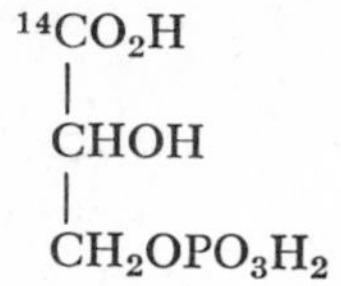

$$\begin{array}{l} ^{14}CO_2H \\ | \\ CHOH \\ | \\ CH_2OPO_3H_2 \end{array}$$

phosphoglyceric acid

The analytical procedure employed in establishing this as the major product of the five-second reaction involves extraction of the algae with a hot alcohol solution. The soluble products are then separated by two-dimensional paper chromatography and, due to the radioactive carbon, can be identified by preparing radioautographs on photographic plates (see page 210).

Calvin further observed that if a longer reaction time for the dark reaction was employed, appreciable amounts of labeled triose phosphates, hexose monophosphates, and hexose diphosphates were produced by these green algae.

At first, a search was made for a two-carbon fragment that combines with $^{14}CO_2$ to form the labeled phosphoglyceric acid as an early product of the dark reaction. It has now been established that $^{14}CO_2$ first combines with a ketopentose diphosphate, namely, ribulose-1,5-diphosphate,

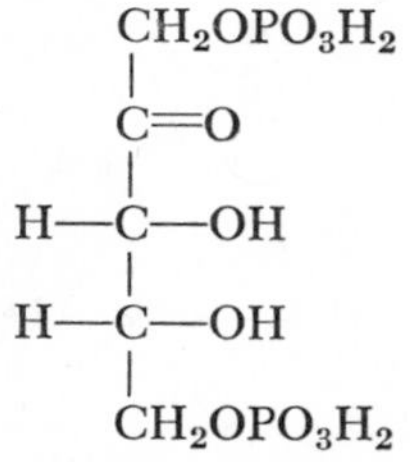

$$\begin{array}{c} CH_2OPO_3H_2 \\ | \\ C{=}O \\ | \\ H{-}C{-}OH \\ | \\ H{-}C{-}OH \\ | \\ CH_2OPO_3H_2 \end{array}$$

ribulose-1,5-diphosphate

formed by the "phosphorylation" of ribulose-5-phosphate. The product of the addition reaction probably has the following structure:

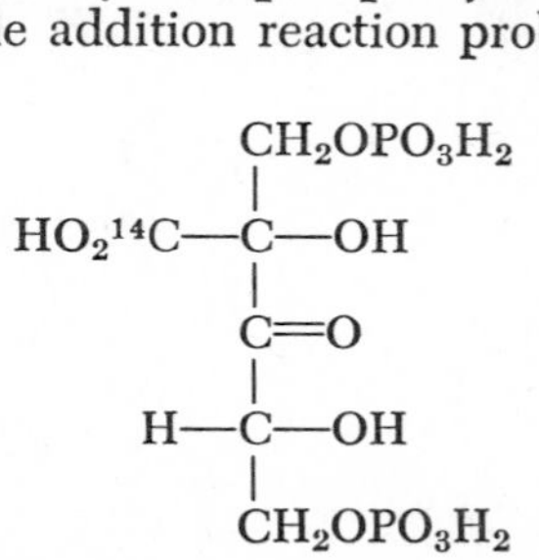

$$\begin{array}{c} CH_2OPO_3H_2 \\ | \\ HO_2{}^{14}C{-}C{-}OH \\ | \\ C{=}O \\ | \\ H{-}C{-}OH \\ | \\ CH_2OPO_3H_2 \end{array}$$

This readily cleaves by hydrolysis into two molecules of phosphoglyceric acid.

TABLE 28.1 THE FATE OF CARBON IN PHOTOSYNTHESIS

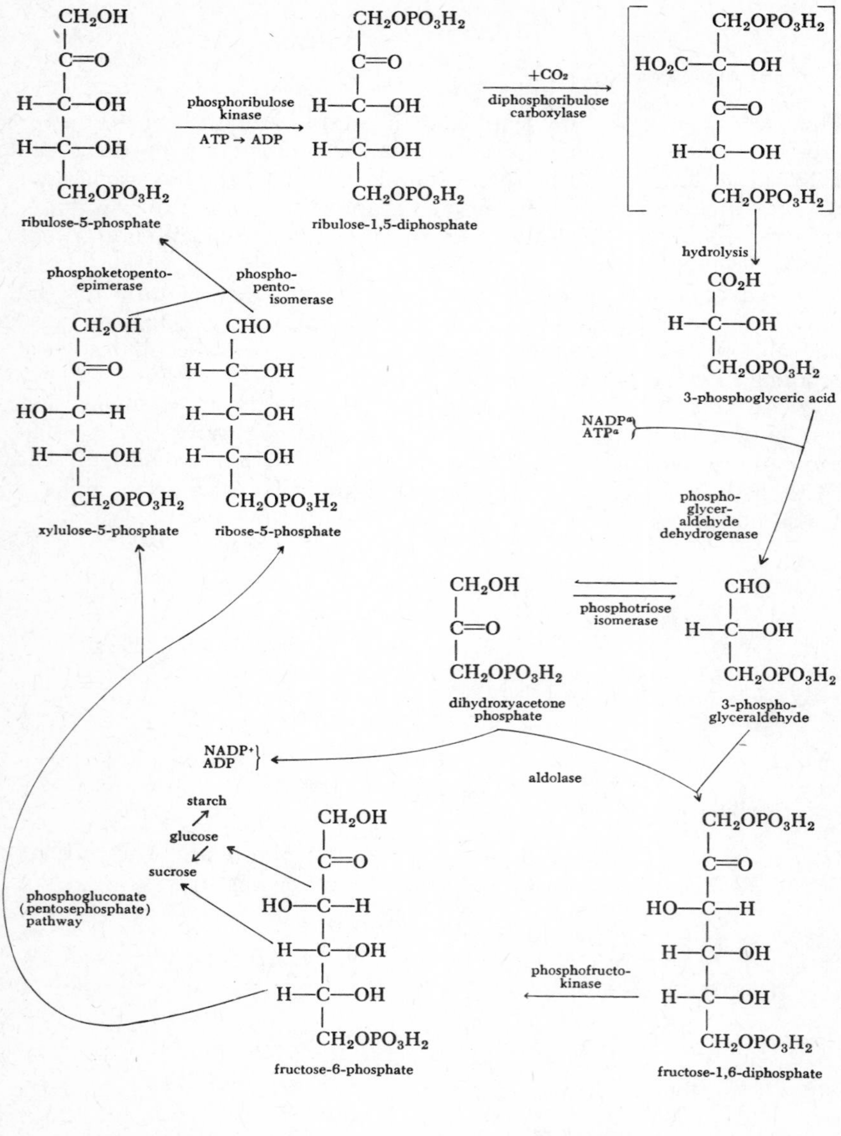

CH_2OH—C=O—H—C—OH—H—C—OH—$CH_2OPO_3H_2$
ribulose-5-phosphate

phosphoribulose kinase
ATP → ADP

$CH_2OPO_3H_2$—C=O—H—C—OH—H—C—OH—$CH_2OPO_3H_2$
ribulose-1,5-diphosphate

$+CO_2$
diphosphoribulose carboxylase

[$CH_2OPO_3H_2$—HO_2C—C—OH—C=O—H—C—OH—$CH_2OPO_3H_2$]

hydrolysis

CO_2H—H—C—OH—$CH_2OPO_3H_2$
3-phosphoglyceric acid

NADP[a]
ATP[a]

phospho-glycer-aldehyde dehydrogenase

CHO—H—C—OH—$CH_2OPO_3H_2$
3-phospho-glyceraldehyde

phosphotriose isomerase

CH_2OH—C=O—$CH_2OPO_3H_2$
dihydroxyacetone phosphate

aldolase

NADP+
ADP

$CH_2OPO_3H_2$—C=O—HO—C—H—H—C—OH—H—C—OH—$CH_2OPO_3H_2$
fructose-1,6-diphosphate

phosphofructo-kinase

CH_2OH—C=O—HO—C—H—H—C—OH—H—C—OH—$CH_2OPO_3H_2$
fructose-6-phosphate

starch
glucose
sucrose

phosphogluconate (pentosephosphate) pathway

CH_2OH—C=O—HO—C—H—H—C—OH—$CH_2OPO_3H_2$
xylulose-5-phosphate

CHO—H—C—OH—H—C—OH—H—C—OH—$CH_2OPO_3H_2$
ribose-5-phosphate

phosphoketopento-epimerase

phospho-pento-isomerase

[a]Produced in the light reaction.

The resulting phosphoglyceric acid is reduced to phosphoglyceraldehyde by the NADPH and ATP produced in the light-catalyzed photolysis of water to oxygen (Table 28.1).

$$\text{phosphoglyceric acid} + \text{NADPH} + \text{H}^+ + \text{ATP} \longrightarrow$$
$$\text{phosphoglyceraldehyde} + \text{NADP}^+ + \text{ADP} + \text{phosphate}$$

In a reversal of the glycolytic cycle (page 402), the following changes then occur:

1. Some of the phosphoglyceraldehyde is rearranged to dihydroxyacetone phosphate by the enzyme phosphotriose isomerase.
2. Interaction of phosphoglyceraldehyde and dihydroxyacetone phosphate in the presence of aldolase produces fructose-1,6-disphosphate.
3. The fructose-1,6-diphosphate is, in part, converted to glucose phosphates and glucose. The latter serves as a source of sucrose and starch in the plant.

It is to be noted that for each six molecules of fructose-1,6-diphosphate formed at this point, only two of the molecules represent newly synthesized hexose phosphate from the carbon dioxide utilized.

Some mechanism must be operative in the plant to regenerate ribulose-5-phosphate, which, as noted above, is a necessary intermediate for the fixation of carbon dioxide in the photosynthetic process. To do this, part of the fructose-1,6-diphosphate synthesized combines with glyceraldehyde phosphate and is converted through several successive steps not indicated in Table 28.1 to two pentose phosphates, xylulose-5-phosphate and ribose-5-phosphate, in the pentose phosphate pathway that has been found to be a minor metabolic pathway competitive with the glycolytic degradation of hexoses. For more information concerning this alternative metabolic pathway, the student is referred to any one of the modern texts in biochemistry. The xylose-5-phosphate and the ribose-5-phosphate are both isomerized to ribulose-5-phosphate to complete the cycle (see Table 28.1).

STUDY QUESTIONS

1. What is the name of the process by which carbohydrates are synthesized from carbon dioxide and water in the green plant?
2. What condition must exist and what materials are required for the photosynthetic process? What important inorganic substance is produced in this process?
3. Discuss the importance of the photosynthetic process to both plants and animals.
4. Sunshine that reached the earth in geologic times past now heats our homes. Explain.
5. What must be the role of chlorophyll in photosynthesis?
6. Discuss the structure of chlorophyll a and chlorophyll b.

7. Describe the organization of that portion of a cell of a green plant in which the photosynthetic process occurs.
8. On what experimental basis has it been established that the oxygen liberated in photosynthesis originates from water molecules rather than from the carbon dioxide molecules that are converted to carbohydrates?
9. In photosynthesis, what chemical changes occur in the "light reaction"?
10. What substance formed in the light reaction is responsible for the conversion of carbon dioxide to carbohydrates in the so called dark reaction?
11. What substance initially combines with carbon dioxide in the dark reaction of photosynthesis? What is the nature of this addition product? Describe the steps by which it will be converted to fructose-1,6-disphosphate.
12. In the dark reaction, one of the initial products formed when carbon dioxide combines with ribulose-1,5-diphosphate is phosphoglyceric acid. Upon what experimental evidence did Melvin Calvin establish this as an early significant product of the dark reaction?
13. Describe briefly (a) the steps by which fructose-1,6-diphosphate can be converted to glucose and starch and (b) how ribulose-5-phosphate, which is required to initiate the dark reaction, is regenerated from it.
14. Explain why the initial formation of six molecules of fructose-1,6-diphosphate in the dark reaction (See Table 28.1) represents the synthesis of only two molecules of hexose diphosphate.

chapter 29

---

the structure and biological role of the nucleic acids

29

Introduction

Nucleic acids, as well as such simple proteins as the histones or protamines, are hydrolysis products of those conjugated proteins known as *nucleoproteins* (page 352). The nucleic acids themselves are high-molecular-weight polymeric substances that are degraded stepwise through the following types of products, which will be described below:

nucleic acids
↘
nucleotides
↘ (→)
nucleosides and phosphoric acid
↘ (→)
sugars and nitrogen bases

Those nucleic acids that upon hydrolysis produce the pentose sugar deoxyribose occur in the chromatin of cell nuclei and are called *deoxyribonucleic acids,* or DNA's. Small amounts of DNA occur in the cytoplasm of the cell. At one time it was believed that nucleic acids occurred exclusively in cell nuclei; hence their class name.

Those nucleoproteins that are hydrolyzed to ribose and are known as *ribonucleic acids,* or RNA's, are most abundant in the cytoplasm of the cell, although small amounts of these occur in the cell nuclei.

Although our knowledge of the nucleoproteins and their hydrolysis products, the nucleic acids, is far from complete, enough is now known concerning these substances to indicate their outstanding importance in biological systems. The self-reproducing chromosomes present in cells appear to possess the property of transmitting hereditary factors. They are primarily, if not exclusively, nucleoprotein in nature. Certain nucleic acids appear also to play a most important role in the fabrication of the proteins of the body.

Viruses consist of nucleoprotein material covered by a protective coating of proteins. They can reproduce only in the host cell but otherwise appear to have the typical characteristics of living material. A number of viruses have been isolated as crystalline solids.

A number of theories concerning the nature and mode of transmission of cancer are at present being considered. Bittner has described an agent, a nucleoprotein, which is present in milk, mammary tissue, blood, and other tissues of certain strains of mice susceptible to mammary tumors and which can be transmitted by their milk to their offspring or to other strains of mice susceptible to this type of tumor.

The final hydrolysis products of the nucleic acids

Complete hydrolysis converts nucleic acids to phosphoric acid, heterocyclic nitrogen bases known as purines and pyrimidines, and a sugar, usually ribose or deoxyribose.

The pyrimidines are six-membered heterocyclic compounds that contain two nitrogen atoms in the ring, separated by a carbon atom. The parent compound pyrimidine,

```
       H
       C
     //4\
   N3    5CH
   |2    6||
   HC     CH
     \\1 /
       N
```

is of little importance. Cytosine and uracil, the 4-amino-2-hydroxy and the 2,4-dihydroxy derivatives, respectively, of pyrimidine are obtained as hydrolysis products of the ribonucleic acids (RNA's), while cytosine and thymine, the 2,4-dihydroxy-5-methyl derivative, are the two pyrimidines usually obtained when deoxyribonucleic acids (DNA's) are hydrolyzed.

```
       NH2               OH                 OH
       C                 |                  |
     //  \               C                  C
   N      C—H          //  \              //  \
   |      ||         N      C—H         N      C—CH3
   C      C—H        |      ||          |      ||
  / \\   /           C      C—H         C      C—H
HO    N             / \\   /           / \\   /
                  HO    N            HO    N

    cytosine          uracil             thymine
```

The purines are bicylic compounds in which two nitrogen atoms occur in each of the two rings. Adenine and guanine, the 6-amino- and the 2-amino-6-hydroxy derivatives, respectively, of purine are hydrolysis products of both the ribonucleic and the deoxyribonucleic acids. Purine, their parent compound, is of little importance.

```
           H
           C
        // 6 \      N
      N1      5C  7  \\
      |        ||     8CH
      C2      4C  9  /
     /  \\  3 /    N
    H     N        H

         purine
```

```
            NH2
             |
             C         N
          //   \     /   \\
        N        C         CH
        |        ||       /
        C        C      N
       /  \\   /    \  /
      H     N         H

           adenine
```

```
            OH
             |
             C         N
          //   \     /   \\
        N        C         CH
        |        ||       /
        C        C      N
       /  \\   /         H
     H2N    N

           guanine
```

Note in the pyrimidine and purine formulas above the typical method employed for numbering these ring systems.

Adenine is also the hydrolysis product of adenosine triphosphate, ATP (page 401). The energy resulting from the oxidation of monosaccharides and lipids in the body is stored in this latter substance, and when needed for muscular contractions it is regenerated by hydrolysis.

Uric acid, caffeine, and theobromine are also derivatives of purine. Uric acid (page 459) is one of the end products of the metabolism of the nucleoproteins in humans, and it occurs in small amounts in the blood and in urine. In gout, the uric acid content of the blood is abnormally high and it deposits in certain joints in the body. Caffeine and theobromine are stimulants and diuretics, the former occurring in coffee and tea, the latter in cocoa.

As previously noted (page 438), the pentoses ribose and deoxyribose

```
       1                      1
      CHO                    CHO
      | 2                    | 2
  H—C—OH               H—C—H
      | 3                    | 3
  H—C—OH               H—C—OH
      | 4                    | 4
  H—C—OH               H—C—OH
      | 5                    | 5
     CH2OH                  CH2OH

     ribose               deoxyribose
```

are both hydrolysis products of the nucleoproteins. Note that in deoxyribose, the alcohol group present on the #2 carbon atom of ribose is replaced by a hydrogen atom. In the nucleic acids, these two pentoses occur in cyclic structures that can be represented as

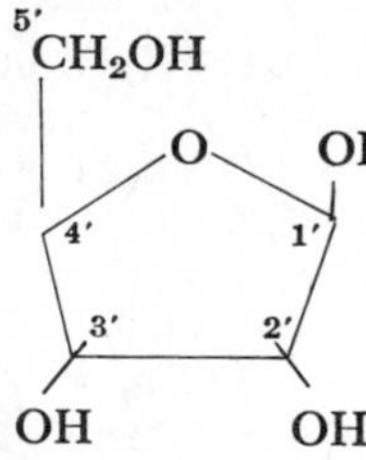

ribose and deoxyribose

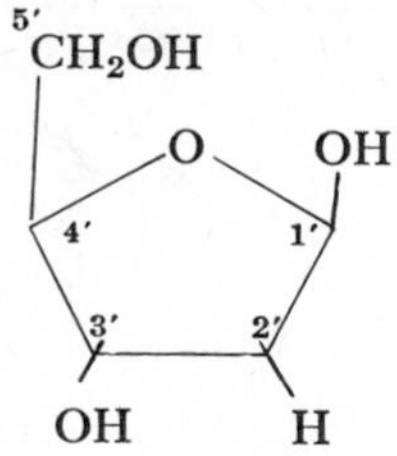

In nucleic acids, nucleotides, and nucleosides, these rings are assigned primed numbers so that there will be no confusion with the numbered substituents of the pyrimidines and purines in these structures. Note that ring carbons and most hydrogen atoms directly attached to the ring are omitted in the above cyclic formulas for these sugars.

Nucleosides

Hydrolysis of a nucleoside produces a nitrogen base (a purine or pyrimidine) and a sugar (usually ribose or deoxyribose). Union of these two types of substances will be through the #1 nitrogen atom of the pyrimidine or #9 nitrogen atom of the purine and the 1′ carbon atom of ribose or deoxyribose. Note that this is a β-glycosidic linkage (page 309). Those nucleosides derived from ribose are called *ribosides,* whereas those derived from deoxyribose are known as *deoxyribosides.*

The ribosides of cytosine and adenine, called *cytidine* and *adenosine,* have the following structures:

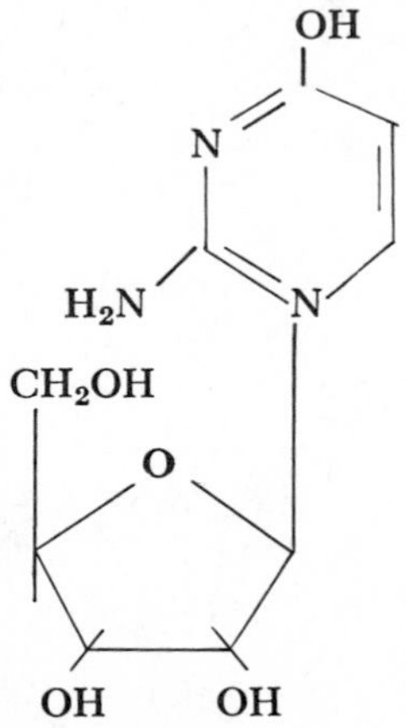

cytidine

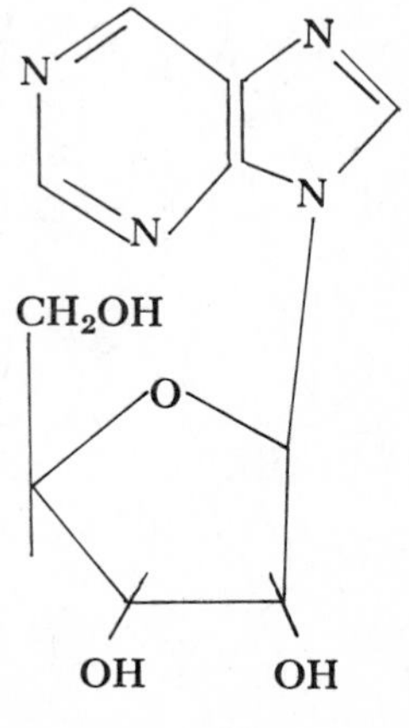

adenosine

TABLE 29.1 SOME NUCLEOSIDES AND NUCLEOTIDES

Pentose Sugar	*Pyrimidine*	*Purine*	*Nucleoside*	*Nucleotide*
Ribose	Cytosine		Cytidine	Cytidylic acid
Ribose	Uracil		Uridine	Uridylic acid
Ribose		Guanine	Guanosine	Guanylic acid
Ribose		Adenine	Adenosine	Adenylic acid
Deoxyribose	Cytosine		Deoxycytidine	Deoxycytidylic acid
Deoxyribose	Thymine		Thymidine[a]	Thymidylic acid[a]
Deoxyribose		Guanine	Deoxyguanosine	Deoxyguanylic acid
Deoxyribose		Adenine	Deoxyadenosine	Deoxyadenylic acid

[a]Because ribonucleic acids are not hydrolyzed to the corresponding nucleosides and nucleotides, the prefix *deoxy* is deleted.

The names of a number of nucleosides, as well as their constituents, are indicated in Table 29.1.

Nucleotides

These are phosphate esters of the nucleosides. In the nucleotides derived from ribose, phosphoric acid may be esterified with the alcohol groups at positions 2′, 3′ or 5′, whereas in the deoxyribose nucleosides, esterifiable alcohol groups are present only on the 3′ and the 5′ carbon atoms.

The nucleotide adenosine-5′-monophosphate, also known as *adenylic acid,* or AA, which can be obtained by the hydrolysis of adenosine triphosphate (ATP), has the structure

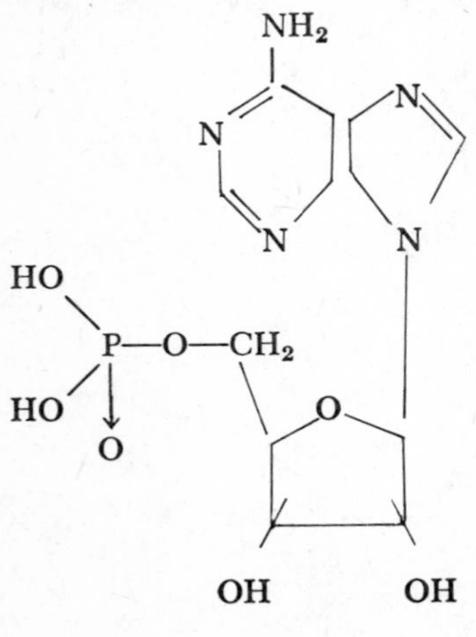

NH_2
N N
N N
HO
P—O—CH_2
HO O
O
OH OH

A number of nucleotides obtained by hydrolysis of nucleic acids are indicated in Table 29.1.

Structure of the nucleic acids

The nucleic acids are polymers whose monomers are the nucleotides. In these polymeric substances, the linkage between the monomer units is through a phosphate diester group that links the 3′ and the 5′ carbon atoms of ribose or deoxyribose to form a linear chain in which phosphate groups alternate with sugar molecules.

$$\mathrm{HO{-}\overset{\overset{\displaystyle O}{\uparrow}}{\underset{\displaystyle OH}{P}}{-}OR} \qquad\qquad \mathrm{RO{-}\overset{\overset{\displaystyle O}{\uparrow}}{\underset{\displaystyle OH}{P}}{-}OR}$$

a phosphate monoester — a phosphate diester

The purines and pyrimidines are attached to the sugar molecules and project out from the linear chain, thus:

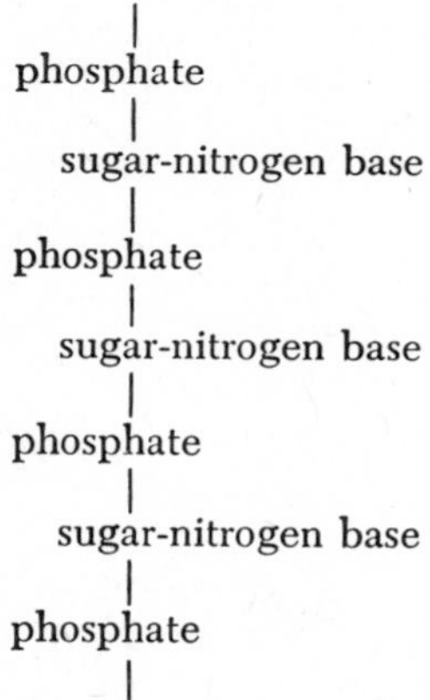

In the structure proposed by Watson and Crick for the DNA's, two parallel polynucleotide chains are wrapped around each other in a double helical structure (see Figures 29.1, 29.2, and 29.3). Such a structure appears to be stabilized by hydrogen bonding between the nitrogen bases that extend inward from the "backbone" of the chains. In such a structure, the purine adenine is hydrogen bonded (page 440) to the pyrimidine thymine, whereas the purine guanine is hydrogen bonded to the pyrimidine cytosine (Figure 29.4). In support of this, adenine and thymine, as well as cytosine and guanine, occur in a 1:1 ratio in the DNA's.

Although less is known about the structure of the RNA's, they appear to be smaller than the typical DNA's and, like the latter, consist of linear polynucleotide chains.

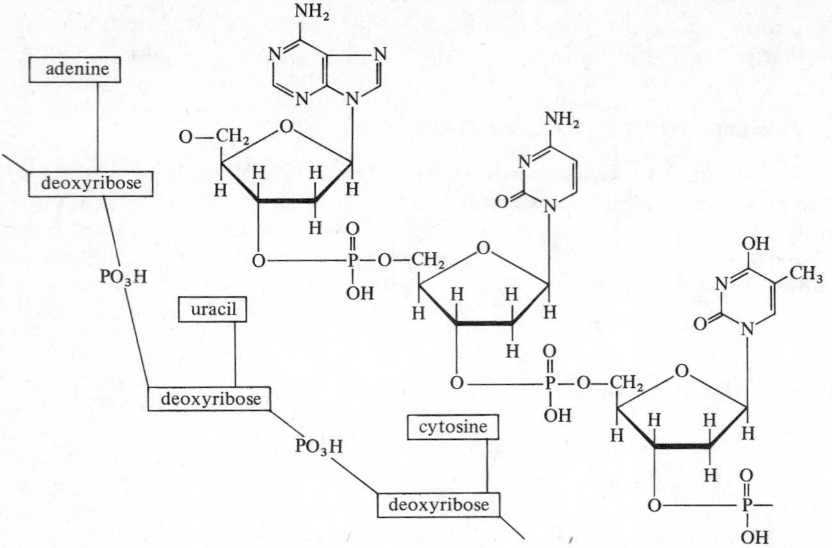

Figure 29.1 A portion of a deoxyribonucleic acid.

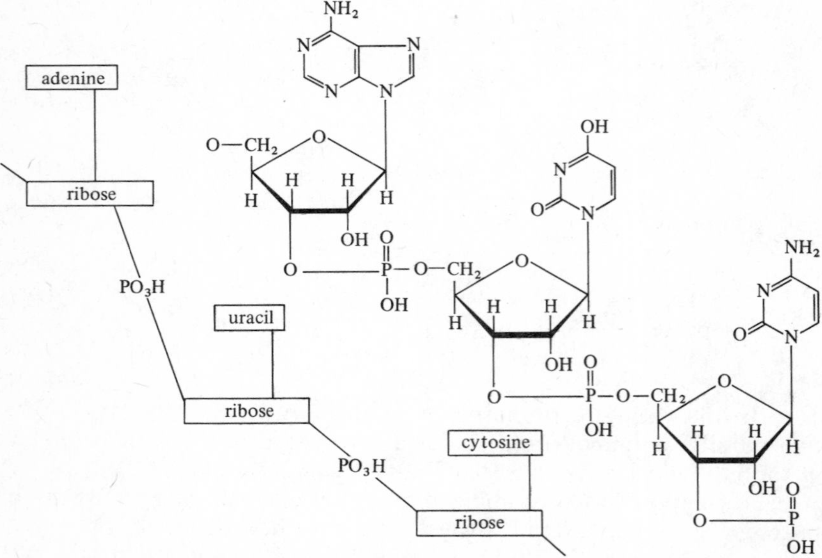

Figure 29.2 A portion of a ribonucleic acid.

DNA replication

Because every cell contains a full complement of DNA, replication of DNA must take place when cells divide (mitosis). This replication appears to occur through a temporary uncoiling of the

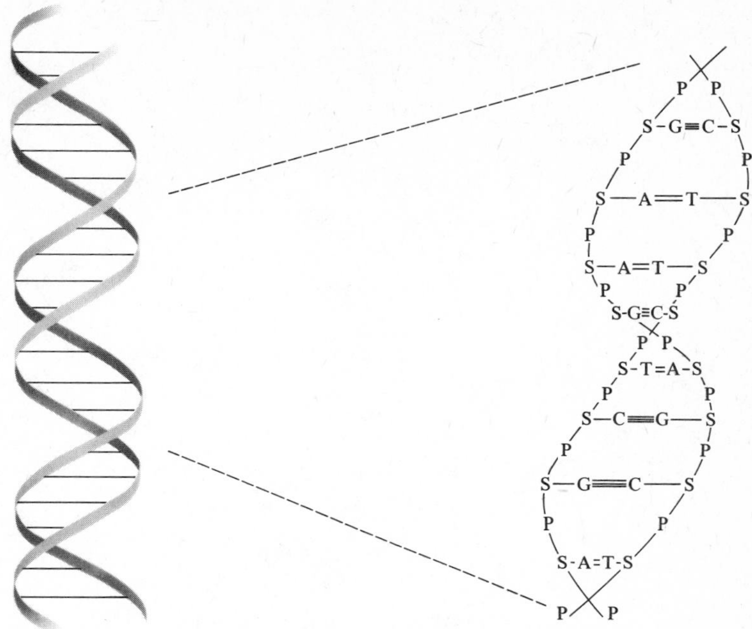

Figure 29.3 Structure of DNA. P, Phosphate; S, sugar; G, guanine; C, cytosine; A, adenine; T, thymine. (From G. I. Sackheim and R. M. Schultz, Chemistry for the Health Sciences, *Macmillan, N.Y., 1969.)*

two chains of the DNA double helix. Each of the bases of the resulting chains will combine through hydrogen bonding with a nucleotide to which is attached a complementary base, that is, adenine in the parent chain can attract and hold a nucleotide with thymine, while cytosine can attract and hold a guanine nucleotide. The linking up of these nucleotide units will produce a new helical strand identical with the other, untwined DNA strand of the original double helix. Thus, utilizing the two untwined DNA strands, two daughter DNA double helices are produced that are identical in every respect with the parent DNA structure (Figure 29.5).

When an ovum combines with a sperm cell at conception, a new cell is produced which has a nucleus in which the parent DNA molecules intermingle. Division of such a cell produces new cells, which contain equal amounts of the genetic material from each parent. This division of nuclear material accounts for the transfer of hereditary characteristics from parent to child.

DNA and the genetic code

Because all DNA molecules have the same "backbone" consisting of alternating deoxyribose and phosphate units, the arrange-

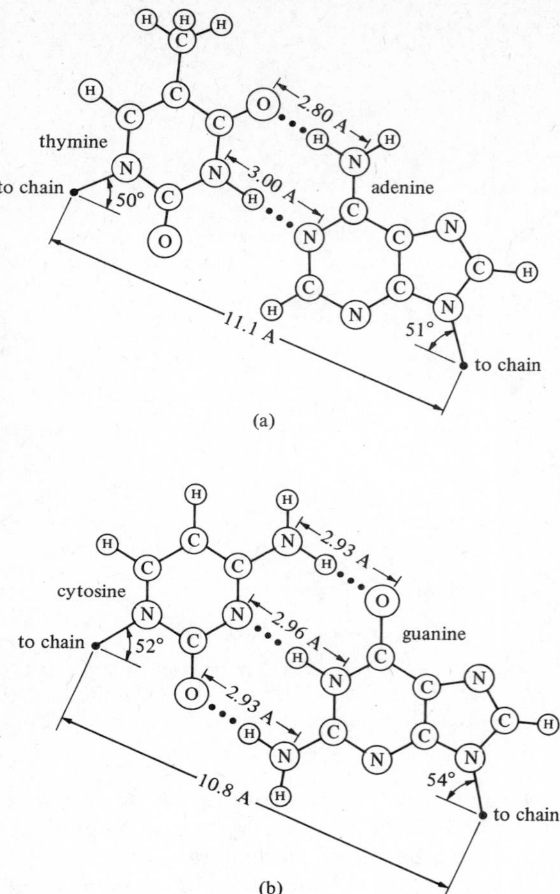

***Figure 29.4** Drawing showing (a) how molecules of adenine and thymine may form a complementary pair held together by two hydrogen bonds, and (b) the same sort of relationship for cytosine and guanine held by three hydrogen bonds. 1 Å is equal to 10^{-8} cm. (From L. Pauling and R. B. Corey,* Arch. Biochem. Biophysics., *65, 164 (1956).)*

ment of the purine and pyrimidine bases attached to the DNA strand must constitute the genetic code. Thus four bases, adenine, thymine, guanine, and cytosine, which can be represented by the letters A, T, G, and C, respectively, are utilized in forming the code. In a DNA molecule, these bases appear to be assembled in groups of three, such as A, C, T; A, T, T; C, C, C; and so on.

The chromosomes of cell nuclei appear to be strands of DNA and proteins. The gene appears to be that region of a DNA molecule that contains a code that directs the synthesis of a specific protein or enzyme.

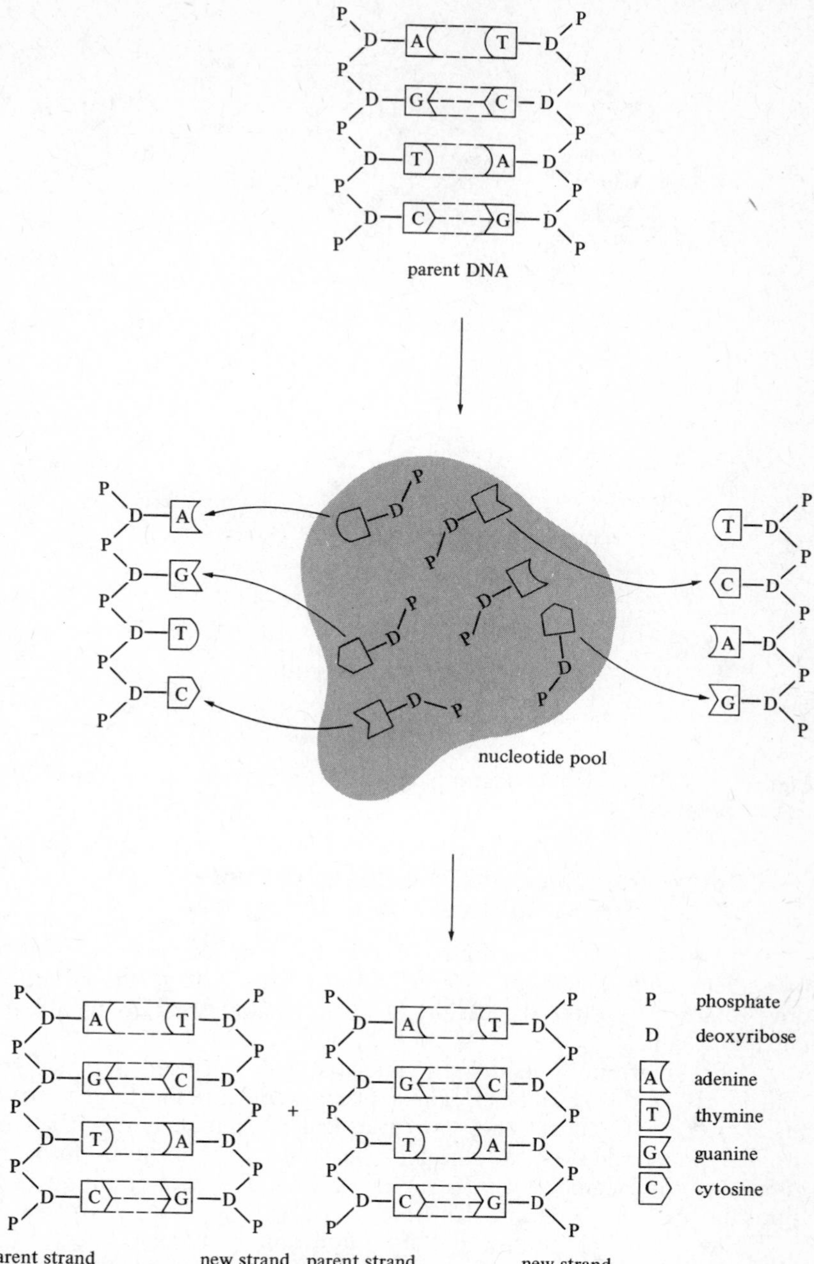

***Figure 29.5** Replication of DNA. (From R. J. Ouellette,* Introductory Organic Chemistry, *Harper & Row, N.Y., 1971.)*

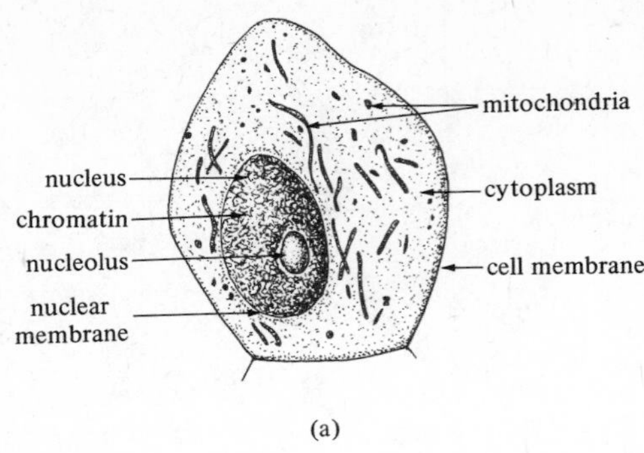

(a)

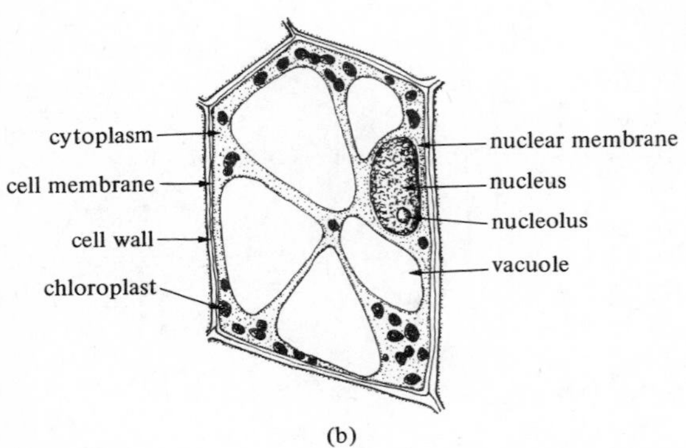

(b)

Figure 29.6 The essential features of (a) a typical animal cell and (b) a typical plant cell.

Cell physiology and the chemical role of certain structures in a living cell

In order to understand more clearly some discussions relating to nucleic acids that appear later in this chapter, it seems advisable now to insert a brief discussion of the physiology of living animal and plant cells.

That material outside the nucleus and enclosed by the cell wall is called the *cytoplasm.* The *mitochondria* are the largest structures in the cytoplasm. They are spherical or rodlike bodies that are usually surrounded by a double membrane that projects into the interior of the mitochondria (see Figure 29.6). The sacs resulting from these inward folds are called *cristae.*

Enzymes that control certain important cellular processes are arranged in a very definite pattern on the surface of the cristae. The fluid that appears between the membrane contains essential coenzymes. Such oxidative processes as the citris acid cycle of carbohydrate metabolism, the oxidation of lipids, the metabolism of proteins, and oxidative phosphorylations occur within the mitochondria.

It is noteworthy that all cells, even those that are carrying out specialized functions, possess certain similarities in structure (Figure 29.6). The central portion of the cell, called the *nucleus*, may contain one or more spheres called *nucleoli*, suspended in a coarse matrix of particles called the *chromatin*. During cell division, the nucleoli disappear and the chromatin particles become separated into thread-like structures called *chromosomes*. The chromosomes are long strands of DNA coated in part with proteins. A gene consist of a certain group of nucleotide bases that constitute a code along a chromosomal strand. The great importance of the chromosomes in the transfer of genetic material from the mother to the daughter cell and in directing the biosynthesis of proteins as noted on pages 449–452.

In all but bacterial cells, the cytoplasm is generally extensively interlaced with interconnected tubular structures, the *endoplasmic reticulum*, which are enclosed by a membrane (Figure 29.7). Attached to the surface of this reticulum are beadlike particles made up mostly of RNA and called *ribosomes*. On these ribosomes, mRNA and tRNA-amino acid complexes act to form new protein molecules (Figure 29.8). Bacterial cells appear to be devoid of the endoplasmic reticulum. In these cells, the ribosomes are dispersed in the cytoplasm and there carry out the usual role in the biosynthesis of proteins. The biosynthesis of lipids and carbohydrates may also occur in the endoplasmic reticulum.

The cytoplasm of cells in green leaves contains ellipsoidal structures called *chloroplasts* (Figure 29.6). These contain such pigments as the chlorophyll and carotenes and must be the locale for the photosynthetic conversion of carbon dioxide and water to the sugars. See Chapter 28.

Transfer of information to RNA molecules and the biosynthesis of proteins

Information coded on a DNA strand is transferred from the nucleus to the cytoplasm by an RNA molecule called messenger-RNA, or mRNA. In the formation of mRNA, the DNA molecule unzips into two DNA strands. The mRNA molecule is synthesized on one of these strands. This process, in which the message is transcribed to an mRNA molecule, is called *transcription*. The mRNA carries a code complementary to the one present on the mother DNA strand except that in this new strand, uracil replaces thymine and ribose replaces deoxyribose.

After synthesis, the mRNA migrates from the nucleus and becomes attached to the ribosomes, the nucleoprotein particles present in the endoplasmic reticulum (see above) of the cytoplasm. Small RNA molecules, which have been called soluble RNA, or sRNA, but are also designated as transfer RNA, or tRNA, are present in the cytoplasm. At the terminal end of all tRNA's, there always appears the same sequence of three nucleotides, cytidylic, cytidylic, and adenylic acids, so it might be more proper to designate tRNA as

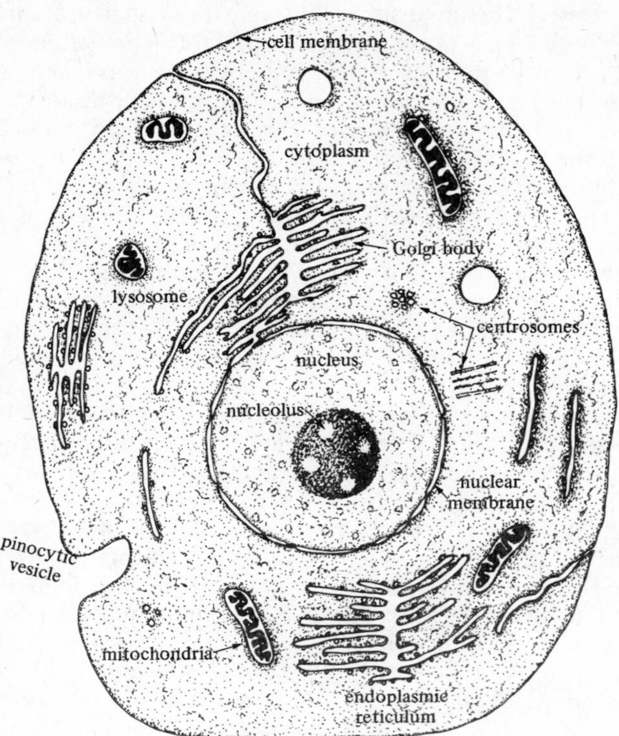

Figure 29.7 Diagram of a typical cell. The oxidative reactions that furnish the cell with energy occur in the mitochondria. Protein synthesis occurs on the ribosomes, which are represented by the dots that line the endoplasmic reticulum. The lysosomes are probably areas of intracellular digestion. It is suspected that the Golgi bodies are concerned with storage, transport, and secretion of newly synthesized macromolecules. One of the pair of centrosomes is shown in cross section (circles) while the other is shown in longitudinal section (rods). In cell division, these part to form poles of apparatus in which two duplicate sets of chromosomes are separated.

tRNA-C-C-A. At a certain location on the strand of a tRNA molecule, there is a triplet unit of four nucleotide bases coded in such a fashion that this RNA molecule can be attached to a single specific amino acid that has previously been activated by combination with a specific enzyme for this amino acid. This activation requires a molecule of ATP. The resulting tRNA-amino acid complexes attach themselves to the mRNA on the ribosomes according to the arrangement of three adjacent nucleotide groups on a certain portion of the mRNA strand. This arrangement of nucleotide bases is called a *codon.* The codons proposed for specific amino acids are indicated in Table 29.2. The arrangement of the three adjacent nucleotide bases on the tRNA strand that codes for a specific amino acid is called an *anticodon.*

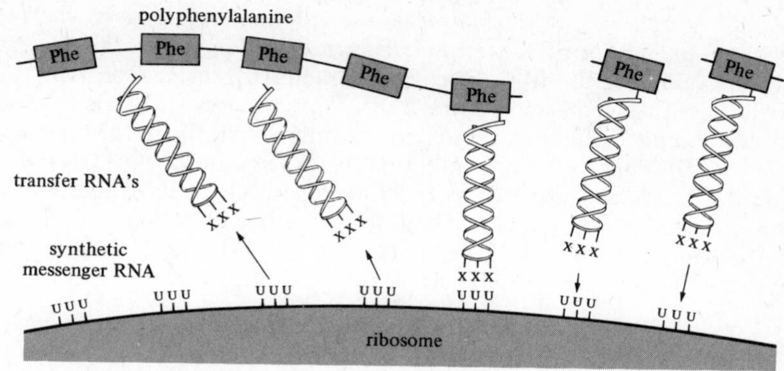

Figure 29.8 The discovery that a messenger RNA synthesized to contain only uracil directs the synthesis of a synthetic protein, polyphenylalanine, containing only one amino acid, phenylalanine, was the first break in the genetic code. Many of the bases in transfer RNA that respond to code words in the messenger RNA are unknown and are indicated by X's.

TABLE 29.2 TRIPLE NUCLEOTIDE BASE CODE OR CODONS FOR mRNA FOR THE SPECIFIC AMINO ACIDS LISTED

Codons or Code Words for mRNA	*Amino Acids*
CCG, CUG, CAG, GCU	Alanine
GUC, CGC, AGA, CGC	Arginine
CUA, UAA, CAA	Asparagine
GUA, GCA	Aspartic acid
UGU, GGU, UGC	Cysteine
GAA, GAG, AUG	Glutamic acid
ACA, UAC, CAG	Glutamine
CGG, UGG, AGG	Glycine
CAC, AUC	Histidine
UAU, AAU, CAU	Isoleucine
CUC, UGU, UAU	Leucine
AAA, AAG, AUA	Lysine
UGA	Methionine
UUU, UUC	Phenylalanine
CCC, CAC, CUC	Proline
UCG, CCU, UCU	Serine
ACA,UCA, CAC	Threonine
UGG	Tryptophan
UAU, ACU	Tyrosine
GUG, UUG, GUA	Valine

Source: Routh, *Introduction to Biochemistry,* Saunders, Philadelphia, 1971.

The tRNA-amino acid complexes are lined up according to instructions coded on the mRNA strand. Enzymes induce the "zipping up" of these amino acids in a structure characteristic of the protein that is being synthesized (see Figure 29.8). This process, in which a protein containing a definite sequence of amino acids is synthesized, is called *translation.* As the protein is being synthesized, it, as well as the tRNA, is detached from the ribosomes. The tRNA again seeks out a molecule of the activated amino acid for which it is coded, and the resulting complex can attach itself to the mRNA–ribosome complex to continue the biosynthesis of the desired protein.

Mutations and genetic diseases

Any error in the reading or copying of the genetic code in a DNA or RNA molecule can lead to a mutation. This may occur due to the addition or deletion of a code letter, a substitution of one code letter for another, or a rearrangement of the coding. Although some mutations may lead to the improvement of a species, many usually produce a weakening and death of the mutant. Sickle cell anemia appears to result from the presence of an abnormal hemoglobin in which the amino acid valine has replaced a single glutamic acid unit in normal hemoglobin.

An error in the coding of DNA may result in the lack of an essential enzyme in the body and a typical dysfunction. In phenylketonuria, the enzyme phenylalanine hydroxylase is not available for the normal conversion of phenylalanine to tyrosine. In such a situation, phenylalanine is converted to phenylpyruvic acid,

$$C_6H_5CH_2COCOOH$$

An accumulation of this latter acid damages the nervous system and frequently produces feeblemindedness. Other genetic diseases are pentosuria (page 461), albinism, hemophilia (page 470), and galactosemia.

Viruses

These appear to consist primarily of large molecules of nucleoproteins, which frequently can be obtained crystalline.

Viruses, although consisting almost exclusively of nucleoproteins, appear to be living organisms. They can infect the cell nuclei of a host and can induce it to reproduce their kind. This usually leads to the rapid multiplication of the virus and the death of the host.

Whereas the nucleic acids of the bacterial viruses, such as the bacteriophage that destroy bacteria (Figure 29.9), appear to consist primarily of DNA, plant viruses, such as the tobacco mosaic virus, are ribonucleoproteins. Animal viruses contain either DNA or RNA. Measle virus appears to belong to the former group, whereas influenza virus belongs to the latter.

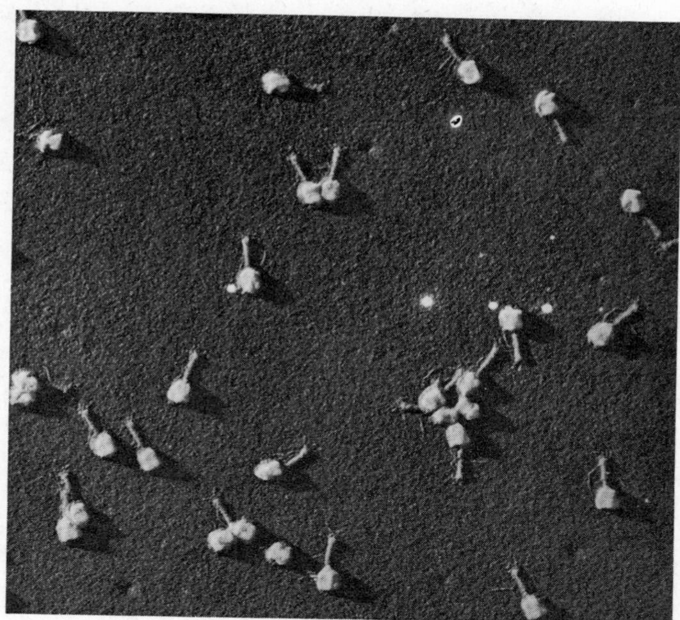

Figure 29.9 An electron micrograph of T2 bacteriophage. (Courtesy The Virus Laboratory, University of California, Berkeley. From Mahler and Cordes, Basic Biological Chemistry, *Harper & Row, N.Y., 1968.)*

STUDY QUESTIONS

1. What are the initial products of the hydrolysis of the nucleoproteins?
2. What types of products are intermediate hydrolysis products of the nucleic acids?
3. Distinguish between a DNA and an RNA. In what portion of the cell do these predominately occur?
4. Describe the products that are formed when a deoxyribonucleic acid is completely hydrolyzed.
5. Which of the hydrolysis products of the nucleic acids may be considered as glycosides of a nitrogen base? Are these α- or β-glycosides?
6. Which of the hydrolysis products of the nucleic acids are esters of phosphoric acid?
7. Describe the characteristic structure of the heterocyclic bases that are obtained when nucleic acids are completely hydrolyzed. These are considered as derivatives of which two heterocyclic nitrogen bases?
8. What purine derivatives are obtained when RNA's are hydrolyzed?
9. What pyrimidine derivatives are obtained when DNA's are hydrolyzed?
10. What are the cause and symptoms of gout?
11. Write structural formulas for uracil, guanine, deoxyribose, deoxycytidylic acid, adenosine triphosphate, thymidine, uridine, and guanylic acid.
12. Describe the system used in numbering the deoxyribose and the adenine portion of the nucleoside adenosine.

13. What role do the following play in the physiology of the cell: mitochondria, chloroplast, genes, endoplasmic reticulum, chromosomes, ribosomes?
14. What is the role of the following types of nucleic acids: mRNA, ribosomal-RNA, DNA, tRNA (sRNA)?
15. Describe the nature of the main chain of a DNA molecule. Where are the nitrogen bases located in a DNA chain?
16. Why and how are two linear chains of a DNA stabilized in a double helical structure?
17. In a DNA double helix, which nitrogen base is hydrogen bonded to thymine? To guanine?
18. Describe the process that is most likely involved in "replicating" a DNA molecule. How does the transcription of an mRNA molecule on a DNA strand differ from the replication of a new DNA strand?
19. How is genetic information transferred from parent to offspring? What changes occur in an ovum when it is fertilized by a sperm cell? What role is played by genes and chromosomes in the transmission of genetic information?
20. What are codons and anticodons? Describe the role of the codons, anticodons, mRNA, tRNA, and ribosomal RNA's in the biosynthesis of a cellular protein. Which type of nucleic acid carries the code that establishes the sequence of amino acids in the protein synthesized?
21. Describe some of the changes in codons that could lead to mutations.
22. What information can you give concerning the following genetic diseases: sickle cell anemia, pentosuria (page 461), phenylketonuria, hemophilia (page 470)?
23. What is the major chemical constituent of a virus? Why must viruses be considered living organisms?
24. Describe the changes that must occur in a normal cell when it is attacked and invaded by a virus.

chapter 30

the urine in health and disease

30

The waste products of the body are excreted through the following channels: the lungs, the skin, the intestines, and the kidneys. Water and carbon dioxide are eliminated by the lungs. Water composes the major portion of the sweat that escapes from the skin; also present are a small amount of water-soluble nitrogenous products, such as urea, a small amount of inorganic salts, lipoidal material, and lactic and butyric acids. The bile pigments and cholesterol are excreted into the intestines by the gall bladder. Besides these two substances, the feces, the excretory product of the intestines, contain undigested particles of food, salts of fatty acids and inorganic iron, and calcium compounds excreted by the wall of the intestines. Water, water-soluble organic nitrogenous products, and inorganic salts are the main excretory products of the kidneys.

Formation of urine in the kidneys

Blood enters the kidneys by means of the renal arteries and passes through capillary nodules called *glomeruli.* An ultra-filtrate of the blood passes through the walls of the glomeruli and into *tubules* (Figure 30.1). This filtrate has the same composition as the blood except that colloidal materials, such as the proteins, are not present. (Ultra-filtrates are filtrates formed by passing a fluid medium through a suitable membrane. Most colloidal substances are unable to pass through the membrane and will be absent in the resulting filtrate.) In the tubules, nearly all of the glucose and amino acids are reabsorbed, as well as large amounts of water and chloride and bicarbonate ions. The renal threshold of a substance is the concentration level in the plasma, above which the substance will appear in the urine. For most individuals, the renal threshold for glucose will be some concentration between 140 and 200 mg of this sugar in 100 ml of the blood plasma. The blood filtrate is reduced to about 1 percent of its original volume in the tubules. Substances such as ammonium salts and hippuric acid (page 460) are possibly formed in the cells lining the walls of the tubules and are excreted into the urine directly. These small tubules of the kidneys join to form the *ureter,* a tube that leads from the kidneys to the bladder (Figure 30.2).

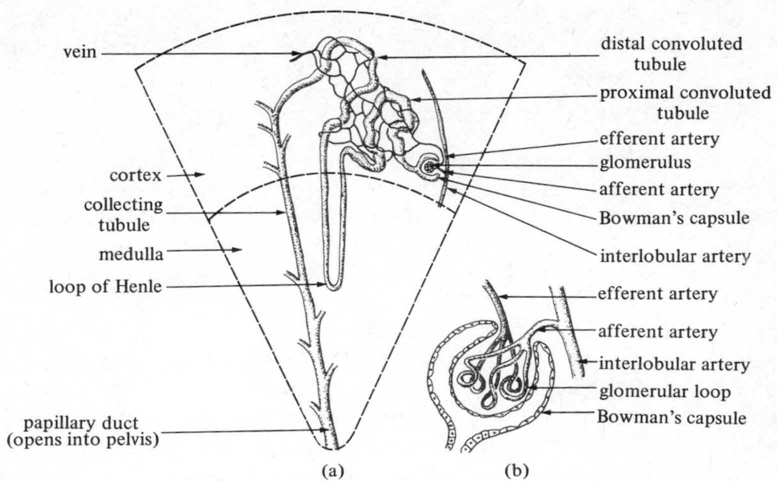

***Figure 30.1** (a) Diagram shows the relation of a uriniferous tubule to the glomerulus and the ultimate renal capillary system. Many such tubules join together to form a series of terminal canals that discharge at the tips of the pyramids into the pelvis of the kidney. (b) Diagram of a renal corpuscle, much enlarged. (From Guyer,* Animal Biology, *4th ed., Harper & Row, N.Y., 1948.)*

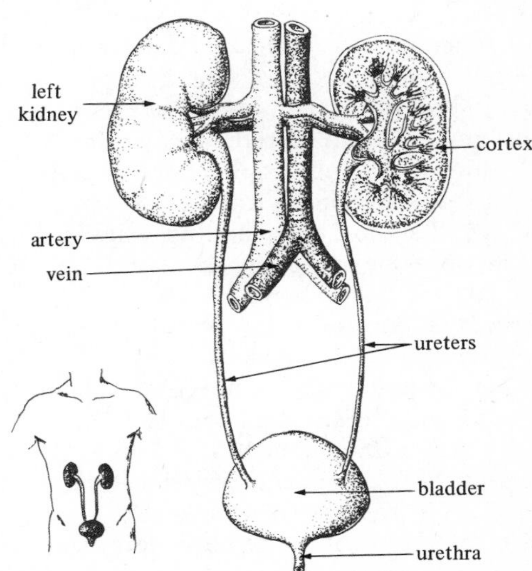

***Figure 30.2** Human excretory system. The ureters carry the urine from the kidneys to the bladder.*

For a more detailed description of the structure and the function of the kidney, the student should consult a textbook of physiology.

The physical properties of the urine

The color of normal urine varies from light yellow to amber and appears to be dependent upon the volume excreted in a given period of time. The larger the amount of urine voided, the lighter the color of urine. The color of urine is due to the presence of small amounts of pigments, the most important of which is urochrome (page 393). A red to dark brown color may indicate the presence of blood. In obstructive jaundice, the urine is often colored dark brown due to the presence of the bile pigment bilirubin (page 392). Black pigments, called *melanins,* are excessively produced by melanotic tumors, and the excretion of these pigments may produce a urine black in color. In *alcaptonuria,* a rare disorder in which tyrosine (page 320) is not properly metabolized in the body, the urine of patients is normal in color when voided but becomes dark brown upon standing. This is due to the oxidation of homogentisic acid that is present in the urine of such individuals.

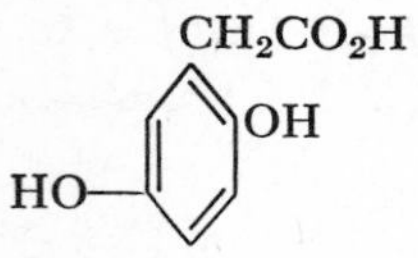

CH_2CO_2H

OH

HO—

homogentisic acid

Ingested drugs, dyes, and pigments may modify the color of urine.

Urine when voided is clear and usually contains no sediments. On standing, it usually becomes cloudy, and sediments may form due to the precipitation of inorganic salts, such as the calcium phosphates.

The volume of urine excreted in a 24 hour period varies between 0.5 and 1.5 liters. It is increased by the ingestion of large quantities of water and is decreased by the loss of water by perspiration and from the lungs. This loss may be appreciable in hot weather. Drugs such as caffeine that cause an increased flow of urine are called *diuretics.*

A decreased flow of urine is called *oliguria,* a condition that usually accompanies fevers. *Anuria* is the complete absence of a flow of urine. Both oliguria and anuria accompany mercuric chloride poisoning and acute cases of nephritis. *Polyuria* is an abnormally large excretion of urine, a condition observed in diabetes mellitus and diabetes insipidus. Patients suffering from the latter disorder may excrete as much as 20 liters of urine a day.

Fresh urine has a slight aromatic odor. On standing, it may have an ammoniacal odor due to hydrolysis of urea to ammonia. Certain drugs and foods may modify the odor of urine. Asparagus imparts the disagreeable odor of methyl mercaptan, CH_3SH. The urine of patients suffering from diabetes mellitus or other disorders that are accompanied by ketosis will have an acetonelike odor. Urine has a salty taste due to the presence of sodium chloride.

The specific gravity of urine is closely related to the percentage of solids present in it. Normal urine usually has a specific gravity between 1.015 and 1.025. This variation in the specific gravity of urine is due to variations in the volume excreted in a given period. In diabetes insipidus, the specific gravity will be nearly unity due to the large volume of urine excreted. In diabetes mellitus, the specific gravity will be abnormally high due to the large concentration of sugar in the urine. In certain types of nephritis, the urine will have nearly the same specific gravity as an ultra-filtrate of blood, namely 1.010.

Humans who subsist on a mixed diet of meats and vegetables excrete a slightly acid urine. A diet limited to fruits and vegetables will produce an alkaline urine. Following a meal, the urine of a normal individual generally becomes alkaline in reaction due to the retention and secretion of hydrochloric acid by the stomach, because an equivalent of base ($NaHCO_3$) will be introduced into the blood plasma for each equivalent of hydrochloric acid secreted into the stomach. This is referred to as the *alkaline tide.* In patients suffering from acidosis, the acidity of the urine can be expected to be abnormally high.

Because there are frequent fluctuations in urine composition, it is generally preferable to determine physical properties on a 24 hour sample of urine, discarding the first voiding in the morning and collecting the first voiding of the next morning.

Normal constituents of the urine

Urea, NH_2CONH_2, is the most abundant solute in the urine. It is the end product of the metabolism of proteins in the body and is synthesized from carbon dioxide and ammonia in the liver (page 418). From 25 to 30 g is excreted daily in the urine of adults, although the amount varies widely and depends upon the protein content of the diet. Urea is a diuretic, and an increased production of urea is accompanied by an increased flow of urine.

About 0.7 g of uric acid is excreted daily in the urine of humans and is a product of the metabolism of nucleoproteins in the body. About 0.2–0.5 g of uric acid is formed from the breakdown of the cellular material of the body.

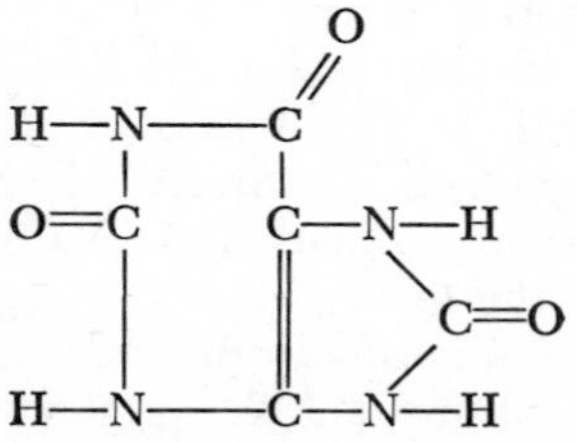

```
                O
               //
H—N——C
   |       |
O=C    C—N—H
   |       ||     \
   |       ||      C=O
   |       ||     /
H—N——C—N—H
```

uric acid

The remaining portion of the uric acid originates from the metabolism of the nucleoproteins present in the diet. A diet rich in glandular meats, such as liver and pancreas, produces an increased excretion of this acid. Uric acid is usually present in the urine as the ammonium and potassium salt, and when urine is acidified with mineral acids, the free acid precipitates as a reddish crystalline deposit. Whereas there is an increased excretion of uric acid in leukemia, the excretion of this acid may be increased or decreased in gout.

The excretion of creatinine in the urine is nearly constant for an individual and amounts to 1 to 1.5 g daily. The *creatinine coefficient* is the number of milligrams of creatinine excreted in the urine of an individual in a 24 hour period for each kilogram of body weight. It is related to the muscular development of the individual and is usually higher for men than for women and children. It is believed that most of the creatinine of the urine is produced from the creatine present in the muscular tissue of the body.

NH_2 | $HN{=}C$ \ $N{-}CH_2{-}CO_2H$ | CH_3

creatine

H / N \ $C{=}O$ | $HN{=}C$ | CH_2 / N | CH_3

creatinine

However, creatinine that is present in the diet is also excreted as such in the urine. Creatinine is the anhydride of creatine. Creatine does not appear in appreciable amounts in the urine of adult males but does occasionally appear in the urine of females and children. This condition is called *creatinuria.*

Hippuric acid, or benzoylglycine, occurs to the extent of about 0.7 g daily

$C_6H_5{-}C({=}O){-}NH{-}CH_2{-}CO_2H$

hippuric acid

in the urine. It is a detoxification product of benzoic acid and other aromatic compounds that are oxidized to this acid in the body. In the liver, benzoic acid is combined with the amino acid glycine, NH_2CH_2COOH, to form the harmless product hippuric acid. Prunes and cranberries are especially rich in benzoic acid and lead to an increased production of hippuric acid in the urine.

Some of the other organic constituents normally present in urine in small amounts are allantoin, indican, amino acids, pigments such as urochrome, vitamins, and hormones.

The sodium, ammonium, potassium, magnesium, calcium, and iron salts of hydrochloric, phosphoric, carbonic, and sulfuric acids occur in the urine. Sodium chloride is the most abundant of these salts, and 10 to 15 g is usually excreted daily. The sulfates and phosphates are formed mostly by the oxidation of the sulfur and phosphorus present in metabolized proteins. The latter may also originate from phospholipids and other phosphate esters. The phosphates and bicarbonates, as well as the ammonium ions, are important buffers in the blood and aid in maintaining the proper pH of the body fluids. The surplus of these are excreted in the urine.

Pathological constituents of the urine

The appearance of certain abnormal constituents in the urine of an individual usually indicates the existence of some pathological condition. It is the usual practice to examine a 24 hour sample of urine. The first voiding in the morning is discarded, but all of the following voidings, including the first voiding the next morning, are collected together. To avoid bacterial action, some preservative (such as toluene) is added to the urine sample in small amounts.

The following pathological constituents are frequently determined in a sample of urine: sugars, proteins, blood, acetone bodies, indican, and bile.

The occurrence of sugars in the urine is called *glycosuria.* Glucose is the most common sugar found in urine and its presence in urine is called *glucosuria.* If the normal amount of glucose is present in urine, there is not a sufficient amount to give a positive test with Benedict's reagent. The following pathological conditions will usually be accompanied by glucosuria: diabetes mellitus, renal diabetes, alimentary glycosuria, and some cases of severe liver damage. The appearance of glucose in the urine thus cannot be interpreted as due in all cases to diabetes mellitus. A determination of the blood-sugar concentration is a better index of the presence and progress of this type of diabetes. The use of Clinitest tablets for the detection of glucose in the urine is described on page 297 and in the laboratory manual.

Lactose is sometimes present in the urine of lactating females, a condition called *lactosuria.* It can be distinguished from true glucosuria by the inability of yeast to destroy the reducing power of the urine of individuals with this condition. *Pentosuria,* a condition in which the pentose xylulose occurs in the urine, appears in some individuals (page 298).

Although the presence of proteins in the urine can be called *proteinuria,* it is more commonly called *albuminuria,* because albumin is the usual protein found in pathological urines. The presence of proteins in the urine is usually an indication of kidney damage or nephritis. Proteins may, however, appear in many infectious diseases, in heart trouble, and in other disorders. In some individuals, proteins will appear in the urine if the person stands for some time but will disappear if he lies down. This is called *orthostatic albuminuria,* and no harmful effects accompany this condition. Although albumin is the most common protein that occurs in proteinuria, it is sometimes accompanied by globulins. Because the globulins are larger molecules, their presence in the urine indicates a more extensive damage to the kidneys than if albumin is found exclusively.

The acetone or ketone bodies are acetone, acetoacetic acid, and β-hydroxybutyric acid. For a discussion of the acetone bodies and the accompanying ketosis and acidosis, see pages 411–414.

Small amounts of indican occur in normal urine.

$C—OSO_3K$
‖
CH
N
H

indican

Larger amounts usually indicate an excessive amount of putrefaction in the intestines, because indican is the product of the bacterial decomposition of tryptophan (page 321).

The presence of hemoglobin in the urine, a condition called *hemoglobinuria,* is due to the hemolysis (page 113) or rupture of red blood cells. The hemoglobin liberated from these cells is excreted in the urine. In *hematuria,* the whole red blood cells appear in the urine. This latter condition is usually the result of a hemorrhage in the kidney or urinary tract. Red blood cells are usually detected by a microscopic examination of the urine.

In obstructive jaundice, the bile pigments are not properly excreted through the intestinal tract. Their concentration in the blood increases, and they are excreted in large amounts in the urine.

The reagents and procedures employed in testing for pathological constituents of the urine, including rapid tablet and tape methods, are discussed in the laboratory manual.

STUDY QUESTIONS

1. What organs and tissues are responsible for the disposal of the waste products of the body?
2. Discuss the various excretory products of the body from the standpoint of their composition.
3. What is an ultra-filtrate? The urine may be considered as a modified ultra-filtrate of what body fluid? What changes occur in the ultrafiltrate from the glomeruli as it passes through the kidney tubules?
4. Discuss the color, odor, volume, specific gravity, and acid or base reaction of normal urine.
5. What pathological conditions are indicated by urine that is (a) dark brown in color, (b) black, (c) a normal color that changes to black on standing?

6. Why does a sediment usually form in normal urine when it stands for some time?
7. What are polyuria, anuria, and oliguria? What pathological conditions may these indicate?
8. How does urine change in specific gravity in diabetes mellitus? In diabetes insipidus? In certain cases of nephritis?
9. What is meant by the term "alkaline tide"? How can the occurrence of this condition following a meal be explained? Will the urine of a vegetarian be acidic or basic in reaction?
10. What are the four most abundant organic constituents of normal urine? What is the origin of each of these in the urine? Write a structural formula for each. About how much of each of these substances is excreted normally in a 24 hour urine sample?
11. What types of food increase the uric acid content of the urine? Why?
12. What is meant by the creatinine coefficient? Upon what is the value of this coefficient dependent? What is creatinuria? How are creatine and creatinine related?
13. How and where is benzoic acid detoxified in the body? What foods are rich in benzoic acid?
14. What inorganic salts are present in the urine? Which of these is the most abundant? What is an important source of the sulfates and phosphates of the urine?
15. What pathological conditions are indicated by the presence of (a) glucose and acetone bodies, (b) albumin, (c) hemoglobin, (d) bilirubin, and (e) large amounts of indican in the urine?
16. Describe the collection and preservation of a 24 hour sample of urine.
17. What is glycosuria? Lactosuria? Pentosuria? Glucosuria? How can lactosuria and pentosuria be distinguished from true glucosuria?
18. What test is to be considered as a more reliable test for diabetes mellitus than the presence of glucose in the urine? Why?
19. Why does the presence of globulins in urine indicate a more severe kidney damage than the presence of albumins exclusively?
20. Explain the occurrence of acetone bodies in the blood and urine. What is ketosis? Why does acidosis usually accompany ketosis?
21. Distinguish between hemoglobinuria and hematuria.
22. After reference to a suitable laboratory manual, describe and explain the clinical tests that are used to detect the presence of glucose, albumin, acetone bodies, bile pigments, and blood in the urine.

chapter 31

the chemistry of blood and respiration

31

Composition of the blood

The blood consists of a fluid called the *plasma,* in which several types of particles called the *formed elements* of the blood are suspended. These consist of the *red blood cells,* or *erythrocytes,* the *white cells,* or *leucocytes,* and the *blood platelets* (Table 31.1). The red blood cells contain the hemoglobin (Figure 31.1) of the blood, and their main function is to carry oxygen from the lungs to the body cells. They also play an important role in controlling the *p*H of the blood and removing of carbon dioxide to the lungs. The white cells serve as defenses against infection, because they are able to attack and destroy harmful microorganisms. The presence of blood platelets is essential for the clotting of shed blood.

About 92 percent of the plasma is water and 8 percent is solids. The proteins, which are nearly all synthesized in the liver, are the most abundant and important of the plasma solids. The four major types of protein present in the plasma are *serum albumin, serum globulin, fibrinogen,* and *prothrombin.* The globulins and albumins are colloidal and cannot diffuse into the cells. They produce a slightly higher osmotic pressure in the blood than in the cells, thus preventing the flow of an excessive amount of fluid into the cells. Certain pathological conditions (such as loss of albumin from the blood due to kidney failure, malnutrition, and some liver diseases) cause the blood osmotic pressure to drop below that of the cells, causing an abnormal amount of fluid to enter the cells. This condition is known as *edema.* Because osmotic pressure is influenced by particle size, albumin with a smaller molecular weight will be more effective in preventing edema than will the other plasma proteins. The antibodies of the blood are present in the globulin fraction of the blood. α- and β-globulins assist in the transport of carbohydrates, lipids, and metal ions.

Fibrinogen and *prothrombin* play important roles in the *clotting of the blood.* In clotting, fibrinogen is changed into a denatured protein called *fibrin.* Fibrinogen resembles the globulins in most of its chemical properties. It, like the globulins, is precipitated by a half-saturated solution of ammonium sulfate; but fibrinogen is precipitated by a half-saturated solution of sodium chloride, whereas the precipi-

TABLE 31.1 CONCENTRATION OF FORMED ELEMENTS OF THE BLOOD PER CUBIC MILLIMETER

Erythrocytes, or red blood cells	about 5,000,000 cells
Leucocytes, or white cells	5,000–10,000 cells
Blood platelets	200,000–400,000 cells

tation of true globulins requires a saturated solution of sodium chloride.

Blood serum lacks the formed elements of whole blood, as well as the fibrinogen present in whole blood and blood plasma.

Blood plasma normally contains about 100 mg of glucose in 100 ml of plasma. After a meal, the amount of glucose is slightly increased. Abnormal increases of glucose in the blood occur in certain pathological conditions, such as diabetes mellitus. Other substances present in small amounts in plasma are amino acids; lipids; lactic acid; nitrogenous waste products such as urea, uric acid, creatine, and creatinine; small amounts of acetone bodies; pigments; enzymes; vitamins; and hormones. The amount of acetone bodies increases considerably in diabetes mellitus (see Table 31.2) and other conditions that produce ketosis. The following inorganic ions are present in the blood plasma: Na^+, Ca^{2+}, Cl^-, HCO_3^-, SO_4^{2-}, Mg^{2+}, $H_2PO_4^-$, and HPO_4^{2-}. The red corpuscles contain a relatively high K^+ concentration and a low Na^+ concentration (page 481).

The blood has a pH of about 7.4, with a possible extreme normal range of 7.3 to 7.5.

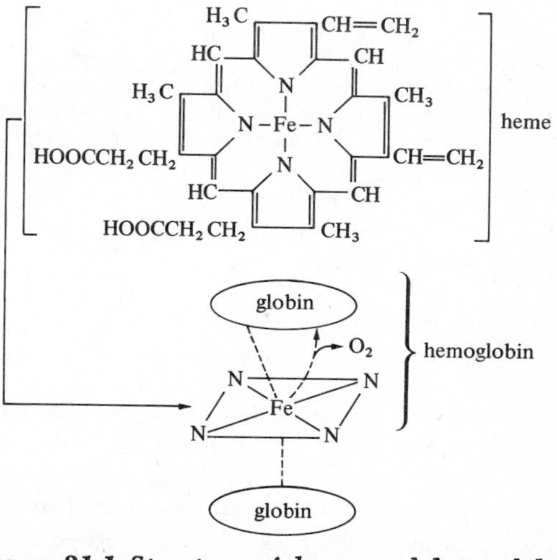

Figure 31.1 Structure of heme and hemoglobin.

TABLE 31.2 COMPOSITION OF NORMAL HUMAN PLASMA[a]

	Average or Representative Value, mg/100 ml	Range, mg/100 ml
Inorganic constitutents		
Water	93,600	92,400–94,400
Chloride	365	355–381
Sodium	316	300–330
Bicarbonate (as $NaHCO_3$)[b]	226	205–280
Potassium	17.2	12.1–25.4
Phosphate (as P) (inorganic)[c]	3.2	2.6–5.4
Calcium (serum)	10	8.2–11.6
Silica (as SiO_2) (whole blood)	9.0	
Sulfur, total nonprotein	3.38	2.95–3.75
Magnesium (serum)	2.0	1.7–2.3
Zinc	0.21	0.12–0.48
Copper	0.12	0.086–0.161
Iron	0.105	0.028–0.210
Carbohydrates		
Glucose, fasting capillary whole blood	93	
Pentose, total	2.55	
Polysaccharides (serum) (as hexose)	102	73–101

Source: H. Downes, *The Chemistry of Living Cells,* 2nd ed., Harper & Row, New York (1962). Data in this table are adapted from the full tables given in H. A. Krebs, "Chemical Composition of Blood Plasma and Serum," *Ann. Rev. Biochem.* **19**, 409 (1950).

[a]Note that the figures for silica and glucose are for whole blood. Except where indicated, all others are for plasma.

[b]All bicarbonate ion present is reported as $NaHCO_3$.

[c]Does not include organic acid-soluble phosphate esters.

[d]It is now believed that all the blood lipids are present in lipoprotein complexes in which the various constituents are found in various ratios.

[e]The six groups listed are the main fractions into which human plasma proteins are separated by electrophoresis.

Functions of the blood

The following are some of the important roles that the blood plays in the proper functioning of the body:

1. It carries digested food products and oxygen to the cells.
2. It removes waste products produced during the metabolism of nutrients in the cells of the body, such as carbon dioxide, urea, and uric acid. Most of the carbon dioxide is carried by the blood to the lungs, where it is excreted. Urea, uric acid, etc., are, for the most part, separated from the blood by the

TABLE 31.2 (*Continued*)

	Average or Representative Value, mg/100 ml	*Range mg/100 ml*
Nonprotein nitrogen compounds		
Amino acids, total (as N) (ninhydrin method)	4.1	3.4–5.5
Creatine (serum)	1.07	0.76–1.28
Creatinine (serum) (male)		1.05–1.65
Uric acid (serum)	4.0	2.9–6.9
Urea (male)	27.1	
Lipids[d]		
Fatty acids (as stearic acid)		200–450
Fats, neutral		0–150
Cholesterol, free and esterified		150–260
Phospholipids		150–250
Proteins[e]	g/100 ml	
Albumin	4.04	
α_1-Globulin	0.31	
α_2-Globulin	0.48	
β-Globulin	0.81	
γ-Globulin	0.74	
Fibrinogen	0.34	
Total	6.72	

kidneys, which excrete these substances in the urine. Some of these products are also excreted by the intestines and through the skin.

3. Through the action of the white corpuscles, antibodies, and other factors, the blood protects the body from invasion by pathological microorganisms.
4. The presence of buffer systems in the blood helps to maintain the blood and many other body fluids at a nearly constant pH.
5. Internal secretions called *hormones,* as well as vitamins and some enzymes, are rapidly transported to their seat of action by the blood.
6. The blood is an essential substance in the heat regulation of the body. This control depends upon the shifting of blood to or from the surface of the body. Heat can be dissipated through the lungs and skin.
7. It maintains the water content of the body tissues.

Clotting of blood

Clotting of blood depends upon a change of the native, salt-soluble protein fibrinogen to a coagulated protein fibrin, which forms a threadlike mass or *reticulum* that enmeshes the formed elements of the blood. The total mechanism of blood clotting is complex, and there is still much uncertainty concerning certain of its phases. Our viewpoint concerning the clotting of blood can be expected to change as new information becomes available to us through research.

The conversion of fibrinogen to fibrin is catalyzed by the enzyme *thrombin.* However, unshed blood is prevented from clotting because thrombin is present as an inactive proenzyme, called *prothrombin,* which is probably a glycoprotein. Calcium ions, *thromboplastin,* and a number of other factors apparently are necessary activators for prothrombin.

If calcium ions are removed from blood, it will not clot. This can be accomplished by adding sodium citrate or sodium oxalate to freshly shed blood. Because oxalates are toxic, sodium citrate is used as an anticoagulant for blood to be used in transfusions.

Thromboplastin, which is apparently a complex conjugated protein, is not present in plasma but is present in many tissues, such as the lungs and the brain. The blood platelets appear to disintegrate when blood is shed and liberate either thromboplastin or some other coagulation activator. Two other factors in plasma that may also be needed to convert prothrombin to thrombin are an *accelerator globulin,* designated often as Ac-globulin, and an *antihemophilic globulin.* All necessary factors except the latter appear to be present in the uncoagulable blood of those afflicted with *hemophilia.*

One unanswered problem is why blood does not coagulate in the blood vessels but does coagulate when shed. One explanation may be the fact that thromboplastin and the platelet factor are introduced into the plasma only when tissue is damaged and platelets disintegrate.

The mechanism of the clotting of blood may be summarized as follows:

$$\text{prothrombin} \xrightarrow[\text{platelet factor, accelerator globulin, antihemophilic factor, etc.}]{Ca^{2+},\ \text{thromboplastin}} \text{thrombin}$$

$$\text{fibrinogen} \xrightarrow{\text{thrombin}} \text{fibrin}$$

Vitamin K is essential for the formation of prothrombin in the liver.

Three substances, heparin, hirudin, and dicumarol, will prevent the coagulation of blood. Heparin is probably produced by the liver and is present in small amounts in the blood and many tissues. It possibly acts by preventing the conversion of prothrombin to thrombin. Hirudin is present in the saliva of the medicinal leech. The anticoagulant dicumarol occurs in sweet clover.

Apparently, serum also contains an inactive form of a proteolytic enzyme called *plasmin* or *fibrinolysin.* Upon activation, this enzyme slowly destroys the blood clot by dissolving fibrin.

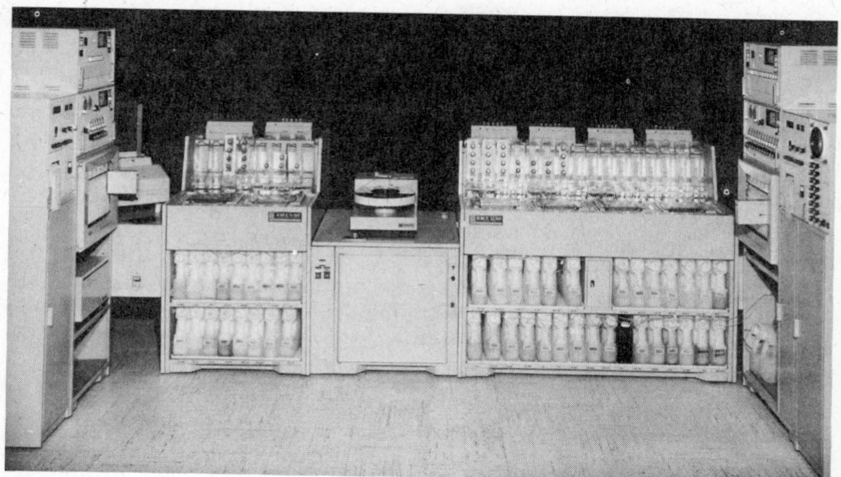

Figure 31.2 A high-speed computer-controlled biochemical analyzer. It is capable of producing 1080 test results an hour. (Courtesy Technicon Instruments Corporation.)

Changes in the composition of the blood in certain diseases

During good health, the chemical composition of the blood remains rather remarkably constant. In certain diseases, the amount of certain constituents increases or decreases greatly, and analysis of blood samples is of assistance in diagnosing and following the course of such a disease. In clinical laboratories, glucose, urea nitrogen, nonprotein nitrogen, uric acid, creatinine, and chlorides are usually determined. Folin and Wu have developed a method of analysis in which the concentration of all of these may be determined on a single blood sample. Automatic procedures are now coming into use for the analysis of blood samples (Figure 31.2).

In *nephritis*, kidney tissue is damaged or destroyed, and colloidal substances, such as the proteins, escape into the urine. In mild kidney damage, the serum albumins appear in the urine, but in severe damage, serum globulins also appear. In very mild cases of nephritis, little change occurs in the composition of the blood. In more severe cases, nonprotein nitrogenous substances, such as urea, uric acid, and creatinine, are retained by the kidneys, whereupon the percentage of urea and total nonprotein nitrogen increases well above the normal level in the blood. In very severe cases of nephritis, the large amount of these waste materials that accumulate in the blood produce *uremic poisoning*, or *uremia*. The ingestion of salts of mercury can also produce nephritis.

Changes in the composition of blood due to diabetes mellitus and other disorders, such as starvation, diarrhea, and liver damage, are discussed elsewhere in this text. In particular, see the discussion of carbohydrate and fat metabolism in Chapter 27.

If the number of red blood cells or the amount of hemoglobin is reduced appreciably below the amount in normal blood, *anemia* results. In *nutritional anemias,* insufficient iron is ingested in the food, due either to an insufficient diet, to lack of absorption of iron, or to a diet low in iron content. This type of anemia may usually be corrected by adding certain iron compounds to or correcting the diet. Because milk is deficient in iron, those on milk diets, particularly infants, may be troubled with this disorder. Anemia may also accompany certain types of jaundice, hemorrhage, cancer, and other infections. This type of anemia will usually improve when the primary cause is corrected.

In the above types of anemia, there is a reduced production of hemoglobin. In *pernicious anemia,* the activity of the blood-building tissues is reduced, with the resulting decrease in the number of red cells in the blood. This type of anemia may be due to an insufficient amount of vitamin B_{12} in the diet or to the absence of the intrinsic factor in the gastric juice. The intrinsic factor is necessary for the absorption of vitamin B_{12} by the cells in the walls of the intestines. In the past, diets high in liver or injection of liver extracts have been used in the treatment of this type of anemia. Injections of vitamin B_{12} are a much more effective treatment.

In *polycythemias,* the number of red blood cells is increased above the normal count.

Blood normally contains a small amount of bilirubin. In *jaundice,* the amount of bilirubin in the blood is greatly increased, and the skin becomes colored yellow-brown. Jaundice may be caused by obstruction of the bile duct (obstructive jaundice) or by destruction of liver cells that normally excrete the bile pigments into the gall bladder. Liver cells may be destroyed by toxic substances such as chloroform. In obstructive jaundice, the stool will be clay-colored (page 393) because no bile pigments can escape into the intestinal tract. The severity of jaundice is often measured by determining the number of times a sample of blood serum must be diluted in order to match the color of a 0.01 percent solution of potassium dichromate. This is called the *icterus index.* Because normal blood must be diluted four- to sixfold to match the standard, it is said to have an index of 4 to 6. In jaundice, the index is much higher.

THE CHEMISTRY OF RESPIRATION

Respiration includes all the processes by which oxygen is absorbed by the blood in the lungs and is transported to the cells for metabolic purposes, and those processes by which carbon dioxide, produced in the cells, is transported by the blood to the lungs and is excreted.

Oxygen transportation by the blood

Oxygen in the inspired air rapidly diffuses through the extremely thin capillary walls in the lining of the lungs. It is then carried by the arterial blood to the tissues, such as those of the muscles and heart.

Most of the oxygen carried by arterial blood is combined chemically with the hemoglobin of the erythrocytes as oxyhemoglobin,

$$\underset{\text{hemoglobin}}{Hb} + O_2 \longrightarrow \underset{\text{oxyhemoglobin}}{HbO_2}$$

and only a small portion (about one-sixtieth) is dissolved physically in the blood plasma.

Oxygen diffuses into the cells from the blood due to the decomposition of oxyhemoglobin. The oxyhemoglobin decomposes when it approaches the cell, because oxygen is being utilized by the cell and its concentration is greatly reduced in this area.

Carbon dioxide transportation by the blood

The products of metabolism in the cell include carbon dioxide and water. The concentration of carbon dioxide in the cell will be much greater than in the blood that comes in contact with the cell, and this gas will diffuse from the cell through the lymph and plasma into the red blood cells. The carbon dioxide combines with water in the red cells to form carbonic acid. This reaction, as well as the decomposition of carbonic acid, is catalyzed by an enzyme *carbonic anhydrase* present in the red cells. Because oxyhemoglobin is a stronger acid than hemoglobin, its decomposition will liberate some alkali ions in the red cells. The increasing concentration of carbonic acid will accelerate the rate of decomposition of the oxyhemoglobin. The carbonic acid will, in part, dissociate into bicarbonate ions, HCO_3^-, which combine with the alkali ions to form alkali bicarbonate salts.

Some carbon dioxide is dissolved physically in venous blood, while a somewhat larger amount is combined with hemoglobin as carbaminohemoglobin:

$$\underset{\text{hemoglobin}}{HbNH_2} + CO_2 \longrightarrow \underset{\text{carbaminohemoglobin}}{HbNHCOOH}$$

This reaction is readily shifted to the left when hemoglobin is converted to oxyhemoglobin in the lungs, thus liberating this combined CO_2. However, the major portion of the carbon dioxide is transported in the blood as the alkali bicarbonate (Figure 31.3).

Other products of metabolism in the cells besides carbon dioxide and water include phosphoric, sulfuric, and organic acids, such as uric and lactic acids. Whereas carbon dioxide is transported to the

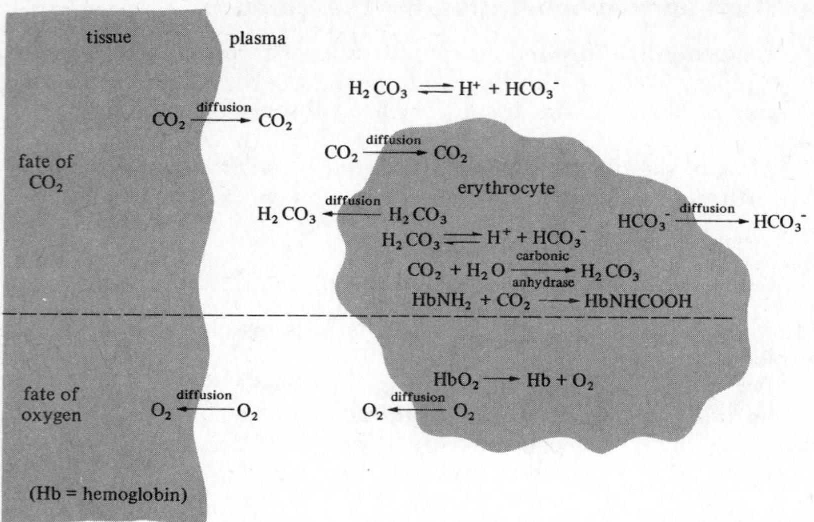

***Figure 31.3** Diagram illustrates the mode of transportation of oxygen from the erythrocytes to the tissue cells and of carbon dioxide from the tissue cells to the erythrocytes.*

lungs, the blood transports these other metabolic products to the kidneys, where they are usually excreted as alkali metal salts. In the blood, these acidic substances assist in the decomposition of oxyhemoglobin to oxygen and hemoglobin.

Venous blood, which is poor in oxygen but rich in carbonic acid and bicarbonates, is transported to the lungs. Because the concentration of oxygen in the inhaled air is greater than in the venous blood, oxygen diffuses into the plasma and finally into the red blood cells. In these cells, oxygen converts hemoglobin into oxyhemoglobin, a stronger acid than the former substance. The oxyhemoglobin reacts with the bicarbonate salts to form the alkali salts of oxyhemoglobin and liberate carbonic acid. Because there is a low concentration of carbon dioxide in the inspired air, there will be a diffusion of carbon dioxide from the blood and into the lungs. The arterial blood, which is now enriched with oxygen and which has lost a portion of its carbon dioxide, is pumped to the cells and the cycle is repeated.

ACID–BASE BALANCE

The student is advised to review pages 154–162 before studying the following material.

Carbon dioxide and other substances, including basic substances such as the alkali carbonates and bicarbonates, are added to the blood by the combustion of foods. In the human, acidic metabolic products usually exceed the basic ones, unless the individual is on a diet made up of fruits and vegetables exclusively.

Although acidic and basic substances are being introduced continuously into the blood, its pH remains unusually constant at pH 7.3–7.5. This constancy of the pH of the blood is maintained in several ways, such as by the action of the buffers of the blood, the retention or excretion of carbon dioxide by the lungs, the excretion or retention of acidic or basic substances by the kidneys, and an increased or decreased production and excretion of ammonia via the kidneys (page 419).

The buffers of the blood include: (1) carbonic acid and the bicarbonates, (2) mono- and dihydrogen phosphate ions, (3) proteins and their alkali salts, (4) hemoglobin and its potassium salt, and (5) oxyhemoglobin and its potassium salt. Buffers (1) and (2) above are present in both the plasma and red cells, (3) primarily in the plasma, (4) and (5) in the red cells. Of the above, the carbonic acid–bicarbonate buffer is the most important one in the blood. Hemoglobin and the proteins are next in importance, whereas phosphates are minor buffers in the blood due to their low concentration.

To maintain the blood at a pH of 7.4 will require a ratio of carbonic acid to bicarbonates of 1 to 20, whereas a ratio of $H_2PO_4^-$ and HPO_4^{2-} of 1 to 4 will be required.

The role of the lungs in maintaining acid–base balance

Addition of such acids to the blood as sulfuric, phosphoric, lactic, uric, acetoacetic, or β-hydroxybutyric will convert some of the bicarbonates present to carbonic acid. The resulting change in the ratio of bicarbonate ions and carbonic acid would tend to decrease the pH. If the concentration of either H^+ or H_2CO_3 is increased, respiration is stimulated, carbonic acid is more rapidly decomposed to carbon dioxide, and the latter is expelled in larger amounts from the lungs. This should bring the ratio of HCO_3^- and H_2CO_3 back to about 20:1 and thus maintain the blood at or near the normal pH value of 7.4.

Less frequently, bases will be introduced into the blood, and the bicarbonate concentration of the blood will increase, with a resulting tendency for the pH also to increase. Under such conditions, respiration is decreased resulting in an increased concentration of carbonic acid in the blood. The net result will then be an increase in both HCO_3^- and H_2CO_3, which will be adjusted to a ratio of 20:1 to return the pH to about 7.4.

The role of the kidneys in maintaining acid–base balance

The kidneys have the ability to excrete or reabsorb either acidic or basic substances in order to maintain acid–base balance in the body. Phosphates are concentrated in the kidneys, and their excretion plays one of the most important roles of the kidneys in maintaining the near constant pH of the blood. If acidic substances are entering the blood, these will be carried to the kidneys, and appreciable amounts of monohydrogen phosphate ions will be converted to dihyrogen phosphate ions. The later will be preferentially eliminated in the urine, whereas HPO_4^{2-} will be reabsorbed in the kidney tubules.

Introduction of bases into the blood will generally result in conversion of $H_2PO_4^-$ ions by bicarbonate ions into HPO_4^{2-}:

$$H_2PO_4^- + HCO_3^- \longrightarrow HPO_4^{2-} + H_2CO_3$$

Compensation for the added base will occur by a decreased elimination of carbon dioxide by the lungs, an increased excretion of HCO_3^- and HPO_4^{2-}, and a retention of $H_2PO_4^-$ by the kidneys.

A slower but effective mechanism of the kidneys to compensate for a large increase in the acidity of the blood involves formation of ammonia (likely from glutamine, the amide of glutamic acid) in the kidney tubules and its union with carbonic acid to form ammonium bicarbonate. The "alkaline reserve" of the body is then maintained in the following way: The ammonium bicarbonate reacts with the sodium salt of some acid (previously neutralized by $NaHCO_3$). The ammonium salt of the acid is excreted in the urine while the $NaHCO_3$, simultaneously formed, is absorbed in the kidney tubules and is returned to the blood.

$$NH_4^+HCO_3^- + Na^+A^- \longrightarrow Na^+HCO_3^- + NH_4^+A^-$$

Therefore, each equivalent of ammonia formed and excreted conserves an equivalent of base ion with an equivalent of bicarbonate ion, thus restoring the bicarbonate level of the blood.

In alkalosis, the production of ammonia by the kidneys is repressed.

ACIDOSIS AND ALKALOSIS

Acidosis is that condition that exists in the body due to a retention of acids or an abnormal loss or excretion of bases, while *alkalosis* results from an unusual retention of bases or an increased excretion or loss of acids. If the acidosis or alkalosis is not severe, the mechanisms of the body will be able to maintain its pH at or near 7.4, and in such cases the term *compensated acidosis,* or *alkalosis,* is used. In extended severe acidosis, the mechanisms of the body may be unable to maintain the pH constant. The latter may drop appreciably below 7.4, a condition known as *uncompensated acidosis,* which may be accompanied by coma or death. In *uncompensated alkalosis,* the pH will increase noticeably above 7.4, and tetany and death may result.

Although acidosis may be due to a nonvolatile acid excess, as in diabetes mellitus, or to an alkali deficit, as in infantile diarrhea, it may also be caused by a primary carbon dioxide excess. This can be caused by a decreased respiration or hypoventilation of the lungs due to the use of certain drugs such as morphine or to an obstruction to respiration that may occur in pneumonia. Acidosis due to an excess of carbon dioxide is called *respiratory acidosis.*

Alkalosis may be caused by an excess of alkali, such as may occur when an overdose of sodium bicarbonate is used in the treatment of an acid stomach, or by the loss of acids from the body, such as may result from extended vomiting. Another condition leading to alkalosis is primary carbon dioxide deficit, or *respiratory alkalosis,* usually a result of a hyperventilation of the lungs due to fever, low oxygen atmospheres at high altitudes, or undue anxiety of an individual.

Carbon dioxide combining power of the blood plasma

Acidosis, because it neutralizes the bicarbonates present in the blood, decreases the ability of the blood to carry carbon dioxide to the lungs. A clinical test has been developed to determine the carbon dioxide combining power of the blood. The results of this test will usually indicate whether a condition of acidosis or alkalosis exists in the body. In this test, blood is drawn into a special tube under oil and is saturated with carbon dioxide. Upon acidification and evacuation in a special instrument, carbonic acid and the bicarbonates are converted to carbon dioxide, which is expelled from the blood fluid. The volume of carbon dioxide gas expelled is measured. In a normal individual, the carbon dioxide combining power of the blood will be equivalent to 50 to 60 ml of carbon dioxide per 100 ml of plasma. A value of 70 ml or above indicates alkalosis. Values of 40 ml or less usually indicate acidosis. If the value is below 30 ml, a very serious condition of acidosis is indicated.

STUDY QUESTIONS

1. Differentiate between whole blood, plasma, and serum.
2. What are the formed elements of the blood, and what functions do they perform?
3. What are the four most important types of proteins in the blood plasma? What role does each of these perform in the blood?
4. What are some of the nonprotein constituents of the plasma?
5. What is the normal pH range of the blood?
6. Enumerate some of the functions of the blood.
7. What inorganic ion is necessary for the clotting of blood?
8. What physical changes occur when blood clots?
9. What is the nature of prothrombin, thrombin, and fibrinogen?
10. What happens to the substances in Question 9 when blood clots?
11. What are the sources and action of hirudin and heparin on the blood?
12. Discuss the chemical changes that occur when blood clots.
13. What vitamin appears to be necessary for the clotting of blood, and what role does it play in the same?
14. What is citrated blood?

15. What substances are most frequently determined in blood analyses?
16. What pathological conditions may be indicated by the following changes?

    (a) an increase of glucose concentration in the blood
    (b) a decrease in the concentration of glucose in the blood
    (c) an abnormally low pH
    (d) an abnormally high pH
    (e) the appearance of proteins in the urine
    (f) a low concentration of hemoglobin in the blood
    (g) an increased amount of bile pigments in the blood
    (h) a very large white blood cell count

17. Describe the various types of anemias and their causes.
18. What is jaundice? Distinguish between obstructive and hemolytic jaundice. What change in the stool occurs in obstructive jaundice?
19. What is the icterus index? Why and how is it determined? What changes occur in this index in jaundice?
20. Describe the changes that may be expected in the composition of the blood in diabetes mellitus, nephritis, anemia, jaundice, polycythemia, and edema.
21. What role does hemoglobin play in the transport of oxygen in the body? What is oxyhemoglobin? What change in the hemoglobin occurs when it is in contact with carbon monoxide? Carbon dioxide?
22. How is the carbon dioxide combining power of the blood plasma determined? What is the value of the carbon dioxide combining power of normal blood plasma? How would you interpret low and high values for the CO_2-combining power of blood plasma?
23. Compare the acid strength of hemoglobin and oxyhemoglobin. Relate the acidity of these substances with their ability to decompose bicarbonates to carbon dioxide.
24. Discuss the transport of carbon dioxide from the cells of the body to the lungs.
25. How is the greater portion of carbon dioxide transported in the blood?
26. What is carbaminohemoglobin, and what roles does it play in the transportation of carbon dioxide in the body?
27. Discuss the properties of the enzyme carbonic anhydrase.
28. What is meant by respiration?
29. List some of the buffers of the blood.
30. Discuss the role of (a) the lungs and (b) the kidneys in maintaining the acid–base balance of the body.
31. What are some of the causes and symptoms of acidosis? Of alkalosis?
32. What is meant by compensated acidosis, uncompensated alkalosis, respiratory acidosis, respiratory alkalosis?

chapter 32

inorganic metabolism

32

A number of mineral elements are present in small amounts in the body that are of great importance in its proper functioning. Some of these are iodine, iron, copper, and cobalt. Although our diet usually contains sufficient amounts of potassium, phosphorus, magnesium, sulfur, copper, and cobalt, a deficiency of calcium, iron, and iodine frequently occurs.

Much knowledge concerning the need for small amounts of minerals has been discovered by feeding experiments using small laboratory animals. Diets deficient in any one of the essential elements will lead to failure of growth, wasting, and in extreme cases death in these experimental animals. Addition of the deficient element to the diet of these animals usually induces a rapid recovery followed by normal growth.

The percentage concentrations of the various elements in the human body are indicated in Table 32.1.

Sodium

The sodium chloride content of unseasoned foods will not be sufficient to meet the needs of the body for these two elements, but the use of salt in seasoning food more than compensates for this deficiency. The diet of the herbivora is usually inadequate in salt, and these animals frequently travel great distances and expose themselves to danger to obtain salt from the salt licks or salt deposits.

Sodium salts, particularly the chloride, are needed to maintain the osmotic pressure of the body fluids. Sodium chloride serves as the source of hydrochloric acid present in the gastric juice of the stomach. Sodium is also present in the body as sodium bicarbonate. This salt acts with carbonic acid as a buffer and is essential in maintaining the acid–base balance of the body fluids. Sodium bicarbonate also plays an important role in the transportation of carbon dioxide from the cells to the lungs (page 473).

Sweating depletes the sodium chloride concentration of body fluids. Miners and others working in a hot environment, as well as those exercising vigorously in hot weather, may have a certain type of heat stroke due to the depletion of salt in body fluids. Under such conditions, salt tablets or dilute salt solutions should be ingested.

TABLE 32.1 THE CONCENTRATION OF ELEMENTS IN THE HUMAN BODY

Element	*Symbol*	*Percentage*
Oxygen	O	65.00
Carbon	C	18.00
Hydrogen	H	10.00
Nitrogen	N	3.00
Calcium	Ca	2.00
Phosphorus	P	1.00
Potassium	K	0.35
Sulfur	S	0.25
Sodium	Na	0.15
Chlorine	Cl	0.15
Magnesium	Mg	0.05
Iron	Fe	0.004
Fluorine	F	Trace amounts
Silicon	Si	
Zinc	Zn	
Copper	Cu	
Cobalt	Co	
Manganese	Mn	
Iodine	I	
Boron	B	

Source: Data from George I. Sackheim and Ronald M. Schultz, *Chemistry for the Health Sciences,* Macmillan, New York (1969).

In Addison's disease, a hypofunctioning of the adrenal glands (page 519), the excretion of sodium chloride is increased. The depletion of sodium salts is further aggravated because the body will have a lessened capacity to excrete potassium salts, and these will tend to accumulate. In this condition, a high salt diet or administration of cortical hormones in indicated.

Sodium chloride aggravates edema (page 466) because of the accompanying increase is osmotic pressure. Patients with edema are usually maintained on salt-free diets or on diets that are quite restricted in their salt content.

Potassium

Potassium ions are important constituents of the red blood cells and tissue cells. The fluids that bathe these cells are very poor in potassium but rich in sodium ions.

Potassium is present in abundance in the diet, especially in fruits and vegetables, and there is little opportunity for a deficiency of this element.

Potassium salts assist in maintaining the osmotic pressure of the body fluids. Potassium, as well as sodium, ions maintain the irritability of nerves and the movement of muscles. Both of these ions are antagonistic to the depressing action of calcium and magnesium ions. Potassium ions exert a relaxing effect on the heart muscles between heartbeats.

Calcium

A proper balance of calcium, phosphorus, and vitamin D must be present in the body for the proper formation of bones and teeth. The greater portion of the calcium in the body is present in these structures. Calcium is also needed in the clotting of blood fibrinogen, the precipitation of milk casein in the stomach, the proper rhythm of the heartbeat, and for the formation of milk in the lactating mammal.

Milk is a good source of calcium in a form that is readily absorbed in the digestive system. Leafy green vegetables serve as an important source of calcium, but calcium is not easily assimilated from this source because it may be present as insoluble salts, such as calcium oxalate.

Infants and children need 1 to 1.5 g of calcium daily, which may be safely satisfied by including a quart of milk per day in the diet. Adults require 0.5 to 0.75 g of this element daily, and this requirement should be satisfied by a pint of milk daily. It must be remembered that pregnant and lactating females will require a greater amount of calcium.

In the absence of satisfactory amounts of vitamin D and phosphorus, calcium will not be properly deposited in the bones and teeth, and a condition known as *rickets* will occur in chidren. If the diet of the adult does not meet the need for calcium, the calcium of the bones will be mobilized to maintain the calcium content of the blood serum, and abnormalities and bending of the bones of the skeleton will result. This is called *osteomalacia.*

Calcium metabolism is regulated by the parathyroid glands (page 523), which are small glands located near the thyroid glands of the neck. If these glands are removed, calcium is excreted and the serum level of this ion drops to a low value. This condition is known as *hypoparathyroidism.* Because calcium is a depressing ion, this low level of calcium in the blood serum may produce convulsions and tetany. High serum calcium levels will produce depression and coma.

Phosphorus

Besides its important role in the formation of bones and teeth, phosphorus is needed in the diet for the following purposes:

1. to suppy phosphate salts to the blood and urine. These salts are used to maintain the acid–base balance in the body fluids, because the excretion of the acid phosphate salts conserves the reserves of acids and bases in the body

2. to form hexose phosphates that are essential for the metabolism of the carbohydrates and the storage of glycogen in the liver
3. to form phosphocreatine and adenosine phosphates that are essential for muscle contraction.

Phosphoric acid is the important inorganic moiety in the nuclei of the cells, where it occurs as a nucleoprotein. The phospholipid cephalin is required in blood clotting (page 469). Another group of phospholipids, the lecithins, appear to play an important role in fat metabolism.

The rate of deposition of phosphorus in the bones and teeth is dependent upon the concentration of calcium and vitamin D in the blood serum.

The phophoproteins and nucleoproteins, which occur in such foods as milk, cheese, and meats, are the most important sources of phosphorus, although the dietary phospolipids and organic and inorganic phosphates serve as other sources of phosphorus.

Magnesium

In experiments, a deficiency of magnesium in the diet of animals has produced convulsions and death. It appears to be essential in the transmission of impulses along the nerves, and in muscle contraction. Magnesium is a depressant, and injection of its salts into the bloodstream produces anesthesia. This treatment has been used to reduce convulsions.

Because magnesium compounds are not readily absorbed through the intestinal wall and tend to pull water into the intestines, they are used as purgatives (Epsom salts). The presence of magnesium ions is required for many enzymatic reactions, particularly in the metabolism of carbohydrates.

Chlorine

Chlorine is essential to life. Its main source in the diet is sodium chloride. It is closely associated with sodium and potassium in metabolism. Chlorides are required for the formation of the hydrochloric acid of the gastric juice.

Iron

Although the human body contains only 3 to 4 g of iron, it is a very essential element. It is the characteristic element of hemoglobin and is needed in its formation. It appears to be responsible for the union of hemoglobin with oxygen and to be essential for many enzyme systems that catalyze oxidation-reduction reactions in the tissues of the body.

Meat and green leafy vegetables are important sources of iron, while milk and bread are poor sources. A certain reserve of iron is furnished the infant by the mother before birth. Most of this iron is stored in the liver and spleen in part as ferritin (page 392) and usually meets the blood-forming need for iron in the first six months of life. If the infant is maintained exclusively on a milk diet beyond this time, a nutritional anemia may develop because of this lack of iron. Anemias caused by lack of sufficient iron in the body may be due to bleeding. These anemias will usually respond to the addition of iron salts to the diet. Although the adult male usually obtains sufficient iron in the normal diet, the female may need an increased iron intake due to menstruation and parturition.

Iron in the form of hematin, a decomposition product of hemoglobin, is not easily absorbed in the intestines, but the iron may be utilized if this compound is decomposed during the process of digestion. It is claimed that ferric salts are not as readily absorbed as ferrous salts from the intestines; however, the former should be readily reduced to the divalent state in the intestines.

Iodine

There is no longer any doubt that iodine is essential in the diet. Its lack produces a condition known as *hypothyroidism,* an insufficient production of the thyroid substances by the thyroid gland (page 515). In this condition, the thyroid gland is usually enlarged, a condition known as *simple* or *endemic goiter.*

In areas where the diet has an insufficient iodine content, small amounts of iodine may be added to the water supply or to table salt (iodized salt). In certain areas, such as the Great Lakes region of the United States, the Pyrenees in Spain, and the Alps in Switzerland, soil and water supplies contain little iodine, and simple goiters are quite common. In regions close to seawater, where iodine is added to the soil by ocean spray carried inland by the wind and sea food is abundant, simple goiters occur infrequently.

Most of the iodine of the body is present in the thyroid gland, but small amounts are also present in the blood. About half the iodine present is thyroglobulin, which is a conjugated protein containing the iodinated amino acid thyroxine.

Because not all enlargements of the thyroid gland are due to lack of iodine, the danger of self-medication with iodine is great.

Fluorine

Most of the fluorine of the body is present in bone and teeth, with the highest concentration being present in the enamel of the teeth. If the water supply contains a small amount of fluorides, children will show a much lower incidence of tooth decay (dental caries). For this reason, many cities are now adding small amounts of fluorides to their water supply.

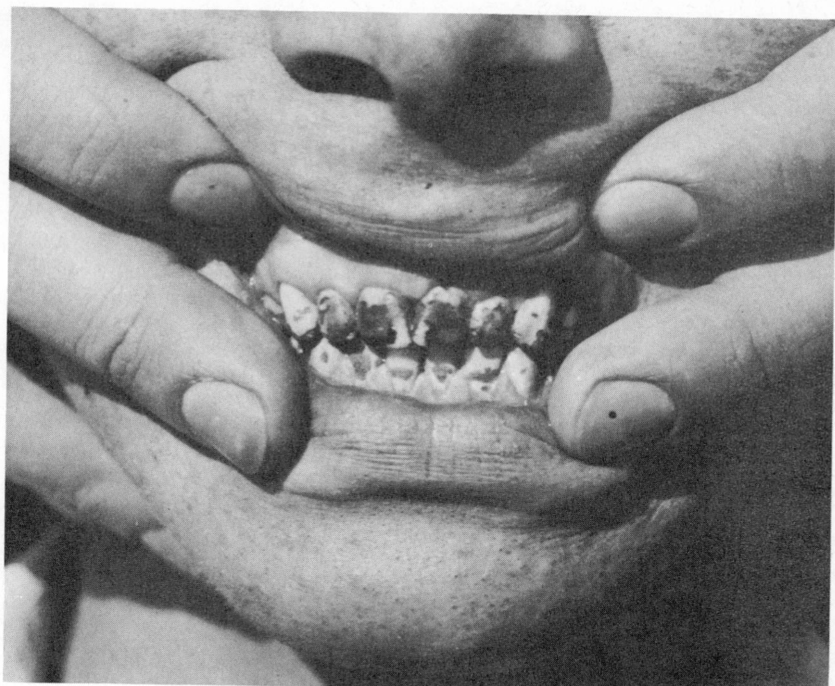

Figure 32.1 Mottled teeth due to an excessive concentration of fluorides in drinking water. Mottling will not occur when the water supply contains less than two parts fluorine per million. (Courtesy Dr. H. Trendley Dean [deceased], formerly of the National Institute of Dental Research.)

If the concentration of fluorides in drinking water is very high, the enamel of the teeth will show characteristic discolored patches, a condition known as *mottled teeth* (Figure 32.1). The teeth are harder but more brittle than normal teeth and are easily damaged.

Sulfur

There should be no deficiency of sulfur in the diet if the diet contains sufficient amounts of the sulfur-containing amino acids, such as methionine and cystine.

Sulfur is present in many proteins and in such important constituents of the body as the bile salts; certain enzymes and coenzymes; keratin, a protein present in hair, nails, and skin; and insulin.

The sulfur of the body is eliminated through the kidneys as inorganic sulfates or as ethereal sulfates in which the sulfate radical is combined with organic radicals. Some toxic substances are combined with sulfate radicals and are thus detoxified and eliminated through the kidneys.

Copper

Although copper is not present in hemoglobin, this element appears to be needed in small amounts in its formation. The blood of anemic test rats has shown no increase in hemoglobin when pure iron salts were supplied in their milk, but a definite improvement was observed when a small amount of copper was also included. Small amounts of copper may be needed for the synthesis of some of the oxidases and other enzymes. Copper is stored in small amounts in the liver, and liver and oysters are usually good sources of this element. It should be remembered that when copper is ingested in appreciable quantities, it is toxic.

Cobalt and manganese

The lack of cobalt has produced anemia in some cattle and sheep. The importance of trace amounts of cobalt in the formation of hemoglobin is indicated by its presence in vitamin B_{12} (page 501). This vitamin is proving of great value in the treatment of pernicious anemia. Larger quantities of cobalt may produce a polycythemia (increased production of red blood cells). Small amounts of manganese may be needed to activate certain enzymes.

Water

Water is one of the most abundant and most important of all compounds present in the body. Nearly two-thirds of the adult body is water, and the water content of the infant is still greater.

Water is an excellent solvent, and in the body it serves as a medium of reactions (most of which will occur only in an aqueous medium). Water serves to regulate the temperature of the body, is a very good heat transfer agent, and is responsible for maintaining an even temperature throughout the body. Because heat is being generated due to oxidation of the digested foodstuffs in the cells and especially during vigorous exercise, it must be dissipated to avoid an appreciable rise in temperature. Because water has a high heat of vaporization, the dissipation of this heat depends upon the evaporation of water by way of the lungs and skin. Even if no perceptible sweating is occurring, much water is evaporated from the lungs and by the sweat glands of the skin.

Water is needed to transport food and food products in the intestines and from the intestines to the cells and to transport the waste products of the body to the skin, lungs, kidneys, and intestines.

WATER AND ELECTROLYTE BALANCE

The water and electrolyte balances in the body are closely interrelated and are related to the acid–base balance. These balances are to a great extent under hormonal control. Thus the vasopressin of the pituitary has an antidiuretic effect, while cortical hormones, such as cortisone and aldosterone, exert a control of sodium ion concentration in the plasma and other body fluids.

Minor variations from normal concentrations of water, electrolytes, and the pH of the body fluids are usually quickly compensated by body mechanisms. However, conditions that lead to an extensive imbalance of water, electrolytes, acids, or bases will usually lead to severe pathological changes in the body, and immediate therapeutic measures must be instituted to avoid dire results.

Many mechanisms are utilized in maintaining the proper balance of water, electrolytes, and acids and bases in the body—a state of balance now frequently referred to as *homeostasis*.

Water is taken into the body in ingested fluids, in water present in food, and as metabolic water (produced by oxidations in the cells of the organic hydrogen-containing compounds of the food). It is excreted from the lungs, skin, intestines, and kidneys. However, the regulation of water balance depends primarily upon the kidneys, for only enough water is excreted by the lungs and skin to control body temperature and by the intestines to maintain the proper consistency of the intestinal contents. In a normal individual, about 400 ml of water is lost in the expired air and 600 ml is lost by perspiration daily. The obligatory water loss through the kidneys is about 500 ml daily. Any intake of water beyond the obligatory amounts specified above will be excreted as an additional volume of urine by the kidneys.

In a normal individual, thirst is a regulator of body water. The human body cannot distinguish between a water and a salt deficiency. Although there is a daily obligatory loss of water and potassium ions by the kidneys, there is no similar obligatory loss of sodium and chloride ions. If salt is removed from the diet, the amount of salt ions in the urine will decrease rapidly, and the kidneys, in a short period of time, will be excreting a nearly salt-free urine.

Normally, the concentration of electroyltes in blood, interstitial, and intracellular fluids in the body is maintained constant. Venous blood will have a slightly higher concentration of electrolytes than arterial blood because of the higher concentration of bicarbonates in the former. Interstitial fluids can be expected to have an electrolyte concentration similar to the plasma of the blood. Because protein concentration in intracellular fluids is higher than in the interstitial fluids, the former can be expected to have a lower concentration of electrolytes if the two fluids are to maintain nearly equal osmotic pressures. It is noteworthy that in extracellular fluids, sodium and chloride ions are quite predominant, whereas in intracellular fluids, potassium and phosphate ions are in much greater abundance.

Osmotic pressures and volumes of body fluids will be affected by changes in water and electrolyte content of these fluids. An increase or decrease of fluid in the body will usually affect the volume of interstitial fluid with little change in blood volume. However, in cases of extended water deficiency, plasma volumes will also be decreased.

Pathological conditions accompanying water and electrolyte imbalance

Although *dehydration* of the body may be due to an inadequate intake of water, a number of pathological conditions are frequently accompanied by dehydration, such as persistent vomiting, intestinal obstruction, diarrhea in infants, hemorrhage, and diabetes mellitus and insipidus. Treatment of dehydration due to these conditions usually involves the replacement of the water and the electrolytes that have been lost by the body.

Persistent vomiting may produce an alkalosis, due to loss of acid from the stomach. Addition of ammonium chloride to the replacement fluid will correct this condition. Addition of sodium lactate to replacement fluid will correct the acidosis that frequently accompanies diarrhea.

Potassium ions will leave the intracellular fluids and enter the interstitial and plasma fluids if there is an appreciable loss of sodium ions from the body fluids. If kidney function is normal, much of this displaced potassium will be excreted. If the kidneys are malfunctioning, this potassium will be retained in the extracellular fluid of the body, and serious heart damage may result. In sodium ion loss from the body, it is usually wise to include some potassium ions in the replacement fluids.

Excessive perspiration leads to large fluid and salt losses and is accompanied by muscle cramps, weakness, and collapse. To avoid these symptoms, dilute salt solutions (about 0.1 to 0.15 percent), which are quite palatable, or salt tablets should be used. Fever leads to similar losses of water and electrolytes, and these losses may be made up by use of dilute salt solutions.

Shock due to trauma and burns usually results in a loss of potassium by the damaged cells and an extensive change in concentration of electrolytes in most body fluids. Shock is usually accompanied by fluid transfer from interstitial to intracellular fluids. In shock, administration of isotonic sodium lactate is suggested.

In adrenalectomies and in Addison's disease, large amounts of sodium ions are excreted (page 519). This results in edema, and potassium ions move from the cells to the interstitial fluids of the body and then are generally partially excreted. Although administration of salt solutions gives some relief, the use of cortical hormones that control electrolyte concentrations, such as cortisone, is the preferred treatment. In edema, replacement solutions containing salt should be used with caution.

STUDY QUESTIONS

1. What are some of the elements essential for the proper growth and functioning of the body?
2. The following pathological conditions may be due to the deficiency of which of the elements: anemias, goiters, rickets, osteomalacia?

3. How is the essential nature of a chemical element in the diet determined experimentally?
4. What are some of the sources of water utilized by the body?
5. What are the functions of water in the body?
6. What are some of the causes of dehydration of the body, and how is this condition usually corrected?
7. When will there be water retention by the body? This retention of water will usually appear as what symptom?
8. Discuss the effect of acidosis on the salt content of the body, the effect of alkalosis on the salt content of the body.
9. How does the body utilize sodium chloride? What role does sodium bicarbonate play in the body?
10. Why must patients with edema be placed on salt-free diets?
11. What is the effect of hypofunctioning of the adrenal glands on the sodium and potassium content of body fluids? What is the name of this disorder?
12. What roles do potassium ions play in the body?
13. What hormone controls the concentration of potassium salts in the body?
14. Discuss the role of sodium bicarbonate in the transportation of carbon dioxide in the body; in the control of pH of the body.
15. How is carbon dioxide excreted from the body?
16. Discuss calcium and phosphorus metabolism from the following standpoints: (a) sources of these elements, (b) role that these elements play in the body, (c) effect of a deficiency of these elements on the body, (d) daily requirements of the body for these elements.
17. What glands control the calcium content of the body? What effect on calcium metabolism does an underactivity of these glands produce?
18. Why does magnesium sulfate act as a strong purgative?
19. Because milk is a poor source of iron, what is the chief source of iron for the newborn infant? When is it necessary to supplement the diet of the infant with iron?
20. Why do human females exhibit anemias more frequently than males?
21. What form of iron is most readily utilized by the body?
22. Iron is needed for the formation of what essential substances of the body?
23. Why are goiters more prevalent in certain geographic regions than in others?
24. What role does iodine play in the functioning of the thyroid gland?
25. How are therapeutic doses of iodine furnished to the body?
26. What is thyroglobulin? (See page 515.)
27. What foods serve as good sources of iodine?
28. What beneficial effects are exhibited by small amounts of fluorides present in the water supply?
29. Describe mottled teeth. How is this condition produced?
30. What essential substances of the body contain the element sulfur? What are the sources of sulfur in the diet?
31. Name the amino acids that contain sulfur. Which of these is an essential amino acid?
32. Discuss the excretion of sulfur from the body.
33. Discuss the role of copper in the formation of hemoglobin. How else may copper be utilized in the body?
34. What metallic element is present in vitamin B_{12}? Discuss the importance of small amounts of this vitamin in the body.

35. What metallic ions have a depressing effect on the body?
36. What is meant by homeostasis?
37. Discuss the role of the kidneys in the regulation of the water balance in the body. What is the usual daily loss of water from the body by respiration and perspiration? What is the usual daily obligatory water loss from the kidneys?
38. What symptoms may accompany excessive perspiration?
39. What changes in the body fluids occur due to the shock that usually accompanies trauma and burns?
40. Is there an obligatory loss of (a) water, (b) sodium ions, (c) potassium ions, and (d) chloride ions from the body?
41. What changes in the electrolyte concentration of extracellular and intercellular fluids will occur when sodium ions are removed from extracellular fluids?

chapter 33

vitamins

33

If test animals are fed purified diets that contain only the essential fats, carbohydrates, proteins, water, and mineral elements, they will cease to grow, will lose weight, and will exhibit symptoms characteristic of the deficiency diseases. Besides the constituents enumerated above, very small amounts of substances called *vitamins* are necessary for the proper functioning of the body. Because the human body cannot synthesize these substances, it must depend upon plants and other animals for them.

In the late nineteenth century, it was noted that the addition of rice polishings to the diet of many Oriental people that were subsisting on a diet of polished rice usually prevented the occurrence of the deficiency disease beriberi, which was quite prevalent in Java and other parts of Asia. In 1911, Funk isolated a cystalline substance, now called *thiamin*, from these rice polishings. This substance cured beriberi in man and a similar disease, *polyneuritis*, in pigeons. He found an amine group in this curative substance, and because it appeared to be essential to life, he called it a *vitamine*. This name was later shortened to *vitamin* and has been extended to all the accessory food factors that have been described. Because many of these substances contain no nitrogen, the name is a misnomer.

Autotrophic organisms, such as green plants, if furnished energy as sunlight, can synthesize all the materials they require for growth and reproduction starting with such simple substances as carbon dioxide, water, nitrogen, and nitrogen-containing salts. Certain of the simpler heterotrophic organisms, such as free-living bacteria and yeast, require, besides these materials, some few simple organic compounds, such as lactic acid, to meet all their needs, because they are able to synthesize all the complex materials they require from these materials. Many heterotrophic organisms, however, require much more complex organic materials for their growth and usually depend on such autotrophic organisms as plants or other heterotrophic organisms for these. Many heterotrophic organisms, finding that a preformed essential substance is available in their diet, have lost the ability to synthesize this substance. Thus milk-souring bacteria, having found a ready source of riboflavin in milk, have lost the ability to biosynthesize this substance, and it must be considered an essential vitamin for these lactic acid bacteria. It is to be presumed that man has lost the ability to or cannot synthesize sufficient amounts of the vitamins discussed in this chapter to meet the demands of his body.

Vitamin A

Two forms of this vitamin are known. Vitamin A_1 occurs abundantly in the livers of saltwater fish, whereas vitamin A_2 occurs in the livers of freshwater fish. Both substances are soluble in fats and fat solvents but insoluble in water. Both are highly unsaturated alcohols and are represented by the following general formula:

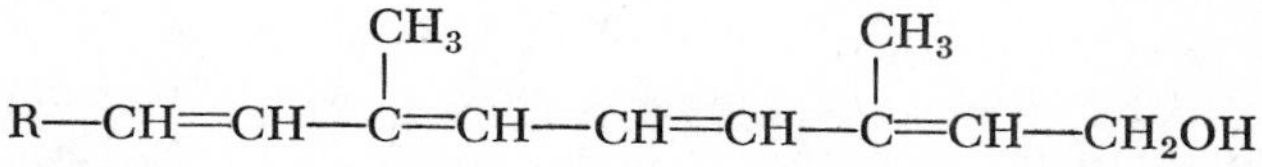

$$R{-}CH{=}CH{-}\overset{\overset{CH_3}{|}}{C}{=}CH{-}CH{=}CH{-}\overset{\overset{CH_3}{|}}{C}{=}CH{-}CH_2OH$$

In vitamin A_1, the R group, which contains a six-membered ring and two carbon atoms joined by a double bond, is

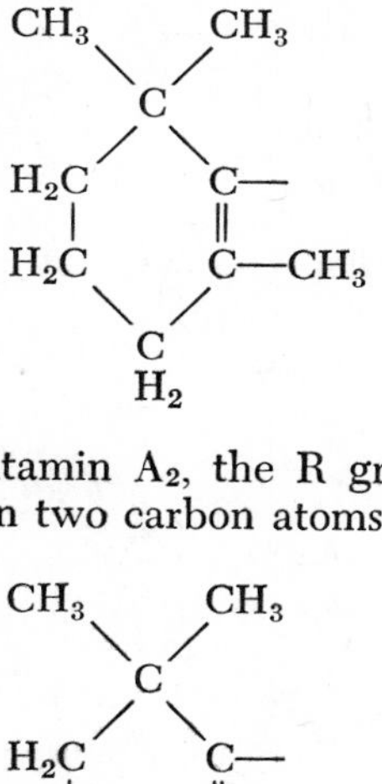

```
CH3     CH3
   \   /
     C
   /   \
H2C     C—
 |      ||
H2C     C—CH3
   \   /
     C
     H2
```

In vitamin A_2, the R group contains an additional double bond between two carbon atoms.

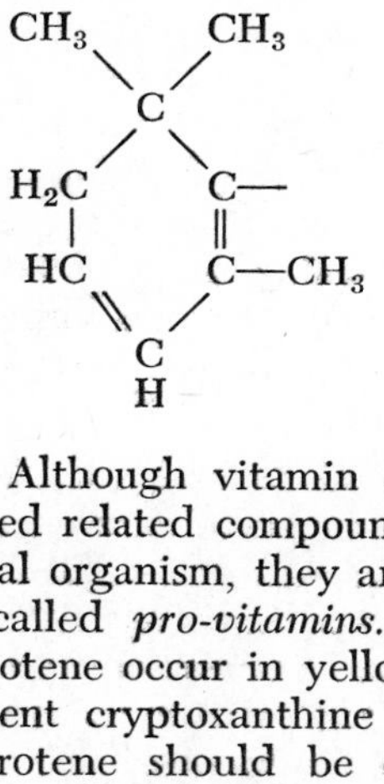

```
CH3     CH3
   \   /
     C
   /   \
H2C     C—
 |      ||
HC      C—CH3
   \\  /
     C
     H
```

Although vitamin A does not occur in plants, several highly colored related compounds do, and when these are ingested into the animal organism, they are converted into vitamin A. Such substances are called *pro-vitamins.* The three colored carotenes, α-, β-, and γ-carotene occur in yellow vegetables such as carrots and yams. The pigment cryptoxanthine can also act as a precursor of vitamin A. β-Carotene should be cleaved into two molecules of vitamin A, whereas the other three plant pigments just mentioned should produce one molecule of vitamin A for each molecule of the pigment utilized. The conversion in the body, however, is much less than 100 percent efficient.

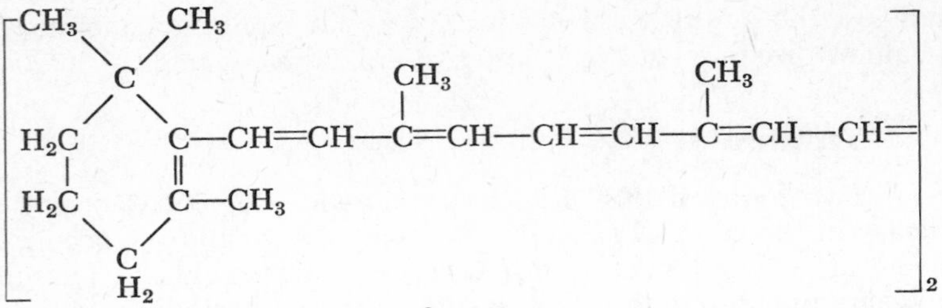

```
 ┌ CH3      CH3                                                          ┐
 │    \    /                                                             │
 │     C                  CH3                    CH3                     │
 │    / \                 |                      |                       │
 │ H2C    C—CH=CH—C=CH—CH=CH—C=CH—CH=                                   │
 │ |      ||                                                             │
 │ H2C    C—CH3                                                          │
 │    \  /                                                               │
 │     C                                                                 │
 └     H2                                                                ┘2
```

β-carotene

As with other vitamins and essential substances, a lack of vitamin A in the diet will induce a slowing of the growth rate or a complete failure of growth. A lack of this vitamin will also produce a shrinking and hardening of the epithelial tissues, such as the membranes of the eyes, of the gastrointestinal tract, skin, mouth tissues, the respiratory tract, and the genito-urinary system. This hardening of the tissues is called *keratinization.* The tear glands of the eyes frequently become *keratinized* and will no longer secrete the tears that wash the eye. Under such conditions, bacteria are able to attack the membranes of the cornea of the eye and produce an infection called *xerophthalmia.* In this disease, the cornea becomes clouded, and light is not transmitted through it.

The change of the respiratory tract tissues produced by a deficiency of vitamin A makes them much more susceptible to respiratory infections. This is usually overcome by the addition of satisfactory amounts of vitamin A to the diet. It must be remembered, however, that vitamin A will have little beneficial effect on respiratory ailments that are not caused by this deficiency. Animals fed diets deficient in this vitamin often have deformed teeth.

The rods, certain cells in the retina of the eye, are sensitive to small amounts of light. This sensitivity has been traced to the presence of a pigment called *rhodopsin,* or *visual purple,* present in the rods. This pigment appears to be a product formed by the union of a protein called *opsin* with the aldehyde *retinal,* which is the product resulting from the oxidation of the primary alcohol group of vitamin A. If there is a deficiency of this vitamin, rhodopsin is not formed in sufficient amounts, and the afflicted individual will not be able to discern objects in dim light, a condition called *night blindness,* or *nyctalopia.* Supplementing the diet with carotenes or vitamin A will usually cure this type of night blindness.

The best sources of vitamin A are the fish liver oils, such as cod liver oils, and halibut liver oil. The yellow and green leafy vegetables are excellent sources of the carotenes. Eggs and dairy products are other good sources of vitamin A.

The vitamin B complex

Soon after it was discovered that the extract of rice polishings would relieve the symptoms of beriberi, it was recognized that several factors were present in this extract, because the antiberiberi factor (thiamin or vitamin B_1) was heat labile (destroyed by heat), yet the resulting material retained some heat stable factors. Riboflavin, or vitamin B_2, has been recognized as one of these factors. More recently, other associated substances in the vitamin B complex, such as niacin, pantothenic acid, and biotin, have been isolated and described. At first these vitamins were identified as vitamin B_1, B_2, B_3, etc. Because there has been much confusion concerning this type of nomenclature, it is common to associate a chemical name with those substances whose chemical structure has been established. The B vitamins appear to be essential constituents of certain coenzymes.

Vitamin B_1, or thiamin

This is a water-soluble, white, crystalline product. It contains an amine group and is usually available as the hydrochloride salt. In neutral or slightly acid solution, or in the solid state, prolonged heating is necessary for the decomposition of this vitamin. In mild alkaline solutions, it is rapidly deactivated. Vitamin B_1, or thiamin, is composed of two substituted heterocyclic rings, a pyrimidine and a thiazole ring

```
N═C              C────N
|  |             ‖    ‖
C  C             C    C
‖  ‖              \  /
N─C                S

pyrimidine ring   thiazole ring
```

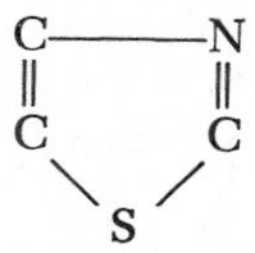

joined by a methylene (CH_2) group. Its structure has been established as

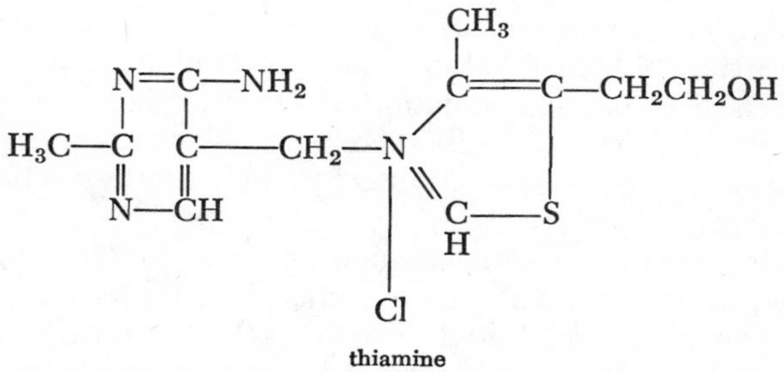

```
                              CH3
                              |
     N═C─NH2                  C══C─CH2CH2OH
     |  |                    /   |
H3C─C  C────CH2─N           |
     ‖  ‖              | \\      |
     N─CH              |  C────S
                       |  H
                       |
                       Cl
```

thiamine

It has been synthesized by R.R. Williams and associates, and the synthetic vitamin is now commercially available.

A deficiency of this vitamin will arrest growth and will produce a loss of appetite, which is frequently accompanied by a loss in weight. Its deficiency in humans produces the symptoms of beriberi (Figure 33.1). A similar disease in birds is called *polyneuritis*. The early symptoms of beriberi involve a degeneration of certain nerves. Great pain is noted when these nerves are placed under pressure. In time, the muscles served by these nerves become stiff and atrophy. The heart becomes enlarged, and death is frequently due to heart

Figure 33.1 The effect of a thiamine-deficient diet (left) and an adequate diet (right). (Courtesy Agricultural Research Service, U.S. Department of Agriculture.)

failure. In cases of beriberi, an edema will develop. Birds with polyneuritis can neither stand, walk, nor fly.

Thiamin is usually found in living tissue combined with two molecules of phosphoric acid as the coenzyme thiamine pyrophosphate. In conjunction with lipoic acid,

```
        CH2
       /   \
    CH2     CH—(CH2)4—COOH
       \   /
        S—S
```

this coenzyme activates the enzyme pyruvic oxidase, which is required to decarboxylate and oxidize pyruvic acid to acetic acid, in the aerobic phase of carbohydrate metabolism. If there is a deficiency of thiamin in the diet, pyruvic acid accumulates, and carbohydrates are not properly metabolized. The accumulation of pyruvic acid may be responsible for the symptoms of beriberi. In yeast, thiamine pyrophosphate (cocarboxylase) acts as a coenzyme for the decarboxylation of pyruvic acid to acetaldehyde.

The following foods are usually good sources of this vitamin: yeast, meat, milk, eggs, nuts, and whole grains. Vegetables and fruits are rather poor sources of this vitamin, while white bread, polished rice, and some breakfast foods are deficient in thiamin content if they are not fortified with the synthetic vitamin.

Riboflavin

This orange-red, crystalline solid has also been called vitamin B_2. It consists of a three-ring heterocyclic molecule, dimethylisoalloxazine, combined with a pentahydroxyl alcohol ribitol, which is a reduction product of the pentose sugar ribose. Riboflavin has been assigned the following structure:

```
              CH2OH
              |
             (CHOH)3
              |
              CH2
              |
              N        N
       ⁄‾‾\ ⁄   \   //   \
  CH3 /    \     C         C=O
     |      |    |         |
  CH3 \    /     C         NH
       \__/ \  //  \     /
              N       C
                      ||
                      O
```

riboflavin

Although there is no characteristic disease associated with a deficiency of this vitamin, on a diet lacking riboflavin rats decrease in weight, tend to lose their hair, and the skin becomes scaly; dogs will react with vomiting, diarrhea, loss in weight, and finally death. In man, there is an inflammation of the skin (dermatitis) and a cracking of the skin in the corner of the mouth. In both rats and dogs, cataracts of the eyes appear.

Riboflavin is required to form the flavin coenzymes, flavin mononucleotide (FMN) and flavin adenine dinucleotide (FAD) (page 374). These are coenzymes for certain dehydrogenases and have the ability to accept hydrogen and to pass it on to the cytochromes or to oxygen. As an example, the flavin nucleotides activate succinic dehydrogenase, an essential enzyme for the conversion of succinic acid to fumaric acid in the citric acid cycle. The lack of riboflavin should seriously interfere with metabolic processes.

Milk is a very good source of this vitamin. Other sources are yeast, liver, lean beef, eggs, cheese, and whole grains.

Nicotinic acid

This is the antipellagra factor. It has also been called the P-P factor and has been confused with riboflavin as vitamin G. Nicotinic acid, or niacin, and niacinamide, to which it is readily transformed in the body, have the following formulas:

```
        H                                H
        C                                C          O
     //   \                           //   \       //
  HC        C—CO2H                 HC        C—C—NH2
  |         ||                     |         ||
  HC        CH                     HC        CH
    \\     /                         \\     /
       N                                N
```

nicotinic acid, or niacin

niacinamide

This vitamin relieves the symptoms of the deficiency disease of man called *pellagra* (Figure 33.2). In this disease, a dermatitis (skin rash) and an inflammation of the mouth appear, and diarrhea may also occur. There is usually a thickening and pigmentation of the skin. The latter stages of this disease may involve nervous and mental disturbances. The disease of dogs called *black tongue* must also be due to a deficiency of this vitamin, because it can be cured by addition of niacin to the diet. A deficiency disease resembling pellagra has been induced in pigs, rats, and mice. Pellagra was formerly prevalent in the South, where some people have subsisted almost exclusively on corn products and fat pork. Due to the improvement of the diet, pellagra is of infrequent occurrence in the United States.

Niacinamide is essential and is a component of the phosphopyridine nucleotides NAD (also designated as DPN, or coenzyme I) and NADP (or TPN, or coenzyme II). These are hydrogen carriers and are thus coenzymes for a number of dehydrogenases, such as those required to dehydrogenate 1,3-diphosphoglyceraldehyde, lactic acid, and isocitric acid of the citric acid cycle. This vitamin is therefore quite essential for proper carbohydrate metabolism. Lean meat, yeast, liver, and milk are good sources of it.

Pyridoxine, pyridoxal, and pyridoxamine

These all have vitamin B_6 activity. Like niacin, they are substituted pyridines:

```
          CH2OH                                   CHO
           |                                       |
           C                                       C
         //  \                                   //  \
  HO—C          C—CH2OH                   HO—C          C—CH2OH
      |         ||                            |         ||
 H3C—C          CH                       H3C—C          CH
       \\     /                                 \\     /
          N                                        N
      pyridoxine                                pyridoxal

          CH2NH2
           |
           C
         //  \
  HO—C          C—CH2OH
      |         ||
 H3C—C          CH
       \\     /
          N
      pyridoxamine
```

A lack of these substances in the diet will produce nervousness, irritability, and weakness in the human. Rats on pyridoxine-deficient diets develop a dermatitis on the paws, lips, cheeks, chin, and on the side of the face; pigs and dogs develop an anemia. Rats and dogs on

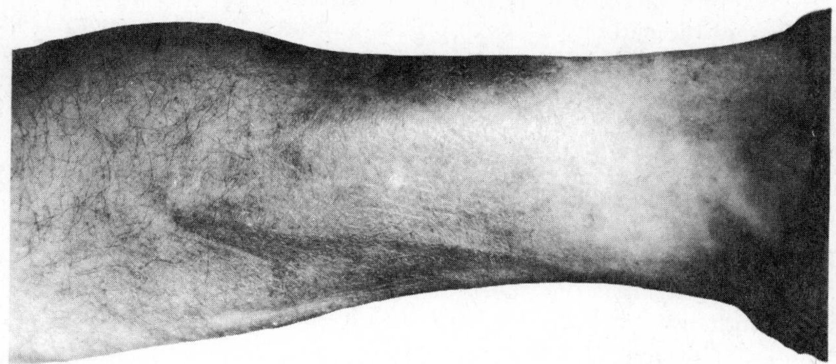

Figure 33.2 The characteristic dermatitis that frequently accompanies the deficiency disease pellagra. (Courtesy Dr. A. C. Bailey.)

a deficient diet may develop epileptic fits, and chicks will not grow on a diet deficient in pyridoxine.

In the body, pyridoxine is probably converted to pyridoxal (the CH_2OH group of pyridoxine is oxidized to the aldehyde group) and to pyridoxamine (in which the OH group in position 4 of pyridoxine is replaced by an amino group). A coenzyme pyridoxal phosphate is required to decarboxylate amino acids to amines. Pyridoxal phosphate and pyridoxamine phosphate are probably coenzymes for the transamination reactions of the amino acids. Pyridoxamine may also be essential for the metabolism of unsaturated acids and tryptophan and for the conversion of the latter to nicotinic acid.

Egg yolks, grains, milk, meat, and fish are good sources of this vitamin.

Pantothenic acid

This is another water-soluble, heat stable substance which was first isolated in 1940. It has been assigned the formula

$$HOCH_2-\overset{\overset{\displaystyle CH_3}{|}}{\underset{\underset{\displaystyle CH_3}{|}}{C}}-CHOH-\overset{\overset{\displaystyle O}{\|}}{C}-NH-CH_2-CH_2-COOH$$

pantothenic acid

It is not known if this vitamin is necessary for human nutrition, although it appears to be necessary in the nutrition of chicks, rats, dogs, and certain microorganisms, such as yeast and bacteria. Deficiency of pantothenic acid in dogs produces upset of the gastrointestinal tract, prostration, and convulsions, whereas rats exhibit a graying of the hair, failure in growth, damage to the adrenal cortex, and dermatitis. Chicks also exhibit a dermatitis on a deficient diet.

Pantothenic acid appears to be essential for the proper metabolism of the carbohydrates, proteins, and fats, for it is a component of coenzyme A, which is required for the union of acetic acid or acetyl groups with oxaloacetic acid to form citric acid in the citric acid cycle (page 403). It is also required in the biosynthesis of cholesterol.

Lean meat, skim milk, eggs, liver, buttermilk, and some green vegetables are good sources of this vitamin.

Biotin

Biotin was isolated by Kögl and Tonnis in 1936, and Du Vigneaud established its structure in 1942.

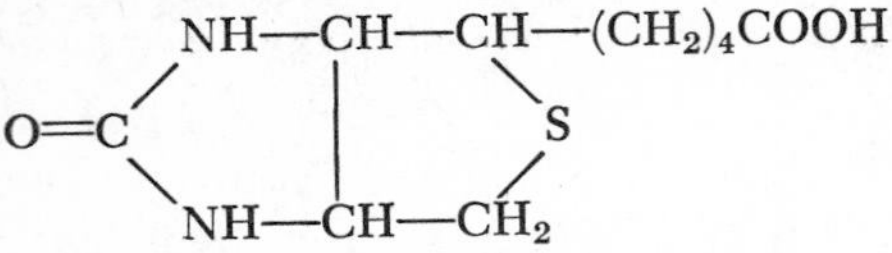

```
       NH—CH—CH—(CH2)4COOH
      /    |    \
  O=C      |     S
      \    |    /
       NH—CH—CH2
```

biotin

Its lack prevents the growth of certain microorganisms, such as yeast and bacteria. Rats fail to grow in its absence, and they also develop skin disorders. The symptoms of this deficiency may be produced in the rat by feeding large amounts of egg white, which apparently prevents the absorption of biotin by the digestive tract. Typical biotin deficiency symptoms have been produced experimentally in humans.

Biotin is probably a component of coenzymes that are necessary for the decarboxylation and oxidation of oxalosuccinic acid to ketoglutaric acid in the citric acid cycle, and also for the union of pyruvic acid with carbon dioxide to form oxaloacetic acid. It appears to be required for the biosynthesis of fatty acids.

Liver, kidney, and milk are excellent sources of biotin.

Pteroylglutamic acid

This has also been called *folic acid* and has the following structure:

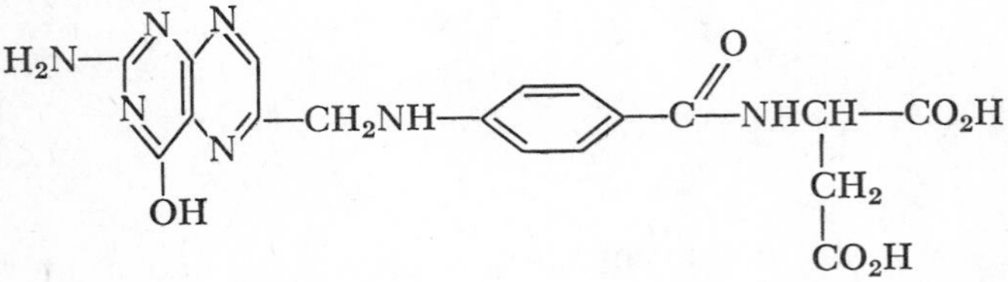

```
       N  N                                   O
H2N—(pteridine ring)—CH2NH—(benzene ring)—C—NHCH—CO2H
       N   N                                     |
       |                                         CH2
       OH                                        |
                                                 CO2H
```

It contains three constituents, a heterocyclic molecule called a *pterin*, *p*-aminobenzoic acid, and glutamic acid. Some other related substances in which a number of glutamic acid molecules are attached to the glutamic acid portion of the molecule by peptide linkages also exhibit folic acid activity.

In the absence of pteroylglutamic acid, laboratory animals do not produce red and white blood cells in sufficient amounts. Thus it has found use in several types of anemia, although it gives only temporary relief in pernicious anemia. Diarrhea and several human diseases, including *sprue* and *leucopenia,* are relieved by the addition of this vitamin to the diet. It appears to be necessary for the growth of certain bacteria. Chicks do not grow and feather on a diet deficient in folic acid.

Pteroylglutamic acid appears to be necessary in certain syntheses that require the addition of a one-carbon-atom residue to certain substances. It may be needed for the synthesis of thymine, a pyrimidine (page 439), and other components of the nucleoproteins, as well as choline and the amino acid methionine.

Good sources of this vitamin are green leafy vegetables, yeast, liver, and cauliflower.

p-Aminobenzoic acid is necessary for the synthesis of pteroylglutamic acid.

Vitamin B_{12}, or cyanocobalamin

This is a complex molecule that contains cobalt, cyanide groups, and phosphorus. A heterocyclic molecule, dimethylbenzimidazole, is obtained upon its decomposition (Figure 33.3).

Vitamin B_{12} has also been called the animal protein factor and is possibly identical with the *extrinsic factor* of the *antipernicious anemia principle.* Vitamin B_{12a} differs from vitamin B_{12} in having cyanide groups replaced by hydroxyl groups and is therefore called *hydroxocobalamin.*

Vitamin B_{12} cannot be absorbed from the intestinal tract in the absence of the intrinsic factor (page 390). Injections of a very small amount of vitamin B_{12} will produce a great improvement in pernicious anemia. It must be essential for the synthesis of red blood cells and appears to be necessary for the transfer of one carbon atom fragments in the body, especially in the synthesis of purines. Vitamin B_{12} improves the appetite and growth of chicks, mice, and rats that have been fed on a diet deficient in this vitamin.

Although vitamin B_{12} is actually inactive in enzymatic reactions, there are active agents that are vitamin B_{12} coenzymes, called *cobamides.* In the best known of these, the cyanide group attached to a cobalt atom in the vitamin is replaced by the nucleoside deoxyadenosine (page 442).

Many bacteria are capable of synthesizing vitamin B_{12}. Its richest animal source is liver.

Vitamin C, or ascorbic acid

This is another water-soluble vitamin; it produces remarkable recoveries from the deficiency disease called *scurvy.* It is a white, crystalline powder that is stable in the dry state but readily decom-

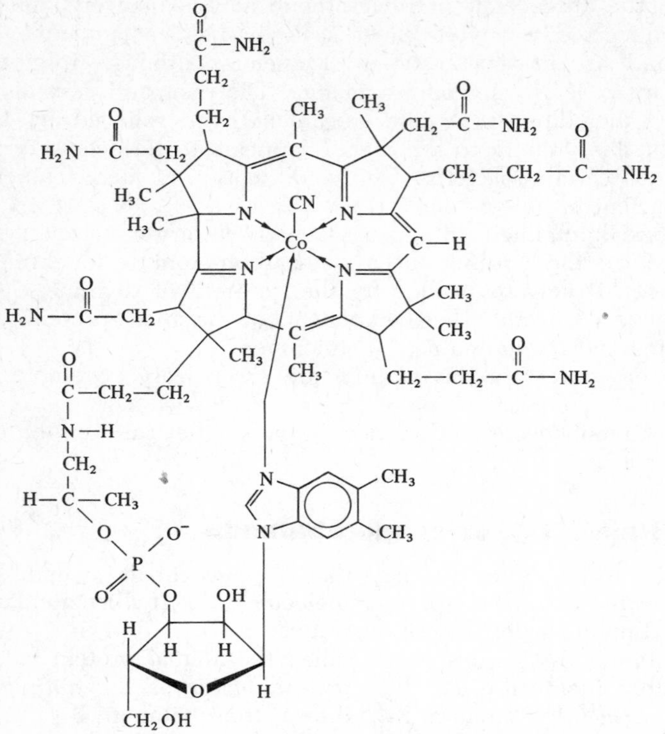

Figure 33.3 Structure of vitamin B_{12}.

poses in aqueous solutions, especially if they are alkaline. It is a strong reducing agent. Small amounts of copper catalyze its oxidation, and for this reason the use of copper cooking vessels should be avoided.

Ascorbic acid has been synthesized and is assigned the following formula:

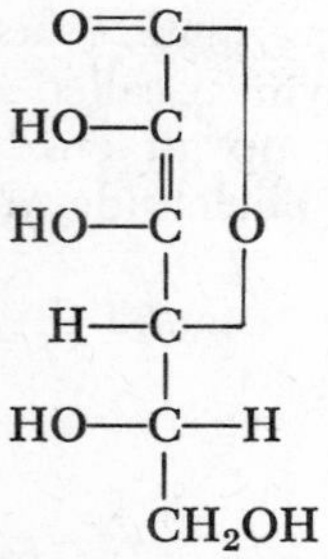

vitamin C, or ascorbic acid

Ascorbic acid appears to be necessary for the replacement of intercellular substances, that is, those chemical substances that cement

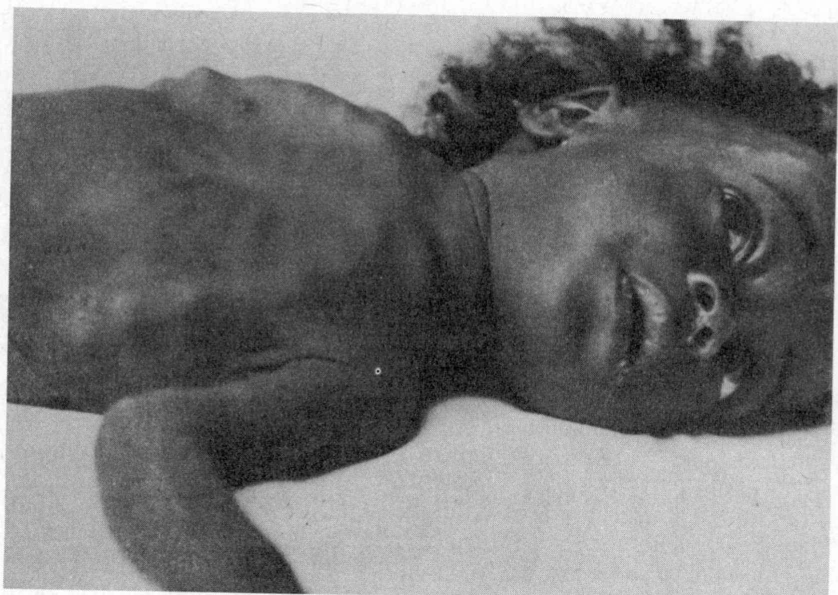

Figure 33.4 A baby suffering the typical effects of scurvy. (From Lester V. Bergman & Assoc.)

the cells of the body together. Ascorbic acid, because of its ease of oxidation, appears to act as a component of coenzymes needed for certain oxidation-reduction reactions that occur in the cells of the body. An ascorbic acid-containing coenzyme is probably required in the metabolism of the amino acid tyrosine.

Slight deficiencies of this vitamin result in few if any observable symptoms, but the individual appears to be more susceptible to infectious diseases and exhibits a slower growth rate. If there is a very extensive deficiency of this vitamin in the diet, scurvy will appear. It was discovered as early as the eighteenth century that foods that are now known to be rich sources of this vitamin would cure scurvy when added to the diet. In 1795, limes were included in the rations of British sailors; hence their nickname "limeys."

The symptoms of scurvy are swelling, reddening, and ulceration of the gums; hemorrhaging from the skin, the muscles, the mucous membrane of the mouth, and the digestive system; pains in the joints; and decalcification of the bones. Because there is damage to the bone marrow that produces the red blood cells of the body, this disease is frequently accompanied by anemia (Figure 33.4). The teeth may be greatly affected by a deficiency of vitamin C, and extensive changes occur in the enamel and the dentine. All of these changes may be due to the nonproduction of the intercellular substances.

Citrus fruits, such as limes, lemons, oranges, and grapefruits, are excellent sources of this vitamin. Other good sources are tomatoes, cabbage, potatoes, strawberries, peppers, and green leafy vegetables that are properly prepared. Extended cooking or the addition of sodium bicarbonate to the cooking water should be avoided. Storage and drying of vegetables may have a deleterious effect on their vitamin C content.

Liver is the best animal source of vitamin C, whereas mother's milk is only a fair source. Cow's milk may also be a fair source, but the quantity of this vitamin in milk is markedly influenced by the diet of the cow. Much of the vitamin may be destroyed in pasteurized or canned milk, which is one reason for adding orange or tomato juice to the diet of infants.

Vitamin D, or calciferol (the antirachitic vitamin)

There have been a number of forms of vitamin D described, but only two, *vitamin* D_2 and *vitamin* D_3, are of importance in nutrition. All of the known forms of Vitamin D are closely related to the sterols previously described (page 281).

Vitamin D_2 is called *ergocalciferol* and can be prepared by the irradiation of ergosterol with ultraviolet light. Ergosterol is a sterol that occurs commonly in fungi, like the ergot in rye, and in yeast. A commercial vitamin preparation produced by the irradiation of ergosterol is sold under the trade name Viosterol. Vitamin D_2 has been assigned the following structure:

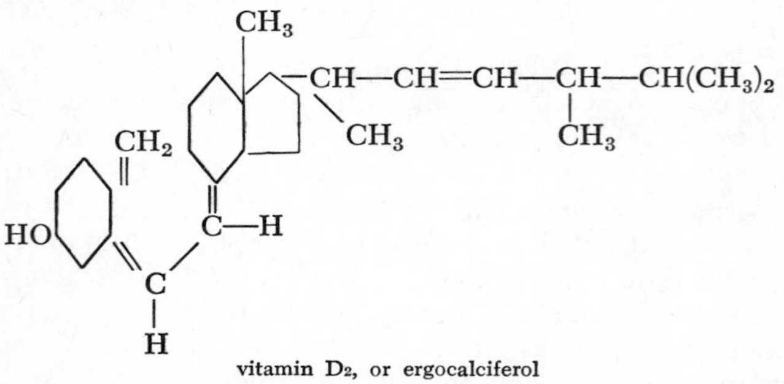

vitamin D_2, or ergocalciferol

Vitamin D_3 is produced by the irradiation of the sterol 7-dehydrocholesterol that occurs in the skin of animals. This sterol is related to the important animal sterol, cholesterol, and contains a double bond and two fewer hydrogen atoms than the latter.

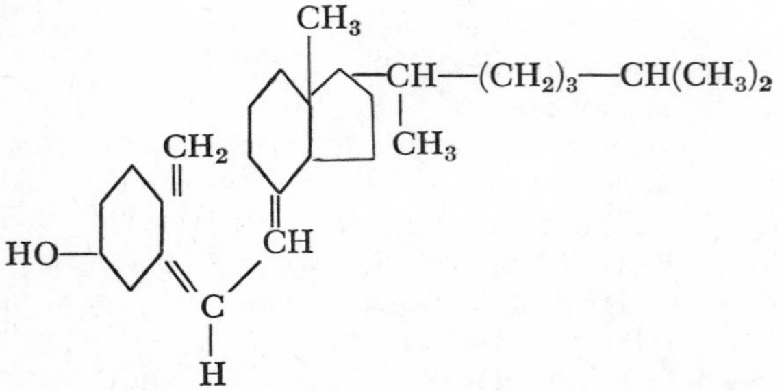

vitamin D_3, or cholecalciferol

The various forms of vitamin D are insoluble in water but soluble in fats and fat solvents. They are found in the nonsaponifiable portion of the fats, seem to be stable to heat, and are resistant to oxidation.

Vitamin D appears to be necessary for the proper absorption of calcium from the intestinal tract and for the proper deposition of calcium phosphate in the bones and teeth. A deficiency of phosphorus, calcium, or vitamin D in the diet of children will produce rickets. In rickets, the cartilage of the epiphyseal line of the bone continues to grow, but there is no deposition of calcium salts in this area. This produces an enlargement of the wrists, knees, ankles, and rib joints. The knobby appearance of the rib joints is called *rachitic rosary*. The growth of the child with rickets ceases, and these children are usually quite irritable. A bulging abdomen is common due to the relaxation of the muscles. The development of the teeth is delayed, and there is a greater evidence of cavities (dental caries).

In adults, deficiencies of phosphorus, calcium, or vitamin D in the diet can produce *osteomalacia*, also called *adult rickets*, or *osteoporosis*. In the former, mineralization of the matrix of the bone does not occur, resulting in a softening and deformation of the bones. In osteoporosis, there is an inadequate formation of the bone matrix, and a decalcification and softening of the bones occur.

The best sources of vitamin D are the fish oils, such as halibut, cod, and shark liver oil, and the flesh of oily fish, such as mackerel, salmon, and sardine. Milk and meat are rather poor sources. It is now the practice to increase the vitamin D content of milk by irradiation or addition of this vitamin.

Vitamin E, or tocopherol

Vitamin E is also called the *antisterility factor*. There are several compounds that possess vitamin E activity, but the most active of these has been called α-tocopherol, a colorless to pale-yellow oil that is soluble in fats and fat solvents but insoluble in water. It is readily oxidized and is destroyed when fats containing it become rancid. α-Tocopherol, now commercially available, has the following formula:

```
           CH3
           |
           C       O   CH3            CH3              CH3
         //  \   /   \ |              |                |
H3C—C          C        C—(CH2)3—CH—(CH2)3—CH—(CH2)3—CH(CH3)2
    |          ||       |
HO—C           C        CH2
     \\      /   \    /
        C         CH2
        |
        CH3
```

α-tocopherol

A deficiency of vitamin E in male rats produces sterility because the nuclei of the cells that produce the male sex cells, or spermatozoa, die and are absorbed. In female rats, the embryo will form and develop normally for several weeks and then die and be reabsorbed. If the degeneration of the male organ has progressed, the addition of the vitamin will have no curative effect. It will usually cure the sterility of the female rat that has been on a vitamin E deficient diet. Many young laboratory animals become paralyzed when they subsist on vitamin E deficient diets. The chief role of vitamin E in cells may be to act as an antioxidant.

This vitamin occurs so abundantly in many foods that there is little opportunity for a deficiency of it in the human diet. A vitamin E deficiency disease has not been produced experimentally in a human subject. Some investigators have claimed that the addition of vitamin E to the diet has cured some cases of sterility in humans, but the value of this vitamin as a cure for human sterility is still uncertain. Administration of vitamin E preparations to women with a history of spontaneous abortions has given variable and unpredictable results.

This vitamin is widely distributed in plant and animal tissues. The richest source of vitamin E is wheat germ oil. Green leafy vegetable, plant oils, such as peanut, palm, or cottonseed oil, whole grains, meat, egg yolks, fish liver oils, and milk are other good sources.

Vitamin K

This is a fat-soluble vitamin and is called the *antihemorrhagic factor,* because it is necessary for the clotting of blood.

Vitamin K_1, which occurs in the leaves of the alfalfa plant, is a pale yellow oil that crystallizes at low temperatures. It has the following formula:

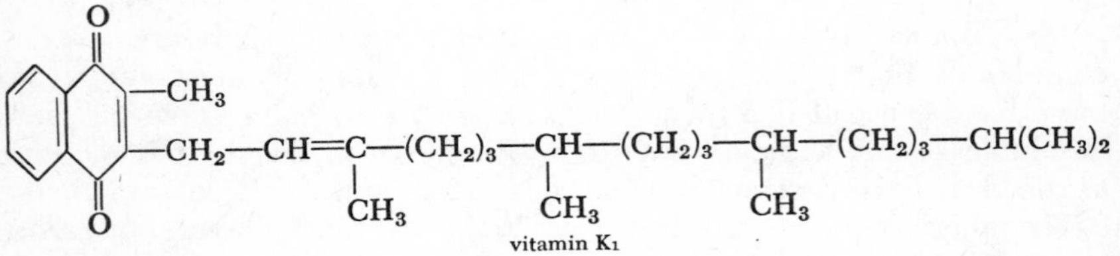

vitamin K_1

Another pale yellow, crystalline compound with vitamin K activity is synthesized by bacteria and is called vitamin K_2.

Other substances have been synthesized that have vitamin K activity. The most potent of these is the water-soluble *menadione,* or 2-methyl-1,4-naphthoquinone, which is about three times as effective a blood coagulant as the naturally occurring vitamin K_1.

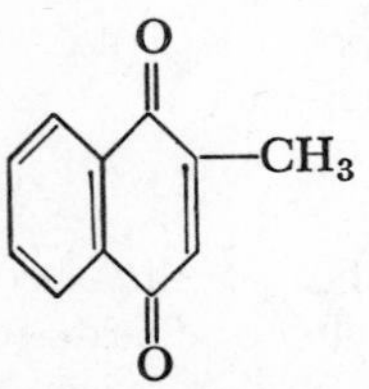

menadione

Vitamin K is required for the formation of prothrombin in the liver. Prothrombin is necessary for the normal clotting of the blood (page 469).

In the absence of bile, vitamin K cannot be readily absorbed from the intestinal tract. In obstructive jaundice and some related diseases, bile is not able to reach the intestinal tract, and patients with such diseases tend to bleed easily on slight injury. Addition of bile to the diet gives some relief from this condition, but response is best on the addition of both bile and vitamin K to the diet or intravenous injection of vitamin K_1. Bile and vitamin K are frequently administered to patients before surgery to prevent profuse and extended bleeding. Infants are sometimes born with low blood prothrombin levels and are then prone to bleeding. To avoid this tendency for bleeding in infants, vitamin K should be administered to the mother for several days before parturition.

This vitamin is found in green alfalfa and in other green leafy vegetables. Another good source is putrefied fish meal, where it has possibly been synthesized by bacteria. Bacteria probably synthesize it in the intestinal tract. Other good sources of this vitamin are tomatoes, liver, cheese, and eggs.

Other accessory food factors

Several substances that are biologically essential for the human are not considered vitamins, because they can be synthesized readily in the body. Among these are *p*-aminobenzoic acid, choline, and inositol.

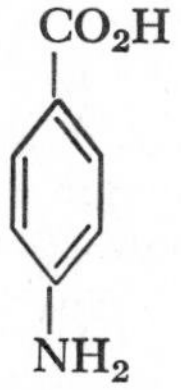

CO_2H — (benzene ring) — NH_2

p-aminobenzoic acid

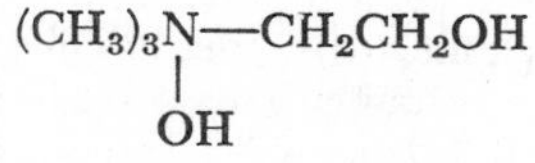

$(CH_3)_3N(OH)—CH_2CH_2OH$

choline

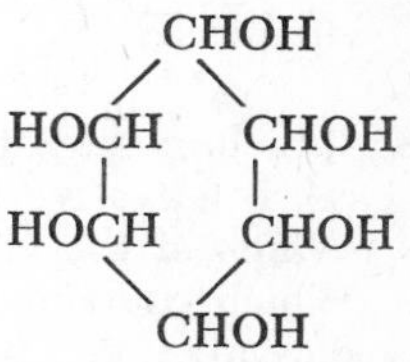

CHOH / HOCH CHOH / HOCH CHOH / CHOH (ring)

inositol

p-Aminobenzoic acid is necessary for the growth of rats, chicks, and certain bacteria. Its lack will produce a graying of the hair and also prevent the lactation of rats. The sole role of this substance may be its use in the synthesis of pteroylglutamic acid (page 500). Choline is required in the synthesis of phospholipids, such as the lecithins (page 285). On a choline deficient diet, dogs, rabbits, and rats develop a fatty liver; chicks develop slipped tendons; and in young rats, kidney hemorrhage and an enlarged spleen develop. In the absence of inositol, mice lose their hair and cease to grow, while rats develop a swelling and a loss of hair about the eyes, a condition called "spectacled eyes."

In bacteria, sulfanilamide appears to be antagonistic to *p*-aminobenzoic acid (page 336). It appears to replace the latter in some essential coenzyme, whereupon a normal metabolic process is blocked in the bacteria, thus stopping the growth or causing the death of the bacteria.

STUDY QUESTIONS

1. Which of the vitamins was first to be discovered? How and by whom was it discovered?
2. What characteristics must a substance possess to be classed as a vitamin?
3. Indicate the role of vitamins in metabolic processes.
4. What type of experiments must be conducted to determine if a substance is an "accessory food factor"?
5. Classify the vitamins according to their water and oil solubility.
6. Which of the vitamins allay the symptoms of the following diseases?

scurvy
rickets
osteomalacia
night blindness
polyneuritis
black tongue
beriberi
xerophthalmia
pellagra
osteoporosis
pernicious anemia

7. What are the symptoms of a deficiency of the following vitamins in the diet?

ergocalciferol
ascorbic acid
thiamin
α-tocopherol
vitamin K_1
vitamin A
nicotinic acid
pyridoxine
vitamin B_{12}

8. Which of the vitamins is (a) a highly unsaturated alcohol; (b) a steroid; (c) a derivative of pyridine; (d) a peptidelike compound?
9. Which of the vitamins contains both a pyrimidine and a thiazole ring? Which contains a naphthalene ring? Which of the vitamins contains a derivative of ribose? Which contains cobalt?
10. How are the carotenes related to vitamin A? Why are these substances called pro-vitamins? Discuss the occurrence of vitamin A and the carotenes.
11. What role does vitamin A play in the proper functioning of the eyes? What is rhodopsin, or visual purple?
12. Describe the best sources of thiamin, riboflavin, nicotinic acid, biotin, ascorbic acid, vitamin D, the tocopherols, and vitamin K in our diet.
13. The ingestion of large amounts of egg white will induce the symptoms of a deficiency of which of the vitamins?
14. What explanation has been given for the bacteriostatic properties of sulfanilamide?
15. Why are British sailors sometimes called "limeys"?
16. Why should copper vessels be avoided in the processing and cooking of vegetables?
17. How can ergosterol and 7-dehydrocholesterol be converted to products with vitamin D activity?

18. What is Viosterol? Ergocalciferol?
19. Identify "rachitic rosary."
20. Which of the vitamins prevents sterility in animals? Have you any information concerning its value in the treatment of human sterility?
21. What role does vitamin K play in the clotting of blood?
22. Why is vitamin K frequently administered to pregnant women shortly before parturition?
23. What substitute for vitamin K has been synthesized in the laboratory? In what way is it superior to natural vitamin K?
24. When taken by mouth, vitamin K is supplemented by what substance? What discomfort may result from this treatment?

chapter 34

hormones

34

There are a number of ductless glands (organs) in the body, such as the thyroid gland, the adrenal glands, the pituitary, and the parathyroid glands, that are called the *glands of internal secretion*, or the *endocrine glands* (Figure 34.1). The secretions of these glands, called *hormones*, pass directly into the bloodstream and are carried by the blood to those tissues or organs of the body that they stimulate. It is possible that hormones have no effect on the glands that produce them.

The hormones resemble the vitamins in their action. They are effective in very minute amounts and appear to be essential in enzyme systems that are necessary for the proper functioning of the body. Hormones are produced in the body itself, whereas vitamins are introduced into the body in the diet.

Hormones of the gastrointestinal tract

There is some evidence that the following hormones are produced in the gastrointestinal tract. Some of these appear to be polypeptides or low-molecular-weight proteins.

Gastrin. This substance is produced in the mucosa of the pylorus and in the upper portion of the duodenum. It is a true hormone, for it is absorbed from the intestine into the bloodstream, which carries it to the cells in the stomach wall. It produces a greater secretion of hydrochloric acid in the gastric juice and a greater motility of the stomach.

Secretin and pancreozymin. These two hormones are produced in the mucosa of the small intestine. Secretin increases the volume and the bicarbonate content, while pancreozymin increases the enzyme content of the pancreatic juice.

Cholecystokinin. This is produced by the intestinal mucosa. It increases the contractions of the gall bladder. The production of this enzyme is increased by the entrance of fats into the duodenum. It has now been definitely established that pancreozymin and cholecystokinin are identical.

Enterogastrone. This is a *chalone* rather than a hormone, for it is an antagonist to gastrin. It is produced in the intestinal mucosa and reduces the flow of gastric juice and the motility of the stomach.

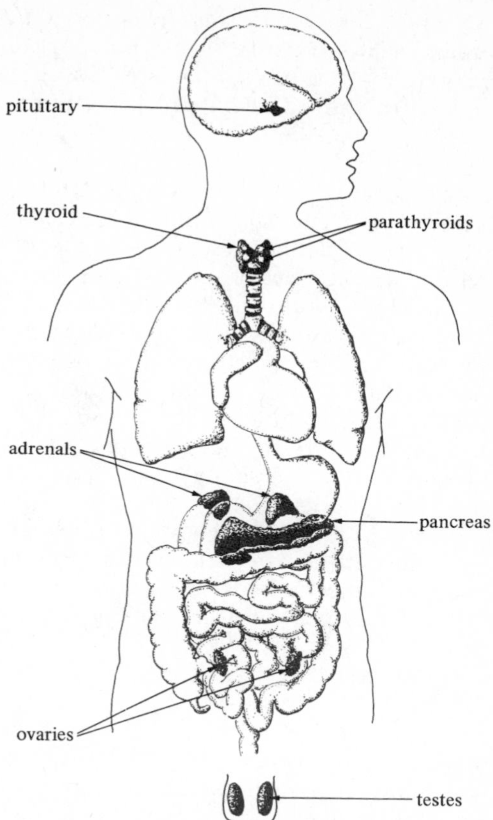

Figure 34.1 Location of the endocrine glands of human beings.

The entrance of fats and carbohydrates into the small intestine increases the production of enterogastrone and thus prolongs the digestive process of the stomach.

Insulin, the hormone of the pancreas

Besides the pancreatic juice produced by the pancreas, a hormone, *insulin,* is elaborated by clusters of cells within the pancreas called the *islets of Langerhans.* Insulin has a great influence on carbohydrate metabolism. Its presence in cells increases the rate of oxidation of carbohydrates and reduces the concentration of glucose in the blood. It also increases the rate of deposition of glycogen in the liver. It is uncertain whether insulin directly increases the rate of conversion of glucose to glycogen or produces a decreased rate of decomposition of glycogen to glucose in the liver.

An active extract of insulin was separated from the pancreas of animals in 1921 by Banting and Best, and in 1926, Abel obtained it as a colorless, crystalline solid. Sanger has elucidated the structure of insulin and has found that it consists of four polypeptide chains joined by disulfide (—S—S—) groups (see Figure 24.2). This protein has a molecular weight of about 12,000. Because insulin is destroyed by the digestive juices, it cannot be given by mouth but is injected subcutaneously.

An underactivity or degeneration of the islets of Langerhans causes diabetes mellitus. Removal of the pancreas of animals will also produce this disease. In diabetes mellitus, the rate of oxidation of glucose in the body is decreased and the glucose concentration of the blood increases. There is also an increased flow of urine. Faulty carbohydrate metabolism is usually accompanied by faulty fat metabolism and by ketosis and acidosis. If the acidosis becomes severe, coma and death may occur. Injection of insulin will produce a marked and rapid recovery from these conditions, but it cannot be considered a cure because if no more is given, the symptoms of diabetes mellitus will reappear in a short time. Usually the affected individual must take insulin the rest of his life.

Protamine insulin, produced by combining insulin with a protein called *protamine*, is substituted for or is used as an adjunct to regular insulin in subcutaneous injections. It is absorbed more slowly in the body than insulin, and larger doses can be given at less frequent intervals.

Hyperactivity or an abnormally large production of insulin is usually due to a tumorous growth of the islets of Langerhans. A hypoglycemia, accompanied by a variety of symptoms, will usually follow such a condition. The blood-sugar level will drop to a very low level. The patient may experience weakness, depression, cold sweating, tremors, and finally collapse. If an overdose of insulin is given to a diabetic patient, similar symptoms will be observed. This is called *insulin shock*. In insulin shock, the patient should be given sugar or other sweet substances immediately.

The synthesis of insulin has been accomplished by Merrifield in the United States and by a group of Chinese chemists.

The pancreas also produces a hyperglycemic factor, called *glucagon*. Glucagon produces a breakdown of liver glycogen and a rise in blood sugar. Thus this hormone is antagonistic to insulin. Its structure has been established, and it is a polypeptide with a molecular weight of 3,500.

Recently, several oral hypoglycemic agents, such as orinase (Tolbutamide) and Diabinese (chloropropamide) have been introduced as antidiabetic agents. Unlike insulin, which is of protein nature, they are not hydrolyzed by the digestive juices and are taken orally. Although generally effective in cases of "adult diabetes," "juvenile diabetes" is usually resistant to their use.

Thyroxine, the hormone of the thyroid gland

The *thyroids* are a pair of glands located in the front of the neck just below the larynx. The two glands are joined by a strip of tissue and have a combined weight of about 25 g. They are quite fibrous and contain many small vesicles that are filled with a colloidal substance usually called *colloid*. The latter is composed mostly of water and a protein called *thyroglobulin*. If thyroglobulin is hydrolyzed, two amino acids, *thyroxine* and in smaller amounts *triiodothyronine*, can be isolated.

$$HO-C_6H_2I_2-O-C_6H_2I_2-CH_2-\underset{NH_2}{\underset{|}{CH}}-CO_2H$$

thyroxine

$$HO-C_6H_3I-O-C_6H_2I_2-CH_2-\underset{NH_2}{\underset{|}{CH}}-COOH$$

triiodothyronine

Triiodothyronine has five times the hormonal activity of thyroxine. Thyroglobulin is apparently degraded to these two amino acids in the thyroid gland, and these acids are then transported by the blood to the site of their action. It is uncertain whether thyroxine, triiodothyronine, or thyroglobulin is the true active principle of the thyroid gland.

The hormone of the thyroid gland has a remarkable effect on the growth and development of the body and in increasing the rate of cell oxidation or metabolism.

An enlargement of the thyroid gland is called a *goiter* (Figure 34.2) and may be due to a deficiency of iodine in the diet. The following symptoms usually accompany an underactivity of the thyroid gland, a condition called *hypothyroidism:* sluggishness, gain of weight, loss of appetite, slower heartbeat, reduced metabolic rate, slower mental processes, and a cold feeling in the extremities of the body.

In infants, severe hypothyroidism produces a condition called *cretinism*. Hypoactivity may be due to a congenital absence of the thyroid gland (no development of this gland in the embryo) or to a serious deficiency of iodine in the diet of the mother during pregnancy. Cretins are pitiful abnormal dwarfs, short in stature but with broad bodies. They have thick skin, thick lips, protruding mouth, and a very coarse dry hair. They are usually mentally retarded, that is, idiotic, and are usually sexually underdeveloped. In early infancy, there is usually a rapid response to the administration of thyroid preparations, and if treatment is continued, the child will show a nearly normal development. If treatment is not started early, cretins will be slow to respond, and a complete elimination of symtoms will be difficult.

A severe hypothyroidism in adults will frequently result in *myxedema*, a condition which may be due to atrophy of the thyroid gland or to a severe deficiency of iodine in the diet. Besides a slow

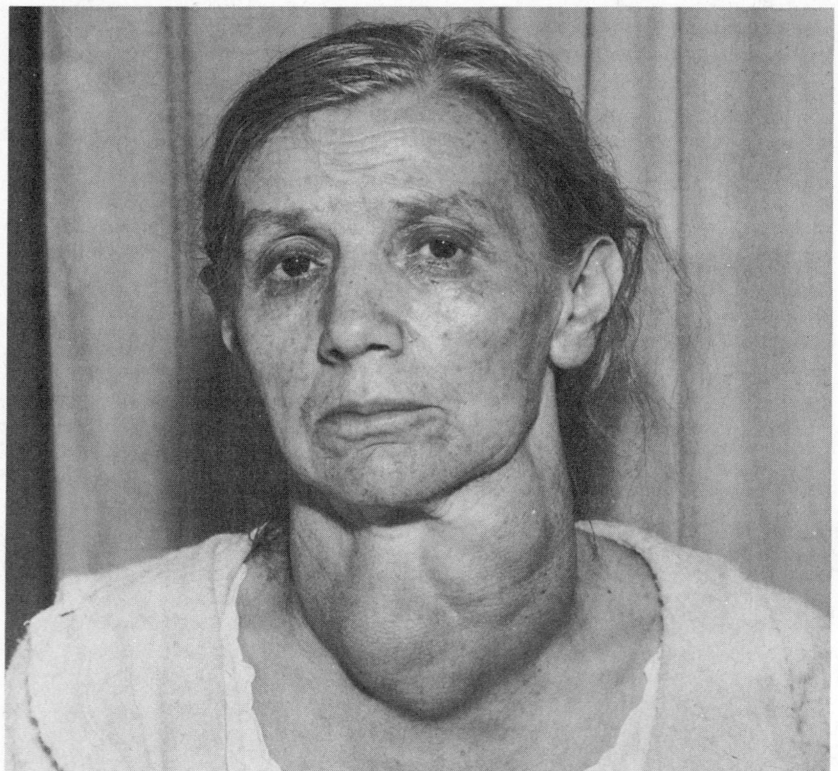

Figure 34.2 Woman with nodular hyperthyroid goiter. (Lester V. Bergman & Assoc.)

pulse rate, a low body temperature, a thickening and drying of the tongue, an increase in weight, and a loss of appetite, the skin becomes filled with water and a proteinlike substance. This produces a very noticeable puffiness about the face and eyes; hence the name of this disorder.

An overactivity of the thyroid gland, or *hyperthyroidism,* is called *Graves' disease;* in this condition the following symptoms are usually observed: extreme nervousness, excitability, increase in appetite, and loss in body weight. The heart rate will be faster and irregular, the metabolic rate will be abnormally high, and the subject will feel hot even in cold weather. Hyperthyroidism may be due to a toxic adenoma (enlargement of the thyroid gland due to the growth of a tumor). In *exophthalmic goiter,* a protrusion of the eyeballs occur.

Hyperthyroidism due to the enlargement of the thyroid gland may be benefited by the surgical removal of part of the gland, the destruction of part of the thyroid tissue by X rays, radium, or radioactive iodine, or by the ingestion of "goitrogenic" substances, such as thiourea or thiouracil. 5-Vinyl-2-thiooxazolidone, which has been identified in turnips, is also goitrogenic.

An increased concentration of cholesterol in the blood usually accompanies hypothyroidism. As thyroid activity increases, cholesterol concentrations decrease. Administration of thyroid extracts may prove valuable in decreasing the incidence of *atherosclerosis* in our older citizens, a condition apparently resulting from an abnormal deposition of cholesterol in the blood vessels of the body.

The hormone calcitonin is produced by the thyroid glands and to a lesser extent by the parathyroids (page 523). This hormone is antagonistic to parathormone; that is, it reduces the calcium content (hypocalcemia) and the phosphate content (hypophosphatemia) of the blood. This hormone accelerates the rate of calcium deposition from blood to the bones and inhibits calcium resorption from the bones.

Epinephrine, or adrenaline, the hormone of the medulla of the adrenal glands

The adrenal glands are two small bodies attached to the top of the kidneys. Each consists of an inner portion called the *medulla*, covering which is a portion of the organ called the *adrenal cortex*.

The medulla of the adrenals elaborates a simple chemical substance called *epinephrine*, or *adrenaline*. This substance contains phenolic hydroxyl groups and an amino group.

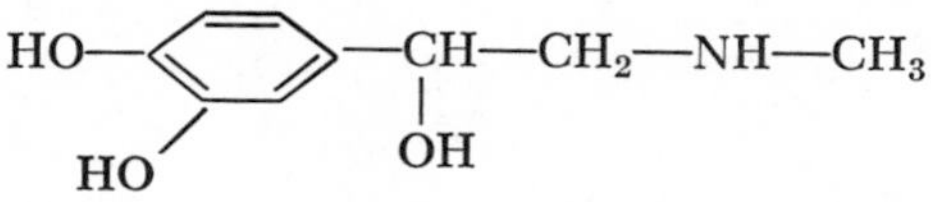

$$\mathrm{HO}\text{—}\langle\text{benzene ring}\rangle\text{—}\underset{\displaystyle\mathrm{OH}}{\overset{}{\mathrm{CH}}}\text{—}\mathrm{CH_2}\text{—}\mathrm{NH}\text{—}\mathrm{CH_3}$$
$$\text{(second HO on the ring)}$$

epinephrine, or adrenaline

Injections of epinephrine in small amounts produce a profound increase in blood pressure. This increase in pressure is due to the constriction of the arterioles. Epinephrine is used when the blood pressure drops suddenly to a very low value, and, because it contracts the arterioles, it is sometimes used in surgery to control bleeding. It is often administered with local anesthetics to increase the time they are effective, because the contracted arterioles will prevent the rapid diffusion of the drug from the localized area.

This drug has a stimulating effect on the heart muscle, and injections of it into the heart have revived heart action when it has stopped during anesthesia. It is frequently used to start the hearts of newborn infants.

Because epinephrine, ephedrine (a plant product), and benzedrine are bronchodilators, they are used to relieve the symptoms of asthma, hay fever, and rhinitis (blocking of the nasal passages).

Epinephrine appears to be antagonistic to insulin. It increases the rate of glycogenolysis in the liver and increases the blood-sugar concentration. Anxiety or fear will increase the production of epinephrine by the adrenal glands. In the latter situation, as well as when the drug is injected, the blood-sugar level may rise above the renal threshold, and glucose will appear in the urine. Epinephrine also produces an increase in the metabolic rate of the body and increases blood circulation and respiration.

Norepinephrine, a compound in which the methyl group on the nitrogen atom of epinephrine is replaced by a hydrogen atom, occurs in small amounts with epinephrine produced by the adrenal glands. It does not have the intense physiological effect of epinephrine. Unlike epinephrine, norepinephrine does not relax smooth muscle, nor does it affect carbohydrate metabolism.

Hormones of the adrenal cortex

Twenty-nine crystalline compounds have been isolated from the extract of the cortex of the adrenal glands. Chemically, these compounds are classed as steroids, because they closely resemble in structure such steroids as the sterols, sex hormones, vitamin D, bile salts, and digitalis. Only six of the compounds isolated appear to have the hormonal activity characteristic of the adrenal cortex. The three that have become best known are corticosterone, desoxycorticosterone (which possesses one less oxygen than corticosterone), and 17-hydroxy-11-dehydrocorticosterone (which contains a hydroxyl group on carbon atom #17 of the dehydrocorticosterone molecule). The name of this 17-hydroxy compound has been shortened to *cortisone.*

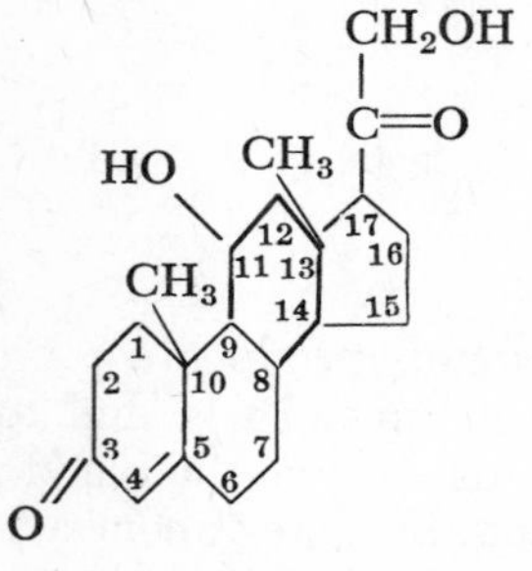

corticosterone

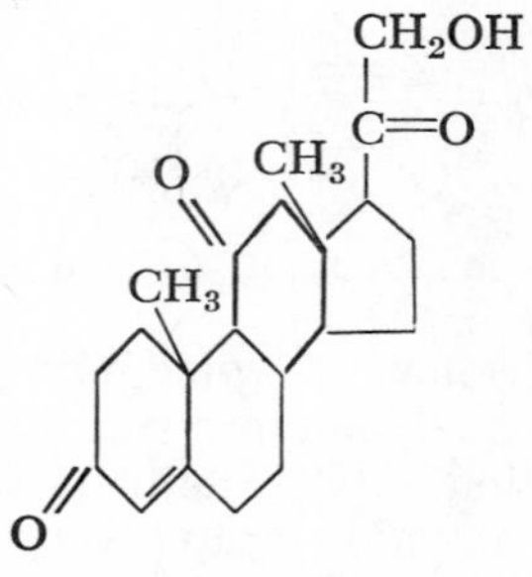

11-dehydrocorticosterone

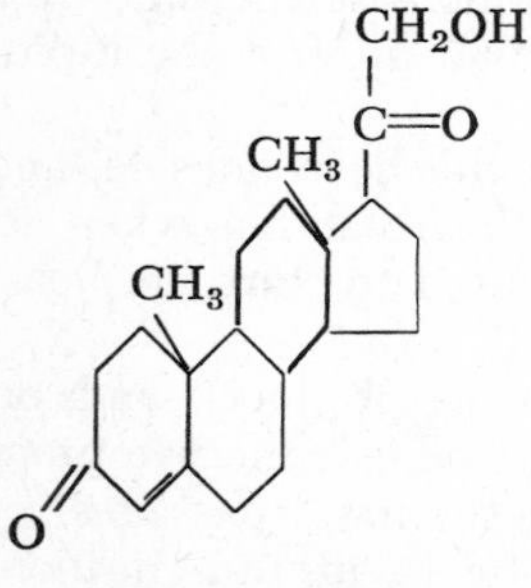

11-desoxycorticosterone

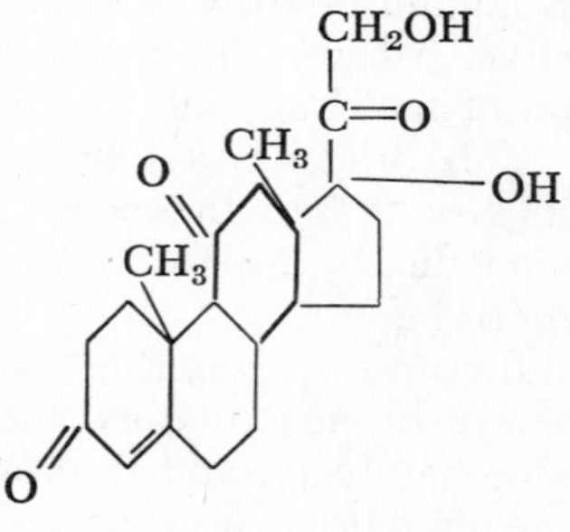

cortisone

Although no serious damage to the body results from the removal of the medulla of the adrenals, the cortex must be of great importance, because death will result within a few weeks after its removal from the animal. The life of the animal from which the adrenals have been removed surgically may be extended for considerable time by the use of diets rich in salt or by the administration of cortical extracts.

Certain of the cortical steroids have a great influence on the salt and water balance of the body. For this reason, they have been called the *mineralocorticoids.* Desoxycorticosterone is an active mineralocorticoid. The twenty-ninth crystalline steroid isolated from the adrenal cortex extract *aldosterone* is a much more potent mineralocorticoid than desoxycorticosterone. It has the following structure:

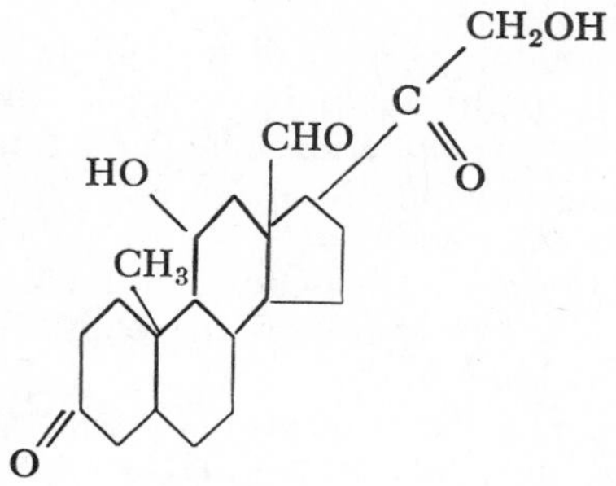

In the absence of these steroids, increased amounts of sodium and chloride ions are excreted, while potassium and phosphate ions are retained and increase in concentration in the body fluids. Most of the water of the body flows to the cells, while there is an increased excretion of water by the kidneys. The blood becomes concentrated due to the loss of water, and the efficiency of the kidneys is thereby reduced. The excretion of such metabolic products as urea, uric acid, and creatinine is reduced, and the concentration of these nitrogen compounds increases markedly in the blood.

The symptoms enumerated above appear in Addison's disease, which results from the partial or complete destruction of the adrenal cortex. A common cause of destruction of the cortical tissue is tuberculosis of the adrenals. Other symptoms of this disease are depression, muscular weakness, vomiting, decreased sexual development, and a characteristic bronzing of the skin due to a deposition of brown pigments. Except for the bronzing of the skin, a salt-rich diet will give some relief from the disorders of this disease, although the use of cortical extracts or of the synthetic acetate of desoxycorticosterone is the preferred treatment.

The *glucocorticoids* influence the carbohydrate balance in the body. In their absence, much of the glycogen of the liver and muscles is converted to glucose, proteins are not properly converted to carbohydrates, and the blood-sugar level falls to a very low value. Corticosterone appears to be active as a glucocorticoid.

Cortisone has now been synthesized and is an effective agent in relieving many of the symptoms of rheumatoid arthritis.

A hyperactivity or overfunction of the adrenal cortex is observed when there is a tumorous growth of the adrenal glands. In children, this hyperactivity will produce precocious growth and the development of sexual characteristics at an early age. In adult females, feminine characteristics decrease and masculinity increases. In adult males, an increased masculinity and sexual desire is noted.

Small amounts of estrone (page 524) and of three androgenic hormones (page 526) are also present in the adrenal cortical extract.

Hormones of the pituitary gland, or hypophysis

The *pituitary gland*, or *hypophysis*, is a small oval body lying just below the brain and attached to the brain by a small amount of tissue. It consists of three lobes: the *anterior lobe*, or *adenohypophysis*, the *posterior lobe*, or *neurohypophysis*, and a lobe that lies between these two, called the *pars intermedia*.

Because the pituitary gland appears to influence the functioning of all the hormone-producing tissues and glands of the body, it has been called the master gland of the body.

The pituitary appears to be under the control of the *hypothalamus* of the brain. This is probably due, in part, to the fact that the hypothalamus produces certain hormones that regulate the activity of the pituitary. Some of the hormones whose synthesis has been ascribed to the pituitary may actually be produced in the hypothalamus and may be stored in the pituitary until utilized.

Two hormones have been isolated from the posterior lobe. One of these, *vasopressin* (pitressin), produces in animals a temporary rise in blood pressure. It also appears to be an antidiuretic, because its absence will permit the production of a very large flow of urine, a disorder called diabetes insipidus. The other hormone, *oxytocin* (pitocin), causes smooth muscle to contract and is used to stimulate the uterine contractions at childbirth. It also appears to stimulate milk secretion.

V. Du Vigneaud and associates have established the structure of and synthesized both vasopressin and oxytocin. Oxytocin is a cyclic octapeptide (a molecule of this hormone can be hydrolyzed to eight molecules of amino acids) with a molecular weight of about 1,000 and consists of the following eight amino acids: leucine, isoleucine, tyrosine, proline, glutamic acid, aspartic acid, glycine, and cysteine, as well as three molecules of ammonia. These amino acids are linked in a ring structure by the disulfide group (—S—S—) of cysteine. Vasopressin has been isolated as *lysine-vasopressin* from hog pituitary and *arginine-vasopressin* from the corresponding beef glands. The vasopressins have structures quite similar to oxytocin. In arginine-vasopressin, phenylalanine replaces isoleucine, and arginine replaces the leucine of the oxytocin structure, while in lysine-vasopressin, lysine replaces leucine.

The oxytocin and the vasopressins synthesized by Du Vigneaud, except for the arginine-vasopressin, which exhibits a lower biological activity, are chemically and physiologically identical to the products isolated from the pituitary glands.

The six hormones that have been isolated in a highly purified form from the anterior lobe of the hypophysis or the adenohypophysis are as follows:

1. A *growth-stimulating*, or *somatotrophic, hormone* (GH), which has a great influence on skeletal growth and the formation of soft tissue. Underactivity or damage to the pituitary gland in children produces a type of dwarfism. Unlike the cretins, these dwarfs do not usually experience mental retardation. Sexual development may, however, be retarded. If there is an overactivity of the anterior lobe, frequently produced by a tumorous growth of the lobe, the child may grow to an unusual height. In adults, the long bones grow in length and are misshapen. A curvature of the spine may also develop. Due to the fibrous growth of tissue, the nose, lips, eyelids, and fingertips thicken, a condition known as *acromegaly* (Figure 34.3).

   The determination of the structure and the synthesis of human pituitary growth hormone (HGH) has been accomplished by Choh Chao Li and associates at the University of California. This protein contains 188 amino acid units and has a molecular weight of 21,500.
2. A *thyrotropic hormone* (TTH) that induces the growth of the thyroid gland and the production of thyroxine. If the anterior lobe is removed, the thyroids shrivel and thyroxine is produced in insufficient amounts.
3. The *adrenocorticotrophic hormone* (ACTH), which appears to regulate the proper functioning of the adrenal cortex and the production of cortical hormones. Dr. Klaus Hofmann and associates at the University of Pittsburgh have synthesized a polypeptide made up of 23 amino acids which has properties identical to ACTH.
4. A *lactogenic hormone* called *prolactin*, which induces the flow of milk in the mammary glands of the mother at childbirth. Also, in the presence of other suitable hormones, it induces the growth of the mammary glands.
5. A *follicle-stimulating hormone* (FSH), a *gonadotrophin*, which stimulates the growth of the graffian follicle in the female (page 523). In the male, it promotes the development of the testes and the production of sperm cells.
6. A second gonadotrophin, the *luteinizing*, or *interstitial, cell-stimulating hormone* (ICSH), which stimulates the development of the ovaries and testes and increased production of testosterone in the testes. In the ovaries, ICSH induces an increased production of the corpus luteum (page 524).

   Gonadotrophins are hormones that stimulate the gonads, that is, the ovaries in females and the testes in males.

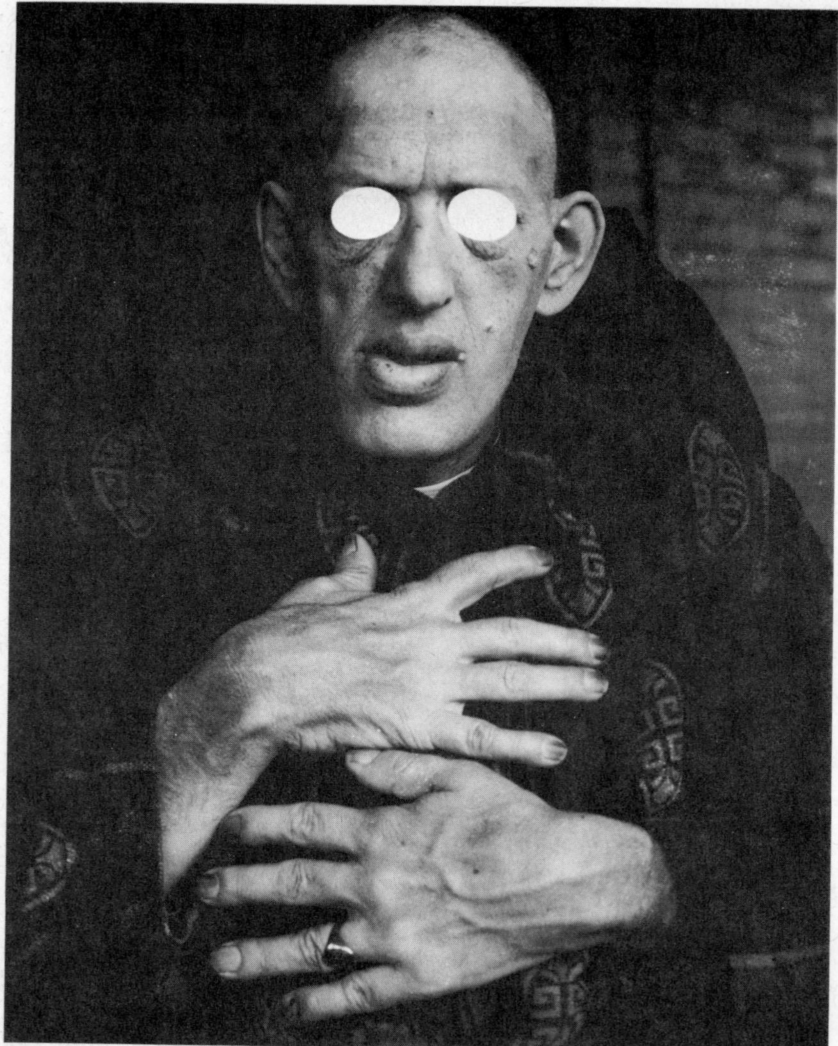

Figure 34.3 A patient with acromegaly. The characteristic appearance is due to a continued growth of the long bones and a fibrous growth and thickening of many tissues. (Courtesy Dr. Warden Ayer, Syracuse, N.Y.)

The diabetogenic effect of the anterior pituitary or adenohypophyseal extract may be identical to either the somatotrophic or the adrenocorticotrophic hormone.

The pars intermedia of the pituitary secretes the melanocyte-stimulating hormone (MSH), or melanotropin. This hormone most likely exists as two different polypeptides designated as α-MSH and β-MSH. The melanotropins are responsible for the darkening of the skin of amphibians and fish and can cause a darkening of the skin of man. The typical bronzing of the skin in Addison's disease (page 519) may be caused by an excess production of this hormone.

The hormone of the parathyroid glands

The *parathyroid glands*, four or more in number, are attached to and are frequently imbedded in the tyroid gland. These glands are very essential to life, and their removal will cause death. In early thyroidectomies (surgical removal of the thyroid gland) on cats and dogs, many deaths occurred because the parathyroid glands were inadvertently removed.

The parathyroid glands produce a hormone called *parathormone*, which controls the calcium and phosphorus balance of the body. Little is known about the chemistry of this hormone, but it appears to be protein in nature.

A hypofunctioning or removal of the parathyroid glands produces a low level of calcium in the blood but a high phosphorus concentration. There is also a decreased excretion of calcium by the kidneys. In extreme cases of underactivity of these glands, tetany and convulsions may be observed. Administration of calcium compounds or parathormone extracts will give relief from these symptoms.

Hyperactivity of the parathyroid glands produces an increase in the calcium content of the blood and an increased elimination of calcium in the urine, while an extensive decalcification and deformities of the bones occur. The blood-phosphorus concentration drops below normal, and coma may result. Enlargement of the parathyroid glands, particularly if an adenoma (a certain type of cancer) is present, is usually responsible for this overactivity. Relief from these symptoms will usually be obtained by surgical removal of the adenoma.

The hormone calcitonin, produced by both the thyroid and parathyroid glands, is antagonistic to the action of parathormone (page 516).

The biologically active portion of human parathormone has recently been synthesized in the laboratories of Ciba-Geigy, Ltd., in Switzerland.

The female sex hormones

There are two groups of female sex hormones. One group is secreted from the cells called *follicles* that line the cavities of the ovaries. This group of hormones is called the *follicular,* or *estrogenic hormones.* The follicles are filled with a fluid called the *follicular fluid.* When an ovum begins to develop in the ovary, it attaches itself to the wall of the follicular cavity. About four weeks are required for the ovum to develop, during which time the follicle moves to the

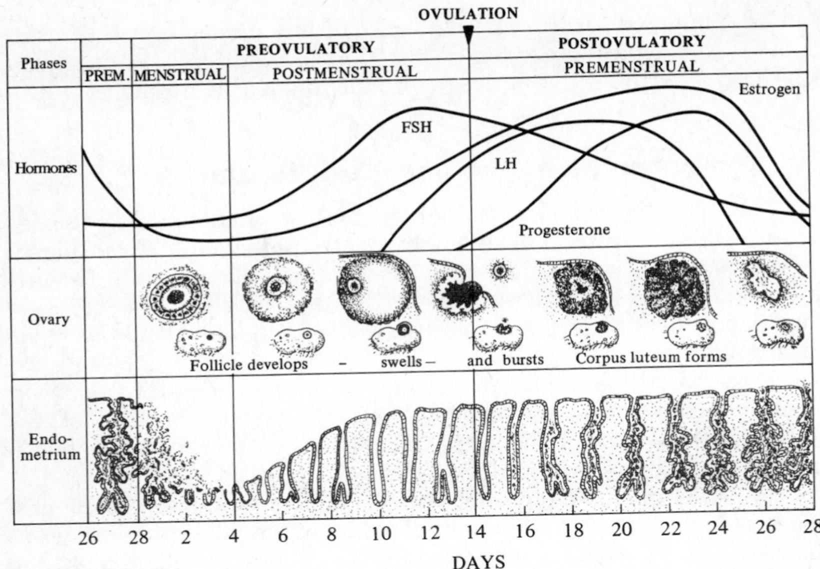

Figure 34.4 The cycle of menstruation and ovulation in the human female. (Based on a drawing in Young, Stebbins, and Brooks, Introduction to Biological Science, *Harper & Row, N.Y., 1956.)*

periphery of the ovary and the ovum is then discharged into the uterus through the *fallopian tube.*

Two hormones, *estrone* and *β-estradiol,* have been isolated from the follicular fluid, the β-estradiol being the more active. In the body, the latter is converted to *estriol,* which possesses little hormonal activity and, along with β-estradiol and estrone, is excreted in the urine. Estrone has also been called theelin; β-estradiol has been known as dihydrotheelin; and estriol, as theelol.

The estrogenic hormones are responsible for the appearance of *estrus,* or the period of heat in females of many mammals. Estradiol is responsible for the development of the vagina, the uterus, and secondary sexual characteristics, such as the enlargement of the mammary glands, the appearance of hair in the pubic region, and the high-pitched voice of the maturing female. The estrogenic hormones are also responsible for the regeneration of the endometrium of the uterus following *menstruation.* The endometrium is a layer of cells that line the inner cavity of the uterus and that slough off during the menstrual period (Figure 34.4).

Estradiol has been used therapeutically to develop the female characteristics in *sexual infantilism.* It has also been used to relieve the distressing symptoms of the *menopause,* or change of life, and in the treatment of cancer of the prostate in the male.

The second group of female sex hormones is produced in the *corpus luteum.* After the matured ovum is discharged into the uterus, the follicle fills with a blood clot that soon changes to a yellow mass called the corpus luteum (literally *yellow body*). The corpus luteum produces a hormone *progesterone*, which is responsible for the thickening of the endometrium of the uterus; thus the uterus is prepared to receive the ovum and maintain it. Progesterone appears also to be responsible for the development of the *placenta*, the tissue that joins the fertilized ovum to the wall of the uterus and through which nutrient passes from the mother to the developing fetus. Progesterone appears to inhibit ovulation during the period that the ovum is present in the uterus. The corpus luteum remains active for six to nine months of the pregnancy. If the ovum is not fertilized, the corpus luteum regresses rapidly and is replaced by scar tissue, the endometrium breaks down, and menstruation follows. Progesterone is frequently employed in the treatment of *amenorrhea*, the absence of menstruation in young women.

The female sex hormones are all steroids and resemble cholesterol (page 287) in their structure. There is a close relationship between female and male sex hormones, both of which are also related to the adrenal hormones. Several of the female sex hormones have been isolated in pure form, and their structures have been established.

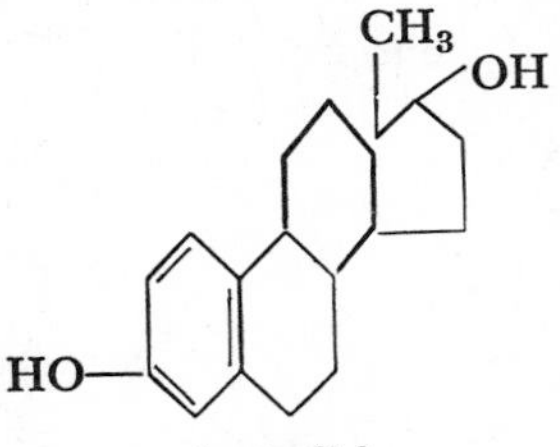

CH_3 OH HO—

β-estradiol

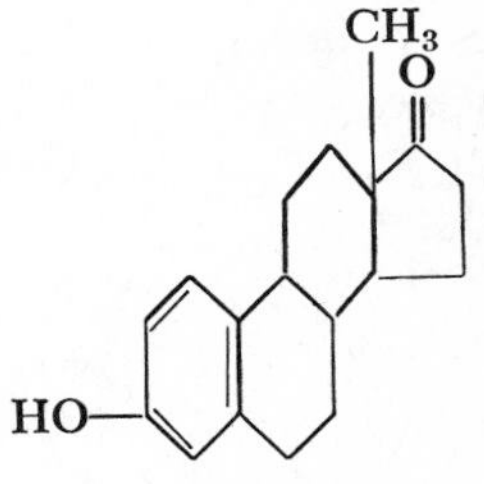

CH_3 O HO—

estrone

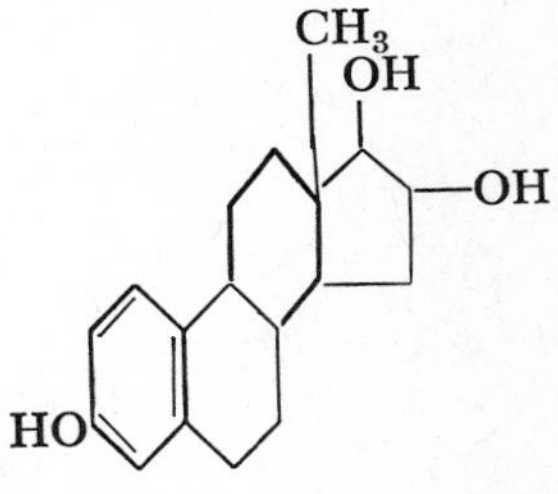

CH_3 OH —OH HO

estriol

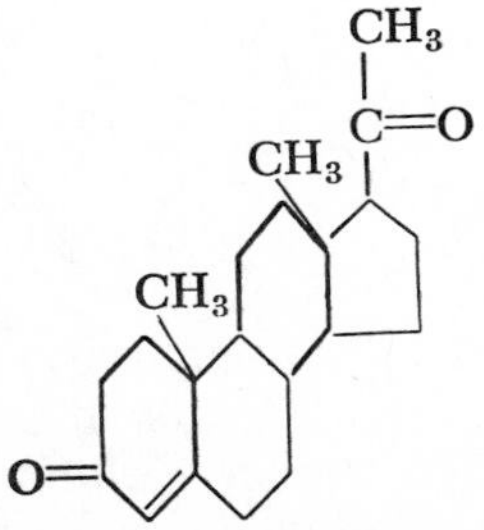

CH_3 C=O CH_3 CH_3 O=

progesterone

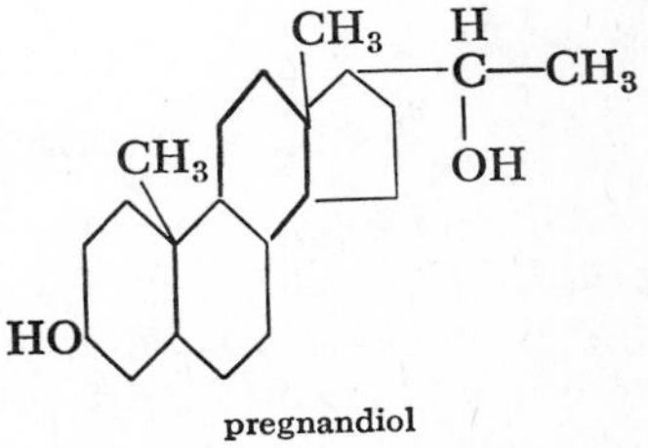

CH_3 H C—CH_3 OH CH_3 HO

pregnandiol

A group of synthetic substances which are not chemically related to the sterols and which are called *stilbesterols* produce nearly the same effects on the female body as estradiol. The most active of these is diethylstilbesterol.

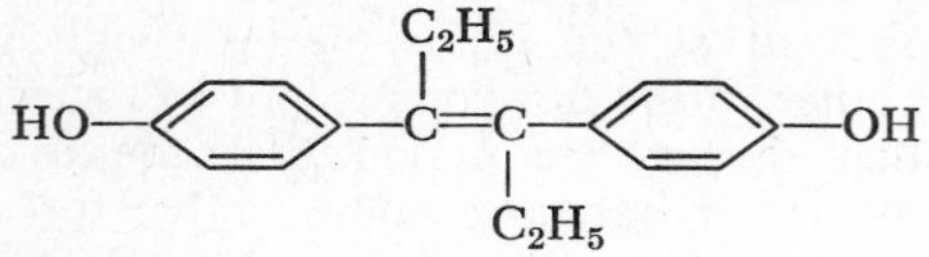

diethylstilbesterol

Because stilbesterols are not destroyed by the digestive system, they may be administered by mouth.

A number of synthetic steroidal substances have now been prepared which exhibit a biological activity equaling or exceeding progesterone and which, unlike the latter, may be taken orally. Of the popular pills, Enovid contains norethynodrel and Orthonovum contains norethindrone. In these oral contraceptives, an orally active estrogenic substance, mestranol, is also included. If taken from the fifth through the twentieth day of the menstrual cycle, ovulation will be inhibited, but if the use of these pills is discontinued on the twentieth day, menstruation will generally occur normally.

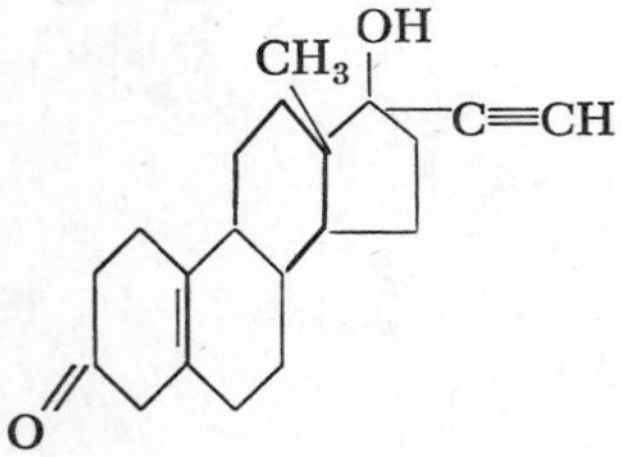

norethynodrel

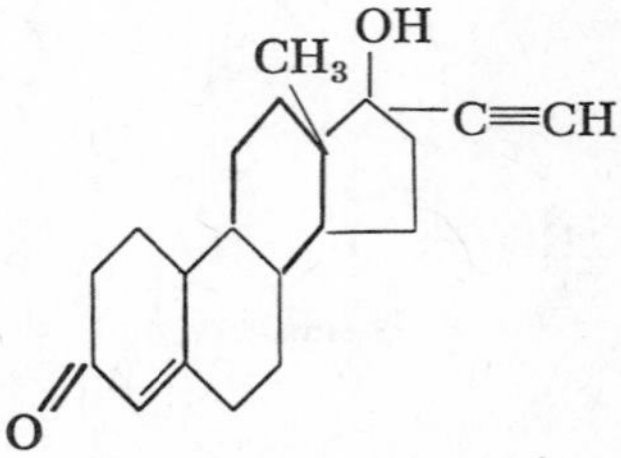

norethindrone

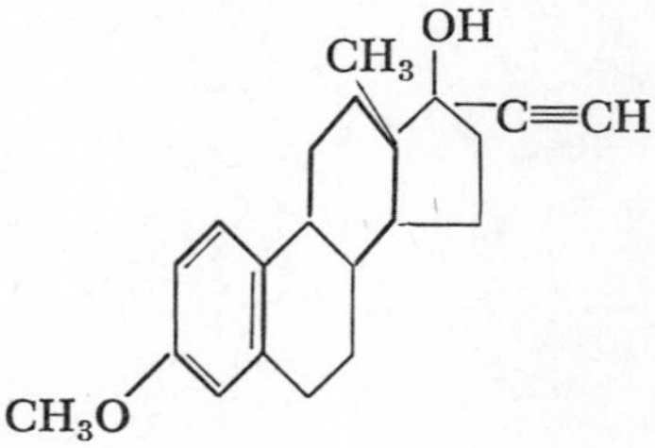

mestranol

The male sex hormones

The hormone *testosterone* is produced by certain cells in the testes of the male. In the body, this hormone is changed to several other hormones, the most important of which is *androsterone*. Both these hormones have been isolated from the urine of males, and their structures have been determined. Both are steroids and are closely related chemically to the female sex hormones.

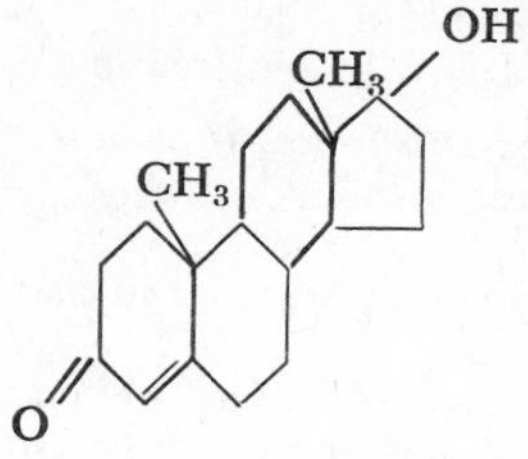

testosterone

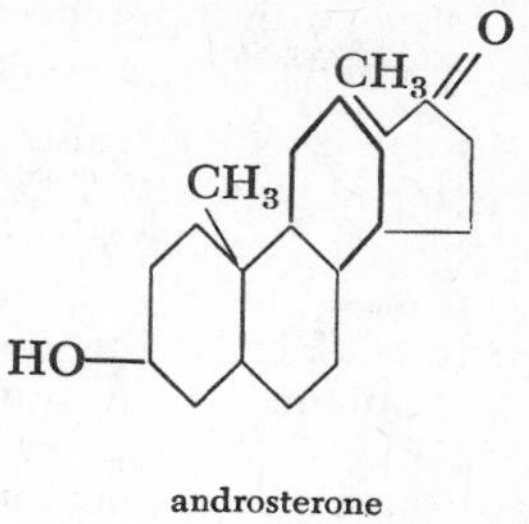

androsterone

In the male, these hormones are responsible for the development of the sexual organs and secondary sexual characteristics, such as the deep voice of the male, the beard, body hair, and the male skeletal structure. The injection of testosterone into capons (castrated male chicks) will produce growth of the wattles, comb, and spurs. Administration of large quantities of male sex hormones to females usually accentuates masculine characteristics while subduing female sex characteristics.

In males in which the testes do not function or are underdeveloped, female characteristics tend to develop. If testosterone is given to a male thus afflicted, the secondary male characteristics will develop, and such an individual may be able to father children if there is some functioning tissue in the testes.

The interrelationship of the hormones

From the preceding discussion of hormones, especially those produced by the pituitary glands, it is to be noted that many body functions are frequently controlled by the harmonious functioning of several of the hormones. Thus the thyrotropic hormone of the pituitary exerts a great influence on the activity of the thyroid, the adenocorticotrophic hormone (ACTH) on that of the adrenals, and the gonadotrophins on that of the sexual organs. Some hormones are antagonistic to others. Examples previously cited are the glucagon of the pancreas and epinephrine of the medulla of the adrenal glands; these are hyperglycemic and are antagonistic to the insulin of the pancreas, which exerts a hypoglycemic effect.

It is not surprising that many hormonal diseases are due to the malfunction of several of the endocrine glands.

STUDY QUESTIONS

1. What are hormones? What are some of the organs that produce them? How do vitamins differ from hormones?
2. What may be the role of hormones in the essential chemical reactions of the body?

3. It is conjectured that a number of hormones are produced by the organs of the digestive tract. Identify the following by name:
   (a) a hormone that produces an increased flow of gastric juice
   (b) a hormone that produces an increased flow of bile
   (c) the hormone that produces an increased flow of pancreatic juice
4. What organ produces insulin? What other substance is produced by this organ? What part of the organ produces insulin?
5. What role does insulin play in metabolism?
6. What disease usually accompanies a hypoactivity of the islets of Langerhans? How is this disease treated?
7. What symptoms would you expect from an overactivity of the islets of Langerhans? What is insulin shock? Why may the pancreas produce an abnormally large amount of insulin?
8. What is protamine insulin? Why is it superior to insulin?
9. Why can insulin not be given by mouth?
10. Why must the diabetic continue the use of insulin?
11. Describe the thyroid gland. Where is it located?
12. Distinguish between thyroxine, triiodothyronine, colloid, and thyroglobulin. What is the chemical composition of thyroxine and triiodothyronine?
13. What physiological effects does thyroxine produce in the body?
14. Describe the symptoms of an underactivity of the thyroid gland; of an overactivity.
15. Describe the cretin and the treatment that is usually prescribed for these individuals. What is myxedema?
16. What is the most common cause of simple goiter? Of an exophthalmic goiter? How are exophthalmic goiters treated surgically? What nonsurgical treatments are giving promise in the treatment of hyperthyroidism?
17. Describe the location and structure of the adrenal glands. What portion of these glands produces epinephrine? What portion produces the cortical hormones?
18. Discuss the composition of epinephrine. By what trade name is it commonly known?
19. What is the effect of epinephrine on carbohydrate metabolism in the body? What other effects does it have on the body?
20. What two drugs are closely related to epinephrine? Discuss some of their uses, as well as those of epinephrine in present-day medical practice.
21. What role does epinephrine play in the almost superhuman power that many individuals will exhibit in grave emergencies?
22. What is the chemical nature of the cortical hormones? What important medical uses are made of the cortical hormones?
23. Describe the results of the surgical removal of the cortex of the adrenal glands.
24. Describe the cause and the symptoms of Addison's disease. Does administration of cortical extracts to an individual afflicted with Addison's disease produce complete remission of its symptoms?
25. Discuss the cause and symptoms of an overactivity of the adrenal cortex.
26. Describe the physical characteristics and location of the parathyroid glands. What precautions are necessary in the surgical removal of the thyroid?

27. What hormone is produced by the parathyroids? It controls the metabolism of what two elements? What is calcitonin?
28. Describe the symptoms of hypoparathyroidism; of hyperparathyroidism.
29. Which gland is called the master gland of the body? It is located in what position in the body? Name the three portions of the gland.
30. Name and describe the functions of the hormones produced by the posterior lobe of this organ; by the anterior lobe.
31. Name the hormone of the pituitary gland that (a) controls growth of an individual, (b) induces the growth and functioning of the thyroid gland, (c) increases the activity of the adrenal cortex, (d) induces the production of milk in the mammal, (e) induces the contraction of smooth muscle, (f) induces the functioning of the sex organs, (g) is an antidiuretic.
32. What is the chemical nature and the action of vasopressin and oxytocin?
33. What are the causes of symptoms of acromegaly?
34. What are the follicular hormones? Which is the most important of these hormones? What influence do these hormones exhibit on the activity of the female organs?
35. By what other name are the follicular hormones known?
36. Which of the follicular hormones is used in the treatment of the menopause?
37. What is the corpus luteum? What hormone is produced by the corpus luteum?
38. What role does the corpus luteum hormone play in pregnancy?
39. Which hormone is responsible for the secondary sexual characteristics of the female?
40. What synthetic substance has an effect on the female similar to β-estradiol?
41. What is the source of androsterone and testerone in the male?
42. What effect will androsterone and testosterone exhibit if administered to a eunuch or to a male who is sexually underdeveloped?
43. To what other substances are the male and female hormones closely related chemically?

answers to problems and selected study questions

ANSWERS TO PROBLEMS IN CHAPTERS 2, 5, 6, 7, 8, 10 AND 27

Chapter 2, page 23

1. (a) 1.83 ml; (b) 183 cm
2. 943 ml
3. (a) 68 kg; (b) 68,000 g
4. 90 ml
5. 10.16 cm; 2.54 m; 35.2 lb; 0.8 oz or 0.05 lb
6. 19.548 g
7. 37° C
8. −112° F
9. −40° C
10. 24° C
11. 1218° F
12. 9350 cal
13. 8200 cal
14. 1.3 g/ml
15. 3,780 g
16. 1.28
17. 117 g
18. 230 ml

Chapter 5, pages 82–84

1. 15.8 tons
2. 54 g
3. (a) 261; (b) 26; (c) 180; (d) 60; (e) 53.5; (f) 76
4. (a) 80%; (b) 52%; (c) 11.3%; (d) 92%; (e) 40%; (f) 62%
5. (a) 40%; (b) 15%; (c) 53.9% (d) 35%; (e) 20%; (f) 15%; (g) 23%; (h) 64.1%
6. 8.3 g
7. 757 g
8. 2.1 g
9. 30.9 g or 15.8 liters of CO_2; 14.2 g of H_2O
10. 734 g
11. 108 g
12. $235 \times \frac{715}{840}$ or 200 ml
13. $1155 \times \frac{775}{760}$ or 1180 ml
14. $500 \times \frac{313}{293}$ or 535 ml
15. $15 \times \frac{373}{293}$ or 19 liters

16. $225 \times \frac{273}{308}$ or 200 ml
17. $415 \times \frac{263}{283} \times \frac{735}{745} = 384$ ml
18. $405 \times \frac{710}{775} \times \frac{293}{308}$ or 351 ml
19. 22.4 liters $\times \frac{295}{273} \times \frac{760}{755}$ or 24.4 liters
20. $355 \times \frac{273}{288} \times \frac{760}{780}$ or 314 ml
21. $5.5 \times \frac{1350}{800} \times \frac{348}{373}$ or 8.2 liters
22. Empirical formulas: (a) CH; (b) C_2H_6O
    Molecular formulas: (a) C_6H_6; (b) C_2H_6O
23. 11.2 liters of CO_2; 11.2 liters of O_2
24. 19.4 g
25. 150 ml
26. 10 liters of CO_2 and 5 liters of O_2
27 (a) 30; (b) 1.84 g; (c) 28; (d) 2.06 g; (e) 2.87 g

Chapter 6, page 99

1. 19 g
2. 7.7 g of Mg; 62 g of $MgSO_4$
3. 99.8 g
4. 14 g
5. 37 percent
6. 441 g
7. 19 ml
8. 50 ml of O_2
9. 20,000
10. 9,150,000 cal or 9,150 kcal
11. 14,500 cal
12. 890 g

Chapter 7, pages 114–116

1. 150 g
2. 2.7 g
3. 6.3 g
4. (a) 8 g; (b) 73.5 g; (c) 794 g; (d) 29 g; (e) 1035 g
5. 17.5 g
6. Ca, 20; PO_4, 31.7; Al, 9; Cu^+, 63.5; Cu^{2+}, 31.8; K, 39; Fe, 28; Sn, 30
7. (a) 9; (b) 5; (c) 32; (d) 35.5; (e) 20
8. (a) 20; (b) 5.3; (c) 28; (d) 18.7
9. 7
10. NaOH, 40; H_3PO_4, 32.7; H_2SO_4, 49; KBr, 119; $Ca(NO_3)_2$, 82; HNO_3, 63; $Ba(OH)_2$, 85.5; $(NH_4)_2CO_3$, 48; $MgSO_4$, 60
11. (a) 11.2 g; (b) 311 g; (c) 3.4 g; (d) 225 g; (e) 4 g
12. (a) 3 *N*; (b) 28.6 *N*; (c) 6.58 *N*; (d) 2.4 *N*; (e) 0.11 *N*
13. 1 liter
14. 0.29 *N*
15. 0.14 *N*
16. sugar 1.09°C; 0.304°; urea, 6.1°, 1.72°; glycerol 4.04°, 1.13°; glycol 6.0°, 1.68°; ethyl alcohol 8.1°, 2.16°; methyl alcohol 11.4°, 3.22°
17. −3° C, 26.6° F
18. 334 ml
19. 375 ml
20. (a) 0.25 meq; (b) 355 mg

Chapter 8, pages 129–130

1. $\frac{1}{9}$, or 0.11
2. 40
3. 0.1 mole of NH_3
4. 0.3 mole of O_2

Chapter 10, page 163

1. $1000 = 10^3$; $1/100 = 10^{-2}$; $100,000 = 10^5$; $1/10,000 = 10^{-4}$
2. (a) 10^7, (b) 10^{-5}, (c) 10^{-9}, (d) 10^{-1}, (e) 10^{13}, (f) 10^{-10}, (g) 10^{13}, (h) 10^{-7}
3. 6
4. 9
5. 3
6. 10^{-9} mole/liter
7. 10,000
8. 3
9. (a) 4.74; (b) 9.25; (c) 7.15; (d) 3.60
10. 1.6×10^{-4}
11. 8.1×10^{-5}
12. $[(CH_3)_2NH_2^+] = [OH^-] = 2.8 \times 10^{-3}$ mole/liter
13. $[H_3O^+] = [NO_2^-] = 2 \times 10^{-2}$ mole/liter

Chapter 27, pages 425–427

1. +66.4°
2. −92°
3. +8.4°

ANSWERS TO SELECTED STUDY QUESTIONS

Chapter 2, page 22

8. Based on the law of conservation of matter, the quantity of reactants must equal the quantity of products; therefore, 58.5 g of sodium chloride will be formed.
11. The following are chemical changes: burning of coal, digestion of food, souring of milk, and ripening of fruit. The others are physical changes.
22. The approximate atomic weight of bromine is 80, while that of carbon is 12.

Chapter 3, pages 48–51

8. Isotopes are atoms of the same atomic number; only their atomic weights and number of neutrons are different.
12. (b) and (e) are isotopes.
27. (a) −1; (b) +2; (d) +3; (e) +2; (f) −2; (g) +2; (h) +1; (i) +3; (j) +1; (k) −1; (l) +4; (m) +2; (n) +3; (o) +3
28. (a) −1; (b) −3; (c) −2; (d) +2; (e) +1; (f) −1; (g) −3; (h) −3; (i) +1; (j) −1; (k) +2; (l) −2; (m) −1; (n) −1; (o) −2; (p) −4
29. (a) +5; (b) +7; (c) +5; (d) +5; (e) +3; (f) +5; (g) +4; (h) +4
30. (a) +6; (b) +4; (c) −2; (d) +5; (e) +5; (f) +3; (g) +6; (h) +4

32. (a) $AlBr_3$; (b) CdI_2; (c) $AlPO_4$; (d) $Sr(NO_3)_2$; (e) Ag_2CO_3; (f) $MgSO_3$; (g) $Zn_3(PO_4)_2$; (h) $Ba(ClO)_2$; (i) $Ca(NO_3)_2$
33. A is a metal; C has six valence electrons; B is chemically inert; the formula of the compound is A_2C_3.
37. Co (58.9) and Ni (58.7), Te (127.60) and I (126.90), and A (39.95) and K (39.10)
55.

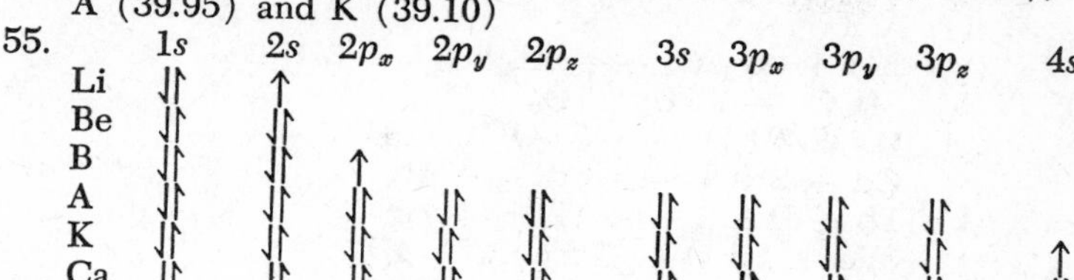

Chapter 4, pages 62–63

2. $2HgO \longrightarrow 2Hg + O_2$
   $2KClO_3 \longrightarrow 2KCl + 3O_2$
   $2H_2O \longrightarrow 2H_2 + O_2$
20. $C_5H_{12} + 8O_2 \longrightarrow 5CO_2 + 6H_2O$
21. $2Mg + O_2 \longrightarrow 2MgO$
    $3Fe + 2O_2 \longrightarrow Fe_3O_4$
    $S + O_2 \longrightarrow SO_2$
    $C + O_2 \longrightarrow CO_2$
    $P_4 + 5O_2 \longrightarrow P_4O_{10}$
22. $CO_2 + H_2O \longrightarrow H_2CO_3$
    $SO_2 + H_2O \longrightarrow H_2SO_3$
    $P_4O_{10} + 6H_2O \longrightarrow 4H_3PO_4$
    $CaO + H_2O \longrightarrow Ca(OH)_2$
    $K_2O + H_2O \longrightarrow 2KOH$
    $MgO + H_2O \longrightarrow Mg(OH)_2$
    $SO_3 + H_2O \longrightarrow H_2SO_4$
35. (a) $Zn + 2KOH \longrightarrow H_2ZnO_2 + H_2$
    (b) $2NaHCO_3 + H_2SO_4 \longrightarrow Na_2SO_4 + H_2O + CO_2$
    (c) $3Zn(OH)_2 + 2H_3PO_4 \longrightarrow Zn_3(PO_4)_2 + 6H_2O$
    (d) $PI_3 + 3H_2O \longrightarrow 3HI + H_3PO_3$
    (e) $2H_2S + SO_2 \longrightarrow 3S + 2H_2O$
    (f) $Fe_3O_4 + 4C \longrightarrow 3Fe + 4CO$
    (g) $3Hg(NO_3)_2 + 2Al \longrightarrow 2Al(NO_3)_3 + 3Hg$
    (h) $2Fe + 6HC_2H_3O_2 \longrightarrow 2Fe(C_2H_3O_2)_3 + 3H_2$
    (i) $C_2H_4 + 3O_2 \longrightarrow 2CO_2 + 2H_2O$
    (j) $CaC_2 + 2H_2O \longrightarrow Ca(OH)_2 + C_2H_2$
    (k) $CaCO_3 + 4C \longrightarrow CaC_2 + 3CO$
    (l) $2Na + 2H_2O \longrightarrow 2NaOH + H_2$
    (m) $2Na_2O_2 + 2H_2O \longrightarrow 4NaOH + O_2$

Chapter 5, page 82

2. 80
10. $\frac{1}{273}$; $\frac{10}{273}$
20. The numerical value of the GMV is 22.4 liters, it is the volume of a mole of a gas at standard conditions.

Chapter 6, pages 96–98

6. only Sn, Zn, Mg, Fe, and Cd

8. (a) $CuO + H_2 \longrightarrow Cu + H_2O$
(b) $2Na + 2H_2O \longrightarrow 2NaOH + H_2$
(c) N.R.
(d) $2AgNO_3 + Fe \longrightarrow Fe(NO_3)_2 + 2Ag$
(e) $2H_2 + O_2 \longrightarrow 2H_2O$
(f) N.R.
(g) $2Al + 3H_2SO_4 \longrightarrow Al_2(SO_4)_3 + 3H_2$
(h) $2H_2O \longrightarrow 2H_2 + O_2$
(i) $Ca + 2H_2O \longrightarrow Ca(OH)_2 + H_2$
(j) $Cd + 2HCl \longrightarrow CdCl_2 + H_2$
(k) $Pb + HgCl_2 \longrightarrow PbCl_2 + Hg$
(l) $3Fe + 4H_2O \longrightarrow Fe_3O_4 + 4H_2$
(m) $FeSO_4 + Zn \longrightarrow ZnSO_4 + Fe$

Chapter 7, page 114

7. Gram-atomic weight/valence = gram-equivalent weight
10. f.p. = −3.72° C; b.p. = 101.04° C
11. All will lower to the same extent.

Chapter 8, pages 128–129

12. (a) formation of unionized water
(b) formation of a volatile gas
(c) formation of a volatile gas
(d) formation of an insoluble solid
(e) formation of an insoluble solid
16. It is independent of the concentration but dependent upon the temperature.
17. The concentration of product appears in the numerator.
18. $R_{\rightarrow} \propto [N_2] \times [O_2]$
$R_{\rightarrow} = k_R \times [N_2] \times [O_2]$
$R_{\leftarrow} = k_L \times [NO]^2$

19. $K_{eq} = \dfrac{[NO]^2}{[N_2] \times [O_2]}$

20. $K_{eq} = \dfrac{[CH_3CHO_2] \times [H_2O]}{[HCHO_2] \times [CH_3OH]}$

$K_{eq} = \dfrac{[HBr]^2}{[H_2] \times [Br_2]}$

$K_{eq} = \dfrac{[NO]^2 \times [O_2]}{[NO_2]^2}$

$K_{eq} = \dfrac{[NH_3]^2}{[N_2] \times [H_2]^3}$

$K_{eq} = \dfrac{[SO_3]^2}{[SO_2]^2 \times [O_2]}$

Chapter 9, pages 148–152

4. Magnesium chloride ionizes into a magnesium ion and two chloride ions.

6. (a) $HNO_3 \rightleftharpoons H^+ + NO_3^-$
   (b) $CaBr_2 \rightleftharpoons Ca^{2+} + 2Br^-$
   (c) $KI \rightleftharpoons K^+ + I^-$
   (d) $MgSO_4 \rightleftharpoons Mg^{2+} + SO_4^{2-}$
   (e) $(NH_4)_2CO_3 \rightleftharpoons 2NH_4^+ + CO_3^{2-}$
14. In concentrated sulfuric acid, much of the H_2SO_4 is present in a covalent structure.
28. (b) The metal must be above hydrogen in the activity series.
    (c) The metal must be above the metal of the salt in the activity series.
29. (a) $K_2O + H_2O \longrightarrow 2KOH$
    (b) $Zn + Cu(NO_3)_2 \longrightarrow Zn(NO_3)_2 + Cu$
    (c) N.R.
    (d) $BaO + 2HNO_3 \longrightarrow Ba(NO_3)_2 + H_2O$
    (e) $FeCl_3 + 3NH_4OH \longrightarrow Fe(OH)_3 + 3NH_4Cl$
    (f) $NaNO_3 + H_2SO_4 \longrightarrow NaHSO_4 + HNO_3$
    (g) $MgCO_3 + HC_2H_3O_2 \longrightarrow Mg(C_2H_3O_2)_2 + H_2O + CO_2$
    (h) $3Sr(OH)_2 + 2H_3PO_4 \longrightarrow Sr_3(PO_4)_2 + 6H_2O$
    (i) $AgNO_3 + NH_4I \longrightarrow AgI + NH_4NO_3$
    (j) $Cu + Cl_2 \longrightarrow CuCl_2$
    (k) N.R.
    (l) $As_2O_3 + 3H_2O \longrightarrow 2H_3AsO_3$
    (m) $2Al + 3H_2SO_4 \longrightarrow Al_2(SO_4)_3 + 3H_2$
    (n) $MgO + SO_2 \longrightarrow MgSO_3$
    (o) $2KClO_3 \longrightarrow 2KCl + 3O_2$
31. $2W(OH)_3 + 3H_2X \longrightarrow W_2X_3 + 6H_2O$
    $3YCl + Ag_3Z \longrightarrow 3AgCl + Y_3Z$
    $BaX + Y_2SO_4 \longrightarrow BaSO_4 + Y_2X$
    $Y_2O + H_2X \longrightarrow Y_2X + H_2O$
    $WCl_3 + 3YOH \longrightarrow W(OH)_3 + 3YCl$
36. (a) basic; (b) acidic; (c) neutral; (d) acidic; (e) basic; (f) acidic; (g) basic; (h) acidic; (i) acidic
40. $Zn(OH)_2 + 2HCl \longrightarrow ZnCl_2 + 2H_2O$
    $Zn(OH)_2 + 2NaOH \longrightarrow Na_2ZnO_2 + 2H_2O$
44. chloric, sulfuric, phosphoric, nitric, carbonic, silicic, arsenic, bromic, iodic, manganic, and chromic acids, respectively.
45. $HClO$ | $FeSO_4$
    H_2SO_3 | $AgNO_3$
    HBr | $BaSO_3$
    HIO_4 | Mg_3N_2
    H_2SO_5 | $Fe(NO_3)_3$
    $HClO_2$ | $HgCl_2$
    H_3AsO_3 | $Ca(ClO)_2$
    $NaBr$ | $Zn(OH)_2$
    $KBrO_2$ | NH_4F
46. iodic acid
    arsenious acid
    zinc nitrite
    stannic oxide
    ammonium monohydrogen phosphate
    lithium carbonate
    sulfur trioxide
    basic lead chloride
    hydriodic acid
    permanganic acid
    mercuric oxide
    magnesium bicarbonate
    aluminum bromide
    strontium sulfide
    ferrous chloride
    sodium dihydrogen phosphate

47. iodite
phosphide
sulfite
bicarbonate
hypobromite
silicate
stannic
cupric
chlorite
nitrate
dihydrogen phosphate
arsenite
ferrous
mercurous

48. $(NH_4)_2CO_3$
$Al_2(SO_4)_3$
$Fe(ClO_3)_3$
K_2S
Na_3PO_4
$CaSO_4$
$Cr(OH)_3$
$Ba(OH)_2$
$Sn_3(PO_4)_4$
$Zn(NO_3)_2$
Ag_3BO_3
$SrSO_4$
$Mg_3(BO_3)_2$
$Fe_2(CO_3)_3$

50. $$K_a = \frac{[H+] \times [C_3H_5O_2-]}{[HC_3H_5O_2]}$$

$$K_a = \frac{[H+] \times [NO_2-]}{[HNO_2]}$$

$$K_a = \frac{[H+] \times [C_7H_5O_2-]}{[HC_7H_5O_2]}$$

$$K_a = \frac{[H+] \times [CN-]}{[HCN]}$$

$$K_a = \frac{[H+] \times [ClO-]}{[HClO]}$$

51. $$K_b = \frac{[CH_3NH_3+] \times [OH-]}{[CH_3NH_2]}$$

52. (a) $K_2 = \frac{[H+] \times [SO_3{}^{2-}]}{[HSO_3-]}$ (b) $K_2 = \frac{[H+] \times [HAsO_4{}^{2-}]}{[H_2AsO_4-]}$

(c) $K_2 = \frac{[H+] \times [C_2O_4{}^{2-}]}{[HC_2O_4-]}$ (d) $K_2 = \frac{[H+] \times [CO_3{}^{2-}]}{[HCO_3-]}$

(e) $K_2 = \frac{[H+] \times [HPO_3{}^{2-}]}{[H_2PO_3-]}$

53. chloroacetic acid > formic acid > acetic acid > picolinic acid > hypochlorous acid > hydrocyanic acid
54. aniline < cinchonine < ammonia < methylamine < dimethylamine

Chapter 10, pages 162–163

7. Vinegar has a pH less than 7, limewater has a pH greater than 7.
12. (c), (e), and (g) should exhibit buffer action.
18. an acetate salt, NH_4OH, and K_2HPO_4, respectively

Chapter 11, pages 179–181

7. (a) All are strong acids except HF, H_2SO_3, and H_3PO_4.
(b) HI, HBr, and H_2SO_3 are weak reducing agents.

(c) None is a good oxidizing agent; H_2SO_4 is a weak oxidizing agent.

16. $CaCO_3 + 2HCl \longrightarrow CaCl_2 + H_2O + CO_2$
17. $MgCO_3 + H_2SO_4 \longrightarrow MgSO_4 + H_2O + CO_2$
26. (a) $HC_3H_5O_3 + NaHCO_3 \longrightarrow NaC_3H_5O_3 + H_2O + CO_2$
    (b) $NaH_2PO_4 + NaHCO_3 \longrightarrow Na_2HPO_4 + H_2O + CO_2$
30. (a) $NaCl + H_2SO_4 \longrightarrow HCl + NaHSO_4$
    (b) $4P + 5O_2 \longrightarrow P_4O_{10}$
    (c) $S + O_2 \longrightarrow SO_2$
    (d) $P_4O_{10} + 6H_2O \longrightarrow 4H_3PO_4$
    $SO_2 + H_2O \longrightarrow H_2SO_3$
    $SO_3 + H_2O \longrightarrow H_2SO_4$
    $NH_3 + H_2O \longrightarrow NH_4OH$
    (e) $ZnS + H_2SO_4 \longrightarrow ZnSO_4 + H_2S$
    (f) $NH_4OH + HNO_3 \longrightarrow NH_4NO_3 + H_2O$
    (g) $NaNO_3 + H_2SO_4 \longrightarrow HNO_3 + NaHSO_4$
    (h) $3H_2 + N_2 \longrightarrow 2NH_3$

Chapter 12, pages 189–191

6. $Fe^{2+} \longrightarrow Fe^{3+} + 1e$
   $Sn^{2+} \longrightarrow Sn^{4+} + 2e$
   $2I^- \longrightarrow I_2 + 2e$
   $Hg_2{}^{2+} \longrightarrow 2Hg^{2+} + 2e$
   $Cl_2 + 2e \longrightarrow 2Cl^-$
7. $2HgCl_2 + SnCl_2 \longrightarrow Hg_2Cl_2 + SnCl_4$
   $Al_2O_3 + 6K \longrightarrow 3K_2O + 2Al$
   $2KI + Cl_2 \longrightarrow 2KCl + I_2$
   $3PbCl_2 + 2Cr \longrightarrow 2CrCl_3 + 3Pb$
   $2Sb + 3Cl_2 \longrightarrow 2SbCl_3$
9. Carbon, sodium thiosulfate, stannous chloride, sulfur dioxide, oxalic acid, and hydrogen iodide are reducing agents. Nitric acid, chlorine, potassium permanganate, and sodium chlorate are oxidizing agents.
10. $ClO_3{}^- + 6H^+ + 6e \longrightarrow Cl^- + 3H_2O$
    $SO_3{}^{2-} + H_2O \longrightarrow SO_4{}^{2-} + 2H^+ + 2e$
    $NO_3{}^- + 2H^+ + 2e \longrightarrow NO_2 + H_2O$
    $2IO_4{}^- + 16H^+ + 14e \longrightarrow I_2 + 8H_2O$
11. $SO_4{}^{2-} + 2Br^- + 2H^+ \longrightarrow SO_2 + Br_2 + 2H_2O$
    $2NO_3{}^- + 3Cu + 8H^+ \longrightarrow 2NO + 3Cu^{2+} + 4H_2O$
    $2ClO^- + 2Cl^- + 4H^+ \longrightarrow 2Cl_2 + 2H_2O$
    $5Fe^{2+} + MnO_4{}^- + 8H^+ \longrightarrow 5Fe^{3+} + Mn^{2+} + 4H_2O$

Chapter 13, page 198

14. 1 micron (μ) = $\frac{1}{1000}$ millimeter
    1 millimicron ($m\mu$) = $\frac{1}{1000}$ micron
    1 Angstrom unit (Å) = 10^{-8} cm = $\frac{1}{10}$ millimicron

Chapter 15, pages 223–224

14. (d) and (e) are isomers, (a) and (c) are homologs.
15. (a) (b) and (f) or (c) and (f)
    (b) (b) and (c)
    (c) (b), (c), and (f)
    (d) (e)

(e) (d) and (h)
(f) (a) and (g)

16. (a) is alicyclic, (b) is aliphatic, (c), (d), and (e) are heterocyclic.

Chapter 16, pages 241–244

1. $CH_3CH_2CH_2COONa + NaOH \longrightarrow C_3H_8 + Na_2CO_3$
4. $2C_4H_{10} + 13O_2 \longrightarrow 8CO_2 + 10H_2O$
   $2C_6H_{14} + 19O_2 \longrightarrow 12CO_2 + 14H_2O$
6. There are two dichloroethanes, of structures $CH_2Cl—CH_2Cl$ and $CH_3—CHCl_2$. There are two trichloroethanes, of structures $CCl_3—CH_3$ and $CHCl_2—CH_2Cl$.
14. $CH_2{=}CH_2 + Cl_2 \longrightarrow CH_2Cl—CH_2Cl$
    $CH_2{=}CH_2 + Br_2 \longrightarrow CH_2Br—CH_2Br$
    $CH_2{=}CH_2 + HI \longrightarrow CH_3—CH_2I$
15. $HC{\equiv}CH + Cl_2 \longrightarrow CHCl{=}CHCl$
    $CHCl{=}CHCl + Cl_2 \longrightarrow CHCl_2—CHCl_2$
17. $HC{\equiv}CH + HBr \longrightarrow CH_2{=}CHBr$
    $CH_2{=}CHBr + HBr \longrightarrow CH_3—CHBr_2$
22. (d) and (f) are isomers, (e) and (g) are isomers.
23. (a) 2-methylbutane
    (b) 3,3-dimethylhexane
    (c) 2,3-dimethylpentane
    (d) 2-methyl-2-pentene
    (e) 2,3-dimethyl-2-pentene
    (f) 2-methyl-3-heptyne
    (g) 3-methyl-1-pentyne
24. (a) 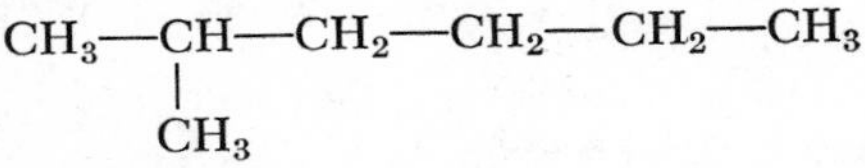 $CH_3—CH(CH_3)—CH_2—CH_2—CH_2—CH_3$

    (b) 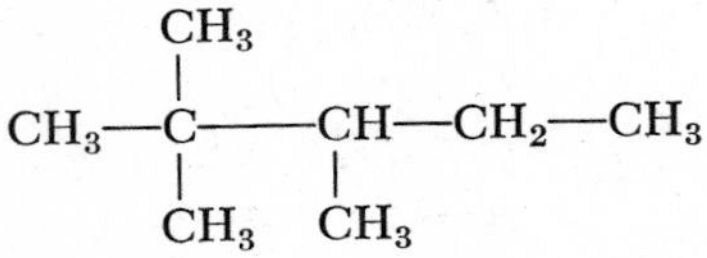 $CH_3—C(CH_3)_2—CH(CH_3)—CH_2—CH_3$

    (c) $CH_3—CH_2—CH{=}CH—CH_2—CH_3$

    (d) 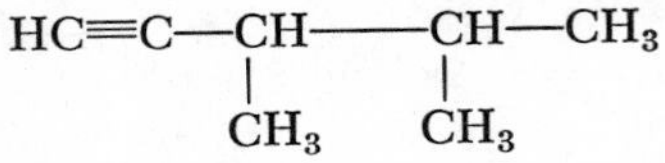 $HC{\equiv}C—CH(CH_3)—CH(CH_3)—CH_3$

    (e) 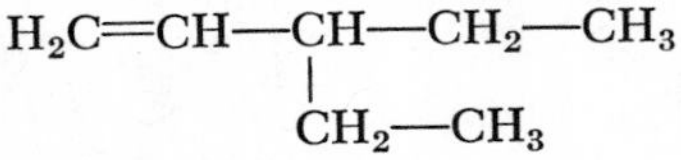 $H_2C{=}CH—CH(CH_2—CH_3)—CH_2—CH_3$

Chapter 17, pages 258–260

3. (a) secondary; (b) primary; (c) primary; (d) tertiary; (e) primary; (f) secondary; (g) secondary; (h) tertiary
5. $CH_3—CH_2OH + CH_3—COOH \longrightarrow$

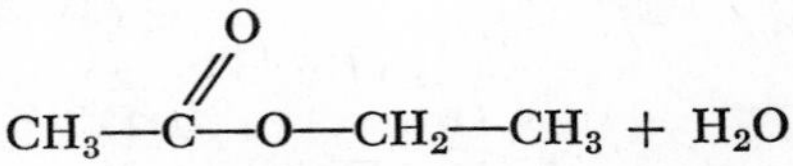

   $CH_3—C({=}O)—O—CH_2—CH_3 + H_2O$

   ethyl acetate

$$CH_3—CH_2—CH_2OH + CH_3—COOH \longrightarrow CH_3—\overset{O}{\overset{//}{C}}—O—CH_2—CH_2—CH_3 + H_2O$$

n-propyl acetate

17. $2CH_3OH + 3O_2 \longrightarrow 2CO_2 + 4H_2O$
    $C_2H_5OH + 3O_2 \longrightarrow 2CO_2 + 3H_2O$
34. The two monochloro-substitution products of *n*-butane have the structures
    $CH_3—CH_2—CH_2—CH_2—Cl$ and $CH_3—CH_2—CHCl—CH_3$
35. The three dichloro-substitution products of isobutane have the structures

$$CH_3—\underset{CH_3}{\underset{|}{CH}}—CHCl_2 \qquad CH_2—\underset{CH_3}{\underset{|}{CCl}}—CH_2Cl$$

$$CH_2Cl—\underset{CH_3}{\underset{|}{CH}}—CH_2Cl$$

37. (a) *n*-butyl alcohol, or 1-butanol
    (b) ethyl alcohol, or ethanol
    (c) *sec*-butyl alcohol, or 2-butanol
38. (a) 2-butanol; (b) 2-methyl-l-butanol; (c) ethanol; (d) 2-methyl-2-propanol; (e) 2,2-dimethyl-1-propanol; (f) 2-propanol; (g) 3-methyl-2-butanol; (h) 3-methyl-3-pentanol

Chapter 18, page 268

3. $CH_3—CH_2—CH_2—CHO$; $CH_3—CO—CH_3$; $CH_3—CH_2—CO—CH_3$; $CH_3—CH_2—CHO$, and $(CH_3)_2CH—CHO$, respectively.
6. (a) propionaldehyde, or propanal
   (b) ethyl methyl ketone, or (2)-butanone
   (c) acetaldehyde, or ethanal
   (d) acetone, or propanone
   (e) *n*-butyraldehyde, or butanal
   (f) diethyl ketone, or 3-pentanone
   (g) methyl *n*-propyl ketone, or 2-pentanone
12. (a) 3-methylbutanal; (b) 2,2-dimethylpropanal; (c) 2-methyl-3-pentanone; (d) 5-methyl-3-hexanone.
13. $CH_3—\overset{H}{\overset{/}{C}}=N—NHC_6H_5$

Chapter 19, pages 277–278

6. (a) $2CH_3—CH_2—COOH + Zn \longrightarrow (CH_3—CH_2—\overset{O}{\overset{//}{C}}—O—)_2Zn + H_2$
   (b) $CH_3—CH_2—COOH + KOH \longrightarrow CH_3—CH_2—CO_2K + H_2O$

(c) $2CH_3—CH_2—COOH + CaO \longrightarrow (CH_3—CH_2—CO_2)_2Ca + H_2O$

(d) $CH_3—CH_2—COOH + Na_2CO_3 \longrightarrow CH_3—CH_2—CO_2Na + H_2O + CO_2$

18. $\frac{2}{3}$ mole of ethyl acetate; $\frac{2}{3}$ mole of water; $\frac{1}{3}$ mole of ethyl alcohol, and $\frac{1}{3}$ mole of acetic acid

24. (a) $CH_3—C(=O)—O—CH_2—CH_3$

(b) $CH_3—CH_2—CH_2—C(=O)—O—CH_3$

(c) $H—C(=O)—O—CH_2—CH_2—CH_3$

(d) $CH_3—C(=O)—O—CH(CH_3)_2$

25. $CH_3—COOH + CH_3—CH_2OH \longrightarrow CH_3—C(=O)—O—CH_2—CH_3 + H_2O$

$CH_3—CH_2—CH_2—COOH + CH_3OH \longrightarrow CH_3—CH_2—CH_2—C(=O)—O—CH_3 + H_2O$

$H—COOH + CH_3—CH_2—CH_2OH \longrightarrow H—C(=O)—O—CH_2—CH_2—CH_3 + H_2O$

$CH_3—COOH + CH_3—CHOH—CH_3 \longrightarrow CH_3—C(=O)—O—CH(CH_3)_2 + H_2O$

26. $CH_3—COOCH_2—CH_3 + NaOH \longrightarrow CH_3—COONa + CH_3—CH_2OH$

$CH_3—CH_2—CH_2—COOCH_3 + NaOH \longrightarrow CH_3—CH_2—CH_2—COONa + CH_3OH$

$H—COOCH_2—CH_2—CH_3 + NaOH \longrightarrow H—COONa + CH_3—CH_2—CH_2OH$

$CH_3—COOCH(CH_3)_2 + NaOH \longrightarrow CH_3—COONa + (CH_3)_2CHOH$

27. (a) 2-methylpentanoic acid
(b) 2,4-dimethylpentanoic acid
(c) 2-hydroxypropanoic acid
(d) 2-chlorobutanoic acid
(e) 2,2-dichloropropanoic acid

Chapter 20, pages 291–292

11.
$$\begin{array}{l} CH_2\text{—}O\text{—}CO(CH_2)_7CH\text{=}CH(CH_2)_7CH_3 \\ | \\ CH\text{—}O\text{—}CO(CH_2)_7CH\text{=}CH(CH_2)_7CH_3 + 3H_2 \longrightarrow \\ | \\ CH_2\text{—}O\text{—}CO(CH_2)_7CH\text{=}CH(CH_2)_7CH_3 \end{array}$$

$$\begin{array}{l} CH_2\text{—}OCO(CH_2)_{16}CH_3 \\ | \\ CH\text{—}OCO(CH_2)_{16}CH_3 \\ | \\ CH_2\text{—}OCO(CH_2)_{16}CH_3 \end{array}$$

35. (a) triolein + 3NaOH $\longrightarrow$
$$\begin{array}{l} CH_2OH \\ | \\ CHOH + 3CH_3(CH_2)_7CH\text{=}CH(CH_2)_7COONa \\ | \\ CH_2OH \end{array}$$

(b) tripalmitin + 3NaOH $\longrightarrow$
$$\begin{array}{l} CH_2OH \\ | \\ CHOH + 3CH_3(CH_2)_{14}COONa \\ | \\ CH_2OH \end{array}$$

36. triolein + $3Br_2$ $\longrightarrow$
$$\begin{array}{l} CH_2O\text{—}CO(CH_2)_7CHBr\text{—}CHBr(CH_2)_7CH_3 \\ | \\ CHO\text{—}CO(CH_2)_7CHBr\text{—}CHBr(CH_2)_7CH_3 \\ | \\ CH_2O\text{—}CO(CH_2)_7CHBr\text{—}CHBr(CH_2)_7CH_3 \end{array}$$

37. Waxes and vegetable oils are simple lipids, while cephalins and glycolipids are compound lipids.
38. See answers to Questions 11 and 35.

Chapter 21, pages 311–314

9. $C_6H_{12}O_6 + 6O_2 \longrightarrow 6CO_2 + 6H_2O$

45. 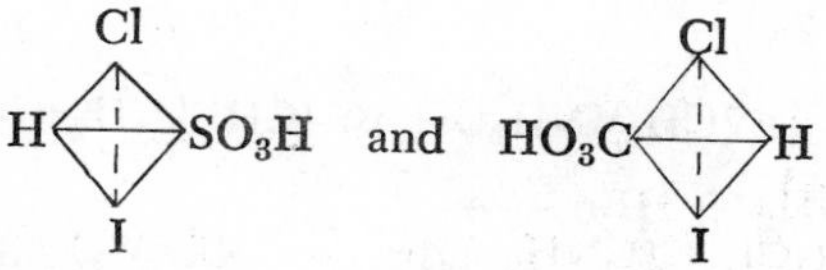 H–C(Cl)(I)–SO_3H and HO_3C–C(Cl)(I)–H

46. (a) $2^3 = 8$; (b) $2^5 = 32$

47. (a)

```
      CHO
       |
   H—C—OH
       |
   H—C—OH
       |
   H—C—OH
       |
      CH2OH
```

(b)

```
      CHO
       |
  HO—C—H
       |
  HO—C—H
       |
  HO—C—H
       |
      CH2OH
```

(c)

```
      CHO
       |
  HO—C—H
       |
   H—C—OH
       |
   H—C—OH
       |
      CH2OH
```

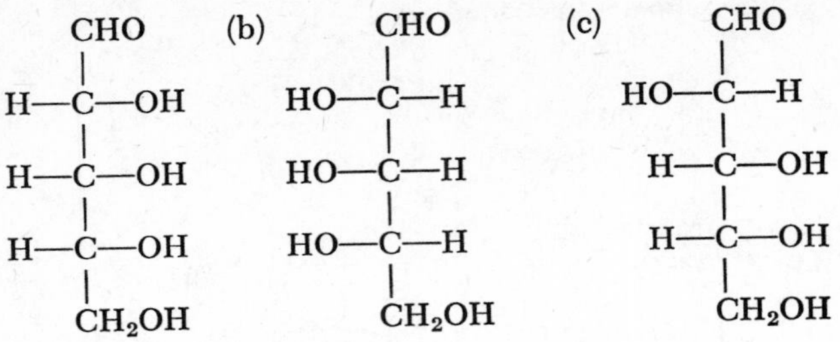

(d)

```
      CHO
       |
   H—C—OH
       |
  HO—C—H
       |
   H—C—OH
       |
   H—C—OH
       |
      CH2OH
```

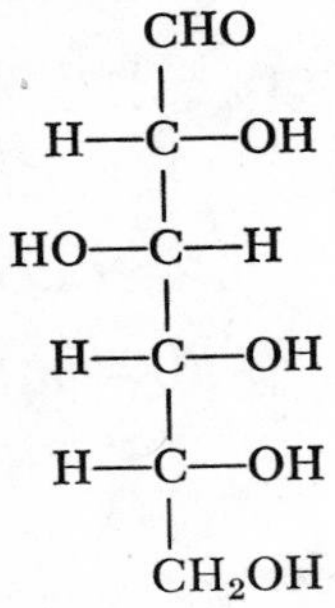

50. See answer to Question 47(d) for structure of (+)-glucose. (+)-Mannose has the structure

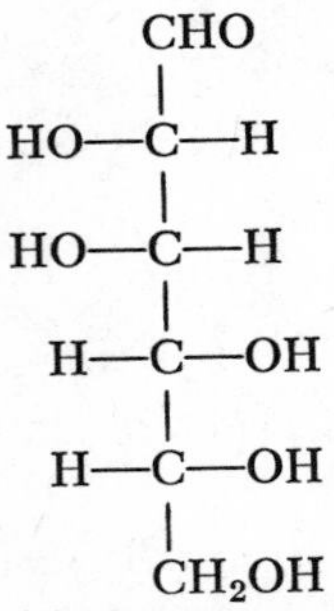

```
      CHO
       |
  HO—C—H
       |
  HO—C—H
       |
   H—C—OH
       |
   H—C—OH
       |
      CH2OH
```

51. (a) 2 asymmetric carbons—2*d*, 2*l*, and 2 racemic modifications
    (b) 4 asymmetric carbons—8*d*, 8*l*, and 8 racemic modifications
    (c) no asymmetric carbons—no stereoisomers
    (d) 1 asymmetric carbon—1*d*, 1*l* and 1 racemic modification
    (e) 3 asymmetric carbons—4*d*, 4*l*, and 4 racemic modifications
    (f) 1 asymmetric carbon—1*d*, 1*l*, and 1 racemic modification
    (g) 2 asymmetric carbons—2*d*, 2*l*, and 2 racemic modifications
52. (a), (c), (e), and (f) represent optically inactive structures.

Chapter 22, pages 322–323

4. $(CH_3)_2NH + H_2O \rightleftharpoons (CH_3)_2NH_2OH \rightleftharpoons (CH_3)_2NH_2^+ + OH^-$
5. $CH_3CH_2CH_2CH_2NH_2 + HBr \longrightarrow$
   $CH_3CH_2CH_2NH_2 \cdot HBr$ or $CH_3CH_2CH_2NH_3Br$
8. 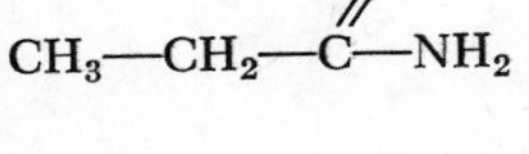 $CH_3—CH_2—\overset{\overset{\displaystyle O}{/\!/}}{C}—NH_2$ propionamide

9. $CH_3{-}CH_2{-}\overset{\overset{\large O}{/\!/}}{C}{-}NH_2 + NaOH \longrightarrow CH_3{-}CH_2{-}COONa + NH_3$

$CH_3{-}CH_2{-}\overset{\overset{\large O}{/\!/}}{C}{-}NH_2 + H_2SO_4 + H_2O \longrightarrow CH_3{-}CH_2{-}COOH + (NH_4)_2SO_4$

15. (a) $(NH_4)_2SO_4 + 2KCNO \longrightarrow 2NH_4CNO + K_2SO_4$

$NH_4CNO \longrightarrow NH_2{-}CO{-}NH_2$

(b) $2NH_3 + CO_2 \longrightarrow NH_2{-}CO{-}NH_2 + H_2O$

Chapter 23, pages 347–348

4. CH_3 (benzene ring) — toluene; CH_3, CH_3 (benzene ring, ortho) — *o*-xylene; CH_3, CH_3 (benzene ring, meta) — *m*-xylene; CH_3, CH_3 (benzene ring, para) — *p*-xylene

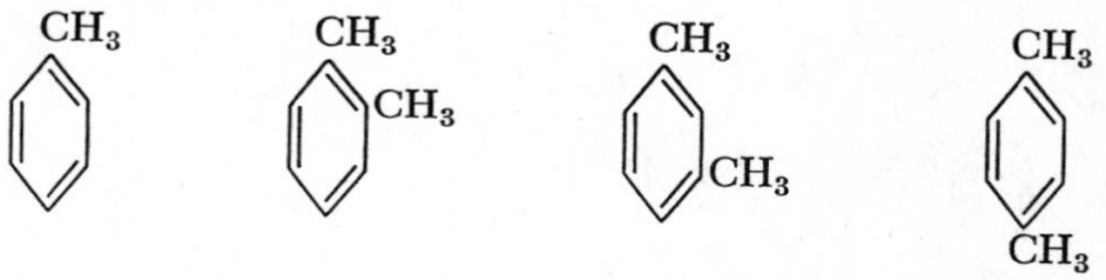

toluene *o*-xylene *m*-xylene *p*-xylene

9. OH (benzene ring) + NaOH ⟶ ONa (benzene ring) + H_2O

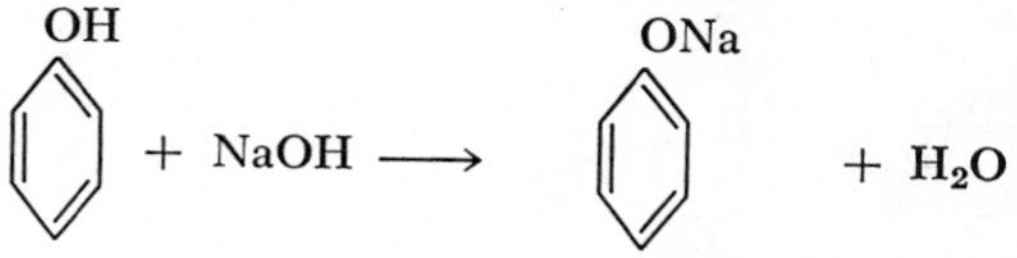

sodium phenoxide

12. CO_2H (benzene ring) + NaOH ⟶ CO_2Na (benzene ring) + H_2O

CO_2H, OH (benzene ring) + NaOH ⟶ CO_2Na, OH (benzene ring) + H_2O

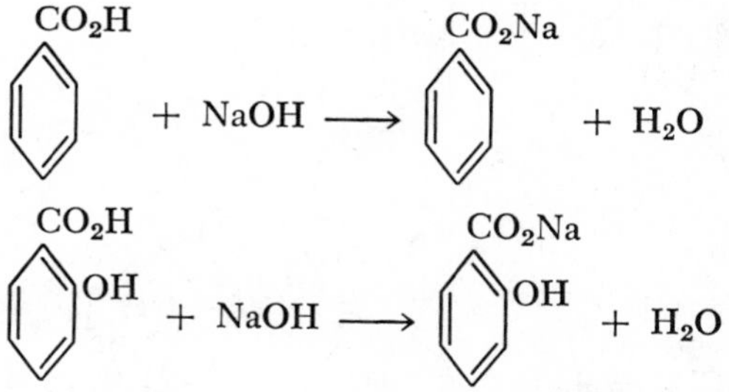

18. There are only one monochlorobenzene and three dichlorosubstitution products of benzene; namely, chlorobenzene and *o*- *m*-, and *p*-dichlorobenzene.

22. CO_2H, OH (benzene ring) + NaOH ⟶ CO_2Na, OH (benzene ring) + H_2O

NH_2 (benzene ring) + HCl ⟶ $NH_2 \cdot HCl$ (benzene ring)

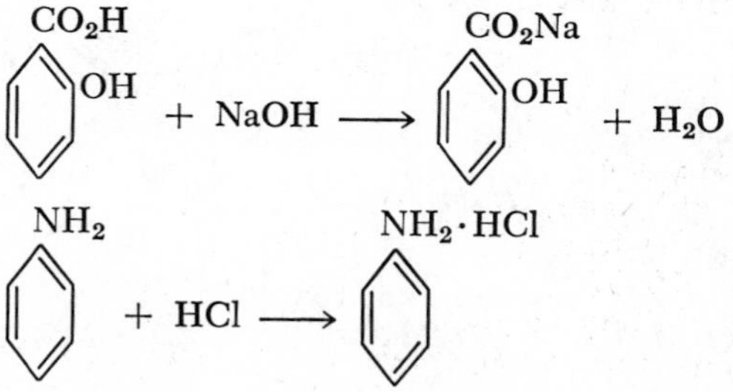

$$C_6H_6 + H_2SO_4 \longrightarrow C_6H_5SO_3H + H_2O$$

$$C_6H_5OH + KOH \longrightarrow C_6H_5OK + H_2O$$

$$C_6H_5CH_3 + HNO_3 \xrightarrow{H_2SO_4} o\text{-}CH_3C_6H_4NO_2 \text{ and } p\text{-}CH_3C_6H_4NO_2$$

$$C_6H_6 + Cl_2 \xrightarrow{Fe} C_6H_5Cl + HCl$$

$$C_6H_5CH_3 + [O] \longrightarrow C_6H_5CO_2H$$

$$p\text{-}NO_2C_6H_4CH_3 + 6[H] \xrightarrow{(Sn + HCl)} p\text{-}NH_2C_6H_4CH_3$$

$$C_6H_5NH_2 + CH_3{-}CO_2H \xrightarrow{\Delta} C_6H_5{-}NHC(=O){-}CH_3 + H_2O$$

25. $CCl_3{-}CHO + 2\ C_6H_5Cl \xrightarrow{H_2SO_4} CCl_3{-}CH(C_6H_4Cl)_2 + H_2O$

Chapter 25, pages 382–384

2. (a) true; (b) false; (c) true; (d) false; (e) true; (f) false; (g) false; (h) true; (i) false; (j) true
13. See pages 370–371.
15. pepsin, 2; salivary amylase, 6.8; trypsin, 8.5
16. (a) carboxy- and aminopeptidases; (b) enterokinase; (c) dehydrogenase

Chapter 27, pages 425–427

14. two
15. 15

Chapter 30, pages 462–463

5. (a) hemoglobinuria, some types of jaundice; (b) presence of melanotic tumors; (c) alcaptonuria

Chapter 31, pages 477–478

16. (a) diabetes mellitus; (b) hypoglycemia, insulin shock; (c) diabetes mellitus, starvation, nephritis, hypoventilation; (d) infantile diarrhea, hyperventilation; (e) albuminuria; (f) an anemia; (g) obstructive jaundice, (h) leukemia

Chapter 32, pages 488–490

2. anemia: deficiency of iron; goiter: deficiency of iodine; rickets: deficiency of calcium; osteomalacia: deficiency of calcium

Chapter 33, pages 508–509

6. scurvy: vitamin C
   rickets: vitamin D
   osteomalacia: vitamin D
   night blindness: vitamin A
   polyneuritis: thiamin
   black tongue: nicotinic acid
   beriberi: thiamin
   xerophthalmia: vitamin A
   pellagra: nicotinic acid
   osteoporosis: vitamin D
   pernicious enemia: vitamin B_{12}
8. (a) vitamin A; (b) vitamin D; (c) nicotinic acid; (d) folic acid or pteroylglutamic acid

Chapter 34, pages 527–529

3. (a) gastrin; (b) cholecystokinin; (c) pancreozymin
31. (a) growth-stimulating, or somatotrophic, hormone (GH)
    (b) the thyrotropic hormone (TTH)
    (c) the adrenocorticotrophic hormone (ACTH)
    (d) prolactin, or lactogenic hormone
    (e) oxytocin
    (f) follicle-stimulating hormone (FSH)
    (g) vasopressin

reference texts and films

General chemistry

BAILAR, J. C., T. Moeller, and J. Kleinberg, *University Chemistry,* Heath, Lexington, Mass. (1965).

BRESCIA, F., J. Arents, H. Meislich, and A. Turk, *Fundamentals of Chemistry: A Modern Introduction,* Acadamic, New York (1966).

GARRETT, A. B., W. T. Lippincott, and F. H. Verhoek, *Chemistry: A Study of Matter,* Ginn, Waltham Mass. (1968).

HERED, W. and W. Nebergall, *College Chemistry,* Heath, Lexington, Mass. (1963).

KEENAN, C. W. and J. H. Wood, *General College Chemistry,* 4th ed., Harper & Row, New York (1971).

KEENAN, C. W., J. H. Wood, and W. E. Bull, *Fundamentals of College Chemistry,* 3rd ed., Harper & Row, New York (1972).

SIENKO, M. J. and B. A. Plane, *Chemistry,* 4th ed., McGraw-Hill, New York (1971).

SISLER, H. H., C. A. Vander Werf, and A. W. Davidson, *College Chemistry,* 3rd ed., Macmillan, New York (1967).

SLABAUGH, W. H. and T. D. Parsons, *General Chemistry,* 2nd ed., Wiley, New York (1971).

TURK, A.. H. Meislich, F. Brescia, and J. Arents, *Introduction to Chemistry,* Academic, New York (1968).

General, organic, and biochemistry

HOFMANN, K. B., *Chemistry for the Applied Sciences,* Prentice-Hall, Englewood Cliffs, N.J. (1963).

HOLUM, J. R., *Elements of General and Biological Chemistry,* 3rd ed., Wiley, New York (1972).

MARIELLA, R. P. and R. A. Blau, *Chemistry of Life Processes,* Harcourt Brace Jovanovich, New York (1968).

MEYER, L. H., *Introductory Chemistry,* Macmillan, New York (1964).

NEVILL, W. A. *General Chemistry,* McGraw-Hill, New York (1967).

ROUTH, J. I., D. P. Eyman, and D. J. Burton, *Essentials of General, Organic, and Biochemistry,* 2nd ed., Saunders, Philadelphia (1973).

SACKHEIM, G. I. and R. M. Schultz, *Chemistry for the Health Sciences,* Macmillan, New York (1969).

REVIEW BOOKS

General, organic, and biochemistry

CHEN, P. S., *Chemistry—Inorganic, Organic, and Biological,* Barnes & Noble (College Outline Series), New York (1968).

Organic and biochemistry

BAUM, S. J., *Introduction to Organic and Biological Chemistry,* Macmillan, New York (1970).

HOLUM, J. R., *Introduction to Organic and Biological Chemistry,* Wiley, New York (1969).

General and organic chemistry

LEE, G. L., H. A. VanOrden, and R. O. Ragsdale, *General and Organic Chemistry,* Saunders, Philadelphia (1971).

Organic chemistry

BONNER, W. A., and A. J. Castro, *Essentials of Modern Organic Chemistry,* Reinhold, New York (1965).

FERGUSON, L. N., *Textbook of Organic Chemistry,* 2nd ed., Van Nostrand Reinhold, New York (1965).

HENDERSON, J. B., D. J. Cram, and G. S. Hammond, *Organic Chemistry,* 3rd ed., McGraw-Hill, New York (1970).

MORRISON, R. T. and R. N. Boyd, *Organic Chemistry,* 3rd ed., Allyn & Bacon, Boston (1973).

NOLLER, C. R., *Textbook of Organic Chemistry,* 3rd ed., Saunders, Philadelphia (1966).

SHIRLEY, D. A., *Organic Chemistry,* Holt, Rinehart & Winston, New York (1966).

Organic chemistry (short course)

DePUY, C. H. and K. L. Rinehart, *Introduction to Organic Chemistry,* Wiley, New York (1967).

FINLEY, K. T. and J. Wilson, *Fundamental Organic Chemistry,* Prentice-Hall, Englewood Cliffs, N.J. (1970).

GRIFFIN, R. W., *Modern Organic Chemistry,* McGraw-Hill, New York (1969).

HART, H. and R. D. Schuetz, *Organic Chemistry,* 4th ed., Houghton Mifflin, Boston (1972).

LINDSTROMBERG, W. W. *Organic Chemistry: A Brief Course,* 2nd ed., Heath, Lexington Mass. (1970).

RICHARDS, J. H., D. J. Cram, and G. S. Hammond, *Elements of Organic Chemistry,* McGraw-Hill, New York (1967).

VAN ORDEN, H. O. and G. L. Lee, *Elementary Organic Chemistry: A Brief Course,* Saunders, Philadelphia (1969).

Biochemistry

TEXTBOOKS

CANTAROW, A. and B. Schepartz, *Textbook of Biochemistry,* 4th ed., Saunders, Philadelphia (1967).

CONN, E. E. and P. K. Stumpf, *Outlines of Biochemistry,* 2nd ed., Wiley, New York (1967).

FRUTON, J. S. and S. Simmonds, *General Biochemistry,* 2nd ed., Wiley, New York (1960).

HAWK, P. B. and B. L. Oser, *Physiological Chemistry,* 14th ed., McGraw-Hill, New York (1965).

JELLINCK, P. H., *Biochemistry: An Introduction,* Holt, Rinehart & Winston, New York (1963).

KARLSON, P., *Introduction to Modern Biochemistry,* Academic, New York (1963).

LEHNINGER, A. L., *Biochemistry,* Worth, New York (1970).

MAHLER, H. R. and E. H. Cordes, *Basic Biological Chemistry,* Harper & Row, New York (1968).

MAHLER, H. R. and E. H. Cordes, *Biological Chemistry,* 2nd ed., Harper & Row, New York (1971).

MAZUR, A. and B. Harrow, *Biochemistry: A Brief Course,* Saunders, Philadelphia (1968).

WHITE, A., P. Handler, and E. L. Smith, *Principles of Biochemistry,* 4th ed., McGraw-Hill, New York (1968).

Biochemistry

SPECIAL TOPICS

BARRY, J. M., *Molecular Biology, Genes, and the Chemical Control of Living Cells,* Prentice-Hall, Englewood Cliffs, N.J. (1964).

CHRISTENSEN, H. N., *Body Fluids and the Acid-Base Balance,* Saunders, Philadelphia (1964).

McELROY, W. D., *Cell Physiology and Biochemistry,* 2nd ed., Prentice-Hall, Englewood Cliffs, N.J. (1964).

PATTON, A. R., *Biochemical Energetics and Kinetics,* Saunders, Philadelphia (1965).

AUDIOVISUAL MATERIALS FOR CHEMISTRY

LIPPINCOTT, W. T., W. R. Barnard, and R. T. Yingling. General chemistry laboratory films. Saunders, Philadelphia.

MASTERSON, W. L. and E. J. Slowinski. Color slides to accompany *Chemical Principles.* 125 two-color 35 mm slides. Saunders, Philadelphia.

O'CONNOR, R. and B. Fowler. *Fundamentals of Chemistry.* 20 films (see science film catalogue CT 34). Harper & Row, New York.

appendix a

mathematical operations

appendix a

Significant figures

When a trained experimentalist determines the density of mercury and indicates it as 13.6 g/ml, one should expect uncertainty in the last digit only. The density of the mercury may actually be 13.5 or 13.7 g/ml but is more likely to be nearer 13.6 g/ml. Thus only the three digits of this number are reliable, and the number of significant figures will be three.

In the table of atomic weights inside the back cover, the atomic weight of mercury is given as 200.59, corresponding to an accuracy of five significant figures. To indicate that there is some uncertainty in the last figure, the expression 200.59 ± 0.01 is frequently used.

In the number 25100, one is not certain whether the two zeros merely locate the decimal point or are significant figures. To indicate an accuracy of measurement of only three significant figures, this number should be expressed as 251×10^2.

Because the position of the decimal point is immaterial in determining the number of significant figures, the numbers

0.0753 0.753 75.3 7.53

are significant to only three figures.

In calculations where it is necessary to round off numbers to obtain one less significant figure, if the last digit to the right is greater than 5, increase the next to last digit by 1, while if it is less than 5, this next to last digit is not changed. Thus 1.357 rounds off to 1.36, while 4.892 rounds off to 4.89. When the last digit is 5, then round off the next to last digit to an even number. Thus, round off 6.345 to 6.34 but 9.575 to 9.58.

In adding or substracting a set of numbers, one should round off to the same number of decimal points. On adding the set of numbers

$$\begin{array}{r} 15.215 \\ 312.6 \\ 27.65 \\ \hline 355.465 \end{array}$$

one should round off to four significant figures, namely 355.5, because some of the numbers have no digits in the second and third decimal places. This is reasonable because if one student weighed some iron fillings to an accuracy of one-tenth gram and another student weighed another sample to an accuracy of milligrams, if both piles of the iron fillings were combined, then the weight of the total quantity of material should only be indicated to the less accurate measure, a tenth of a gram.

In the multiplication and division of numbers, the answer should be rounded off to the number of significant figures that appear in the quantity with the least number of significant figures.

EXAMPLE

Multiply 332 by 16 and divide by 574.

$(332 \times 16)/574 = 9.0800$

Because the figure with the least number of significant figures is 16, with two, then the answer must be rounded off to two significant figures, namely 9.1.

PROBLEMS

1. In carrying out the following calculations, be sure to report the answers to the correct number of significant places:

   (a) $653.216 - 342.6 =$
   (b) $653.216/342.6 =$
   (c) $892.16 + 132.235 + 482.3 =$
   (d) $789.255 \times 4.5 =$

2. Round off the following numbers to indicate three significant figures:

   (a) 631200 (b) 43255
   (c) 3226 (d) 3223
   (e) 0.07875 (f) 8425

The answers to the above problems are

1. (a) 310.6, (b) 1.9, (c) 1506.7, (d) 3541.6
2. (a) 6.31×10^5, (b) 4.33, (c) 3.23, (d) 3.22, (e) 7.88×10^{-2}, (f) 8.42

Exponents and exponential numbers

When manipulating very large or quite small numbers, one will usually find it easier to express these in exponential form. There will be less opportunity for error than when a large number of zeros must be counted.

The exponent indicates the number of times a number, called the *base*, must be multiplied by itself to give a specific number.

Thus 32 can be expressed as 2^5 (2 to the fifth power), that is, $2 \times 2 \times 2 \times 2 \times 2$, which equals 32. For the following discussion, all exponential numbers will be to base 10.

In the following table, some typical numbers are converted to exponential numbers.

Number	*Exponent of Ten*	*Conventional Exponential Number*
10	1	10^1
100	2	10^2
1,000	3	10^3
100,000	5	10^5
10,000,000	7	10^7
0.01	−2	10^{-2}
0.1	−1	10^{-1}
0.00001	−5	10^{-5}
0.0000001	−7	10^{-7}
1	0	10^0

Note from this table that 1 is equal to 10 to the zero power and is expressed exponentially as 10^0.

Such a number as 72184 can be expressed exponentially in the following ways:

7281.4×10^1
728.14×10^2
72.814×10^3
7.2814×10^4
0.72814×10^5

Conventionally, 7.2814×10^4 is the preferred form; that is, the exponential number is expressed as some number between 1 and 10 multiplied by some power of 10. Although 0.00615 could be expressed exponentially as 0.0615×10^{-1} or 0.615×10^{-2}, the expression 6.15×10^{-3} is usually preferred.

For numbers larger than 1, the exponential of the exponential number will be equal to the number of places the decimal point is moved to the left. For numbers less than 1 expressed decimally, the exponent of the exponential number will be negative and will be numerically equal to the number of places the decimal point is moved to the right, or one more than the number of zeros between the decimal place and the first digit.

In multiplying exponential numbers, the decimal numbers must be multiplied but the exponents need only be added.

EXAMPLES

$10^5 \times 10^3 = 10^8$
$10^{-7} \times 10^4 \times 10^8 = 10^5$

$(4 \times 10^{-3})(5 \times 10^{6}) = 20 \times 10^{3} = 2 \times 10^{4}$
$(1.5 \times 10^{5})(3 \times 10^{-9}) = 4.5 \times 10^{-4}$

In dividing, the decimal number in the numerator is divided by the decimal number in the denominator, but the exponent of 10 in the denominator is subtracted from the exponent of 10 in the numerator.

EXAMPLES

$10^{6}/10^{2} = 10^{4}$
$10^{-5}/10^{5} = 10^{10}$
$6 \times 10^{-9}/2 \times 10^{-7} = 3 \times 10^{-2}$
$2 \times 10^{4}/4 \times 10^{-4} = 20 \times 10^{3}/4 \times 10^{-4} = 5 \times 10^{7}$
$1/10^{-6} = 10^{6}$

The last example illustrates the rule that an exponential number may be transferred from the denominator to the numerator, or vice versa, by changing the sign of the exponent.

The following examples illustrate the use of powers and roots of exponential numbers:

EXAMPLES

the square of $10^{3} = 10^{6}$
the cube of $10^{-4} = 10^{-12}$
$(3 \times 10^{-3})^{2} = 9 \times 10^{-6}$
$\sqrt[3]{8 \times 10^{9}} = 2 \times 10^{3}$

PROBLEMS

1. Convert the following numbers to their proper exponential form:

(a) 1,000,000,000 (b) 0.0000001
(c) 66770 (d) 0.0326
(e) 0.695 (f) 113.004
(g) 25.5 (h) 0.00886

2. Carry out the following multiplications:

(a) $10^{6} \times 10^{5} =$
(b) $10^{-9} \times 10^{3} =$
(c) $10^{-8} \times 10^{-2} =$
(d) $(6 \times 10^{-7})(3.2 \times 10^{4}) =$
(e) $(4 \times 10^{4})(7 \times 10^{-6}) =$
(f) $(8.2 \times 10^{-5})(8 \times 10^{-9}) =$

3. Carry out the following divisions:

(a) $10^{8}/10^{5} =$ (b) $10^{4}/10^{7} =$
(c) $10^{-6}/10^{-3} =$ (d) $10^{-3}/10^{-7} =$
(e) $6.4 \times 10^{-4}/4 \times 10^{5} =$ (f) $8.6 \times 10^{9}/4.3 \times 10^{-7} =$
(g) $7.5 \times 10^{-6}/5 \times 10^{-5} =$ (h) $2.5 \times 10^{-3}/5 \times 10^{3} =$

4. Carry out the operations indicated in the following:

(a) $(7 \times 10^{-5})^{2} =$ (b) $(3 \times 10^{-2})^{3} =$
(c) $(2 \times 10^{-3})^{4} =$ (d) $\sqrt{64 \times 10^{-8}} =$
(e) $\sqrt[3]{27 \times 10^{-15}} =$ (f) $\sqrt[3]{6.4 \times 10^{-5}} =$

The answers to the above problems are

1. (a) 10^9, (b) 10^{-7}, (c) 6.677×10^4, (d) 3.26×10^{-2}, (e) 6.95×10^{-1}, (f) 1.13004×10^2, (g) 2.55×10^1, (h) 8.86×10^{-3}
2. (a) 10^{11}, (b) 10^{-6}, (c) 10^{-10}, (d) 1.92×10^{-2}, (e) 2.8×10^{-1}, or 0.28, (f) 6.56×10^{-13}
3. (a) 10^3, (b) 10^{-3}, (c) 10^{-3}, (d) 10^4, (e) 1.6×10^{-9}, (f) 2×10^{16}, (g) 1.5×10^{-1}, or 0.15, (h) 5×10^{-7}
4. (a) 4.9×10^{-9}, (b) 2.7×10^{-5}, (c) 1.6×10^{-11}, (d) 8×10^{-4}, (e) 3×10^{-5}, (f) 4×10^{-2}

Common logarithms

The commonly used logarithms, also called *Briggsian logarithms,* are to the base 10; that is, they are the power to which 10 must be raised to give the number. Thus if $y = 10^x$, then x is the logarithm of the number y (to base 10). Because $1000 = 10^3$, then the logarithm of 1000 is 3.

What is the logarithm of 25.7? Because this number is between 10 (whose logarithm is 1) and 100 (10^2, and whose logarithm is 2), its logarithm will be between 1 and 2. The integral part of this logarithm will be 1 and is called the *characteristic,* while the decimal portion of the logarithm is called the *mantissa.* The mantissa of the number 25.7, which is always independent of the position of the decimal point, will be found in the four-place logarithm table which appears on pages 560 and 561. Look down the column labeled "natural numbers" to 25 and then move over to the column marked 7. You will find that the mantissa of 257 is 0.4099. The logarithm of 25.7 is then 1.4099.

Note that the characteristic of a number greater than 1 will be equal to 1 less than the number of digits to the left of the decimal point.

You will find some difficulty in expressing the logarithm of a number between 0 and 1. In such a case, the characteristic will be negative and numerically will be greater by 1 than the number of zeros between the decimal point and the first integral digit in the number. As an example, 0.62 has a characteristic of −1, while 0.062 has a characteristic of −2; the mantissa of these two numbers is 0.7924. The logarithm of 0.062 can be expressed as $\bar{2}.7924$ $(-2 + 0.7924)$, in which the integral number is negative while the decimal mantissa is positive. To avoid this situation where the two parts of the logarithm are of opposite sign, many prefer to express this logarithm as 8.7924 − 10.

It is frequently necessary to find the number, called the *antilogarithm,* or *antilog,* that corresponds to a certain logarithm. Thus to find the number or antilog corresponding to the logarithm 1.8882, find the number in the four-place log table that corresponds to the mantissa 0.8882. Because this mantissa appears under column 3 of the line of the natural number 77, then the three digits of the number will be 773. The characteristic of the logarithm indicates that there are two digits before the decimal point. Hence the antilog of 1.8882 is 77.3.

Logarithms can be conveniently used to multiply and divide a set of numbers. Roots and powers of numbers are also readily determined using logarithms. The student is referred to an algebra text if he does not know how to carry out these computations utilizing logarithms.

The scales on a slide rule are generally logarithmic. The student is encouraged to use a slide rule in his computations, because these can generally be read to four significant figures (page 554). This is all the precision needed in solving most of the problems in this text.

PROBLEMS

1. What is the logarithm of the following numbers?

(a) 0.0001 (b) 100,000
(c) 672.5 (d) 1.938
(e) 0.0247 (f) 0.749
(g) 0.000062 (h) 1

2. Given the following logarithms, determine their antilog:

(a) 3.8142 (b) 8.7042 − 10
(c) 0.3909 (d) $\bar{3}.9279$
(e) 1.0170 (f) 9.9930 − 10

The answers to the above problems are

1. (a) −4, (b) 5, (c) 2.8277, (d) 0.2874, (e) 8.3927 − 10, (f) 9.8745 − 10, (g) 5.7924 − 10, (h) 0
2. (a) 6520, (b) 0.0506, (c) 2.46, (d) 0.00847, (e) 10.4, (f) 0.984

FOUR-PLACE LOGARITHMS

Natural numbers	0	1	2	3	4	5	6	7	8	9
10	0000	0043	0086	0128	0170	0212	0253	0294	0334	0374
11	0414	0453	0492	0531	0569	0607	0645	0682	0719	0755
12	0792	0828	0864	0899	0934	0969	1004	1038	1072	1106
13	1139	1173	1206	1239	1271	1303	1335	1367	1399	1430
14	1461	1492	1523	1553	1584	1614	1644	1673	1703	1732
15	1761	1790	1818	1847	1875	1903	1931	1959	1987	2014
16	2041	2068	2095	2122	2148	2175	2201	2227	2253	2279
17	2304	2330	2355	2380	2405	2430	2455	2480	2504	2529
18	2553	2577	2601	2625	2648	2672	2695	2718	2742	2765
19	2788	2810	2833	2856	2878	2900	2923	2945	2967	2989
20	3010	3032	3054	3075	3096	3118	3139	3160	3181	3201
21	3222	3243	3263	3284	3304	3324	3345	3365	3385	3404
22	3424	3444	3464	3483	3502	3522	3541	3560	3579	3598
23	3617	3636	3655	3674	3692	3711	3729	3747	3766	3784
24	3802	3820	3838	3856	3874	3892	3909	3927	3945	3962
25	3979	3997	4014	4031	4048	4065	4082	4099	4116	4133
26	4150	4166	4183	4200	4216	4232	4249	4265	4281	4298
27	4314	4330	4346	4362	4378	4394	4409	4425	4440	4456
28	4472	4487	4502	4518	4533	4548	4564	4579	4594	4609
29	4624	4639	4654	4669	4683	4698	4713	4728	4742	4757
30	4771	4786	4800	4814	4829	4843	4857	4871	4886	4900
31	4914	4928	4942	4955	4969	4983	4997	5011	5024	5038
32	5051	5065	5079	5092	5105	5119	5132	5145	5159	5172
33	5185	5198	5211	5224	5237	5250	5263	5276	5289	5302
34	5315	5328	5340	5353	5366	5378	5391	5403	5416	5428
35	5441	5453	5465	5478	5490	5502	5514	5527	5539	5551
36	5563	5575	5587	5599	5611	5623	5635	5647	5658	5670
37	5682	5694	5705	5717	5729	5740	5752	5763	5775	5786
38	5798	5809	5821	5832	5843	5855	5866	5877	5888	5899
39	5911	5922	5933	5944	5955	5966	5977	5988	5999	6010
40	6021	6031	6042	6053	6064	6075	6085	6096	6107	6117
41	6128	6138	6149	6160	6170	6180	6191	6201	6212	6222
42	6232	6243	6253	6263	6274	6284	6294	6304	6314	6325
43	6335	6345	6355	6365	6375	6385	6395	6405	6415	6425
44	6435	6444	6454	6464	6474	6484	6493	6503	6513	6522
45	6532	6542	6551	6561	6571	6580	6590	6599	6609	6618
46	6628	6637	6646	6656	6665	6675	6684	6693	6702	6712
47	6721	6730	6739	6749	6758	6767	6776	6785	6794	6803
48	6812	6821	6830	6839	6848	6857	6866	6875	6884	6893
49	6902	6911	6920	6928	6937	6946	6955	6964	6972	6981
50	6990	6998	7007	7016	7024	7033	7042	7050	7059	7067
51	7076	7084	7093	7101	7110	7118	7126	7135	7143	7152
52	7160	7168	7177	7185	7193	7202	7210	7218	7226	7235
53	7243	7251	7259	7267	7275	7284	7292	7300	7308	7316
54	7324	7332	7340	7348	7356	7364	7372	7380	7388	7396

FOUR-PLACE LOGARITHMS (*Continued*)

Natural numbers	0	1	2	3	4	5	6	7	8	9
55	7404	7412	7419	7427	7435	7443	7451	7459	7466	7474
56	7482	7490	7497	7505	7513	7520	7528	7536	7543	7551
57	7559	7566	7574	7582	7589	7597	7604	7612	7619	7627
58	7634	7642	7649	7657	7664	7672	7679	7686	7694	7701
59	7709	7716	7723	7731	7738	7745	7752	7760	7767	7774
60	7782	7789	7796	7803	7810	7818	7825	7832	7839	7846
61	7853	7860	7868	7875	7882	7889	7896	7903	7910	7917
62	7924	7931	7938	7945	7952	7959	7966	7973	7980	7987
63	7993	8000	8007	8014	8021	8028	8035	8041	8048	8055
64	8062	8069	8075	8082	8089	8096	8102	8109	8116	8122
65	8129	8136	8142	8149	8156	8162	8169	8176	8182	8189
66	8195	8202	8209	8215	8222	8228	8235	8241	8248	8254
67	8261	8267	8274	8280	8287	8293	8299	8306	8312	8319
68	8325	8331	8338	8344	8351	8357	8363	8370	8376	8382
69	8388	8395	8401	8407	8414	8420	8426	8432	8439	8445
70	8451	8457	8463	8470	8476	8482	8488	8494	8500	8506
71	8513	8519	8525	8531	8537	8543	8549	8555	8561	8567
72	8573	8579	8585	8591	8597	8603	8609	8615	8621	8627
73	8633	8639	8645	8651	8657	8663	8669	8675	8681	8686
74	8692	8698	8704	8710	8716	8722	8727	8733	8739	8745
75	8751	8756	8762	8768	8774	8779	8785	8791	8797	8802
76	8808	8814	8820	8825	8831	8837	8842	8848	8854	8859
77	8865	8871	8876	8882	8887	8893	8899	8904	8910	8915
78	8921	8927	8932	8938	8943	8949	8954	8960	8965	8971
79	8976	8982	8987	8993	8998	9004	9009	9015	9020	9025
80	9031	9036	9042	9047	9053	9058	9063	9069	9074	9079
81	9085	9090	9096	9101	9106	9112	9117	9122	9128	9133
82	9138	9143	9149	9154	9159	9165	9170	9175	9180	9186
83	9191	9196	9201	9206	9212	9217	9222	9227	9232	9238
84	9243	9248	9253	9258	9263	9269	9274	9279	9284	9289
85	9294	9299	9304	9309	9315	9320	9325	9330	9335	9340
86	9345	9350	9355	9360	9365	9370	9375	9380	9385	9390
87	9395	9400	9405	9410	9415	9420	9425	9430	9435	9440
88	8445	9450	9455	9460	9465	9469	9474	9479	9484	9489
89	9494	9499	9504	9509	9513	9518	9523	9528	9533	9538
90	9542	9547	9552	9557	9562	9566	9571	9576	9581	9586
91	9590	9595	9600	9605	9609	9614	9619	9624	9628	9633
92	9638	9643	9647	9652	9657	9661	9666	9671	9675	9680
93	9685	9689	9694	9699	9703	9708	9713	9717	9722	9727
94	9731	9736	9741	9745	9750	9754	9759	9763	9768	9773
95	9777	9782	9786	9791	9795	9800	9805	9809	9814	9818
96	9823	9827	9832	9836	9841	9845	9850	9854	9859	9863
97	9868	9872	9877	9881	9886	9890	9894	9899	9903	9908
98	9912	9917	9921	9926	9930	9934	9939	9943	9948	9952
99	9956	9961	9965	9969	9974	9978	9983	9987	9991	9996

appendix b

the calculation of the pH of a buffer

appendix b

The method employed in calculating the pH of a buffer system will be illustrated for a weak acid in combination with a salt with a common ion. The expression for calculating the pH of such a solution, frequently called the *Henderson-Hasselbalch equation,* is derived from the ionization constant of the weak acid (page 141).

For the general acid HA which ionizes in the following fashion

$$HA \rightleftharpoons H^+ + A^-$$

the ionization constant will be (see page 141)

$$K_a = \frac{[H^+] \times [A^-]}{[HA]}$$

The hydrogen ion concentration will then be given by the expression

$$[H^+] = \frac{K_a \times [HA]}{[A^-]}$$

Because HA is slightly ionized in an aqueous solution and this ionization is further repressed by the presence of the common ion A^- (page 159), [HA] may be taken as equal to the original concentration of the acid. Now, because most of the A^- ions will be furnished by the highly ionized salt and few by the ionization of the *weak* acid, $[A^-]$ may be taken as the concentration of the salt added. Then

$$[H^+] = K_a \times \frac{[\text{acid}]}{[\text{salt}]}$$

and then taking the negative logarithms of each member of this equation gives

$$-\log [H^+] = -\log K_a - \log \frac{[\text{acid}]}{[\text{salt}]}$$

and

$$-\log [H^+] = -\log K_a + \log \frac{[\text{salt}]}{[\text{acid}]}$$

Because $-\log [H^+] = \text{pH}$ and $-\log K_a$ is frequently designated as

pK_a, then the usual form of the Henderson-Hasselbalch equation employed is

$$pH = pK_a + \log \frac{[\text{salt}]}{[\text{acid}]}$$

It is to be noted that if equimolar concentrations of the salt and the weak acid are used in preparing a buffer, then

$$pH = pK_a$$

as

$$\log \frac{[\text{salt}]}{[\text{acid}]} = \log 1 = 0$$

ILLUSTRATIVE PROBLEM 1

What is the pH of a solution in which 0.1 mole of formic acid and 0.1 mole of sodium formate are present in each liter of solution?

Because K_a for formic acid is 1.8×10^{-4} (see page 152), then

$$pK_a = -\log 1.8 \times 10^{-4} = 3.7$$

In this solution (see above),

$$pH = pK_a + \log \frac{[\text{salt}]}{[\text{acid}]} = pK_a + \log 1 = pK_a$$

Then the pH of the solution will be 3.7.

ILLUSTRATIVE PROBLEM 2

Calculate the pH of a solution that contains 0.1 mole of H_2CO_3 and 1 mole of $NaHCO_3$ per liter of solution.

Consider that the carbonic acid is acting as a weak acid and that sodium bicarbonate furnishes the common ion HCO_3^-. Then

$$H_2CO_3 \rightleftharpoons H^+ + HCO_3^-$$

Then

$$K_a = K_1 \text{ for carbonic acid} = 4.3 \times 10^{-7}$$

and

$$pK_a = -\log 4.3 \times 10^{-7} = 7 - 0.63 = 6.37$$

Substituting in the Henderson-Hasselbalch equation,

$$pH = pK_a + \log \frac{[\text{salt}]}{[\text{acid}]}$$

$$= 6.37 + \log \frac{1}{0.1}$$

$$= 6.37 + \log 10 = 7.37$$

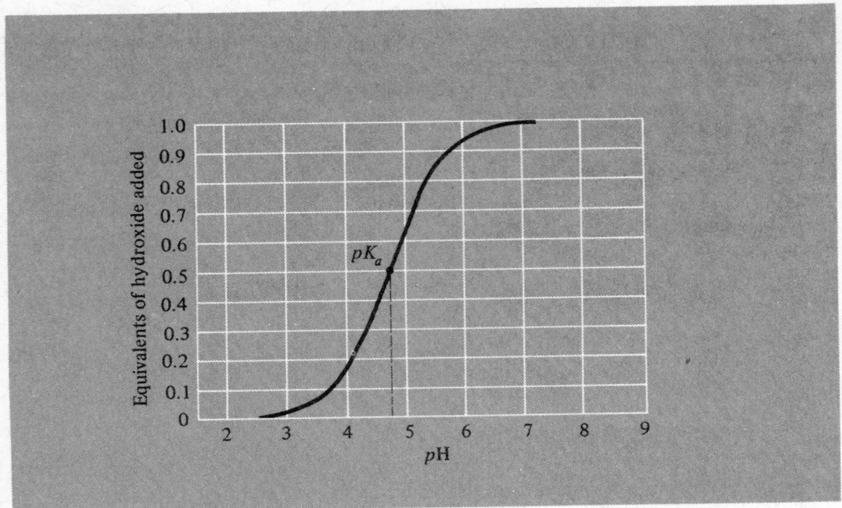

Figure B.1 The titration of acetic acid.

The pK_a of a weak acid may be determined experimentally by measuring the pH of a solution of the weak acid as it is being titrated by a strong base. When the acid is half neutralized, the pK_a will equal the pH of the solution, as indicated by the titration curve in Figure B.1.

STUDY QUESTIONS

1. For what purpose is the Henderson-Hasselbalch equation utilized? Derive this equation from the ionization constant expression for a weak acid. In the Henderson-Hasselbalch equation, why can the concentration of the weak acid present after ionization be taken as equal to the concentration of acid added to the solution? In this equation, why can the concentration of the anion derived from the ionization of the weak acid be disregarded?
2. When will the pK_a of an acid numerically equal the pH of a solution of the acid?
3. How can the pK_a and the ionization constant of a weak acid be determined experimentally?

PROBLEMS

For the numerical value of the ionization constants, K_a, of weak acids, see pages 142 and 152.

1. By the use of the Henderson-Hasselbalch equation, calculate the pH of the following buffers.

   (a) 0.1 mole of H_2CO_3 and 0.1 mole of $NaHCO_3$ per liter
   (b) 0.1 mole of formic acid and 0.01 mole of sodium formate per liter
   (c) 0.1 mole of acetic acid and 0.1 mole of sodium acetate per liter
   (d) 0.01 mole of formic acid and 0.1 mole of sodium formate per liter

2. If the pK_a of acetic acid is 4.7, calculate the concentration of sodium acetate that must be present in a 0.1 *M* solution of acetic acid so that the solution will have a pH of (a) 5.7, and (b) 3.7.

The answers to the above problems are

1. (a) 6.37, (b) 2.7, (c) 4.7, (d) 4.7
2. (a) 1 *M* sodium acetate, (b) 0.01 *M* sodium acetate

appendix c

---

metric-english and other conversion factors

appendix c

Metric-English conversion factors

Length	1 meter (m) = 39.37 inches (in.)
	1 kilometer (km) = 0.62 mile
	1 inch = 2.54 centimeters (cm)
	1 centimeter = 0.394 in.
Mass	1 pound (lb) (avoirdupois) = 454 grams (g)
	1 ounce (oz) (avoirdupois) = 28.4 g
	1 kilogram (kg) = 2.2 lb
Volume	1 liter = 1.06 quarts (qt)
	1 cubic foot (ft^3) = 28.3 liters
	1 milliliter (ml) = 0.06 cubic inch ($in.^3$)
	1 cubic inch = 16.4 ml
	1 gallon (gal)(U.S.) = 3.79 liters
	1 ounce (liquid) = 30 ml (approx.)
Heat measurement	1 calorie (cal) = 0.004 British Thermal Unit (BTU)
	1 British Thermal Unit = 252 cal
	1 Fahrenheit degree = $\frac{5}{9}$ or 0.55 Celsius degree
	1 Celsius degree = $\frac{9}{5}$ or 1.8 Fahrenheit degrees
	$°F = \frac{9}{5}°C + 32$
	$°C = (°F - 32) \times \frac{5}{9}$
Pressure	1 atmosphere (atm) = 760 millimeters (mm) of Hg
	1 millimeter of Hg = 0.019 pounds per square inch (psi)
	1 pound per square inch = 51.7 mm of Hg

Other important conversion factors

Length	1 Angstrom (Å) = 1×10^{-8} cm
	1 micrometer (μm) = 1 micron (μ) = 10^{-6} m
	1 nanometer (nm) = 1 millimicron ($m\mu$) = 10^{-9} m
Mass	1 gram = 1000 milligrams (mg)
	1 gram = 10^6 micrograms (μg)
	1 microgram = 10^{-6} g
Volume	1 liter = 1000 ml = 1000 cubic centimeters (cm^3)
Heat measurement	Absolute (Kelvin) temperature = Celsius temperature + 273
	Celsius temperature = absolute temperature − 273
Pressure	1 millimeter of Hg = 1 torr (0°C at sea level)

Others

1 gram of water at 4°C = 1 ml = 1 cm^3
1 gram-molecular-weight (GMW) of a gas = 1 gram-molecular-volume (GMV) = 22.4 liters at standard conditions
N = Avogadro's number = 6.02×10^{23} molecules

index

index

D

S